Stars and Stellar Evolution

Stars and Stellar Evolution

K.S. de Boer   and   W. Seggewiss

17 avenue du Hoggar
Parc d' activités de Courtabeuf, B.P. 112
91944 Les Ulis Cedex A, France

Cover image: The stellar association LH 95 in the Large Magellanic Cloud showing star formation, young stars and old stars.
HST-ACS image, courtesy of D. Gouliermis and NASA/ESA

ISBN 978-2-7598-0356-9

This work is subject to copyright. All rights are reserved, whether the whole or part of the material is concerned, specifically the rights of translation, reprinting, re-use of illustrations, recitation, broad-casting, reproduction on microfilms or in other ways, and storage in data banks. Duplication of this publication or parts thereof is only permitted under the provisions of the French and German Copyright laws of March 11, 1957 and September 9, 1965, respectively. Violations fall under the prosecution act of the French and German Copyright Laws.

© EDP Sciences, 2008

Contents

1 Introduction   1
    1.1 Historical background 1
        1.1.1 History of the characterization of stars 1
        1.1.2 History of the ideas about the evolution of stars 2
    1.2 Stellar evolution - the importance of gravity 3
    1.3 Relevance of stars for astrophysics 4
    1.4 Elementary astronomy and classical physics 5
        1.4.1 Classical observations 5
        1.4.2 The Planck function 7
        1.4.3 Spectral lines, metallicity, and gas conditions 8
    1.5 The surface parameters of stars 9
        1.5.1 The Hertzsprung-Russell Diagram, HRD 9
            1.5.1.1 Observational HRDs: M_V with SpT or $B - V$ 9
            1.5.1.2 Physical HRD: luminosity L and effective temperature T_{eff} 10
        1.5.2 Spectral energy distributions 12
        1.5.3 Relation between M_V, M_{bol}, and L 12
        1.5.4 Caution with mass - luminosity - temperature relations 12
    1.6 Surface parameters and size of a star 13
    1.7 Names of star types from location in the HRD 14
    1.8 Summary 14

2 Stellar atmosphere:
Continuum radiation + structure   15
    2.1 Introduction 15
    2.2 Radiation theory 16
        2.2.1 Definitions 16
            2.2.1.1 Radiative intensity 16
            2.2.1.2 Mean intensity, radiative flux 17
            2.2.1.3 Radiation density and radiation pressure 17
        2.2.2 The equation of radiation transport 17
        2.2.3 Exploring the equation of radiation transport 18
            2.2.3.1 a: No background intensity: $I_\nu^0 = 0$ 19
            2.2.3.2 b: background intensity: $I_\nu^0 \neq 0$ 19
            2.2.3.3 Graphic representation of the cases 19
    2.3 Thermodynamic equilibrium 19
    2.4 The radiative transfer in stellar atmospheres 20
        2.4.1 Effects of geometry 20
        2.4.2 Including all frequencies 20
    2.5 Continuity equation 20
    2.6 Special cases and approximations 21
        2.6.1 Atmospheres in LTE 21
        2.6.2 Plane parallel atmosphere 21

		2.6.3 Limb darkening	21
		2.6.4 Gray atmosphere; Rosseland mean	23
	2.7	Structure of stellar atmospheres	24
		2.7.1 Temperature structure	24
		2.7.2 Pressure structure	24
	2.8	Opacity and the absorption coefficients	25
		2.8.1 Absorption due to ionization	25
		2.8.1.1 Total absorption cross section for hydrogen	25
		2.8.1.2 Absorption due to ionization of helium	26
		2.8.1.3 Absorption due to ionization of metals	26
		2.8.2 The H^- ion	26
		2.8.3 Absorption due to dissociation	27
		2.8.4 Free-free transitions	27
		2.8.5 Scattering	27
		2.8.6 Total absorption coefficient	28
		2.8.7 Effects of gas density on opacity	29
	2.9	Emission and the emission coefficient	29
	2.10	The spectral continuum and the Planck function	29
		2.10.1 Effects for the CMD	30
		2.10.2 Backwarming, blanketing	31
		2.10.3 Electron density and opacity effects	31

3 Stellar atmosphere: Spectral structure   33

3.1	Spectral lines		33
	3.1.1 Line profile		33
	3.1.1.1 Lorentz profile		33
	3.1.1.2 Pressure broadening		34
	3.1.1.3 Doppler broadening		35
	3.1.1.4 The Voigt profile		35
	3.1.2 Shape and strength of spectral lines and curve of growth		36
	3.1.2.1 Small optical depth in the line ($\tau \ll 1$ and/or $\alpha \ll 1$)		36
	3.1.2.2 Very large optical depth in the line ($\tau \gg 1$ and/or $\alpha \gg 1$)		37
	3.1.2.3 Intermediate α and/or τ		38
	3.1.2.4 Shape of curve of growth		38
3.2	Statistics		38
	3.2.1 Boltzmann statistics and excitation equation		38
	3.2.2 Ionization and Saha equation		39
3.3	Statistics and structure in stellar spectra		40
	3.3.1 Excitation		40
	3.3.2 Ionization		40
	3.3.3 Spectrophotometric methods		40
	3.3.4 Balmer jump and Balmer Series		41
	3.3.5 T_{eff} and $\log g$ from Strömgren photometry		42
	3.3.6 Metallicity from Strömgren photometry		43
	3.3.7 Spectroscopy and the curve of growth		43
	3.3.7.1 Excitation		43
	3.3.7.2 Ionization		44
	3.3.7.3 Depth structure of atmosphere		44
	3.3.7.4 Abundance of elements		45
3.4	Special features		45
	3.4.1 The G-Band		45
	3.4.2 Quasi-molecular absorption: H_2 and H_2^+		45
	3.4.3 Molecular absorption in cool atmospheres		46

	3.5	Magnetic fields and Zeeman effect	47
	3.6	Gravitational settling and radiation levitation	48
	3.7	Stellar rotation	49
	3.7.1	Rotation broadening of spectral lines	49
	3.7.2	Rotation and average surface parameters T, M_V, $B-V$	49
	3.8	Stellar classification: the MKK system and newer methods	50
	3.8.1	Development of stellar classification towards the MKK system	50
	3.8.2	Quality of the MK classification process	51
	3.8.3	New classification methods	51

4 Stellar structure: Basic equations — 53

4.1	Four basic equations for the internal structure		53
	4.1.1	Mass continuity	53
	4.1.2	Hydrostatic equilibrium	53
	4.1.3	Energy conservation	54
	4.1.4	Temperature gradient	55
		4.1.4.1 Radiative energy transport	55
		4.1.4.2 Convective energy transport	56
		4.1.4.3 Conductive energy transport	56
4.2	Stability and time scales		57
	4.2.1	Virial theorem	57
	4.2.2	Kelvin-Helmholtz time scale	57
	4.2.3	Nuclear time scale	58
	4.2.4	Dynamical time scale	58
4.3	Convection versus radiation		59
	4.3.1	Schwarzschild's criterion for convection	59
	4.3.2	Ledoux's criterion for convection	61
	4.3.3	Estimates for $\nabla_{ad} < \nabla_{rad}$	61
		4.3.3.1 Adiabatic gradient ∇_{ad}	61
		4.3.3.2 Radiative gradient ∇_{rad}	61
	4.3.4	Absorption-driven or radiation-driven convection?	62
		4.3.4.1 Large absorption coefficient κ	62
		4.3.4.2 Large flux $F(r)$	62
	4.3.5	Convective overshoot	62
	4.3.6	Mixing length theory	63
4.4	Material functions		63
	4.4.1	Opacity	64
	4.4.2	Equation of state	65
		4.4.2.1 Ideal gas law	65
		4.4.2.2 Radiation pressure	65
		4.4.2.3 Degenerate gas	66
	4.4.3	Energy production functions: nuclear fusion and gravity	67
4.5	Stellar winds		68
	4.5.1	Coronal models	68
	4.5.2	Radiative winds	68
		4.5.2.1 Line driven winds	69
		4.5.2.2 Continuum-driven winds	69
		4.5.2.3 Dust-driven winds	69
	4.5.3	Bi-stability winds: fast and dilute or slow and dense	70
	4.5.4	Winds enhanced due to stellar rotation	70
	4.5.5	Pulsation-driven winds	70

5 Nuclear fusion in stars — 71
- 5.1 Energy production: fusion of H and He — 71
  - 5.1.1 Binding energy of nuclei — 71
  - 5.1.2 Estimates for the occurrence of hydrogen fusion — 72
  - 5.1.3 The Gamow peak — 72
  - 5.1.4 proton–proton chain — 73
  - 5.1.5 CNO cycle — 74
  - 5.1.6 Temperature dependence of H-fusion energy production — 75
  - 5.1.7 He fusion: the triple alpha process — 75
- 5.2 Nucleosynthesis — 76
  - 5.2.1 Carbon and oxygen burning; α-capture — 76
  - 5.2.2 Nitrogen burning — 77
  - 5.2.3 Fusion to heavier elements — 77
  - 5.2.4 General considerations (NSE); s-, r- and p-process — 78
    - 5.2.4.1 s-process — 78
    - 5.2.4.2 r-process — 79
    - 5.2.4.3 p-process — 79
  - 5.2.5 Nucleosynthesis and the Universe; Yields — 80
  - 5.2.6 The burning of Lithium — 80
- 5.3 Neutrinos — 80
  - 5.3.1 Mean free path for neutrinos — 80
  - 5.3.2 Solar neutrinos — 82
  - 5.3.3 Neutrino experiments — 82
  - 5.3.4 The "solar neutrino problem" — 83
  - 5.3.5 Neutrino oscillations — 83
  - 5.3.6 The Sudbury Neutrino Observatory and solution of the problem — 84
    - 5.3.6.1 Relevant neutrino reactions — 84
    - 5.3.6.2 Advantages of heavy water — 85
    - 5.3.6.3 The solution of the solar neutrino problem — 85
- 5.4 Nobel prize 2002 for neutrino research — 85

6 Stellar structure: Making star models — 87
- 6.1 The equations of state and their complications — 87
- 6.2 Polytropes; Consequences of differing equations of state — 87
  - 6.2.1 The general polytropic equation — 88
  - 6.2.2 Special polytropes — 88
    - 6.2.2.1 Polytrope for ideal gas — 88
    - 6.2.2.2 Completely convective stars — 89
    - 6.2.2.3 Non-relativistic degenerate electron gas — 89
    - 6.2.2.4 Relativistic completely degenerate electron gas — 89
- 6.3 Balance between internal pressure and gravitation — 90
- 6.4 The maximum mass of a normal star — 90
- 6.5 The minimum mass of a star — 91
- 6.6 Methods for solving the differential equations — 93
  - 6.6.1 Numerical solutions — 93
  - 6.6.2 Differential equations against mass shell — 93
  - 6.6.3 Adding stellar evolution — 94
  - 6.6.4 A model using gaussian functions — 94
- 6.7 Vocabulary for stellar structure: definitions — 94
- 6.8 Zero-age-main-sequence star parameters from models — 95
  - 6.8.1 ZAMS: structure as a function of mass shell — 95
  - 6.8.2 ZAMS: parameters along the ZAMS - a star as a leaky ball — 96
    - 6.8.2.1 Similarity along the MS; homology; thermostat; luminosity and mass — 96

               6.8.2.2 A star as a leaky ball: general behaviour, effects of chemical composition 98
    6.9   Internal structure and chemical composition 98
          6.9.1   Consequences of nuclear enrichment for stellar structure 98
          6.9.2   Non-hydrogen stars 99
          6.9.3   Central temperature and density of He and C stars 99
    6.10 Summary 100

7 Star formation, proto-stars, very young stars   101
    7.1   Evidence of star formation, populations, IMF 101
          7.1.1   Signs of present star formation 101
          7.1.2   Star-formation processes and results of star formation 102
    7.2   Molecular clouds: places of star formation 102
          7.2.1   Discovery and importance of interstellar molecules 102
          7.2.2   Characteristics of molecular clouds 104
          7.2.3   Observed phenomena in star forming regions 104
    7.3   Instabilities in the interstellar gas 106
          7.3.1   Gravitational instability (Jeans instability) 106
          7.3.2   Thermal instabilities 108
                  7.3.2.1 Energy input and energy loss 108
                  7.3.2.2 Density fluctuations and their growth 108
          7.3.3   Stability and ambipolar diffusion in molecular clouds 109
                  7.3.3.1 Low efficiency of star formation 109
                  7.3.3.2 Cloud support mechanisms 109
                  7.3.3.3 Ambipolar diffusion 111
    7.4   Theoretical scenario of star formation 111
    7.5   Pre-main-sequence evolution (PMS evolution) 113
          7.5.1   Energy source of PMS stars 113
          7.5.2   Theory of pre main-sequence stars 114
                  7.5.2.1 Contraction along the Hayashi line in the earliest phase 114
                  7.5.2.2 The accretion rate $\dot{M}$ 115
    7.6   Bipolar outflows, jets, Herbig-Haro objects, disks 115
          7.6.1   Definition of bipolar outflows and Herbig-Haro objects 115
          7.6.2   Some physical characteristics of bipolar flows 116
          7.6.3   Circumstellar disks 116
          7.6.4   Origin of outflows 117
    7.7   Very young stars 118
          7.7.1   General characteristics of T Tauri stars 118
          7.7.2   T Tau stars and X-ray emission 119
          7.7.3   T Tauri stars as young objects 120
          7.7.4   Herbig Ae and Be stars 121
    7.8   Summary 121

8 The almost stars: Brown Dwarfs   125
    8.1   Introduction and naming problems 125
    8.2   Nuclear fusion in brown dwarfs 125
          8.2.1   Deuterium burning 125
          8.2.2   Lithium burning 126
    8.3   Evolution and surface parameters of BDs 127
    8.4   How ubiquitous are BDs? 128
    8.5   Deuterium, litium and cosmology 128
    8.6   The limit to giant planets 129
    8.7   Summary 130

9 Stars out of balance: from MS star to red giant — 131

- 9.1 Main-sequence stars 131
  - 9.1.1 Changes in the main-sequence phase 131
    - 9.1.1.1 Evolution due to the changing composition of the interior 131
    - 9.1.1.2 The end of the main-sequence phase 132
- 9.2 Effects of convection on the MS phase 132
  - 9.2.1 Stars without inner convection ($M_{\text{init}} < 1.15\,M_\odot$) 132
  - 9.2.2 Stars with inner convection ($M_{\text{init}} > 1.15\,M_\odot$) 133
- 9.3 Why and how does a star become red giant? 134
  - 9.3.1 A "gedankenexperiment": the gravothermal hysteresis cycle 134
  - 9.3.2 The hysteresis cycle and real stars 136
  - 9.3.3 A second red giant phase 137
- 9.4 The overall stellar thermal equilibrium (STE) 137
- 9.5 Isothermal He core and Schönberg-Chandrasekhar limit 139
- 9.6 Luminosity evolution of red giants 139
  - 9.6.1 Red giant luminosity depends on M_{init} 139
  - 9.6.2 Effects of metallicity 139
- 9.7 The core drives the evolution, the envelope follows 140
- 9.8 Duration of the main-sequence phase 140

10 Stellar evolution: Stars in the lower mass range — 141

- 10.1 Defining the low mass range 141
  - 10.1.1 The MS-mass limit of $\simeq 1.15\,M_\odot$ 141
  - 10.1.2 The MS-mass limit of $\simeq 0.5\,M_\odot$ 143
- 10.2 H shell burning: the red giant phase 143
  - 10.2.1 Evolution of the RG core and of the H-burning shell 144
  - 10.2.2 The RG surface: spectral lines, mass loss and dust 144
  - 10.2.3 The end of the RG phase: He ignition, He flash 145
- 10.3 Core He-burning stars 147
- 10.4 The end of core He burning and on to the AGB 147
  - 10.4.1 General aspects 147
    - 10.4.1.1 The end of core He burning 147
    - 10.4.1.2 Envelope thickness, pulses, dredge-up, hot bottom burning, s-process fusion 147
  - 10.4.2 Low mass core He burners ($M_{\text{init}} < 2\,M_\odot$): Horizontal-Branch stars 148
    - 10.4.2.1 HB stars and the various types 148
    - 10.4.2.2 Metal content and age of HB stars, morphology of HBs 149
    - 10.4.2.3 Evolution of stars on the HB and toward the AGB 149
  - 10.4.3 AGB stars: structure and evolution 151
    - 10.4.3.1 AGB star evolution and the CMD 151
    - 10.4.3.2 He-shell flashes (thermal pulses) and convection 151
    - 10.4.3.3 Third dredge-up: nuclear fusion and s-process 152
    - 10.4.3.4 Flashes and mass loss of fusion enriched material 152
    - 10.4.3.5 All happenings in a very thin layer 153
  - 10.4.4 Higher mass core He burners: blue loop stars and the AGB 153
    - 10.4.4.1 Stars with M_{init} larger than 7 to 8 $M_\odot$ 154
    - 10.4.4.2 Stars with $M_{\text{init}} = 2$ to 7 $M_\odot$ 154
    - 10.4.4.3 AGB stars and hot bottom burning 155
  - 10.4.5 Timescales 155
- 10.5 The end of the AGB phase 155
  - 10.5.1 Massive AGB stars: OH/IR stars and pAGB stars 156
  - 10.5.2 Low mass AGB stars: pAGB stars and planetary nebulae 156

- 10.6 The end phase: white dwarfs 157
  - 10.6.1 Classification of WDs 157
  - 10.6.2 Ultimate fate of WDs 159
  - 10.6.3 Born-again stars 159
- 10.7 Initial to final mass relation for lower MS stars 160
- 10.8 Some special stars 160
  - 10.8.1 Pulsational variables: RR Lyrae, δ Cepheids, PG 1159 and ZZ Ceti stars . 160
    - 10.8.1.1 RR Lyrae stars 160
    - 10.8.1.2 δ Cepheid stars 161
    - 10.8.1.3 PG 1159 stars 161
    - 10.8.1.4 ZZ Ceti stars (pulsating WDs) 161
  - 10.8.2 λ Bootes stars 161
  - 10.8.3 Cool subdwarf stars 162
  - 10.8.4 Blue stragglers 162
- 10.9 Gaps and bumps in the MS, HB, AGB 163
  - 10.9.1 Gap on the main sequence 163
  - 10.9.2 Gaps on the HB 163
  - 10.9.3 The RGB and AGB bumps 163
- 10.10 The Red clump 164
- 10.11 Summary 164

11 Stellar pulsation and vibration   167
- 11.1 Describing a star with oscillations 167
  - 11.1.1 The formalism 167
  - 11.1.2 Oscillations and limiting frequencies 168
  - 11.1.3 The driving forces of oscillations 169
- 11.2 Spherically symmetric radial pulsations 170
  - 11.2.1 Formalism for radial pulsation 170
  - 11.2.2 Atmospheric radial pulsations 171
  - 11.2.3 Details of the κ mechanism 172
- 11.3 Types of pulsational variables 172
  - 11.3.1 The instability strip: δ Cep, W Vir, RR Lyr, δ Sct, DA variables . . 172
    - 11.3.1.1 Cepheids 172
    - 11.3.1.2 RR Lyr 173
    - 11.3.1.3 δ Sct 175
    - 11.3.1.4 DA variables or ZZ Cet stars 175
  - 11.3.2 Main-sequence variables 176
  - 11.3.3 Red variables: Miras 176
  - 11.3.4 Massive variables (LBVs) 176
- 11.4 Vibrations 176
- 11.5 Helioseismology 177
- 11.6 Asteroseismology 178
  - 11.6.1 Doppler imaging and spotted stars 179
  - 11.6.2 Doppler-shift asteroseismology 179
  - 11.6.3 Photometric asteroseismology 180
  - 11.6.4 PG 1159, sdB, and DB variables 181
- 11.7 The Solar cycle of 11 years; effects on climate 181

12 Stellar coronae, magnetic fields and sunspots   183
- 12.1 Stellar coronae 183
- 12.2 Effects of radiation transport 184
- 12.3 Magnetic fields 184
- 12.4 Sunspots 185
- 12.5 Prominences and flares 185

    12.6 Relevance of the structures for stellar evolution 186

13 Stellar evolution:
Stars in the higher mass range   187
    13.1 Defining the high mass range 187
    13.2 Types of high mass stars 187
        13.2.1 The O and Of-type stars 188
            13.2.1.1 Determining the temperature of O stars 188
            13.2.1.2 Determining the mass of O stars 189
            13.2.1.3 Oe/Be stars 190
            13.2.1.4 Summary O type stars 190
        13.2.2 B type stars 190
        13.2.3 Wolf-Rayet (WR) stars 190
        13.2.4 Luminous blue variables: LBVs; P-Cygni stars 191
        13.2.5 Red supergiant stars 192
    13.3 Expanding envelopes, luminous winds 192
        13.3.1 Processes of radiation acceleration 193
            13.3.1.1 Radiative acceleration by the continuum 194
            13.3.1.2 Radiative acceleration through spectral lines 194
        13.3.2 Making a P-Cyg profile 195
        13.3.3 Mass loss 195
            13.3.3.1 Velocity profile 196
            13.3.3.2 Density profile 196
    13.4 Evolution and the HRD 197
        13.4.1 General nature of evolution of high mass stars 197
            13.4.1.1 Evolution of stars of $15 - 25$ $M_\odot$ 199
            13.4.1.2 When does a star evolve with a blue loop? 200
            13.4.1.3 Evolution of a 60 $M_\odot$ star 200
        13.4.2 Evolution and effects of metallicity 202
        13.4.3 Evolution and effects of rotation 203
        13.4.4 See a star evolve: P Cygni 203
    13.5 Nuclear fusion times and endphases of high mass stars 203

14 Rotation and stellar evolution   205
    14.1 General aspects of rotation 205
    14.2 Rotation and effects of deformation 205
        14.2.1 Rotation and variation in T_{eff} 205
        14.2.2 Rotation and effective gravity 206
    14.3 Possible effects of rotation on structure 206
        14.3.1 Rotation and meridional circulation 206
        14.3.2 Rotation driven instabilities 207
            14.3.2.1 Brunt-Väisälä oscillations 207
            14.3.2.2 Solberg-Høiland instability 207
            14.3.2.3 Baroclinic instability 207
            14.3.2.4 Shear instability 208
        14.3.3 Rotation of the Sun 208
        14.3.4 Convective flows will be turbulent 208
    14.4 Braking internal rotation 209
        14.4.1 Stabilizing forces 209
        14.4.2 Redistribution of angular momentum with evolution 209
    14.5 Magnetic field and rotation 209
        14.5.1 Rotation makes a magnetic field stronger 209
        14.5.2 Rotation braking by magnetic fields 210
        14.5.3 Loosing angular momentum 210

14.6 Rotation and mass loss 210
    14.6.1 Mass loss disks 210
    14.6.2 Mass loss and loss of angular momentum 211
14.7 Chemical effects of rotation: mixing 211
14.8 Rotation and mass loss affect high mass star evolution 213
14.9 Rotation and mass accretion affect WD evolution 215

15 The first stars   217
15.1 First stars have very low metal content 217
15.2 Making a star in metal-free gas 217
15.3 Evolution of first stars 218
15.4 Nucleosynthesis in Population III stars 221
15.5 Lithium in first stars 221

16 Models and variation of "free" input parameters   223
16.1 Effects on models and evolution 223
    16.1.1 Complications with convection 223
    16.1.2 Effects of metal content 224
    16.1.3 Effects of mass loss 225
    16.1.4 Effects of rotation 225
16.2 Effects of combined parameters 225

17 Degenerate stars: WD, NS, BH   227
17.1 White dwarfs 227
    17.1.1 Internal structure of WDs 227
    17.1.2 Atmosphere of a WD 228
    17.1.3 Cooling and crystallization of a WD; cooling time 229
    17.1.4 Chandrasekhar limit, maximum mass of a WD 230
    17.1.5 Transfer of mass onto a WD; Eruptions 230
    17.1.6 Can a WD become NS? 230
17.2 Neutron stars 231
    17.2.1 Two ways for stars to become NS 231
    17.2.2 Structure and mass of neutron stars 232
    17.2.3 The surface layers of a NS 232
    17.2.4 Behaviour of neutron stars: pulsars 232
17.3 Strange (quark) stars 233
17.4 Black holes 233
    17.4.1 Schwarzschild radius 234
    17.4.2 Observational evidence for the presence of stellar black holes 235
17.5 Nobel prize 2002 for X-ray astrophysics 235

18 Supernovae   237
18.1 Historical supernovae, supernova rate 237
18.2 Observed types of supernovae 237
18.3 Theories about supernovae 239
    18.3.1 Hydrodynamic (core collapse) supernovae 239
        18.3.1.1 Onset of the collapse 239
        18.3.1.2 The collapse 240
        18.3.1.3 End of the collapse and rebounce 240
        18.3.1.4 The explosion 240
        18.3.1.5 Decay of luminosity 240
        18.3.1.6 Endothermic nuclear reactions and light curve bump . . . 240
        18.3.1.7 Deceleration 241
    18.3.2 Thermonuclear supernovae 241

	18.3.3 Other mechanisms to make SNe	242
18.4	Supernovae and their progenitors	242
18.5	Hypernovae / Gamma-ray bursts	243
18.6	Initial mass of stars becoming super- or hypernova	245
18.7	SN Type Ia and cosmology	245
18.8	SN 1987A in the LMC	246
	18.8.1 SN 1987A itself	246
	18.8.2 Effects of SN 1987A on its environment	246
18.9	Endproduct of first stars: M_{init} to M_{final}	247

19 Evolution of binary stars — 249

- 19.1 Introduction — 249
- 19.2 Equipotential surfaces — 250
  - 19.2.1 Mathematical formulation — 250
  - 19.2.2 Graphical representation of equipotential surfaces — 250
- 19.3 Mass exchange — 251
  - 19.3.1 General case — 251
  - 19.3.2 Conservative mass exchange — 253
  - 19.3.3 Classification scheme for close binary systems — 255
  - 19.3.4 Complications — 256
    - 19.3.4.1 Non-conservative mass exchange — 256
    - 19.3.4.2 Accretion disks — 257
    - 19.3.4.3 Common envelopes; merging stars — 258
- 19.4 Evolution of binary stars — 259
  - 19.4.1 Towards massive X-ray binaries and beyond — 259
  - 19.4.2 Towards low-mass X-ray binaries — 263
  - 19.4.3 Microquasars — 263
  - 19.4.4 Low mass binary systems: towards cataclysmic binaries, SN Ia — 264
  - 19.4.5 WDs and rotation: Nova and SN Ia phenomena — 265
- 19.5 Variety of binary evolution; special objects explained — 266
  - 19.5.1 Multiple branching in binary evolution — 266
  - 19.5.2 Special objects now explained by binary evolution — 267
    - 19.5.2.1 Cataclysmic variables; Novae; Supersoft X-ray sources — 267
    - 19.5.2.2 Type Ia supernovae — 267
    - 19.5.2.3 Type Ib and Ic supernovae — 268
    - 19.5.2.4 X-ray binaries (HMXB, LMXB) — 268
    - 19.5.2.5 Binary pulsars — 268
    - 19.5.2.6 High speed OB stars — 268
    - 19.5.2.7 Merged stars — 268
- 19.6 Summary — 269

20 Luminosity and mass function — 271

- 20.1 The luminosity function — 271
- 20.2 The stellar initial mass function — 272
  - 20.2.1 Power law mass functions; equivalences — 273
  - 20.2.2 Salpeter mass function — 274
- 20.3 Relation between the luminosity and mass functions — 274
- 20.4 Determinations of the mass function — 274
  - 20.4.1 Star clusters — 275
    - 20.4.1.1 Open clusters — 275
    - 20.4.1.2 Globular clusters — 275
    - 20.4.1.3 Mass segregation — 275
  - 20.4.2 Field stars — 276
  - 20.4.3 Completeness of the photometry — 276

20.4.4 Results for mass functions	277
20.4.5 The high-mass end of the IMF	278
20.5 The IMF and its universality	279
20.6 The mass function for the first stars	279

21 Isochrones — 281
21.1 Definition . . . 281
21.2 Examples . . . 281
    21.2.1 Effects of metal content of stars . . . 283
    21.2.2 Transforming (L,T)-isochrones to $(M_V, B-V)$-isochrones . . . 283
    21.2.3 Difference between isochrones and evolutionary tracks . . . 284
21.3 Using isochrones in CMDs . . . 284
21.4 Synthetic CMDs . . . 285
21.5 Special CMD-regions to find the age of star groups . . . 287
21.6 Star formation history (SFH) . . . 288
    21.6.1 Photometric SFH . . . 288
    21.6.2 SFH and synthetic spectral energy distributions . . . 288

22 Stars influence their environment — 289
22.1 Star formation and IS cloud metal content . . . 289
22.2 Effects of first stars . . . 289
22.3 Chemical evolution . . . 290
    22.3.1 Consumption of primordial D, Li, He . . . 290
    22.3.2 Metal production and yield . . . 290
    22.3.3 Radioactive decay and nucleochronometry . . . 291
22.4 What comes of all evolution? . . . 292
    22.4.1 Stars and their light . . . 292
    22.4.2 Stellar remnants . . . 292
    22.4.3 Gas returned to IS space . . . 293

23 Summary; Questions, Constants, Acronyms, Lists — 295
23.1 Stars and their structure . . . 295
23.2 Stars and their evolution . . . 296
23.3 Stellar evolution in comparison . . . 297
23.4 Stars and effects for their environment . . . 300
23.5 List of questions . . . 301
23.6 Acronyms, Constants, Abbreviations . . . 303
23.7 List of Figures . . . 305
23.8 List of Tables . . . 310
    Index . . . 311

Preface

Most of the baryonic mass in galaxies is stored in stars, and stars are the objects we can see easily. Stars come in a large variety of shapes and states, reflecting the different possibilities nature has, as well as the fact that stars evolve in the course of their lives.

The functioning and behaviour of stars is, of course, based on two levels of physics. One level is that of large scale structure. It is governed by gravity, macroscopic gas physics, and the way energy is transported through gas. The other level is that of microphysics. This includes the processes of nuclear fusion, the physical state of the gas (also under extreme conditions), and the effects chemical composition and ionization structure have on the energy transport by radiation and convection. The intimate interplay between the two levels, together with the fusion-driven changes taking place inside the star, make "stars and stellar evolution" a fascinating and very broad topic. Moreover, stars need not exist all by themselves (as the Sun does) but may exist in pairs, which can come to intensive interactions during their evolution.

The topic of the book touches on the questions "From where did the Sun come?" and "What will be its fate?". Yet, the Sun plays only a minor role in this text (except that its mass, size, and luminosity are the basis for the stellar units).

The emphasis is on all stars with all their evolutionary phases. The text does not aim to explain all the intricate physics for and in stars. Most of the standard physics is included, of course; for details of aspects not treated in depth, references to other texts are provided. But essential mechanisms are addressed. Nor does the text pretend to be a full review of the literature. But it gives the reference background as well as access to the specialized literature on the various topics. What has been attempted is to give a description for most (but not all) of the phases of evolution possible in their context, illustrated by numerous forms of the "Hertzsprung-Russell Diagram". The text thus rather aims at all those, the general astronomer and the observer alike, who need to understand where an encountered star can be placed in the vast parameterspace of stellar evolutionary states.

Two chapters deal with the nature of stellar ensembles, addressing interpretational problems encountered with their observations but also the possibilities to test theories about stars and their evolution. Here the application of all stellar evolutionary phases presented comes to bear on understanding large and distant stellar systems.

Numerous figures have been included from the literature. The aim was to include figures of didactical relevance while at the same time trying to have figures from original research works. Finding a balance there is not easy and, ultimately, the choice is always subjective. In several cases we had to adapt the figures to suit the readability or to better suit the didactical use, which in most cases meant enlarging the labelling, sometimes also adjusting the lay-out. We hope the original authors approve of those adaptions.

Various figures have been remade from original data. These are in particular Hertzsprung-Russell diagrams, in order to have them all in the same scale (i.e., the same axis ratios, 4 units in $\log L$ for 1 unit in $\log T_{\text{eff}}$). All these can, after proper overall scaling, be overlain.

The text has its origins in our class on "Stars and Stellar Evolution", taught in german. The class has had its own evolution: the introduction of the "Bonn International Physics Programme" in 1998 provided the impetus to rework and translate the german write-up and to improve the description of many aspects of stars and their evolution. The text thus grew over the years.

The chapters of this book were composed by KSdB and WS with topical contributions from Tom Richtler (TR) and Tim Schrabback (TS) while several students allowed parts of text from their Theses to be included.

We thank Georges Meynet and Allen Sweigart for providing new data of model calculations which allowed us to (re-)make various figures. Georges generously supported our endeavour, gave extensive advice, prevented errors and proposed various improvements of the presentation. We thank Steve Shore for a critical reading of the manuscript and Michel Breger, Alvio Renzini, Detlev Schoenberner and Ed van den Heuvel for that of individual chapters; their suggestions helped to make the text in many places much clearer. Our colleagues Thibaut Decressin, Patrick Eggenberg, Michael Hilker, Norbert Langer, Maria Massi, Klaus Strassmeier, Allan Sweigart, Karel van der Hucht and Klaus Werner provided data and/or advice on various parts of the text. Over the years, numerous students made suggestions and we are grateful for their encouragement. We thank in particular Martin Altmann, Torsten Kaempf, Manuel Metz, Soroush Nasoudi, Jörg Sanner and Philip Willemsen for their contribution of diagrams and/or permission to include textparts of their respective theses.

We of course are indebted to many colleagues for their permission to include figures from their research papers in our text. Many colleagues gave additional advice related with these figures. As said, in many instances figures were adjusted in lay-out for the needs of this text.

Thanks go also to the respective publishers for their permission to reproduce figures from the original publications according to the respective copyright rules. These include: the European Southern Observatory for figures from Astronomy & Astrophysics and Astronomy & Astrophysics Supplement Series (through the Editor-in-Chief); the American Astronomical Society ("reproduced by permission of the AAS") for figures from the Astronomical Journal, the Astrophysical Journal, and the Astrophysical Journal Supplement; the Annual Review Corporation ("Reprinted, with permission, from the Annual Reviews of Astronomy and Astrophysics by Annual Reviews www.annualreviews.org"); the International Astronomical Union for figures from the proceedings of their Symposia and Colloquia; Springer and Kluwer for figures from the Astronomy & Astrophysics Review and some of their other publications ("With kind permission of Springer Science+Business Media"); Blackwell Publishers for figures from the Monthly Notices of the Royal astronomical Society; Wiley-VCH for figures from Reviews of Modern Astronomy; and the Publications of the astronomical Society of the Pacific (here the rights reside with its authors). Of course, with each figure the reference to the source is given.

A few figures have been obtained from sources, whose location has slipped from our memory. We request the authors of these figures to make this known to us so that proper credit can be given at a later stage.

Finally, we thank Marie-Louise Chaix, France Citrini, Jean Fontanieu and Jean-Marc Quilbé of EDPSciences for their essential and technical support.

We hope that all readers will benefit from this text. We will be grateful for all comments and suggestions for improvement.

September 2008

Klaas S. de Boer & Wilhelm Seggewiss
Sternwarte of the
Argelander Institut für Astronomie
Universität Bonn

Chapter 1

Introduction

1.1 Historical background

The brilliance of stars on a clear night at a place not yet affected by light pollution is amazing. The uneven distribution of stars over the sky, their range in brightness, and possibly the recognition that they have different shades of colour make stars fascinating objects.

1.1.1 History of the characterization of stars

With the realization in the 17th century that the Sun is a gigantic source of heat around which the planets revolve came the first thoughts about the nature of the nightly twinkling lights. If they were objects like the Sun, they really must be far away. Christian Huygens (1695; *Kosmotheoros*) tried to calculate the distance to the brightest star by comparing its brightness with the brightness of the Sun as seen through the same telescope[1]. He found from the so estimated intensity ratio that the distance to Sirius would be just a factor of four smaller than the real distance to that star (not knowing about and thus not considering the intrinsic differences between stars...).

Kepler and Galileo derived from planetary motions that the Earth revolves around the Sun. Galileo suggested that the distance of stars might be found using the **annual parallactic shift** of their positions, to implicitly prove that the Earth moves indeed. The first parallax was measured in 1838, independently by Bessel, by Henderson and by Struve. Once sufficient parallactic distances were known it became feasible to intercompare stars in a systematic manner. A **reference distance** was agreed upon: **10 pc** (parsec), the distance of a parallactic shift of 1/10 of an arcsecond. The brightness a star would have at this distance is since called **absolute magnitude**.

In particular Hertzsprung noted (based on parallaxes) at the beginning of the 20th century that red stars came in two kinds, the very luminous ones and the feeble ones. With knowledge of the Planck function (1900!) for radiating bodies, equal colour implied equal temperature and thus equal output of a unit surface area, so that the more luminous star had to have a large total surface, thus had to be big. This led to the type names **giant star** and **dwarf star**.

Spectroscopy was essential too. Using the knowledge of laboratory absorption and emission spectra of flames (late 19th century) as well as of atomic physics (early 20th century), explaining many a spectral line from atoms as well as from ions, one could start to investigate the chemical composition of stars. More importantly, understanding the spectral lines led to the derivation of the surface temperature of stars. Thus it became possible to sort the original spectral classifications (Secchi, Manny, Annie Cannon) into a **spectral sequence** *running parallel with temperature*. Russell combined the new spectral types with the absolute magnitudes of the stars.

---

[1]Huygens designed for that purpose a very small diaphragm to be mounted in front of the telescope which he then used to observe the Sun. In order to get, during day time, his eyes adapted to conditions of nightly observing, he sat a long time with a cloth wrapped around his head in the darkened living room with the telescope ready. Only then did he (with the help of an assistant) look through telescope and diaphragm to the Sun. It required many experiments to get everything (including diaphragm) right! (Story as retold by Andriesse 1994).

Ultimately these investigations resulted in the establishment of a diagram, soon called the **Hertzsprung Russell Diagram (HRD)** after the mayor players of the beginning of the 20th century, of **absolute magnitude against spectral type**.

1.1.2 History of the ideas about the evolution of stars

Stars were originally thought to be eternal lights. They were fixed and stable for all time.

Yet, once in a while something happened in the sky (beyond eclipses or the appearance of a comet). In Chinese annals "guest stars" are mentioned: a star became visible and then faded. Later detailed studies by western astronomers showed these guest stars were novae or supernovae (we still see the remnant of the one of 1054, the Crab Nebula). Also in Europe such phenomena were observed (see Table 18.1). In 1596 Fabricius discovered that the star Mira (o Cet) is variable (meaning not stable, thus not eternal). This led to new speculations about the nature of stars.

Since the beginning of the 17th century, with the availability of the telescope, "double stars" received considerable attention. Using the formalism for gravitational attraction developed by Newton (late 17th century), the observed shifts in relative position meant these double stars were orbiting stars, a real discovery. It did fit to the recognition that further orbiting systems exist beyond the solar system (the planets around the Sun) of which Jupiter with its revolving moons was one of the great discoveries by Galileo. Thus orbital periods and the first stellar masses were estimated. These required, however, that a parallactic distance was known, too (the major axis of the orbit has to be known in length units).

Once spectra could be seen (and later photographed), differences in spectral appearance became known. This led to the spectral classification scheme developed at Harvard. Herschel divided light into the visible and the "invisible" rays. The visible ones were between the violet and the red (the "cool rays"), the invisible ones were beyond the red. These latter ones were called the "hot rays", because when sunlight was concentrated by a lens, these could set things on fire. That the cool rays have more energy than the hot rays became gradually clear and was well established with the formulation of Planck's formula (Planck 1901).

Until the turn of the 20th century, one had no idea about the source of the energy of the stars. Estimates existed for the time span the Sun might burn if it were using some of the known fuels, like oil, gas or coal. It led to a life span of several ten-thousands of years, already longer than the time (as calculated from biblical texts) since the creation of the universe (see, e.g., Toulmin & Goodfield 1965).

It was speculated that stars might accrete H from interstellar space and (chemically) burn it to H_2O. However, this provided too little energy. Or the Sun might, according to Mayer in 1849, accrete meteorites and convert the potential energy into radiation energy. The Sun would thus accrete mass and must have been lighter in the past which led to conflict with knowledge about motions in the solar system (eclipse data). Or the solar energy could just come from contraction, as proposed by Helmholtz in 1853. Kelvin showed in 1862 that this would allow for a solar age of up to 50 million years. From thermodynamics it further followed that the stars must be really hot inside (balance between gravity and internal pressure) and that if they were cooling and contracting (what they naturally should do), they would radiate for only a few million years. These times were much shorter than the then already accepted age of the solar system of the order of 10^9 years, estimated from the geological history of the Earth (see, e.g., Cutler 2003).

When Einstein developed his theories about energy and matter and proposed their 'equivalence' (through $E = mc^2$), a mechanism for energy production offered itself: destruction of matter and conversion into radiation. In this manner stars would radiate and become lighter with time. With long life spans, stars clearly must show *ageing* and thus *evolution*.

Once the HRD was discovered and with it the ordering of stars along the **main sequence**, the idea thus was that stars form with high mass and that they use the mechanism of mass destruction to produce energy (radiation). Since the more massive stars were the intrinsically brighter (as had been derived from the few well investigated double star systems), they would have a high rate of mass destruction. So stars should become progressively less massive, would progressively become less compact and thus would produce less energy. In other words, stars would evolve *down* along the main sequence. The total life span of stars was then calculated to be of the order of 10^{13} years. Again the age of the universe came into play.

A completely different approach to find the age of the universe came from investigations of the kinetic energy of stars. That energy was found to be roughly equal for all stars, as had been reiterated by Seares in 1922. It implied, assuming that with time the kinetic energy of stars in an ensemble would become equal, that stars should be as old as 0.5-$1.0\,10^{13}$ years. The age estimates from mass destruction and from kinematics agreed, thus establishing a consistent theory.

Eddington and others had in the 1920s speculated about the fact that the mass of a He-nucleus is somewhat less than that of 4 H-atoms and that this might play a role in the generation of energy in stars. With the developments of nuclear physics the idea emerged that perhaps nuclei can fuse somehow, in this case 4 H $\rightarrow$ ^{4}He, producing energy through $E = mc^2$ from the mass deficit. It did not provide as much energy as the process of complete destruction of matter, but allowed a star to shine more or less stable for periods of up to 10^{11} years[2]. That time span, however, was much shorter than the age derived from the mean kinematic energy of stars in space.

Ultimately many of the problems of the above theories were solved.
- First, von Weiszäcker (1937) and Bethe & Critchfield (1938) proposed a way for the fusion of H into He, so a process was established. Clearly not all H of a star could be transformed since the fusion required very high temperatures and only the core is hot enough for that fusion.
- Secondly, stars operate as if having a highly sensitive thermostat. They radiate in a very stable manner: if the interior would heat up somewhat and thus have more fusion it would expand, thereby reducing the internal density and thus the rate of the fusions taking place, and vice versa. This ensures that stars stay more or less *at the same location on the main sequence*, until the fuel runs out at which point large structural changes take place.
- Finally, the considerations about the kinetic energy of the stars and the age derived from that had to be revised because, with the discovery of the orderly motions of galactic rotation by Oort in 1924, the assumptions made to derive an age from stellar kinematics showed to be not adequate.

For some of the history of these ideas see, e.g., Beckett (1874), and Jeans (1934, 1st edition; the later editions were revised to include nuclear physics and older ideas were omitted).

Once fusion of hydrogen was established as an astrophysical reality the quest was on to see if further fusion processes can take place. In statistical equilibrium He may start to form berillium which may fuse on to carbon but temperatures required are much higher than known to exist in normal main-sequence stars. Moreover, at such temperatures ^{12}C would soon fuse with He to form oxygen, leaving little of the carbon, in fact much less than the universe appears to contain. Fred Hoyle predicted that there should be a "nuclear resonance level" of ^{12}C at about 7.65 MeV allowing the reaction ^{8}Be + ^{4}He $\rightarrow$ ^{12}C + γ to take place at lower temperatures than based on statistics. This prediction was experimentally confirmed and the success formed the basis for his drive to uncover (with others) the sequence of fusion steps leading to elements heavier than He (for details of the story see Hoyle 1994). The full theory was described in the famous paper of Burbidge, Burbidge, Fowler, & Hoyle (1957), better known as B^{2}FH.

1.2 Stellar evolution - the importance of gravity

Stars are born, live their lives, evolve and come to an end. The life of stars is governed by three important fundamental forces: gravity, the electromagnetic force, and the strong nuclear force. Of these, gravity determines the course of the evolution of a star, i.e., to contract, while the others are, in a way, just modifying the general workings of contraction. A general treatment of questions about the nature of cosmic structures and the relevant physics is given by Celnikier (1989).

Stars form from large and cool gas clouds. Their formation sets in once a sufficiently large amount of material is assembled in a sufficiently small volume so that the self-gravity induces a contraction tending toward an as small as possible sphere. This contraction, which leads to a build-up of pressure, is first hampered by the counteracting forces based on the physics of gases.

---

[2] From radioactive dating techniques the age of the oldest meteorites was established to be in the order of $4 \cdot 10^9$ years. Meteorites are believed to have formed together with the Solar system, so the Sun must be of similar age.

On the other hand, the interaction of pressure (i.e., the kinetic energy of the particles) with the atomic structure leads to transformation of energy into a dissipatable form, allowing radiative cooling. Thus, contraction will continue as long as this energy can be dissipated.

Further contraction is slowed down due to the workings of the electrostatic force. With contraction, the gas is heated and therefore becomes (partially) ionized. Particles with the same charge now repel each other. Oppositely charged particles attract each other but given the high temperature their high-speed collisions effectively do not lead any more to recombination.

The workings of the strong nuclear force allow processes which (in very hot gas) lead to efficient, but not indefinite, counteraction of contraction. In the high-density, high-temperature centres of stars the kinetic energy of a fraction of the (fully ionized) atomic nuclei can overcome the mutual electrostatic potential barrier and the strong nuclear force then allows the two nuclei to fuse. In such reactions energy is liberated providing the heat to keep the internal pressure up and thereby preventing further contraction. There is a delicate balance here because a too large fusion rate will increase the temperature and so induce expansion of the gas making the fusion slower (so a function like a thermostat).

A star tries to be in "overall thermal equilibrium" always. However, energy is continuously radiated away because a star is like a "leaky ball". This loss will be made up by gravity, i.e., by contraction inducing higher density and temperature and so more fusion. One can say that, essentially, the brightness of a star is due to the force of gravity, acting on matter able of nuclear fusion.

Once the supply of material for the actual fusion process is exhausted contraction sets in again, increasing the pressure inside the star. In that case the temperature may increase enough to initiate other nuclear fusion processes, which again halt contraction. The actual sequence of such fusion-based intermissions of the contraction depends on the initial mass of the star.

Summarizing, *gravity pushes the conditions inside a gaseous sphere (via intermediate steps) into a high-density high-temperature regime which may allow fusion ractions (within the limits posed by the nuclear forces) leading to a continuous adjustment of the stellar thermal equilibrium. It also leads to an overall lower energy state because the energy liberated by the fusion leaks away.*

Ultimately, no further energy source is available to halt the persistent workings of gravity and the final contraction sets in. This leads to a very cool and very dense gaseous sphere (again limited by the physics of dense gases) consisting of highly condensed material.

The diverse possibilities of such stages of contraction lead to large ranges in stellar structure. The way of the changes in those structures is the topic of stellar evolution.

1.3 Relevance of stars for astrophysics

Studying stars has many good reasons. These can be grouped into six broader categories.

- **Studying stars for their own sake**

A prime goal is to find out how stars are constructed. What makes a gaseous sphere to be a star? Why are stars round and why does a gaseous sphere have so sharp an edge? Do stars come in varieties, large and small or hot and cold? What is the run of parameters such as density and temperature in a star? What is the source of the energy produced? How do stars evolve? And how do stars form? Do planets come with them? Then, maybe of most immediate relevance for mankind, can we understand the Sun?

- **Studying stars as members of stellar systems**

Given the uneven distribution of stars over the sky, immediately the question about the nature of their spatial distribution presses itself upon us. Moreover, simple observations showed the existence of *clusters of stars*. What are they and how do they behave? With spectroscopic investigations it became clear that stars do not all have the same chemical composition. Is there chemical evolution? Is that related with evolution of the Milky Way as a galaxy? Is there a change in the appearance of stellar systems with time?

- **Stars synthesise heavy elements**

According to the "big bang" model for the universe, in the beginning only the elements hydrogen, helium, lithium and perhaps some boron were created. All heavier elements have been created later in stars through nuclear fusion. In the course of the life of a star but in particular near the end, stellar material enriched in heavy elements is returned to interstellar space. Thus the chemical composition of the interstellar medium (ISM) changes with time and so reflects the chemical evolution of the universe.

- **Stars generate energy for the ISM**

During the life of a star, photons are released which contribute to the energy balance of the ISM. And at the end of a stars life, lots of energy is dumped into the ISM either trough strong stellar winds or through supernova explosions. Thus stars strongly influence the energetic structure of the interstellar matter.

- **Stars as objects to help unravel the structure of the Milky Way**

Once the nature of stars is known from their spectra, their apparent brightness allows to infer their distance. This then is input for studies of the spatial distribution of the stars, and, including radial velocities and proper motions, for studies of the kinematics and structure of the Milky Way.

- **Stars as objects to calibrate cosmological distances**

If stars evolve and if the Galaxy evolves, there must have been a beginning of it all. Since Hubble we know the universe expands and also this points at some origin. In order to find the structure and the evolution of the universe one has to be able to determine accurate galaxy distances. Only the stars are the well defined objects fit to determine distances (gas clouds are diffuse and of non-standard shape and size). Thus stars are at the base of all distance determinations, be it direct through parallaxes or indirect using photometry and/or spectroscopy with the HRD, or using the nature of the variable RR Lyr and δ Cep stars. Furthermore, galaxies can be considered as complete populations in evolution. And at large distances in the universe we observe galaxies in a state they had in the past (as defined by our time), in particular as seen at optical and near-IR wavelengths, at which the radiation is predominantly stellar radiation.

In short, there are many very good reasons to study stars and to explore their structure to arrive at knowledge about their nature and evolution as well as that of other cosmic entities.

The goal of this text is to provide insight in stars and in stellar evolution. Basic formulae of physics will be provided to allow an interpretation of the radiation seen from the gaseous atmospheres of stars. It also will provide formulae about the structure of gaseous spheres. The more exotic aspects of the behaviour of star gas under circumstances not producible on earth is presented, too. Note that the more interesting phenomena of stars and stellar evolution occur when the gases are near a critical state, in general near phase or structure transitions[3].

Modelling stars is a rather complex process and the outcome of the modelling is not always understandable at first sight. Its ultimate aim is to explain the nature and origin of all important kinds of stellar objects known from observations and of all evolutionary sequences.

1.4 Elementary astronomy and classical physics

A brief rehearsal of elementary astronomy is in place here. When observing stars one generally gets information about the stellar surface. The astronomy given in this section therefore deals ONLY with aspects of the stellar surface. Some basic physics is added.

1.4.1 Classical observations

Observations of stars comprise measurement of the brightness in wider or narrower spectral bands or the recording of spectra.

- **Photometry** was, since the 1950s, largely performed with the Johnson filter system (see Tab. 1.1), which actually is the filter set used for early colour television experiments. Since the 1970s a new system became popular, devised by Strömgren. The filter bands were narrower (see Tab. 1.1) and were chosen to lie in parts of the spectral energy distribution with relatively strong differences between star types. The brightness in a chosen wavelength band is given in *magnitudes*.

---

[3]Such critical states show that a substantial portion of changes and happenings is of a stochastic nature and results in power law distributions (Buchanan 2000).

Table 1.1: Names of some photometric filter bands with their wavelength(width) [nm]

Johnson	U 365(90)	B 440(100)	V 550(100)	R 700(150)	I 900(150)
Strömgren	u 350(38)	v 410(20)	b 470(20)	y 550(20)	βN 486(3), βW 486(15)

Numerous other photometric systems have been developed with the same goal in mind. The ones surviving are those with which large amounts of data were/are accumulated, which is the case with the astrophysically useful systems.

Note that the intensity I_0 as arriving at the top of the Earth atmosphere is transformed by several wavelength dependent functions into the signal S measured

$$S(\lambda) = I_0(\lambda) \ A(\lambda) \ O(\lambda) \ F(\lambda) \ Q(\lambda) \tag{1.1}$$

with A being the extinction by the Earth atmosphere, O the absorption by the telescope optics, F the transmission function of the chosen filter, and Q the quantum efficiency of the detector.

Since the late 1960s satellites provided photometry at wavelengths not observable from the ground, such as, e.g., in the UV (the satellites OAO-2 and ANS at 150, 180, 220, 250 and 330 nm) and in the IR (the satellite IRAS, at 12, 20, 60 and 100 μm).

A very useful parameter is the **colour index**. It is given by

$$m_X - m_Y = -2.5 \log \frac{I_X}{I_Y} \quad (+ \text{ some constant}) \tag{1.2}$$

and is, in fact, proportional to the slope of a spectral energy distribution in a double logarithmic diagram ($\log I$ vs. $\log \nu$). Note that the convention is to always have the sequence of *shorter wavelength* X before the *longer wavelength* Y. The best known colour index is $B-V$ but any combination is valid (in the Johnson system[4]. e.g.: $U-B$, $B-V$, $V-I$, etc., and in the Strömgren system: $b-y$, or the more advanced indices $m_1 = (v-b) - (b-y)$, $c_1 = (u-v) - (v-b)$, and $\beta = \frac{\beta W}{\beta N}$). A graphic representation of a simple colour index can be found in Fig. 1.1. Note also that a colour index can be measured irrespective of the distance of the star (with a big enough telescope). But there are effects of interstellar reddening[5].

- **Spectroscopy** allows to classify spectral energy distributions according finer detail, leading to spectral types. The standard MKK system (see Ch. 3.8) has the types O,B,A,F,G,K,M with numeric subdivisions as well as the further types: R,N,C. However, the outcome of the typing depends on the spectral dispersion used as well as on the spectral range considered (see in, e.g., Corbally et al. 1993). It leads only to a *type* and *not directly to physical parameters*. Clearly, for 'normal' stars the types can, after calibration, be translated into temperature and further parameters. In addition there is the *luminosity class*, a crude indication of luminosity (based on spectral line widths) which is not in a simple way related with location of a star in the HRD. Spectral lines of numerous elements can be found in the lists published by Moore (1972).

Spectroscopy at other wavelengths became available from satellites, too. Important is the International Ultraviolet Explorer (IUE) satellite (1978-1997) with low dispersion (7 Å resolution) and high dispersion ($\simeq 25$ km s^{-1} resolution), spectra between 1150 and 3200 Å. The Hubble Space Telescope, HST, records with its instruments shorter stretch spectra at various resolutions from the UV to the IR and with very good signal-to-noise ratio (S/N). IR spectra are available from the Infrared Space Observatory, ISO (portions between 2 and 100 μm).

- **Parallaxes** are measured as small semi-annual shifts of nearby stars against the 'fixed' background star field. The best parallaxes come from fields in which all motions can be checked against extragalactic sources. The measured parallactic angle π leads to the distance of the star

$$d \ [\text{pc}] = 1/\pi \quad \text{with } \pi \text{ in arcsec} \tag{1.3}$$

---

[4] The colour indices are by definition = 0 for stars of spectral type A0; the transformation into cgs units was calibrated later. For details about absolute calibrations see Hayes (1985)

[5] Interstellar dust causes a dimming of light where the shorter wavelengths are more extinguished than the longer; the absorption A_V [mag] is related to the extinction colour change $E(B-V)$ as $A_V \simeq 3.1 \ E(B-V)$.

with 1 pc = 206265 AU = $3.09 \cdot 10^{18}$ cm. The visual brightness $m_V = V$ (brightness in the V-band in magnitudes) of a star at 10 pc distance is called the absolute brightness with **absolute magnitude,** M_V. When the true distance d of a star is known, its V can be transformed into M_V, the brightness it would have at the reference distance (of 10 pc)

$$m_V - V = V - M_V = 5 \log d - 5 \quad . \tag{1.4}$$

Very accurate parallaxes have been obtained in the satellite project HIPPARCOS (= HIgh Precission PARallax COllecting Satellite).

- **Velocities** of stars come in several varieties.

Radial velocities can be derived from the Doppler-shift of spectral lines with respect to their reference values (from laboratory spectra) as $v_{\rm rad} = (\Delta\lambda/\lambda) \cdot c$.

Transverse velocities can be found from the angular motion, called the **proper motion**, of a star with time (astrometry; comparison of photographic plates; requires time intervals of years!) together with its distance. (The farther away a star is, the smaller its proper motion.)

Rotation of a star shows up in the shape of spectral lines.

Temperature and perhaps **turbulence** show up in velocity widths of spectral lines.

Rotation, temperature and turbulence can be derived, if the observations are good enough (spectral dispersion), from the *shape* of the spectral line.

- **Luminosities** can be obtained only if the distance of the object is known, too. The luminosity L of an object is normalized using the luminosity of the Sun. The 'solar constant' forms the basis, with $L_\odot = 3.85 \cdot 10^{33}$ erg s^{-1}. **Note:** astronomy uses predominantly the "astronomical" units such as pc, $M_\odot$, $L_\odot$, etc., which are based on cgs units. For conversions see Table 23.2.

1.4.2 The Planck function

The most important form of continuum emission in physics and astrophysics is the thermal or heat radiation. For a source in **thermodynamic equilibrium**, in which all radiation and collision processes are in balance, the radiation, also called **black body radiation**, is given by Planck's formula (Planck 1901)

$$B_\nu(T) = \frac{2h\nu^3}{c^2} \cdot \frac{1}{e^{h\nu/kT} - 1} \tag{1.5}$$

with h Planck's constant and k the Boltzmann constant. B is the emitted energy [erg] in a unit of frequency [Hz] per unit of time [s] from a unit of surface [cm^2] into a unit solid angle [ster], and thus B [erg s^{-1} cm^{-2} Hz^{-1} ster^{-1}].

This equation can be approximated for the two regimes well away from the maximum of the function. These are:

1) the Wien approximation: for $\frac{h\nu}{kT} \gg 1$ one has $(e^{h\nu/kT} - 1) \simeq e^{h\nu/kT}$, thus

$$B_\nu(T) \simeq \frac{2h\nu^3}{c^2} \cdot e^{-h\nu/kT} \quad . \tag{1.6}$$

2) the Rayleigh-Jeans approximation: with $\frac{h\nu}{kT} \ll 1$ a Taylor series expansion, leading to

$$B_\nu(T) = \frac{2\nu^2 kT}{c^2} \quad . \tag{1.7}$$

The Rayleigh-Jeans approximation is (when plotted logarithmically) a straight line. A collection of Planck functions is shown in Fig. 1.1.

Note that the *intensity* I of radiation is the energy per unit energy band crossing through a unit suface in unit time from a given solid angle. The solid angle may be very small when the intensity from a particular source (a star) is measured. The units of I are the same as those of B, yet these parameters are different in what they stand for.

The temperature of objects may follow just from the slope of the spectral energy distribution, if it is a Planck function (see for the method Fig. 1.1). For stars a colour index measured in the visual can be used (within limits, Ch. 2.10) to find the temperature of the surface.

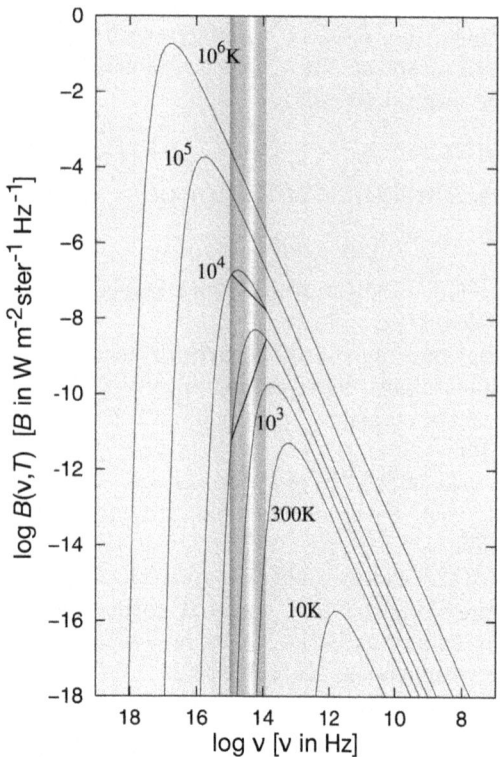

Figure 1.1: Logarithmic representation of the Planck function for black-body radiation. The curves are labelled with the temperature, the unlabelled one is for $T = 6000$ K, close to the temperature of the Sun. The shaded band indicates the visual part of the electromagnetic spectrum. The left side of the Planck function ($h\nu/kT \gg 1$) is the Wien part, the right side ($h\nu/kT \ll 1$) is the Rayleigh-Jeans part (see Sect. 1.4.2).

When performing photometry, one measures the intensity in some wavelength or frequency band. Relating intensities in two such bands leads to a *colour index* (see Sect. 1.4.1). The two straight lines in the graph show the 'slope' between two points on each of two Planck curves representing such 'measurements'. This slope corresponds uniquely to the temperature. When $h\nu/kT$ is small (the temperature of the source is very high with respect to the frequency used for the measurement) the measuring domain lies in the linear Rayleigh-Jeans part (so the same slope for a large range in T) and only a lower limit to the temperature can be derived.

It has to be emphasized that the radiation coming from the surface of a star hardly ever can be described by a Planck function. Stellar spectral energy distributions are based on a Planck function but with considerable deviations due to a large number of opacity effects (see Chs. 2 & 3).

1.4.3 Spectral lines, metallicity, and gas conditions

In spectra, absorption structures are visible caused by absorbing atoms and ions of the gas the light passes through. Extensive tabulations of spectral lines are available by Moore (1972), and of transition probabilities by, e.g., Wiese et al. (1966). The energy levels of atoms and ions, often shown in *Grotrian diagrams*, are determined from quatum mechanics.

The gas in stellar atmospheres is mostly in or near thermodynamic equilibrium leading to some population of the various energy levels in atoms and ions. One can model this and make a comparison with observations.

The line strengths and shapes are governed by the conditions in the gas. Temperature and gas density together set the excitation and ionization state of the gas (Saha equation; Ch. 3.2.2) and thus which spectral lines are present, as well as the strength of the lines. The (observed) level of ionization of a given element can be used to derive the temperature, thus also explaining the spectral types. The gas density is mostly visible in the width of spectral lines (see Fig. 3.1), allowing to find the surface gravity and the luminosity class of the star. Furthermore, the strength of the lines can, in relation with the temperature of the gas, also be used to derive the amounts of absorbing material and thus to find the chemical composition of stellar atmospheres.

Normally, hydrogen makes up 90% of the atmosphere gas (number fraction), helium 10%. Heavier elements, collectively called "metals", make up less than 1/1000 of the atmosphere nuclei. The abundance of an element M is normally given relative to the Solar value as

$$[\text{M/H}] = \log(N(\text{M})/N(\text{H}))_* - \log(N(\text{M})/N(\text{H}))_\odot \quad . \tag{1.8}$$

The total of metals is also known as "metallicity". The abundance of H, He, and the metals is also indicated with the letters X, Y, Z.

1.5. THE SURFACE PARAMETERS OF STARS

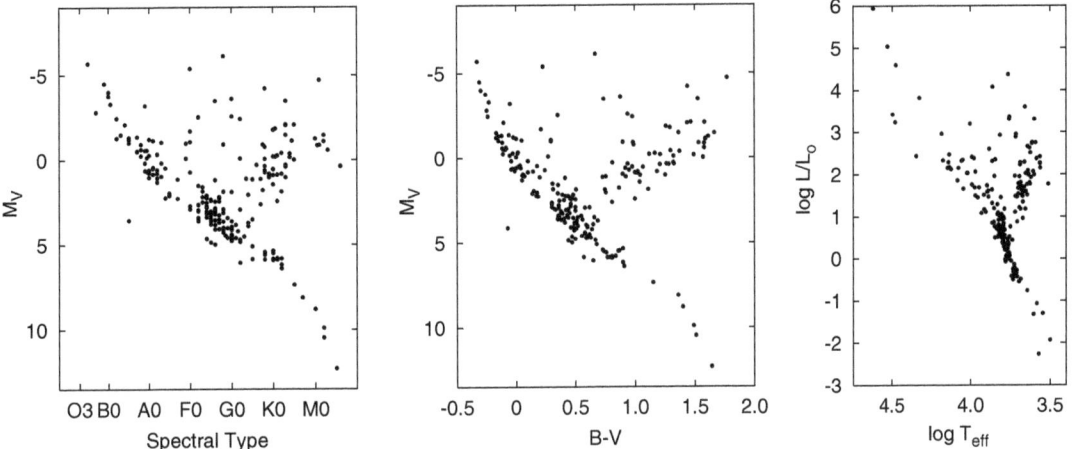

Figure 1.2: The Hertzsprung Russell Diagram (HRD) comes in three types:
    left: M_V vs SpT;   **middle**: M_V vs $B-V$;   **right**: L vs T_{eff}.
Parameters of field stars are shown in the three diagrams. To make an HRD, the distance for each star must be known. (Note, not all stars appear in each of the diagrams.)
The **spectral type diagram** is the original HRD. Spectral types are discrete and are based on spectral lines in relation with temperature. There is no limit to the left or right because of the freedom to define further or refined spectral types.
The **colour magnitude diagram** (CMD) uses a measured colour index (originally $C.I.$, colour index; normally $B-V$) which is also a temperature indicator. There is a blue limit to the colour index, based on the limiting slope of the Planck function (Rayleigh-Jeans limit; Sect. 1.4.2).
The **theoretical HRD** uses $\log L$ and $\log T_{\text{eff}}$. Here the limits are set by the limits of the parameters with which stars can exist. The red giant branch is rather steep (small range in $\log T$ for large range in $\log L$) in contrast to what is seen in the other diagrams where a large range in colour index (Wien part of Planck function at low T) or in spectral type (where lots of classification detail is possible) is present. The ratio of the axes is choosen such, that lines of constant stellar radii run under 45° (see Eq. 1.10 and Fig. 1.3). This choice makes the diagram rather "tall" in comparison with the other 2 diagrams.
The three diagrams can be transformed into each other, but the transformations are not simple.

1.5 The surface parameters of stars

The electromagnetic radiation emanating from the surface is the source of information about stars. Photometric and spectroscopic data can be used to derive in some way information about the physical parameters of the stellar surface.

1.5.1 The Hertzsprung-Russell Diagram, HRD

Combining flux-calibrated star data into the HRD led to a first realization, that stars can have widely different temperatures and luminosities. Note that a HRD (all forms of Fig. 1.2) **shows only surface parameters**.

1.5.1.1 Observational HRDs: M_V with SpT or $B-V$

One observes of a star its spectral type, SpT, and the brightness, like m_V. Only when the distance is known can the absolute magnitude M_V be determined. A diagram of **absolute magnitude** (based on early parallax distances) **versus SpT** is the "original" HRD[6]. It has discrete steps in

---
[6]Russell in 1914; first ideas in this direction in 1907.

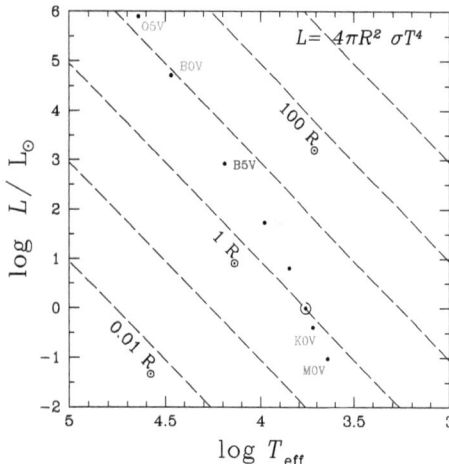

Figure 1.3: The HRD showing the relation between L, T_{eff}, and R in the form of $L = 4\pi R^2 \cdot \sigma T_{\text{eff}}^4$. The axis values are choosen such, that lines of constant R run under 45°. This diagram has the same proportions as Fig. 1.2 right panel and as several figures furtheron in the text. Some main-sequence stars are entered with their spectral type. This diagram can be overlaid on the L vs T HRD.

spectral type (see Fig. 1.2, left). Classically, spectral types ran from O5 (high temperature) to M9 (low temperature). For extensions at the high T end see Ch. 13.2.1.

Making a colour index measurement, one arrives at the **colour-magnitude diagram, CMD**[7]. The morphology of the data point collection is different from that in the HRD (M_V vs. SpT), because the colour index $B-V$ has a blue limit toward high temperature stars (see Fig. 1.2, middle panel). The SpT can accommodate any temperature, since it is based on the occurrence of spectral lines, thus, e.g., those of very highly ionized states (at high temperature).

1.5.1.2 Physical HRD: luminosity L and effective temperature T_{eff}

The name HRD is also used for the so called **physical HRD**. In this diagram the *luminosity* L is plotted against the surface temperature T_{eff} (see Fig. 1.2, right).

Consider a star with radius R. Its surface measures $A = 4\pi R^2$. Radiation is emerging from each unit of surface which, if the material is in thermodynamic equilibrium, is given by the Stefan-Boltzmann formula, being the integral of the Planck function

$$\int B_\nu \, d\nu = S = \int \frac{2h\nu^3}{c^2} \cdot \frac{1}{e^{h\nu/kT} - 1} \, d\nu = \sigma \, T^4 \quad . \tag{1.9}$$

with $\sigma = \frac{2\pi^5 k^4}{15 c^2 h^3}$. Thus the luminosity of a star (parameters of the surface) is given by

$$L \equiv A \, S = 4\pi R^2 \cdot \sigma T^4 \quad . \tag{1.10}$$

This allows to make a diagram (double logarithmic) of $\log L$ versus $\log T$ containing lines of constant R. Note that stellar radii can be determined reliably only using interferometric techniques.

If the radiation of the object considered is *not* following the Planck function (this is the case for most stars!), one may integrate over the observed distance corrected intensity distribution, $I_0(\lambda) \cdot 4\pi \, d^2$, in which d is the distance to the star. One then equates these integrals

$$\int I_0(\lambda) \cdot 4\pi \, d^2 \, d\nu = L = \sigma T_{\text{eff}}^4 \cdot 4\pi R^2 \tag{1.11}$$

in which the equivalent Planck temperature is called the **effective temperature**, T_{eff}, of the star (or, rather, of the stellar surface).

In the physical HRD the temperature used is always T_{eff}. Any difference between T_{eff} and a temperature found from photometry (using, e.g., the colour index as a measure for the slope of the spectral energy distribution as if due to a Planck function; see Fig. 1.1) can be attributed to opacity effects in the gas of the stellar atmosphere. These can be accounted for (see, e.g., Ch. 2.10).

---
[7]Hertzsprung in 1911, first ideas in 1905; using photographic and visual magnitudes: $m_{\text{pg}} - m_{\text{vis}}$.

1.5. THE SURFACE PARAMETERS OF STARS

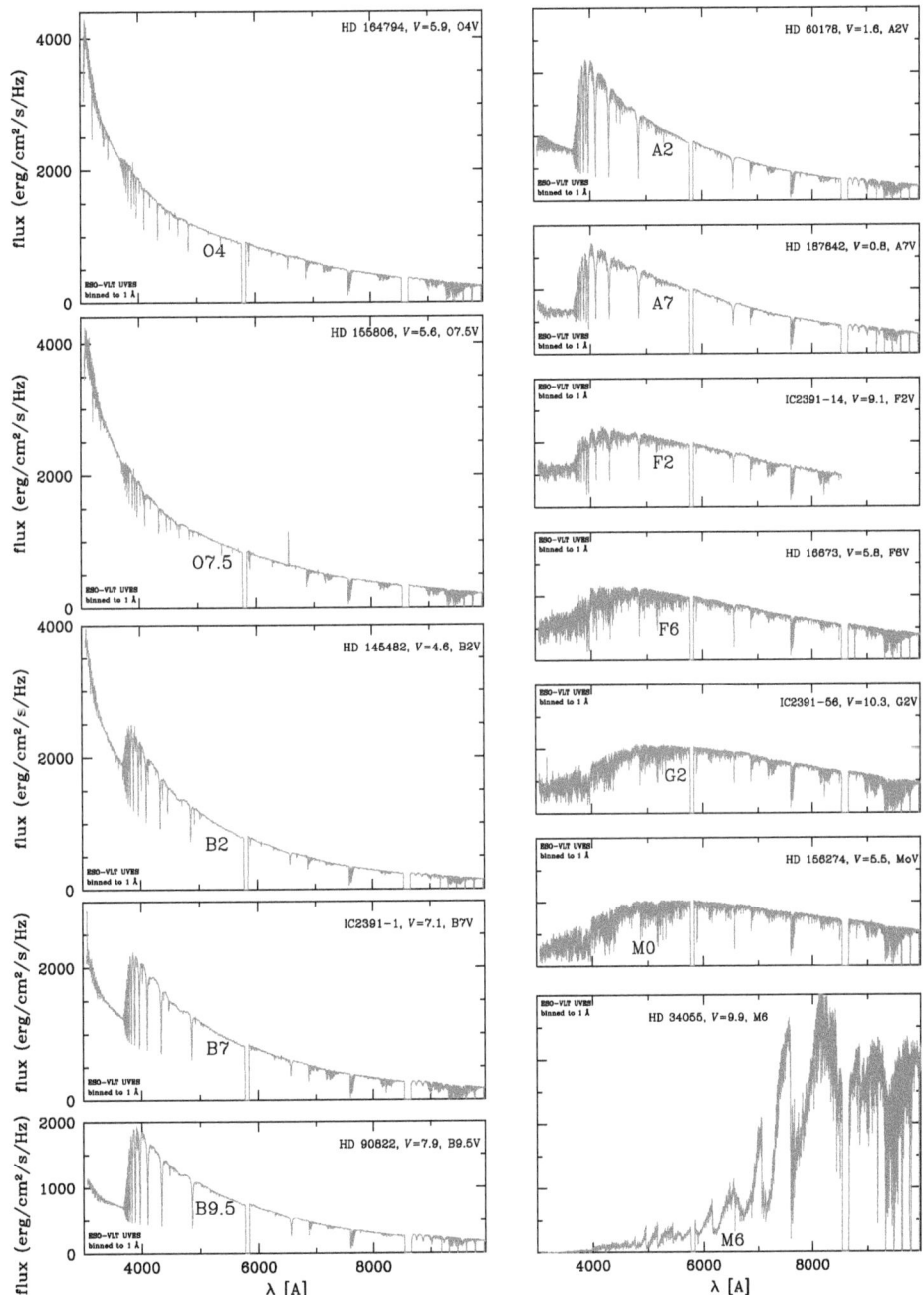

Figure 1.4: Spectra of stars (their names are given in each panel) for a range of spectral types as obtained with the ESO-VLT UVES spectrograph (data from the UVES Paranal Observatory Project ESO DDT Program ID 266.D-5655) as described by Bagnulo et al. (2003).
Each spectrum consists of various parts, each from a high dispersion echelle spectrum; the setup used leaves small gaps near 5750 and 8600 Å. The redmost part of the F2 spectrum is not available. The spectra were binned to 1 Å elements and normalized to (approximately) equal flux ($\simeq$1000) at 5500 Å (except for the M6 spectrum). The plots are linear both in flux and in wavelength.
Note the change in the shape of the spectral energy distribution (and thus in $B-V$) from the hottest (O-type, with $T_{\rm eff} \simeq 50000$ K) to the coolest (M-type, with $T_{\rm eff} \simeq 3000$ K) stars. Note the emergence of the Balmer jump near 3840 Å (Ch. 3.3.4) for stars of type later than O9 and its disappearance for stars later than F6. In the M6 star spectrum strong molecular bands (Ch. 3.4.3) cause severe flux depression at wavelengths smaller than 7000 Å.

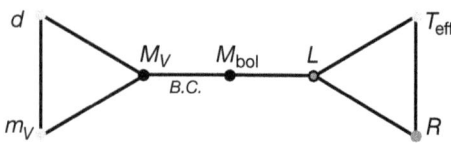

Figure 1.5: Relation between observable and intrinsic parameters of the surface of a star. T_{eff} and m_V are easy to measure, d is more difficult. The equations relating the parameters are, for the parameters from left to right, Eq. 1.4, Eq. 1.13, Eq. 1.14 and Eq. 1.10.

1.5.2 Spectral energy distributions

The overall spectral energy distributions of stars resemble the distribution of the Planck function. However, atoms, ions and molecules in stellar atmospheres may modify this theoretical spectral energy distribution considerably. In particular the hydrogen atom has a big impact with its absorption lines from the **Balmer series**, as well as with the **Balmer jump**. But other elements contribute too. For the numerous processes modifying a simple Planck spectrum see in particular Ch. 2 but also Ch. 3.

Examples of spectral energy distributions are given in Fig. 1.4.

1.5.3 Relation between M_V, M_{bol}, and L

The three versions of the HRD (Fig. 1.2) are interrelated. The (extinction corrected) absolute magnitude, M_V, gives the absolute brightness of a star in the V-band. However, the integrated brightness is more important. It is given as the **bolometric magnitude**

$$M_{\text{bol}} = -2.5 \log \int_0^\infty I_\lambda \, d\lambda \quad , \qquad (1.12)$$

a parameter in magnitude units. Clearly, the relation between M_V and M_{bol} depends on the nature of the spectral energy distribution, thus in particular on the surface temperature of the star. The conversion between M_V and M_{bol} requires to account for the flux not covered in the V-band. This factor is the so called **bolometric correction**, $B.C.$,

$$M_{\text{bol}} \equiv M_V - B.C. \qquad (1.13)$$

which can be determined for all stellar types. $B.C.$ clearly depends on T_{eff}. The use of M_{bol} is gradually decreasing since L is the more relevant parameter.

The physical equivalent of M_{bol} is the luminosity L. Clearly L is some transformation of M_{bol}, in the sense $L = 10^{-0.4 M_{\text{bol}}} + \text{const}$ (from magnitudes to physical units). Using the Sun as a reference (thus for the Sun $B.C. = 0$), one has

$$M_{\text{bol}} = 4.74 - 2.5 \log L \qquad (1.14)$$

with L in units of $L_\odot$. The relations between $V = m_V$, M_V, M_{bol} and L can be given in a schematic way incorporating d, T_{eff}, and R (Fig. 1.5).

1.5.4 Caution with mass - luminosity - temperature relations

From observations and study of binary stars it was found in the 19th century that the bluer a star is, the more mass it has. So it was deduced that this colour sequence is also a mass sequence. It followed that the luminosity of a star is roughly proportional to the **mass** as $L \simeq M^{\simeq 3}$, the exponent depending on what actually the mass[8] of the MS star is. High mass main-sequence stars behave more as $L \simeq M^2$. Using homology considerations (Ch. 6.8.2.1) it is possible to derive a relation between luminosity and surface temperature: $L \simeq T^{7 \pm 1}$. All these **relations are only approximate** and **apply only to main-sequence stars**. They should therefore be used with *great caution* always, and only if one is sure the star considered is of main-sequence type.

---

[8]There is a **risk of confusion** with the letter M: absolute magnitude (Eq. 1.4) or metallicity (Eq. 1.8) or mass.

1.6 Surface parameters and size of a star

The relation $L = 4\pi R^2 \sigma T_{\text{eff}}^4$ for the radiation and surface parameters of a star can easily be used to compare stars. Comparing with the Sun one thus has

$$\frac{L}{L_\odot} = \frac{R^2}{R_\odot^2} \frac{T_{\text{eff}}^4}{T_\odot^4} \tag{1.15}$$

or in logarithmic form (for a star: *)

$$\log\left(\frac{L_*}{L_\odot}\right) = 2\log\left(\frac{R_*}{R_\odot}\right) + 4\log\left(\frac{T_*}{T_\odot}\right) \quad . \tag{1.16}$$

Gravity determines the size of a star. Newton's law indicates $F = G\,\frac{m\,M}{d^2}$ with $F = m\,a$, where m is the mass of some small element on the stellar surface, M is the stellar mass, G is the gravitational constant, d the separation of the objects and a the acceleration. At the surface of a star (thus at distance R) a is normally given by the symbol g. One thus has

$$g = G\,\frac{M}{R^2} \quad . \tag{1.17}$$

This expression can be transformed into one relative to solar values as

$$\frac{g}{g_\odot} = \frac{M}{M_\odot}\left(\frac{R^2}{R_\odot^2}\right)^{-1} \tag{1.18}$$

or in logarithmic form

$$\log\left(\frac{g_*}{g_\odot}\right) = \log\left(\frac{M_*}{M_\odot}\right) - 2\log\left(\frac{R_*}{R_\odot}\right) \quad . \tag{1.19}$$

It is possible to combine Eqs. 1.16 and 1.19 by eliminating the radius so that the combined equation has only the mass, the temperature and gravity, and the luminosity as variables. (*Note that L, R, and M are in astronomy always expressed in solar units; in contrast, for T_{eff} and $\log g$ always the values itself are given, because they can be derived directly from spectral lines showing the physical conditions of the gases in stellar atmospheres.*) The luminosity is, of course, based on knowledge of the distance of a star. Since distance is not easily determined, while the spectral energy distribution is relatively easy to obtain, the luminosity can be split as $L_* = l_* \cdot 4\pi\,d^2$, where l_* is the integral of the extinction corrected spectral energy distribution as measured at the Earth (see Eq. 1.11, $\int I_\lambda d\lambda$; d is the distance of the star in pc, preferably from a parallax). Thus

$$\log\left(\frac{M_*}{M_\odot}\right) = \log g_* - 4\log T_* + \log l_* + 2\log d + 15.11 \quad , \tag{1.20}$$

where the numerical constant is

$$15.11 = \log 4\pi + 2\log\left(3.09 \cdot 10^{18}\right) - \log L_\odot + 4\log T_\odot - \log g_\odot \tag{1.21}$$

with 1 pc $= 3.09 \cdot 10^{18}$ cm, $L_\odot = 3.85 \times 10^{33}$ erg s^{-1}, $T_\odot = 5800$ K, $\log g_\odot = 4.44$ (cgs units).

Eq. 1.20, with the relatively easily determined observationally parameters (T_{eff}, $\log g$ and l_*), can be used to calculate the mass and/or the distance of stars. Both mass and distance are normally difficult to determine.

Two special cases with simple solutions exist. For stars in clusters d can be found from the location of the main sequence in the CMD so that M of the stars can be calculated (see de Boer et al. 1995 for an example of globular cluster stars). For stars of horizontal branch nature in the Milky Way field one knows the value of M from modelling and theory of stellar evolution (or from investigations of globular cluster stars), so that with l and T from observations the distance of the individual stars can be calculated.

1.7 Names of star types from location in the HRD

Once the HRD was developed, stars received names based on their location in the HRD or CMD.

Hertzsprung became aware of the range in nature of red stars based on their large range in luminosity (using early parallaxes; Sect. 1.1.1). Since a radius can be estimated from $L = 4\pi R^2 \sigma T_{\text{eff}}$ he named the luminous red stars "giants" and the not very luminous red stars "dwarfs". The dwarfs established the main sequence (MS), the red giants the red-giant branch (RGB).

Stars below the main sequence thus are "subdwarf (sd) stars", stars on the lower part of the red giant branch are called "subgiants", stars above the RGB "supergiants" (SGs).

Later one discovered a strip of stars lying horizontally in the CMD, stars then called "horizontal-branch stars" (HB stars).

1.8 Summary

The basic observational parameters of stellar astronomy are given: magnitudes, spectra, the Hertzsprung-Russell diagram, and further simple relations for the surface parameters of stars.

Further knowledge of observational stellar astronomy can be obtained from a large variety of textbooks (e.g., Hearnshaw 1996). Understanding stars and their evolution and the large variety of manifestations of stars is what the coming chapters are about.

Numerous textbooks deal with stars and stellar evolution. A few are Cox & Giuli (1968), Böhm-Vitense (1989), and Kippenhahn & Weigert (1990). Others will be mentioned later.

References

Andriesse, C.D. 1994, "Titan kan niet slapen - Een biografie van Christiaan Huygens"; Uitgeverij Contact, Amsterdam

Bagnulo, S., Jehin, E., Ledoux, C., Cabanac, R., Melo, C., Gilmozzi, R., and the ESO Paranal Science Operations Team. 2003, ESO Messenger 114, 10

Beckett, Sir E. 1874, "Astronomy", Soc. Prom. Chr. Knowledge, London

Bethe, H., & Critchfield, C.L. 1938, Phys. Rev. 54, 248

Böhm-Vitense, E. 1989, "Stellar Astrophysics", Cambridge Univ. Press.

Buchanan, M. 2000, "Ubiquity"; Phoenix, London

Burbidge, E.M., Burbidge, G.R., Fowler, W.A., & Hoyle, F. 1957, Rev. Mod. Phys. 29 No. 4 (B^2FH)

Celnikier, L.M. 1989, "Basics of Cosmic Structures", Editions Frontières

Cox, J.P., & Giuli, R.T. 1968, "Principles of Stellar Structure"; Gordon & Breach, New York

Cutler, A. 2003, "The Seashell on the Mountain Top - How Nicolaus Steno solved an Ancient Mystery and created a Science of the Earth"; Plume/Penguin,

Corbally, C.J., Gray, R.O., & Ferguson, G.R. 1993, "The MK Process at 50 years: A Powerful Tool for Astrophysical Insight", Astron. Soc. Pacific Press

de Boer, K.S., Schmidt, J.H.K., & Heber, U. 1995, A&A 303, 95

Hayes, D.S. 1985, in IAU Symp. 111, "Calibration of Fundamental Stellar Quantities", eds. D.S. Hayes et al., Reidel, Dordrecht; p. 225

Hearnshaw, J.B. 1996, "The Measurement of Starlight", Cambridge University Press

Hoyle, F. 1994, "Home is where the Wind Blows - Chapters from a Cosmologist's Life"; Univ. Science Books, Mill Valley

Jeans, J. 1934, "The Universe around us", The Macmillan company, New York, and Cambridge Univ. Press (1st edition)

Kippenhahn, R., & Weigert A. 1990, "Stellar structure and evolution", Springer

Moore, C. 1972, "A Multiplet Table of Astrophysical Interest"; NBS, Washington

Planck, M. 1901, Ann. Physik, 4, 553

Toulmin, S., & Goodfield J. 1965, "The Discovery of Time", Univ. Press, Chicago

von Weizsäcker, C.F. 1937, Physik. Zeitschrift 38, 176,

Wiese, W.L., Smith, M.W., & Glennon, B.M. 1966, "Atomic Transition Probabilities, Part I", NSRDS, Washington; & Wiese, W.L., et al., 1969, "Part II"

Chapter 2

Stellar atmosphere: Continuum radiation + structure

2.1 Introduction

A star is a gaseous sphere in which the inward pull of gravity is balanced by the expanding tendency due to the internal gas pressure gradient. Of stars only the 'surface' is visible. The energy from the interior flows outward and leaves the star as radiation.

In this chapter (and in Ch. 3) the physics of the structure of the visible stellar surface layers, the **stellar atmosphere**, is explored. For that we have to consider the details of the balance of energy flow (which determines temperature and density) with gravity of these surface layers. The major aspect is the transfer of radiation.

Our knowledge of the structure of stellar atmospheres derives from the nature of the radiation detected and from modelling. The observed spectral energy distribution can be thought of as being composed of two parts, the
  a) **smooth radiation continuum**
  b) **smaller scale spectral structure**
The modelling involves all the detailed physics with all the relevant parameters, leading to so called **model atmospheres**.

In a way, describing the structure of the atmosphere is simple. One needs to establish a description for the gravitational balance and the radiation transport:

- the **radiation transport equation** which specifies
  → the emergence of radiative energy at the surface
  → the temperature distribution in the surface layers (thermal balance)

- the **hydrostatic equation** which in a stationary situation describes
  → the pressure/temperature distribution in the surface layers (mechanical balance)

When the description is complete, the model atmosphere can be used to calculate the details of the emitted spectral energy distribution. One thus obtains *the modelled continuum intensity* (this chapter), *the spectral line intensities and the line structure* (Ch. 3), as well as further useful parameters such as *the colour indices* of the spectral energy distribution. Note that the modelling is based on the luminosity L of the star of which the size is set by the energy generation processes in the interior.

A model as indicated above does not describe yet how a real star will appear to us. A complete description must include the geometry of the atmosphere, characterized by the radius of the star.

Models must be tested by comparison with reality. Deviations found will necessarily lead to improvements of the models. One thus has a process leading to consistency:

16 CHAPTER 2. STELLAR ATMOSPHERE: CONTINUUM RADIATION + STRUCTURE

```
                      compare with observations
                               ↗                ↘
      model atmosphere                                    observations
                               ↖                ↙
                          derive parameters
```

The parameters used for the characterisation of a stellar atmosphere model are the
- *effective temperature*, T_{eff}, describing the overall released energy σT_{eff}^4,
- *surface gravity*, g (mostly given as $\log g$), characteristic for the pressure structure,
- *chemical composition*, given by the abundance of the elements, XYZ (with X for hydrogen, Y for helium, and Z for the rest), which governs the overall absorption capabilities (absorption coefficient κ) of the atmosphere gases.

In the following the basic equations describing the structure of a stellar atmosphere are given. Much more on stellar atmospheres can be found in the monographs by, e.g., Unsöld (1955), Mihalas (1978), Böhm-Vitense (1989).

2.2 Radiation theory

2.2.1 Definitions

To characterize radiation a set of parameters has to be defined. These include the *radiative intensity*, the *mean intensity*, the *flux*, the *radiation density* and the *radiation pressure*.

(Note: the proper way to describe radiation theory is in frequency space, where $E_{\text{photon}} = h\nu$. One can, of course, give the formalism also based on wavelength λ. That approach is taken in Böhm-Vitense (1989) and in some other textbooks, because it is felt that wavelength is easier to understand for beginners in astrophysics; Unsöld and Mihalas use the frequency space.)

2.2.1.1 Radiative intensity

The radiative intensity I_ν is defined as the energy dE_ν within a frequency interval $d\nu$ passing per unit of time dt through a surface $d\sigma$ and being directed into a solid angle $d\omega$. Thus

$$I_\nu(\theta, \phi) = \frac{dE_\nu}{\cos\theta \, dt \, d\nu \, d\omega \, d\sigma} \quad (2.1)$$

in which ϕ and θ are the angles of orientation of $d\omega$ with respect to the normal vector to the surface element $d\sigma$ considered (see Fig. 2.1).

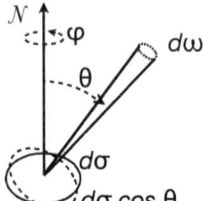

Figure 2.1: The geometry for the definition of **radiative intensity** is shown. The intensity goes through $d\sigma$ but with an angle θ with respect to the normal $\mathcal{N}$ to $d\sigma$.

The radiative intensity may be integrated over all frequencies, leading to the *integrated radiative intensity* $I = \int_0^\infty I_\nu d\nu = \int_0^\infty I_\lambda d\lambda$. Also, one can transform the definition of intensity to that based on wavelength: $I_\lambda = c/\lambda^2 \cdot I_\nu$.

Note that the intensity of radiation can be given in a wide range of units, each based on the traditions in the various wavelength, frequency, or energy ranges. Some examples are:
[W m^{-2} Hz^{-1} sr^{-1}], [erg s^{-1} cm^{-2} Hz^{-1} sr^{-1}], [photons s^{-1} keV^{-1} sr^{-1}].

In the study and description of stellar atmospheres, the direction coordinate normally is choosen positive in the direction away from the centre of the star.

2.2. RADIATION THEORY

2.2.1.2 Mean intensity, radiative flux

The **mean intensity**, $\overline{I_\nu}$, mostly given as J_ν, is defined as the average of I over all solid angles ω

$$J_\nu = \frac{1}{4\pi} \int_{\theta=0}^{\pi} \int_{\phi=0}^{2\pi} I_\nu(\theta, \phi) \cos\theta \sin\theta \, d\phi d\theta = \frac{1}{4\pi} \int I_\nu(\omega) \, d\omega \qquad (2.2)$$

in which the $1/4\pi$ is the normalization (in the case $I_\nu = 1$ the integral gives 4π as a result).

A further parameter is the **radiative flux**, $\vec{F}_\nu$. It is the net energy in the interval $d\nu$ passing each second through a unit area in the direction of the z-axis, the vertical axis,

$$\vec{F}_\nu d\nu = \int I_\nu \, d\nu \, \cos\theta \, d\omega \quad . \qquad (2.3)$$

The radiative flux can be split into two parts: the outward flux F_ν^+ and the inward flux F_ν^-, with $F_\nu = F_\nu^+ + F_\nu^-$, the sum of the outward and inward radiative flux. Note again the directionality in the description of stellar atmospheres: the *positive direction is outward* (with F_ν^+), the negative direction is inward (F_ν^-). In an isotropic radiation field clearly $F_\nu^+ = -F_\nu^-$ so that $F_\nu = 0$.

For a *spherical star* the mean intensity and the radiative flux are related: $J_\nu = \frac{1}{\pi} F_\nu$.

An observer at some distance d from the star measures the intensity in a very small $d\omega = \frac{\pi R^2}{d^2}$ (as seen from the observer) with R the radius of the star. Then the observed flux f_ν (apparent intensity, as in apparent magnitude) clearly is $df_\nu = I_\nu d\omega$ and so $f_\nu = \frac{R^2}{d^2} F_\nu$.

Note: there is some confusion in the literature about the definitions just given. Some schools use for the mean intensity J, others $\overline{I}$, while the factor $1/4\pi$ (as in Eq. 2.2) is sometimes incorporated in the definition of the flux.

2.2.1.3 Radiation density and radiation pressure

One can define the density of the energy of the radiation in a gas. Radiation with energy $dE_\nu = I_\nu \, d\sigma \, dt \, d\omega$ goes through a surface element $d\sigma$. This radiation passes in a time interval dt through a volume element $dV = d\sigma ds$ where $ds = c \, dt$, thereby coming from all directions. The *energy density of the radiation* [W m^{-1}] is thus found by integrating over all solid angles $d\omega$ as

$$U_\nu = \int \frac{dE_\nu}{dV} \, d\omega = \frac{1}{c} \int I_\nu \, d\omega \quad . \qquad (2.4)$$

In the case of *isotropic radiation* the integrals are symmetric and can be solved to $U_\nu = \frac{4\pi}{c} I_\nu$. The concomittant *photon density* (divide U_ν by the photon energy) is $N_\nu = \frac{4\pi}{c} \frac{I_\nu}{h\nu}$. The *total radiation density* (integrate U_ν over frequency) is $U = \frac{4\pi}{c} I$.

Since photons have momentum the radiation represents also **pressure**. The momentum of a photon is $p_\nu = \frac{h\nu}{c}$. At a reflecting surface photons may come in under an angle θ. Then the net incident energy is reduced by the factor $\cos\theta$ and the effect of momentum on the vertical is also reduced by $\cos\theta$. At reflection the total change of momentum is $p_\nu = \frac{2}{c} I_\nu \cos^2\theta$. Thus the pressure exerted by the photons is $P_{\text{rad},\nu} = \int p_\nu d\nu = \int \int \frac{2}{c} I_\nu \cos^2\theta \, d\omega d\nu$, the **radiation pressure**. In the case of isotropy, $P_{\text{rad},\nu} = \frac{4\pi}{3c} I_\nu = \frac{1}{3} U_\nu$. For the importance of radiation pressure see Ch. 4.4.2.2.

2.2.2 The equation of radiation transport

Returning to the definition of intensity and the volume element dV of length ds, one can imagine that the intensity changes over the element ds. There may be *emission* as well as *absorption*.

The **emission coefficient** j_ν is defined as the energy emitted by the volume element dV in a unit of time and frequency into a solid angle ω

$$j_\nu = \frac{dE_\nu}{dt \, dV \, d\nu \, d\omega} \qquad (2.5)$$

in units [W m^{-3} Hz^{-1} sr^{-1}] or [erg s^{-1} cm^{-3} Hz^{-1} sr^{-1}]. For more on j_ν see Sect. 2.9.

The **absorption coefficient** κ_ν is defined by the change in the intensity due to absorption in the material over the path ds (see Fig. 2.2)

$$dI_\nu = -\kappa_\nu I_\nu \, ds \quad . \tag{2.6}$$

Normally, the absorption coefficient κ is the mass absorption coefficient κ_ν [cm^2 g^{-1}]. (Sometimes the atomic absorption coefficient $\kappa_{\nu,\text{numb. of at.}} = \kappa_\nu \cdot \rho/N$ [cm^2] is used.) More on the nature of the absorption coefficient and the kind of processes contributing to κ is given in Sect. 2.8.

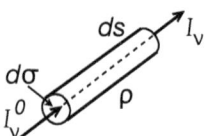

Figure 2.2: The geometry for the definition of the **absorption coefficient** κ is shown. The small volume $dV = d\sigma \cdot ds$ contains material with density ρ and of total particle number N.

In relation with the absorption coefficient one defines the **optical depth** τ_ν as

$$d\tau_\nu = \kappa_\nu \, ds \quad \text{and} \quad \tau_\nu = \int_0^s \kappa_\nu \, ds \quad . \tag{2.7}$$

Note that s is positive in the outward direction (the direction to the observer).

Integrating Eq. 2.6 over the path length ds and solving the differential equation leads to

$$I_\nu = I_\nu^0 \, e^{-\tau_\nu} = I_\nu^0 \, e^{-\int \kappa_\nu \, ds} \quad . \tag{2.8}$$

Three cases need mentioning:
- In the case of $\tau = 1$ it follows that $I_\nu = I_\nu^0/e$.
- For $\tau \gg 1$ one has *large optical depth*, the material is opaque and $I_\nu \ll I_\nu^0$.
- For $\tau \ll 1$ one has *small optical depth*, the gas is transparent and $I_\nu \simeq I_\nu^0$.

The total change of the intensity, covering the absorption and the emission, gives us the **radiation transport equation**

$$dI_\nu = -\kappa_\nu I_\nu \, ds + j_\nu \, ds \tag{2.9}$$

$$\frac{dI_\nu}{\kappa_\nu ds} = -I_\nu + \frac{j_\nu}{\kappa_\nu} \tag{2.10}$$

$$\frac{dI_\nu}{d\tau_\nu} = -I_\nu + S_\nu \tag{2.11}$$

in which the function S_ν is called the **source function**. It describes the source of the change in intensity based on the properties of the material. Clearly, when $S_\nu > 0$ (more emission than absorption) $\frac{dI_\nu}{d\tau} > 0$, while the intensity decreases when there is more absorption than emission.

2.2.3 Exploring the equation of radiation transport

The radiation transport equation can be solved for the case that S_ν is constant along the path considered as

$$I_\nu = I_\nu^0 \, e^{-\tau_\nu} + S_\nu \left(1 - e^{-\tau_\nu}\right) \quad . \tag{2.12}$$

This equation describes in a very recognizable way, that the intensity measured (I), equals the optical depth reduced intensity entering the volume from the rear (the first term on the right side) plus the intensity produced inside the box (S) diluted by the appropriate optical depth relation (the second term on the right side). One can explore the significance and the effective behaviour of this equation by investigating a few realistic cases.

2.2.3.1 a: No background intensity: $I_\nu^0 = 0$

When $I_\nu^0 = 0$ Eq. 2.12 reduces to $I_\nu = S_\nu \left(1 - e^{-\tau_\nu}\right)$.
Case $\tau_\nu \ll 1$ leads through Taylor expansion to $I_\nu = +\tau_\nu \cdot S_\nu$; all radiation produced (source function) can be seen by an outside observer, except for a slight reduction due to absorption (optical depth). This is the case for the Solar Corona or for interstellar emission nebulae.
Case $\tau_\nu \gg 1$ leads to $I_\nu \simeq S_\nu$; the intensity seen equals the source function since, due to the high optical depth, none of the photons produced in the interior of the material can escape (they are immediately scattered or absorbed). If $\tau \to \infty$ then $I_\nu = B_\nu$.

2.2.3.2 b: background intensity: $I_\nu^0 \neq 0$

The non-zero intensity, $I_\nu^0 \neq 0$, situation is in particular applicable to stars. One can rewrite Eq. 2.12 as

$$I_\nu = S_\nu + (I_\nu^0 - S_\nu)e^{-\tau_\nu} \qquad (2.13)$$

and the small and large optical depth cases can easily be approximated.
Case $\tau_\nu \ll 1$ leads to $I_\nu = I_\nu^0 - \tau_\nu(I_\nu^0 - S_\nu)$. If $I_\nu^0 > S_\nu$, we have the case of 'normal' spectral absorption (perhaps as lines; see Ch. 3) of an existing continuum. If $I_\nu^0 < S_\nu$, we have the case of emission (perhaps as lines) superposed on an existing continuum.
Case $\tau_\nu \gg 1$ leads to the same solution as with no background intensity (see above), $I_\nu = S_\nu$.

2.2.3.3 Graphic representation of the cases

One can appreciate all the discussed cases using a graphic form. Adopt a certain behaviour of κ_ν (and thus of τ) and consider what the outcome is on I_ν, both for $I_\nu^0 = 0$ and $I_\nu^0 \neq 0$.

2.3 Thermodynamic equilibrium

Because of the temperature and thus atomic motions in gases there is interaction of the particles amongst each other (e.g., collisions leading to excitation or ionization, as well as reverse processes) and the particles are distributed over the numerous possible energetic states according to statistical properties (see Boltzmann statistics, Ch. 3.2). However, in a situation when all such processes take place so rapidly and frequently that for every excitation there is a de-excitation, for every absorption of a photon there is an emission of the same kind, etc., then also the radiation is isotropic and in balance with the material. All processes are in balance and there are *no effects of time* (no changes with time). This condition is called **thermodynamic equilibrium (TE)**. In that case the radiation intensity is according the **Planck function**. Thus $S_\nu = I_\nu = B_\nu$.

In cases of TE, the isotropy of radiation allows to calculate the radiation density and radiation pressure, as given above (Sect. 2.2.1.3).

Thermodynamic equilibrium will, in fact, nowhere exist in the real universe. Normally there is a source of energy in a star (nuclear fusion) while outside the star the universe is cold, so that energy will flow away from the stellar centre. Yet in small regions inside the star, material is almost in thermodynamic equilibrium, since radiation comes from all sides while the gas-density gradient is very small. In this case one regards the material to be locally in TE, or in **local thermodynamic equilibrium (LTE)**, saying that locally the equilibrium is (almost) realized, but acknowledging that the star as a whole is not in thermodynamic equilibrium at all. If the gas is not in LTE one speaks of non-LTE or (often in publications) NLTE.

In the case of LTE, $S_\nu = B_\nu$ and $\frac{dI_\nu}{d\tau_\nu} = 0$. Thus also $B_\nu = j_\nu/\kappa_\nu$. Also, the expressions for radiation density and pressure (see Sect. 2.2.1.3) are assumed to be valid. One also uses the concept of just *thermal equilibrium* which means a stable temperature structure (also in time).

Note that there is in the concept of TE and LTE *no mention of effects of gravity* since the described TE and LTE just apply locally to the gas.

A description of the total structure of stars has to include gravity. Then the "stellar thermal equilibrium" (STE; Ch. 9.4) becomes important dealing with the overall energy balance of the star.

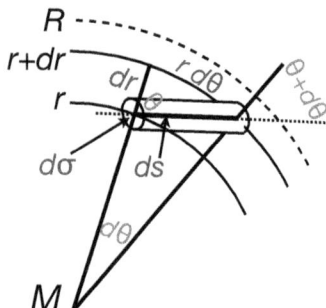

Figure 2.3: Sketch of an element with volume $dV = d\sigma \cdot ds$ at distance r from the centre (M) of a star with orientation θ and $d\theta$ is indicated. The parameters for the geometry of the radiation transport are defined by the triangle formed by dr, $r\,d\theta$, and $d\sigma$. The radiation transport equation is given by Eq. 2.15.

2.4 The radiative transfer in stellar atmospheres

2.4.1 Effects of geometry

Before being able to use the radiation transport equation for the structure of stars and the stellar atmospheres, we have to incorporate explicitly effects of geometry. This is achieved by assuming the volume dV to be at a distance r from the stellar centre, and that the length ds of dV points in the direction θ (see Fig. 2.3). The equation describing the situation is

$$dI_\nu(r,\theta) = -\kappa_\nu \, I_\nu(r,\theta) \, ds + j_\nu \, ds \quad \text{or} \quad \frac{dI_\nu(r,\theta)}{d\tau_\nu} = -I_\nu(r,\theta) + S_\nu \qquad (2.14)$$

and since $dr = ds \, \cos\theta$ and $r\,d\theta = -ds \, \sin\theta$ this can be transformed into the **general radiation transport equation**

$$\frac{\partial I_\nu}{\partial r} \cos\theta - \frac{\partial I_\nu}{\partial \theta} \frac{\sin\theta}{r} = -\kappa_\nu \, I_\nu + j_\nu \quad . \qquad (2.15)$$

2.4.2 Including all frequencies

In a stellar atmosphere the radiation transport equation must be solved for every frequency, ν. Also, all these ν-dependent solutions must work together to make one consistent atmosphere model. This leads to the continuity equation.

2.5 Continuity equation

Until here, the nature of the *energy transport* in stellar atmospheres was left open. It may be just radiation, but it may also include energy transport through, e.g., conduction or convection.

In the special case of *energy transport only by radiation*, it is the radiation transport equation which suffices to describe the transport.

One can obtain insight in the relation of the radiative flux F with depth r. For that one integrates the radiation transport equation (Eq. 2.14) over the whole solid angle $d\omega$ (as in Eq. 2.3) and obtains

$$\frac{1}{4\pi} \frac{dF_\nu}{dr} = \kappa_\nu (I_\nu - S_\nu) \qquad (2.16)$$

for which the integration and differentiation were interchanged ($\int \frac{dI}{d\tau} d\omega = \frac{\int dI d\omega}{d\tau}$). Assuming now energy transport only by radiation, then $F = \int F_\nu d\nu$ must be independent of r ($\frac{dF}{dr} = 0$) or

$$\frac{1}{4\pi} \int_0^\infty \kappa_\nu \, F_\nu \, d\nu = \int_0^\infty \kappa_\nu \, S_\nu \, d\nu \quad . \qquad (2.17)$$

The description of the radiation transport requiring relations for every frequency comes here to a synthesis, in that the equation above involves only integrals over the entire spectrum. It provides the connection between the frequency dependent transport equations and the total radiative energy transport. This equation is called the *continuity equation*.

2.6 Special cases and approximations

The number of variables in the radiation transport equation is not really large, but at least two of the variables are rather complicated functions. These are j_ν as well as $\tau_\nu = \int \kappa_\nu ds$ with $\kappa_\nu \sim \rho(T, XYZ)$. They can generally not be dealt with in an analytic way.

There are four 'simplifications' which allow to derive special aspects of the solutions of the general radiation transport equation (Eq. 2.15). Schematically they are

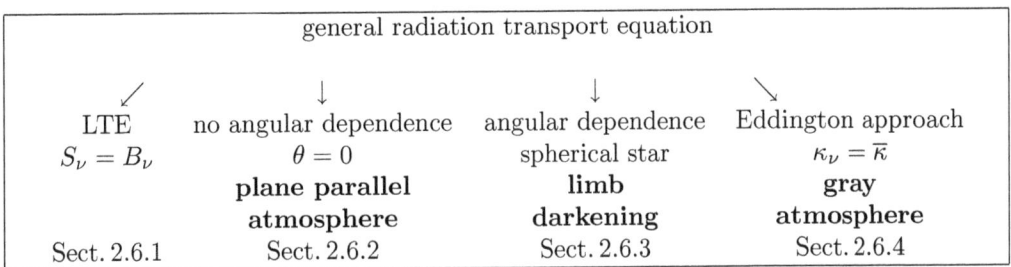

and these will be discussed in the following. In several cases some of these approximations are used in combination.

2.6.1 Atmospheres in LTE

If the atmosphere is in LTE, then the gas and the radiation field behave in all layers at all depths r according to TE. Clearly $S_\nu = \frac{j_\nu}{\kappa_\nu} = B_\nu(T(r))$ (Kirchhofs law) and the radiation is isotropic. Although nearer to the surface this condition is not really met, the LTE approach is used in cases where $T(r)$ varies only very slightly with depth. An entire star is not in TE, but one may, as noted before, for practical purposes (like the calculation of excitation conditions, etc.), assume that locally TE is valid.

The LTE approximation is used in many instances, in part also in the next subsections.

2.6.2 Plane parallel atmosphere

In the case of a stellar atmosphere with very large radius or in order to keep the mathematics simple, one may treat the radiation transport as that in a plane of material in which the successive layers are parallel, a *plane-parallel atmosphere*. In that case $d\theta = 0$ (see Fig. 2.3) and $d\tau_\nu = -\kappa_\nu dr$. The general radiation transport equation (Eq. 2.15) simplifies to

$$\frac{dI_\nu}{d\tau_\nu} \cos\theta = I_\nu - S_\nu \quad , \tag{2.18}$$

the radiative transfer equation for the plane-parallel case, with $S_\nu = j_\nu/\kappa_\nu$.

In plane-parallel atmospheres the total radiative flux F does not depend on the depth r, so that $dF(r)/dr = 0$.

Since here $dF(r)/dt = 0$ and F = constant, one has a case of *conservation of energy*. Therefore, Stefan-Boltzmann's law must hold so that in each and every layer

$$F = \sigma T^4(r) \quad . \tag{2.19}$$

If one were bold enough to think this all holds near the surface of spherical stars too, it would mean $F = \sigma T_{\text{eff}}^4$.

2.6.3 Limb darkening

When considering the edge of a real stellar atmosphere, one has to account for the fact that the radiation field is not isotropic. Also, the angular aspect, θ, of the radiation flow is of relevance.

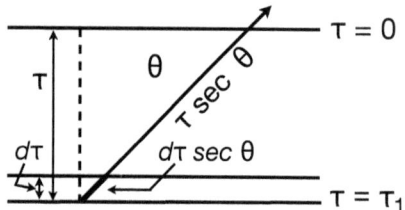

Figure 2.4: Generalized geometry of the outer layers of a stellar atmosphere. At depth τ a thin layer with depth $d\tau$ is considered. For the case of a curved atmosphere this leads to the derivation of the limb darkening effect (Sect. 2.6.3).

Multiplying the radiation transport equation (Eq. 2.11) with $e^{-\tau \sec\theta}$ (see Fig. 2.4) and integrating over $\tau \sec\theta$ leads to

$$I\, e^{-\tau \sec\theta} = -\int_\tau S\, e^{-\tau' \sec\theta}\, d\tau'\, \sec\theta \qquad (2.20)$$

One now can treat the outward and the inward radiation separately.

Outward radiation comes from optical depth τ and directions $0 \leq \theta \leq \frac{\pi}{2}$, leading to the appropriate formula for $I(\tau, \theta)$. From the interior one has the intensity

$$I_\nu(0, \theta) = -\int_0^\infty S_\nu(\tau_\nu) e^{-\tau_\nu \sec\theta}\, d\tau'\, \sec\theta \qquad (2.21)$$

for which $\tau \to \infty$ toward larger depth r.

Inward radiation comes from directions $\frac{\pi}{2} \leq \theta \leq \pi$, leading to the appropriate $I(\tau, \theta)$. At the very edge of the atmosphere one is at $r = 0$ so that $\tau_0 = 0$ and from $\frac{\pi}{2} \leq \theta \leq \pi$ clearly $I(0, \theta) = 0$.

Consider now the two extreme positions as seen by an observer: the very edge and the centre of the stellar disk.
- *At the edge* of the visible disk of the star the direction toward the observer $\theta = \frac{\pi}{2}$ and thus $\sec\theta = 1/\cos\theta \to \infty$. Therefore $I_\nu(0, \frac{\pi}{2}) \to 0$.
- *At the centre* of the disk one has as direction toward the observer $\theta = 0$ and so $\sec\theta = 1$. Then $I_\nu(0, 0) = \int_0^\infty S_\nu(\tau_\nu)\, e^{-\tau_\nu}\, d\tau_\nu$.

Note that the stellar disk can normally only be seen in detail by an observer when the star is the Sun. For all other stars their distance in combination with the spatial resolution of observations does not allow to see this behaviour (except in interferometric measurements).

To even further elaborate this case, one can approximate the depth dependence of the source function at some location $\tau(r)$ in the atmosphere. Eddington and Barbier explored this in the first half of the 20th century. Here the approximation of the source function by Böhm-Vitense (1989, her Ch. 5.4)

$$S_\nu(\tau_\nu) = a_\nu + b_\nu \tau_\nu \qquad (2.22)$$

(Taylor expansion $\frac{dS_\nu}{d\tau_\nu}$) is used. After inserting this into Eq. 2.20 one arrives at

$$I_\nu(0, \theta) = a_\nu + b_\nu \cos\theta \quad . \qquad (2.23)$$

Inserting this into Eq. 2.3 the expression for the flux is

$$\pi F_\nu(0) = 2\pi \int_0^1 (a_\nu + b_\nu \cos\theta) \cos\theta\, d(\cos\theta) = \left(a_\nu + \frac{2}{3}b_\nu\right)\pi \quad . \qquad (2.24)$$

Comparing this with the approximation used one must conclude that

$$F_\nu(0) = S_\nu\left(\tau_\nu = \frac{2}{3}\right) \quad . \qquad (2.25)$$

This is, in fact, the **Eddington-Barbier relation**.

Further simplifications can now be applied. Assuming LTE ($S_\nu = B_\nu$, the Planck function) one obtains $F_\nu(0) = \pi\, B_\nu(T(\tau = \frac{2}{3}))$. Assuming further a gray atmosphere ($\kappa_\nu = \bar\kappa$, see below) one arrives at $F(0) = \sigma T_{\text{eff}}^4$. Then the conclusion is

$$T_{\text{eff}} = T\left(\tau = \frac{2}{3}\right) \qquad (2.26)$$

2.6. SPECIAL CASES AND APPROXIMATIONS

Table 2.1: Limb darkening of the Sun: theory and observation

$\sin\theta = \frac{r}{R_\odot}$	$\cos\theta$	$\frac{2}{5}(1+\frac{3}{2}\cos\theta)$	$[\frac{I(\theta)}{I(0)}]_{obs.}$
0.00	1.00	1.00	1.00
0.20	0.98	0.99	0.99
0.40	0.92	0.95	0.96
0.55	0.84	0.90	0.92
0.75	0.66	0.80	0.83
0.87	0.48	0.69	0.74
0.95	0.31	0.59	0.63
0.97	0.23	0.53	0.55
1.00	0.00	0.40	–

or, in words, the effective temperature of the stellar surface equals the temperature at optical depth 2/3.

To find the limb darkening in a simple manner, one has to proceed with the gray atmosphere approximation.

2.6.4 Gray atmosphere; Rosseland mean

A considerable simplification in the construction of models is achieved when a simplified form of the expression for the absorption coefficient κ_ν (and thus for the opacity τ_ν) can be found (for details about the real structure of κ see Sect. 2.8). Rosseland (1924) adopted $\kappa_\nu \sim \bar{\kappa}$, which allowed to simplify the radiation transport equation drastically. This mean absorption coefficient is a flux-weighted mean opacity, called the **Rosseland mean** after the original investigator. The atmosphere calculated this way is called a *gray atmosphere*.

The mean absorption coefficient $\bar{\kappa}$ is found using $F = \int F_\nu$, with at great depth in the atmosphere $F = \frac{1}{3}\frac{dS}{d\tau} = \frac{1}{3}\frac{dB}{d\tau}$, and defining $d\tau = \bar{\kappa}\,ds$. Then

$$\frac{1}{\bar{\kappa}} = \frac{\int_0^\infty \frac{1}{\kappa_\nu}\frac{dB_\nu}{dT}d\nu}{\frac{d}{dT}\int_0^\infty B_\nu d\nu} \quad . \tag{2.27}$$

The gray atmosphere is really unphysical, but it is an approach very useful to make first estimates about the structure of a stellar atmosphere.

For a gray atmosphere, Eq. 2.12 now becomes

$$\cos\theta\,\frac{dI(\tau,\theta)}{d\tau} = I(\tau,\theta) - S(\tau) \tag{2.28}$$

and Eq. 2.17 is

$$S(\tau) = \frac{1}{4\pi}F(\tau) \quad . \tag{2.29}$$

Combining these two leads to an equation for $I(\tau,\theta)$ which can be solved for the surface of an atmosphere, where $F(0) = F^+$, leading to

$$S(\tau) = \frac{3}{4\pi}F\cdot(\tau + q_\tau) \tag{2.30}$$

in which q_τ in principle is a constant of integration. In reality, rather $q_\tau = q(\tau)$, a numerical function, with $1/2 \leq q(\tau) \leq 1$ and $q(0) = \sqrt{3} = 0.5774$. $q(\tau)$ can only be found numerically and then a reasonable fit function is $q(\tau) \simeq 0.7104 - 0.1331e^{-3.4488\tau}$.

Eddington chose a different approach. He multiplied the gray transport equation with $\cos\theta$ and integrated over all solid angles. This results in

$$S(\tau) = \frac{3}{4\pi}F\cdot\left(\tau + \frac{2}{3}\right) \quad . \tag{2.31}$$

Returning this result into the direction dependent transport equation one obtains

$$I(\tau,\theta) = \frac{3}{4\pi} F \cdot \left(\cos\theta + \tau + \frac{2}{3}\right) \tag{2.32}$$

so that a relation between I and angle of sight can be found. E.g., normalizing to the centre of the stellar surface ($\theta = 0$) leads to

$$\frac{I(0,\theta)}{I(0,0)} = \frac{2}{5}\left(1 + \frac{3}{2}\cos\theta\right) \quad . \tag{2.33}$$

Such values for the Sun are given in Table 2.1.

Using these equations one can obtain insight in the temperature structure of a stellar atmosphere. Note that independent information comes from the use of spectral lines of different excitation and ionization stages of metals. This will be discussed in Ch. 3.3.7.

2.7 Structure of stellar atmospheres

2.7.1 Temperature structure

In the case of LTE the source function equals the Planck function. Considering the outward flux and using Stefan-Boltzmann one then has

$$\pi\, S(\tau) = \sigma\, T^4(\tau) \tag{2.34}$$

so that with the expression for S in the gray atmosphere one arrives at

$$T^4(\tau) = \frac{3}{4} T_{\text{eff}}^4 (\tau + q_\tau) \quad . \tag{2.35}$$

At the surface (limit $\tau \to 0$) one has for T_0 at $q_\tau = 2/3$

$$T_0 = \frac{1}{2^{1/4}}\, T_{\text{eff}} = 0.841\, T_{\text{eff}} \tag{2.36}$$

so that for the Sun with $T_{\text{eff}} = 5780$ one finds in this approximation $T_0 = 4860$ K.

2.7.2 Pressure structure

In order to build a full model for temperature and density of a stellar atmosphere, one now has to combine the expressions for the gravitational structure with the expressions for the physics of the gases. The gas pressure is $P_{\text{gas}} = nkT$. The pressure increases inward from $r + dr$ to r, so that

$$dP_{\text{gas}} = -\rho\, g\, ds \tag{2.37}$$

with gas density ρ and gravitation according $g = G\frac{M}{r^2}$. Dividing by the optical depth one then has

$$\frac{dP_{\text{gas}}}{d\bar{\tau}} = \frac{g}{\bar{\kappa}_m} \tag{2.38}$$

with $\bar{\kappa}_m = \bar{\kappa}_m(T, P, XYZ)$ the temperature-, pressure-, composition-dependent mass absorption coeffiecient. This requires numerical solutions.

Assuming a gray atmosphere and a temperature structure as derived above, and approximating $P_{\text{gas}} = (g/\bar{\kappa}_m)\bar{\tau}$ one can make the transition to the geometric structure

$$\frac{dP_{\text{gas}}}{P_{\text{gas}}} = d\ln P_{\text{gas}} = -\frac{dr}{H_{\text{P}}} \quad \text{with} \quad H_{\text{P}} = \frac{kT}{\bar{\mu}\, g} \tag{2.39}$$

H_{P} being the pressure scale height (equivalence height). The mean molecular weight, $\bar{\mu}$, for fully ionized gases is easy to calculate (Eq. 4.65) but for partially ionized gases (as is normally the case in stellar atmospheres) it is a complicated function of T, P, and XYZ.

Results for the structure of the solar atmosphere are given in Table 2.2.

2.8. OPACITY AND THE ABSORPTION COEFFICIENTS

Table 2.2: Parameters for the outer solar atmosphere. Data from Minnaert (1953)

$\bar{\tau}$	T [K]	$\log P_{\text{gas}}$	$\log P_e$	ρ [g cm^{-3}]	d [km]
0.01	4650	3.74	−0.15	1.4 10^{-8}	−412
0.02	4700	4.01	0.09	2.7	−325
0.06	4790	4.30	0.36	5.0	−230
0.10	4890	4.43	0.50	6.6	−187
0.20	5090	4.60	0.71	9.4	−129
0.32	5290	4.71	0.87	11.7	−89
0.40	5400	4.77	0.96	13.1	−67
0.60	5660	4.86	1.15	15.4	−32
0.80	5870	4.91	1.32	16.6	−12
1.00	6070	4.94	1.48	17.3	0

d = depth considered from the outside, with $d = 0$ at $\tau = 1$.
$P_e = n_e kT$ and can be derived from spectral lines (see Ch. 3).

2.8 Opacity and the absorption coefficients

Absorption is an expression used for the sum of true absorption (κ_ν) and scattering (σ_ν). The dominant form is in most stellar gases the true absorption, and is presented first. There are several forms of absorption: absorption due to ionization of atoms, absorption due to electronic transitions in atoms and ions (see Ch. 3), absorption due to dissociation of molecules, and absorption in free-free transitions. A special case is the absorption due to ionization of H$^-$ (see below).

2.8.1 Absorption due to ionization

Atoms can be ionized by photons having an energy larger than the ionization energy E_{ion} of the actual excitation state of the atom. Excess energy will be transferred to the ion and the electron as kinetic energy $E_{\text{ph}} = h\nu = E_{\text{ion}} + \frac{1}{2}m_e v^2 + \frac{1}{2}m_{\text{ion}} v^2$. The kinetic energy becomes part of the thermal energy of the gas. The process is also described as the bound-free transition (b-f). The reverse process is the recombination (free-bound, f-b), in which photons are produced.

Ionization can take place for frequencies $\nu > \nu_{\text{ion}} = E_{\text{ion}}/h$ and this produces a sharp depression, the **ionization edge**, in the spectral continuum. The absorption coefficient of hydrogen is (Kramers in 1923)

$$a_n = \frac{1}{n^2} \frac{64\pi^4}{3\sqrt{3}} \cdot \frac{1}{\nu^3} \cdot \frac{Z^4 m_e e^{10}}{ch^6 n^3} g \tag{2.40}$$

with g the Gaunt factor (a factor with which deviations from a simple formula are taken care of). Clearly, the absorption coefficient is proportional to ν^{-3} so that, starting at the ionization edge, the absorption due to ionization becomes smaller with larger frequency.

2.8.1.1 Total absorption cross section for hydrogen

Since atoms may be in an excited state, ionization can take place also from this excited state. This leads to another ionization edge at that limit. The frequency of the ionization edges of hydrogen-like atoms (H, He, C^{+5}, etc.) can be given by

$$\nu_{\text{edge}} = R Z^2 \frac{1}{n^2} \tag{2.41}$$

in which R is the Rydberg constant and Z the nuclear charge. For $n = 1, 2, 3...$ the edges are called after Lyman, Balmer, ... (see Fig. 2.5).

For the interpretation of observations of the shape of the absorption edges of hydrogen see, e.g., Ch. 3.3.2. There it is also shown that the level of hydrogen ionization in the stellar atmosphere

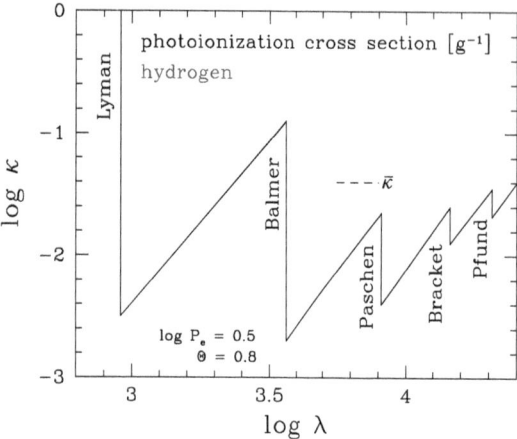

Figure 2.5: The absorption coefficient of hydrogen is plotted in logarithmic scale ($\log \kappa$ [g^{-1}] vs. $\log \lambda$). The absorption edges from the levels $n = 2$ (Balmer), $n = 3$ (Paschen), $n = 4$ (Brackett), and $n = 5$ (Pfund) are marked. The model used has $\Theta = 0.8$ ($T_{\text{eff}} \simeq 6000$ K) and $\log P_e = 0.5$. The Rosseland mean opacity, $\bar{\kappa}$ (see Sect. 2.6.4), is indicated.

plays a role only in a certain temperature range: hydrogen is fully ionized for $T \geq 20\,000$ K and the ionization edges have disappeared, while for $T \leq 6000$ K hydrogen is not even excited to level $n = 2$ so that the Balmer and higher n absorption edges are not present.

2.8.1.2 Absorption due to ionization of helium

The ionization energy of neutral He is 25.4 eV ($\lambda = 506$ Å). In the solar atmosphere the level of excitation of He is vanishingly small, so that He absorption plays no role in the atmosphere itself.

In hot stellar atmospheres He is quite visible. There are twice as many He ionization edges as H edges, and every second edge coincides with an H edge (H and He have nearly equal Rydberg constants). This is immediately clear using Eq. 2.41:

$$\text{He II}: \nu \simeq \frac{4}{n^2} \quad \text{and} \quad \text{H}: \nu \simeq \frac{1}{n^2} \qquad (2.42)$$

He has edges at 228, 911, 2050, 3644 and 5694 Å. The 3644 Å edge is called the Pickering edge. These edges have consequences for the overall opacity in stellar gas (see Fig. 2.8 and Ch. 4.4.1).

2.8.1.3 Absorption due to ionization of metals

With the metals, the more important ones are those of abundant elements. As an example, consider the atmosphere of the Sun, in which the metals are neutral or singly ionized. Fe I has an I.P.= 7.90 eV so the ionization edge is at $\lambda = 1570$ Å. Excitation to higher levels is easy so that ionization can take place at almost any wavelength. Si I has an I.P. = 8.15 eV thus an edge at $\lambda = 1520$ Å. Mg I has I.P. of 7.65 eV, edge at $\lambda = 1630$ Å, and a clear edge from an excited level at $\lambda = 2520$ Å. In hot atmospheres similar considerations can be made for the higher ionic states.

Ionization potentials are listed in "Allen's Astrophysical Quantities" (Cox 2002). Information about further energy levels is available in Moore (1959) or Wiese et al. (1966, 1969).

Note that the metals impose numerous absorption lines onto the spectrum. In many cases, there are so many that individual lines hardly can be recognized and the absorption looks like continuous absorption. An example is given in Fig. 2.6.

2.8.2 The H$^-$ ion

In gas near 5000 to 6000 K most of the metals are singly ionized. This leads to some electron density. In such gas, depending on density and availability of electrons, hydrogen may build the 'radical' H$^-$, the H$^-$ ion. The binding energy of H$^-$ is 0.75 eV and there are no excited states of H$^-$ below that energy. One can apply the Saha equation (Ch. 3.2.2) to calculate the number of H$^-$ ions in the atmospheres of stars. For the Sun one so arrives at $n(\text{H}^-) \simeq 3 \cdot 10^{-8}\, n(\text{H})$. For details see, e.g., Böhm-Vitense (1989, Ch. 7).

2.8. OPACITY AND THE ABSORPTION COEFFICIENTS

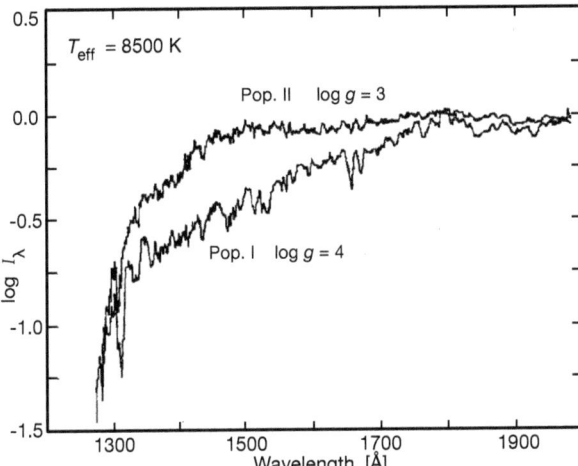

Figure 2.6: The absorption coefficient κ due to metals is mostly due to individual spectral lines. There are, however, so many that the effect of their overlapping is like that of continuum absorption. The average spectral energy distribution in the UV of stars of Population I is shown together with that of stars of Population II, the latter being metal poor by about a factor 10, thus $[M/H]\simeq -1$. The UV intensity of the metal poor stars is higher because of substantially smaller overall opacity. The spectra were normalised in the V-band. Figure from Huenemoerder et al. (1984).

This ion is, of course, easily dissociated which requires photons with an energy of more than 0.75 eV. This radiative dissociation produces a very broadly spread contribution (centered on $\lambda \simeq 16\,500$ Å) to the opacity of the atmospheric gases (see Fig. 2.8, left panel). It leads to a broad and shallow depression in the spectral brightness of the Sun in the near IR part of the spectrum.

2.8.3 Absorption due to dissociation

Molecules can dissociate through absorption by photons having an energy larger than the dissociation energy, $h\nu > E_{\text{diss}}$. This absorption takes place over a large range of frequencies and leads therefore also to continuum absorption. If molecule AB is dissociated into the atoms A and B by a photon $h\nu$, the energy of the photon is not only used for the dissociation ($AB + h\nu_{\text{diss}} \to A + B$) but the excess energy is taken up as kinetic energy or possibly as energy to excite the atoms remaining: $E_{\text{ph}} = E_{\text{diss}} + E_{\text{kin}} + E_{\text{exc}}$.

Dissociations have a certain probability to take place. This is characterized by a dissociation constant K as

$$K_{AB} = \frac{P_A P_B}{P_{AB}} = kT \frac{n_A\, n_B}{n_{AB}} \quad . \tag{2.43}$$

the latter part of the equation assuming $PV = nkT$. The abundance of various molecules in cool stellar atmospheres is shown in Fig. 2.7, some dissociation constants are listed there, too. Examples of spectral line absorption by molecules can be found in Figs. 3.10 and 10.21.

2.8.4 Free-free transitions

At higher temperatures and higher electron densities, electrons passing by ions are accelerated in the Coulomb field and then radiate the *Coulomb-Bremsstrahlung*. This process is also indicated as a free-free transition (f-f), producing free-free radiation. Similarly, energy can be absorbed from the photon field, leading to acceleration. The absorption coefficient for this process, in the case of fully ionized gas (in stellar interiors) and with 80% H and 20% He by mass, is

$$a_{\text{ff}} = 1.32 \cdot 10^{-2} \frac{n_e^2}{T^{3/2}} \frac{1}{\nu^2} g \tag{2.44}$$

again with g as a correction factor ($g \simeq 1$).

2.8.5 Scattering

Some of the photons are scattered by the gas particles. There are several kinds of scattering.

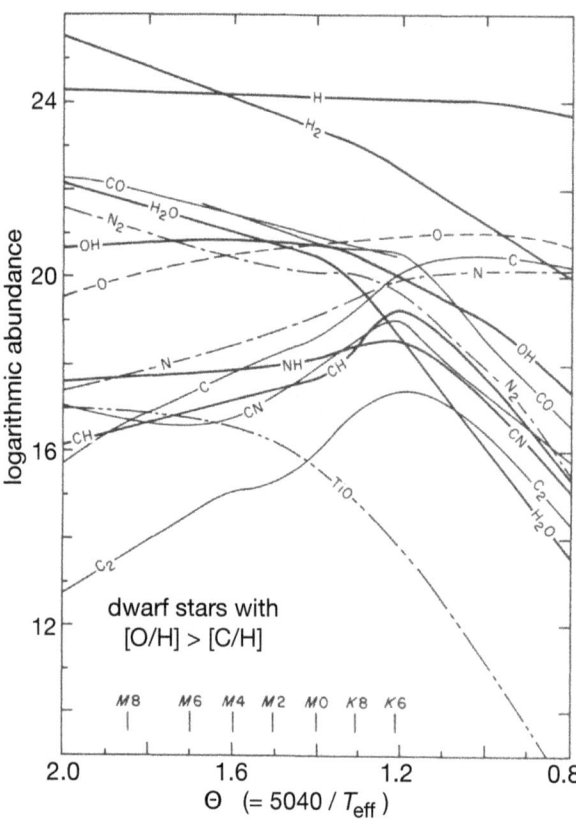

Figure 2.7: The abundance of molecules in cool stellar atmospheres is indicated in a logarithmic scale, along with the abundance of the relevant elements (for an atmosphere richer in O than in C). The temperature is given as Θ (see Ch. 3.2.2) so that the temperature range shown is from $T_{\text{eff}} \simeq 2500$ K at the left to $T_{\text{eff}} \simeq 6300$ K at the right. Figure from Aller (1963).

Molecules may be dissociated, requiring a certain energy depending on the excitation condition of the gas (see Eq. 2.43). Dissociation constants for two molecules are given below.

T [K]	log kT	log K H$_2$	log K TiO
5000	−0.37	7.7	. .
4000	−0.40	6.4	3.7
3000	−0.59	4.4	0.7
2000	−0.76	0.3	−5.3

Dissociation constants for H$_2$ and TiO (as given by Aller 1963)

Photons may have 'resonant' scattering on atoms and ions. This process is rather absorption and instantaneous re-emission of the photon. It takes place at frequencies around the frequency of an electronic transition (an absorption line). The cross section is

$$\sigma_R(\nu) = \frac{8\pi}{3} \frac{e^4}{m_e^2 c^4} \left(\frac{\nu}{\nu_0}\right)^4 N \tag{2.45}$$

with ν_o the resonance frequency of a transition (e.g. 1215 Å for the hydrogen Lyman-α line).

Photons may be scattered by free electrons (*Thomson scattering*) or by a molecule (*Rayleigh scattering*). The cross section is

$$\sigma_e = \frac{8\pi}{3} \frac{e^4}{m_e^2 c^4} n_e = 6.65 \cdot 10^{-25} n_e \tag{2.46}$$

with n_e the electron density. Since in the atmosphere of the Sun the relative number of electrons is small ($n_e/n_{\text{atom + ion}} \simeq 3 \cdot 10^{-3}$) there is hardly any Thomson scattering. In hot atmospheres hydrogen and metals are ionized so $n_e/n_{\text{nuclei}} \simeq 1$ and Thomson scattering is important indeed.

Another scattering process is the *Compton scattering* in which a photon scatters on an relativistic electron and therby gains energy. This may be of relevance in the deep interior of stars.

2.8.6 Total absorption coefficient

The total absorption coefficient of gas in the broader visual domain of spectral energy distributions is presented in Fig. 2.8. Two cases are shown: the cool case ($\Theta = 1$, $T_{\text{eff}} = 5040$ K) and the hot case ($T_{\text{eff}} = 28\,300$ K).

For opacity at higher temperatures (rather pertaining to layers below the atmosphere and the conditions of the stellar interior) see Ch. 4.4.1 and Fig. 4.4.

2.9. EMISSION AND THE EMISSION COEFFICIENT

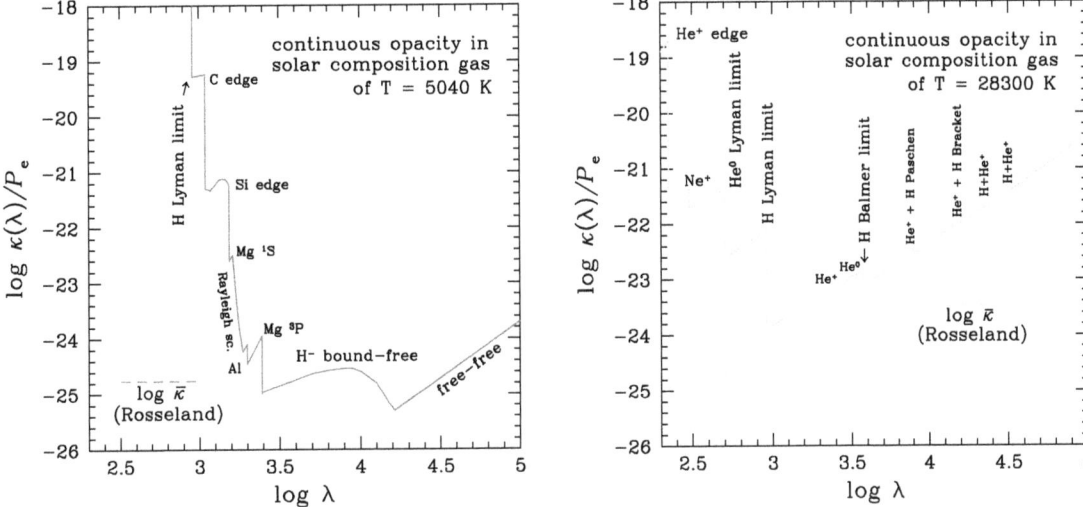

Figure 2.8: The total absorption coefficient κ/P_e [g^{-1}] due to continuous processes (foremost ionization). *Left*: gas at $T_{\text{eff}} = 5040$ K. *Right*: gas at $T_{\text{eff}} = 28\,300$ K. Various ionization edges are explained. Note that the Rosseland mean $\bar{\kappa}$ has different values for different temperatures. The figures are an adaptation of those of Unsöld (1977) who used diagrams from Vitense (1951) for a range of T as input; for sources of metal photoionization cross sections see, e.g., de Boer et al. (1972).

2.8.7 Effects of gas density on opacity

Most of the processes presented have a dependence on gas density. This may be, e.g., due to excitation of the state from which absorption takes place or due to a density dependence of the ionization or dissociation balance (for details on these mechanisms see Ch. 3.2). The opacity thus depends, for a given temperature, also on gas density. The effect is rather indirect and shows up clearly only under certain conditions.

One effect of gas pressure on the opacity is given in Fig. 2.9.

2.9 Emission and the emission coefficient

Radiation set free by hot gas can be described by the Planck function, because the gas is in LTE. Thus inside stars $j_\nu = B_\nu(T)$.

Further sources of emission are:
◦ free-free transitions or Coulomb-Bremsstrahlung (Ch. 2.8.4). Electrons passing by ions are being accelerated and thus emit radiation.
◦ free-bound transitions or recombination radiation. Each recombination produces a photon which, depending on the mean excitation state of such ions, is not easily absorbed.

Such further sources of emission become noticeable individually only when the conditions in the gas start to deviate from LTE (rather near the stellar surface). Remember that LTE is defined as the condition in which all radiation-matter interactions are in balance.

2.10 The spectral continuum and the Planck function

The source function, S_ν, of the gas in the interior of stars is the Planck function, because the gas is in LTE. However, in stellar surface layers the gas is not in LTE which results in an observable spectral energy distribution deviating considerably from the Planck like distribution emerging from the interior layers.

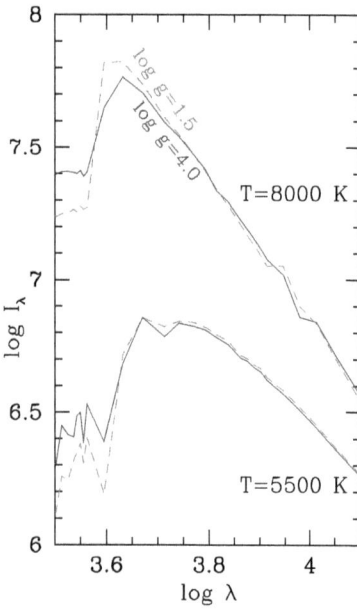

Figure 2.9: A comparison of the spectral energy distributions of cool main-sequence stars with those of red giants of the same temperature shows the effect of different gas density on the strength of the opacity due to H (the Balmer jump) as well as of the H$^-$ ion. In the atmosphere of the red giant ($\log g = 1.5$) the lower gas density leads to less H$^-$ and thus to larger n_e so to more H-atoms excited to the Balmer level. Thus the red giant has, in comparison, a larger Balmer jump (see Ch. 3.3.4 for the explanation of the Balmer level population). This explains the difference in location of the main sequence and the giant branch in an $U-B$ vs $B-V$ diagram. Graph using data from Kurucz (1992).

A simple method to imagine how the stellar surface spectral energy distribution looks like is to carry out a thought experiment. Starting with the Planck function one applies absorption of continuum nature like that given in Fig. 2.8. (Working with logarithmic representations one can just "substract" one curve from the other.) The absorption consists always of absorption by the metals, plus that by H and He depending on the temperature. For $T_{\text{eff}} < 6000$ K there is the contribution by H$^-$. For stars with T_{eff} in the range of 6000 to 20 000 K the hydrogen atom contributes with absorption due to photo-ionization. In the visual part of the spectrum, the contribution of H leads around the Balmer jump (Fig. 2.5 and Ch. 3.3.4) to a temperature effect. Its effect on the spectral energy distribution can be seen clearly in the two-colour diagram, $U-B$ vs. $B-V$ (Fig. 2.10).

The next step is to add the effects of individual absorption lines (Ch. 3). Such are, in the middle temperature range, foremost the lines of the hydrogen Balmer series, but all other lines contribute, too. Especially when the density of absorption lines in the spectrum is large, the effect looks like that of a continuum depression, as in Fig. 2.6. Note that absorption lines are generally much stronger and much more abundant on the Wien side of the maximum than in the Rayleigh-Jeans part of the energy distribution.

2.10.1 Effects for the CMD

The Planck spectral energy distribution and its deviations show up also in the CMD (see Fig. 1.2 middle panel and Fig. 21.3). Going along the main sequence from the cool to the hot end one has the following effects.

Very cool end: When these stars are observed in B and V one measures rather at the short wavelength side of the intensity maximum. The spectral slope is steeply red, and the spectrum is loaded with absorption lines. Stars of slightly higher temperatures already have a considerably increased emitted intensity (in M_V).

Cool end: The B and V bands are still 'on the Wien side of the Planck function', but now $B-V$ changes quickly toward higher temperatures, while M_V changes less drastically. The main sequence levels off. Why does this differ from the behaviour of the very cool stars?

Middle temperature range: Stars are, in the visual, now near the peak of the intensity distribution. In this temperature range the photoionization of hydrogen dominates the structure of the atmospheric opacity causing the above described deviations of the Planck spectrum.

Hot end: The visual part of the spectral energy distribution is approaching that of the Rayleigh-Jeans approximation, and $B-V$ becomes a constant (Rayleigh-Jeans limit). For hotter stars M_V

2.10. THE SPECTRAL CONTINUUM AND THE PLANCK FUNCTION

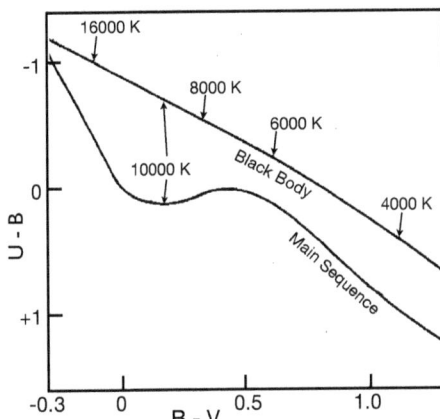

Figure 2.10: The $U - B$ vs. $B - V$ two-colour diagram shows how the colours of the Planck function change with temperature. The run of the colours of main-sequence stars deviates from the Planck colours, largely due to effects of opacity by the hydrogen ionization from the Balmer level (the Balmer jump). For details of the Balmer jump see Ch. 3.3.4. Note that when these effects play no role (toward the lower temperatures, especially $T < 4000$ K) the colour slope is the same as that for the Planck function. The locus of red giants in this diagram is different from that of main-sequence stars, because in their atmospheres density effects play a role for the opacity (see Fig. 2.9).

will be brighter still. This can be understood as being due to the larger stellar radius (more mass) as well as due to the *linear* dependence of M_V on T (Rayleigh-Jeans part of the spectrum!).

2.10.2 Backwarming, blanketing

One important aspect has to be added to the considerations above. When due to continuum opacity (and also spectral line opacity) energy is absorbed, the retained energy leads to an increase in temperature of those gases. This effect is in the literature called *backwarming* or *blanketing*. The higher temperature implies a higher temperature Planck function, so that all of the thought experiment has to be performed anew with that new spectral energy distribution. Clearly, in doing so, the luminosity must stay the same. Another way to phrase the effects is that due to opacity the emergent energy is redistributed over the spectrum. This is, in a way, the fullfillment of the continuity equation (Sect. 2.5). In general, the intensity taken away at the higher energy end of the spectral energy distribution is compensated for by an overall rise in the spectral energy distribution, which therefore is also noticeable in its lower energy part.

2.10.3 Electron density and opacity effects

Also the gas density plays a role in the size of the atmospheric opacity. At lower density, the fewer collisions between electrons and atoms or ions lead to a change in the ionization balance (see Ch. 3.2.2), e.g., in cooler atmospheres to less H in the form of H^-, or to less excitation (see Ch. 3.2). The combination of H^- and electron density is of interest here: with *fewer* H^- there are *more* free electrons thus more excitation of H to the Balmer level and so a larger Balmer jump. The effects for the spectral energy distribution in atmospheres of $T_{\text{eff}} = 8000$ K and 5500 K can be seen in Fig. 2.9.

References

Aller, L.H. 1963, "Atmospheres of the Sun and the Stars", 2nd. Edition; Ronald, New York
Böhm-Vitense, E. 1989, "Stellar Astrophysics, Vol 2: Stellar Atmospheres"; Cambridge Univ. Press.
Cox, A.N. 2002, "Allens Astrophysical Quantities", Springer, Heidelberg
de Boer, K.S., Koppenael, K., & Pottasch, S.R. 1972, A&A 28, 145
Huenemoerder, D.P., de Boer, K.S., & Code A.D. 1984, AJ 89, 851
Kramers, H.A. 1923, Phil. Mag. 46, 836
Kurucz, R.L. 1992, in IAU Symp. 149, "The Stellar Populations in Galaxies", eds. B. Barbuy & A. Renzini; Kluwer, p. 225
Mihalas, D. 1970, "Stellar Atmospheres"; Freeman & Co, San Francisco
Minnaert, M. 1953, in "The Solar System I, The Sun", Ed. G.P. Kuiper, Univ. Chicago Press

Moore, C.E. 1959, "A Multiplet Table of Astrophysical Interest", NBS Techn. Note No 26
Rosseland, S. 1924, MNRAS 84, 525
Unsöld, A. 1955, "Physik der Sternatmosphären"; Springer, Heidelberg
Unsöld, A. 1977, "The New Cosmos"; Springer, Heidelberg
Vitense, E. 1951, Z.f.Ap. 28, 81
Wiese, W.L., Smith, M.W., & Glennon, B.M. 1966, "Atomic Transition Probabilities, Part I", NSRDS, NBS, Washington
Wiese, W.L., Smith, M.W., & Miles, B.M. 1969, "Atomic transition probabilities, Part II", NSRDS, NBS, Washington

Chapter 3

Stellar atmosphere: Spectral structure

The recognition of structure in the spectrum of a star came from letting sunlight fall through a prism. Later Fraunhofer assigned (capital) letters to the prominent lines, lines later were called Fraunhofer lines. A few letters are still in use, such as H and K for the two Ca II lines near 3933 and 3968 Å as well as D (1 & 2) for the Na I lines near 5890 Å. (Some letters turned out to mark absorption produced by the Earth's atmosphere.) The spectral lines detected originate in the tenuous small optical depth layers of a star, the stellar atmosphere.

Spectral lines are identified by the name of the atom or ion and the wavelength. Neutral atoms produce transitions of the first spectroscopic state I, the first ion lines of the second spectroscopic state II, etc. The absorption line from C^{+3} near 154.8 nm is given as C IV λ1548 (λ in Å). The so-called 'forbidden lines' (forbidden according to the selection rules of the Bohr model but possible nevertheless with small probability) are indicated by square brackets around the ion name as, e.g., [O III] λ5305, semi-forbidden lines by just one of the brackets as for C III] λ1907.9. For tabular material on spectral lines see, e.g., Wiese et al. (1966, 1969).

There are numerous books about stellar atmospheres and spectral lines. A very thorough text is that of Unsöld (1955), others are from, e.g., Mihalas (1970) and Böhm-Vitense (1989).

3.1 Spectral lines

Spectral lines have a well defined shape, which is produced by the combined effect of the quantum mechanics (properties of the individual transitions) and the behaviour of the atoms in the stellar atmosphere (bulk properties of gas).

3.1.1 Line profile

3.1.1.1 Lorentz profile

The energy difference between two states can be calculated using quantum mechnics. Since the upper and lower state have a certain fuzziness each, the energy difference is not exactly equal to the simple Bohr model of the energy levels. One can calculate the probability ($f_{\rm lu}$) that a transition from the lower state l to the upper state u takes place with energy $\Delta E = h\nu$, with ν the frequency of the photon absorbed in the transition. This probability then can be incorporated in an expression for the absorption coefficient κ as

$$\kappa_\nu^{\rm line} = \frac{\pi e^2}{mc} n_{\rm l} f_{\rm lu} \frac{1}{\pi} \frac{\gamma/4\pi}{(\nu - \nu_o)^2 + (\gamma/4\pi)^2} = \frac{\pi e^2}{mc} n_{\rm l} f_{\rm lu} \, \Psi(\nu) \qquad (3.1)$$

in which ν_0 is the nominal frequency of the transition and $\nu - \nu_0$ is often given as $\Delta\nu$. $\Psi(\nu)$ is the *Lorentz function* which is normalized such that $\int \Psi(\nu) \, d\nu = 1$.

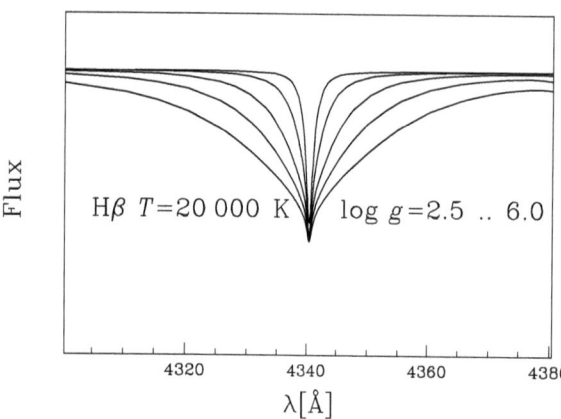

Figure 3.1: The shape of a spectral line depends on the pressure in the stellar atmosphere. At high pressure the lines get **pressure broadened** (see Sect. 3.1.1.2). The spectral line is Hβ modelled for several values of $\log g$ (2.5, 3.0, 4.0, 5.0, 6.0). The intensity scale runs from 0 to 1.2. Stellar surfaces with 'high gravity' have high gas density and thus broadened lines, surfaces with 'low gravity' have less dense gas so little pressure broadening. Figure from model database by Heber, Stw. Bamberg.

The constant γ, known as the damping constant, gives in a thin and collisionless gas the full half-intensity width of the absorption with

$$\gamma = |4\pi \, \Delta\nu| \quad \text{and} \quad \gamma = \frac{8\pi^2 e^2 \nu_o^2}{3 m_e c^3} \;\; \text{s}^{-1} \;. \qquad (3.2)$$

This full half-intensity natural line width is very small: $\Delta\lambda_n = 1.2 \cdot 10^{-4}$ Å. In quantum mechanics γ is interpreted as the reciprocal of the mean life time of the upper state. The parameter γ comes from the theory of radiation damping, in which radiation stimulates excited atoms to make an energy transition.

Classically, the gas containing such atoms was considered as a collection of charged oscillators (Maxwells equations, etc.) with $\mathcal{N}$ the number of oscillators. Using classical theory, the same formula as the one from quantum mechanics was derived with one difference: instead of $n_l f_{lu}$ one had $\mathcal{N} = \mathcal{N}_l$, the number of oscillators doing the absorption. (For details see Unsöld 1955, his Chs. 68 & 71.)

The name for the parameter f_{lu}, the **oscillator strength**, still reflects the classical approach. It provides the conversion constant between the bulk parameter and the physical parameter, $\mathcal{N}_l = N_l f_{lu}$. The oscillator strength f_{lu} is proportional to the transition probability A_{ul}, the Einstein constant for a spontaneous transition, as

$$f_{lu} = \frac{g_l}{g_u} \frac{1}{3\gamma} A_{ul} \;. \qquad (3.3)$$

3.1.1.2 Pressure broadening

Atoms in the process of a transition may be influenced by particles in their vicinity and then Eq. 3.2 is no longer valid. Two theoretical approaches are known in the literature.

The *collision theory* considers the radiating atom when another particle passes close by. The wavetrain in the process of being emitted is disturbed and is now considered as a collection of small sections being emitted in time. The full line profile then follows from Fourier analysis.

In the *statistical approach* one asks for which fraction of time is the spectral line shifted due to the influence of a colliding particle. The statistics leads then to the full profile.

Both approaches are special cases of the general complex of interactions leading to pressure broadening. Details can be found in Unsöld (1955, Chs. 75-83). Pressure broadening is very important. It induces a change in the shape of the spectral lines, making them broader proportional to the gas density. As an example, Fig. 3.1 shows the shape of one spectral line for a range of gas densities (or pressures) characterized by the surface gravity of the star.

The effect of pressure broadening on spectral lines explains why it is possible to derive the *luminosity class* of a star from the spectrum. Stars relatively compact for their mass must have high gas density (thus large $\log g$) at the surface (where $\tau \simeq 1$) and thus pressure broadened

3.1. SPECTRAL LINES

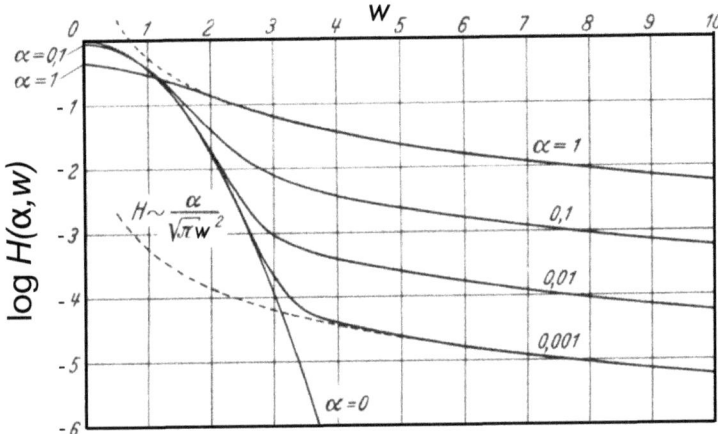

Figure 3.2: The Voigt function $H(\alpha, w)$ for different values of α (see Sect. 3.1.1.4) at a fixed value of $\log g$. When α is small, the line profile is like the Doppler profile and the central portion of H is called the **Doppler core** of the profile. When α is large, strong damping is present and the Voigt profile shows pronounced **damping wings**, extending to large values of w. Note that the shape of H is normalised for the Doppler width w. Note also that the damping wings for $\alpha = 1$ and 0.001 have the same shape (dashed extensions). The shape of an absorption line is obtained through $I = I_0 e^{-\tau}$ with $\tau \sim H(\alpha, w)$; see Eqs. 3.7 & 3.11 and Fig. 3.3. Adapted from Unsöld (1955).

spectral lines, while stars relatively extended for their mass thus having a less dense surface (small $\log g$) have little broadened spectral lines.

3.1.1.3 Doppler broadening

In all gases atoms move because of their temperature related kinetic energy. Also, bulk motions may take place in gas. All these random motions produce a line-of-sight Doppler shift for the wavelength (frequency) of a spectral line. For a gas this leads to broader spectral lines due to 'Doppler broadening'.

Doppler motion in an ideal gas (stellar atmosphere gas is close to that condition) can be given (in frequency space, $v/c = \Delta\nu/\nu$) by the Gauss function

$$\phi(\Delta\nu) = \frac{1}{\Delta\nu_D \sqrt{\pi}} e^{-\left(\frac{\Delta\nu}{\Delta\nu_D}\right)^2} \qquad (3.4)$$

with as width normalization $\Delta\nu_D = \frac{\nu}{c}\sqrt{\frac{2kT}{\mu}}$ (atomic weight $= \mu$). Also this function is normalized: $\int \phi(\Delta\nu) d\nu = 1$. The half width half maximum in velocity units is $b = 2\sqrt{2}\Delta\nu_D(c/\nu_0)$.

3.1.1.4 The Voigt profile

A spectral line as we observe it has a shape governed by the processes described above. The shape of the Lorentz profile as modified by Doppler broadening is given by the convolution of the functions Ψ (see Sect. 3.1.1.1) and ϕ (Sect. 3.1.1.3) as

$$\int_{-\infty}^{+\infty} \Psi(\nu - \Delta\nu) \otimes \phi(\Delta\nu) \, d(\Delta\nu) =$$

$$\frac{1}{\Delta\nu_D \sqrt{\pi}} \frac{\gamma}{4\pi^2} \int_{-\infty}^{+\infty} \frac{exp[-(\frac{\Delta\nu}{\Delta\nu_D})^2]}{(\nu - \nu_0 - \Delta\nu)^2 + (\gamma/4\pi)^2} \, d(\Delta\nu) = \frac{1}{\Delta\nu_D \sqrt{\pi}} H(\alpha, w) \qquad (3.5)$$

in which $H(\alpha, w)$ is the *Voigt function* (Waldemar Voigt, 1850-1919), with $\int_{-\infty}^{+\infty} H(\alpha, w)dw = \sqrt{\pi}$. The two important parameters in the Voigt function have been defined as

$$\alpha = \frac{\gamma}{4\pi\Delta\nu_D} \quad \text{and} \quad w = \frac{\nu - \nu_0}{\Delta\nu_D} \quad . \tag{3.6}$$

One can recognize in α the aspects of damping ($\gamma/4\pi$) and Doppler-broadening ($\Delta\nu_D$). The parameter α gives the relative importance of the two. Fig. 3.2 gives the Voigt function in relation to w, the relative width due to the velocity dispersion.

3.1.2 Shape and strength of spectral lines and curve of growth

The *shape of an absorption line* is defined by the frequency dependent absorption coefficient of the absorption line

$$\kappa_\nu^{\text{line}} = \frac{\pi e^2}{mc} n_l \, f_{\text{lu}} \frac{1}{\Delta\nu_D \sqrt{\pi}} \, H(\alpha, w) \quad . \tag{3.7}$$

Intensity is absorbed out of the continuum intensity I so that the remaining intensity is $I_\nu = I_\nu^{\text{cont}} e^{-\tau}$ with $\tau \sim H(\alpha, w)$. Note that I^{cont} is often given as I^0 or I_0.

Since the absorption is governed by $e^{-\tau}$, and while spectra are always presented in a *linear* intensity scale, the shape of the Voigt profile (Fig. 3.2) is transformed. In the centre of an absorption line the optical depth is always largest, so that little light is left for further absorption. Increasing the optical depth hardly leads to a measurable further loss of intensity in the centre of the line. The spectral line is said to be *saturated*. However, absorption will be stronger in the wings. At modest spectral resolution (which was always the case with instrumentation of the past and is still used for faint objects with small intensity) the spectral line is degraded by the resolution of the spectrograph and all information on the true line shape is lost. One then can retrieve from the data only the total strength of absorption.

The *strength of an absorption line* is defined by the light missing from the spectrum. One integrates over the continuum-normalised spectrum through

$$W_\lambda = \int \frac{I^{\text{cont}} - I_\lambda}{I^{\text{cont}}} d\lambda \tag{3.8}$$

giving the strength of the absorption in units of Å, being the **equivalent width** of absorption. This parameter has its equivalents in frequency or velocity space as

$$\frac{W_\lambda}{\lambda} = \frac{W_\nu}{\nu} = \frac{W_v}{c} \quad . \tag{3.9}$$

Since $I_\nu = I_\nu^{\text{contin}} e^{-\tau_\nu}$, in which τ_ν is based on the line-of-sight integral over the amount of material according to

$$\tau_\nu = \int_{s_1}^{s_2} \kappa_\nu \, ds \tag{3.10}$$

the equivalent width of an absorption line is therefore

$$W_\nu = \int_{\text{line}} 1 - e^{-\tau_\nu} \, d\nu \quad . \tag{3.11}$$

3.1.2.1 Small optical depth in the line ($\tau \ll 1$ and/or $\alpha \ll 1$)

For $\alpha \ll 1$ one has (see definition of α above) $4\pi\Delta\nu_D \gg \gamma$, or, in words, the effect of the Doppler-broadening is much more important than the effect of the damping. The absorption profile shows, in fact, only the central part, the **Doppler core** of the line.

In this case the equivalent width can be simply derived (since $\tau \ll 1$) by

$$W_\nu = \int_{\text{line}} 1 - e^{-\tau_\nu} \, d\nu = \quad \text{(for small } \tau\text{)} \quad = \int \tau_\nu \, d\nu = \int_0^{+\infty} \int_{s_1}^{s_2} \kappa_\nu \, ds \, d\nu \quad . \tag{3.12}$$

3.1. SPECTRAL LINES

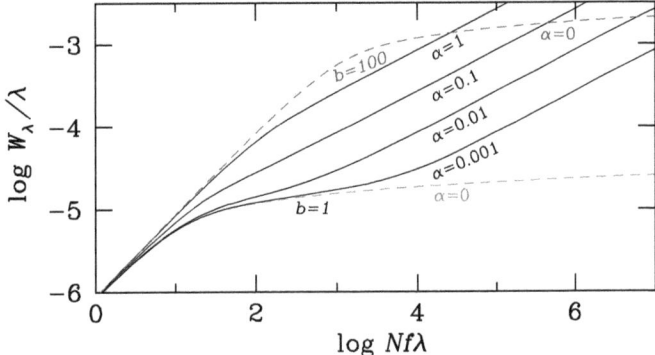

Figure 3.3: The curve of growth (COG) is shown for $b = 1$ km s^{-1} and different values of α (as in Eq. 3.6 and compare with Fig. 3.2). The units are $\log W_\lambda/\lambda$ vs. $\log Nf\lambda$, the latter arbitrarily normalized. For smaller α the strength of the line reaches Doppler saturation and grows only when (at larger N) the absorption wings start growing. For larger α stronger damping leads to wider damping wings (Fig. 3.2) and stronger and stronger lines. To see what a larger b brings, the curve for $b = 100$ km s^{-1} and $\alpha = 0$ is added. Curves for larger b values would have the same shape as the ones shown for $b = 1$ km s^{-1}, but shifted up and to the right to match the appropriate b level (see the $b = 100$ km s^{-1} curve). For pure Doppler broadening ($b \to \infty$), the CoG would be a straight line (the dashed upper-left curve). For no damping ($\alpha = 0, \gamma = 0$) the line saturates (the level in W_λ/λ depending on b) and stays for all $Nf\lambda$ nearly at that level.

Assuming that the nature of the material doing the absorption does not change over the line of sight, the integral can be split in two parts: the integral over the line of sight and the integral over the frequency interval. (Note that the frequency interval runs from 0 to $+\infty$ nominally, but that in practice only the 'small' stretch in frequency around ν_0 is relevant.) Thus

$$W_\nu = \int \tau_\nu d\nu = \int_0^{+\infty} \frac{\kappa_\nu}{n_1} d\nu \cdot \int_{s_1}^{s_2} n_1 ds = \frac{\pi e^2}{mc} f_{lu} \cdot n_1 L = \frac{\pi e^2}{mc} N_1 f_{lu} \qquad (3.13)$$

in which $\int n_1 ds = n_1 L = N_1$, the **column density** of material [cm^{-2}]. Transforming this to the classical wavelength based equivalent width (units of Å, in spectroscopy still common) one arrives at

$$\frac{W_\lambda}{\lambda} = \frac{\pi e^2}{mc^2} N_1 f_{lu} \lambda \qquad (3.14)$$

in which $N_1 f_{lu}$ is again $\mathcal{N}$, the classical number of oscillators. *For small optical depths, the equivalent width is linearly proportional to the amount of material and to the line-constant $f\lambda$.*

3.1.2.2 Very large optical depth in the line ($\tau \gg 1$ and/or $\alpha \gg 1$)

For $\alpha \gg 1$ clearly $\gamma \gg 4\pi \Delta\nu_D$ or, in words, the effects of damping are much more pronounced than the effects of Doppler broadening. The shape of the absorption coefficient shows **wide damping wings** (see Fig. 3.2) and $H(\alpha, w)$ can (in the limit) be simplified to be $H(\alpha, w) \simeq \frac{\alpha}{\sqrt{\pi} w^2}$ so that

$$\tau_\nu = \int_{s_1}^{s_2} \frac{\pi e^2}{mc} n_1 f_{lu} \frac{1}{\Delta\nu_D \sqrt{\pi}} \frac{\alpha}{\sqrt{\pi} w^2} ds \ . \qquad (3.15)$$

After separating again the integration over line of sight from that over frequency one obtains

$$\frac{W_\lambda}{\lambda} = \sqrt{\frac{\pi^2 e^2 \gamma \lambda}{mc^3}} \sqrt{N_1 f_{lu} \lambda} = \frac{\pi^2 e^2}{mc^2} \sqrt{\frac{8}{3\lambda}} \sqrt{N_1 f_{lu} \lambda} \qquad (3.16)$$

indicating a *square root* proportionality of the equivalent width to $Nf\lambda$. The location of the curve of growth for such lines still does depend on λ (because of $\sqrt{8/3\lambda}$).

3.1.2.3 Intermediate α and/or τ

In the middle range, the Voigt function cannot be solved analytically but must be treated numerically. The result is that the equivalent width of a line behaves rather proportional to the *logarithm* of $Nf\lambda$ as

$$\frac{W_\lambda}{\lambda} \sim \log N_l f_{lu} \lambda \quad . \tag{3.17}$$

Since $\frac{W_\lambda}{\lambda}$ usually is plotted logarithmically, the intermediate part becomes nearly a flat line (see Fig. 3.3).

3.1.2.4 Shape of curve of growth

The curve showing how an absorption line grows in strength with increase of optical depth ($Nf\lambda$) is called the curve of growth (**COG**). Such curves are always plotted in a logarithmic scale. Units and normalisation of the parameters may vary. Fig. 3.3 shows the COG with W_λ and $Nf\lambda$ normalised in a way usefull for stellar atmosphere physics. Note, however, that light passing through a stellar atmosphere passes through gases with a range in density and temperature, while the approach above implied a uniform slab of gas. But see Sects. 3.3.7 and 3.3.7.3.

3.2 Statistics

The atoms and ions in a gas will each be in certain energetic states. In thermodynamic equilibrium all processes (absorption, emission, excitation, de-excitation) are in balance. This means that the population of the available energy levels is governed by statistics. This can be used both to learn about the excitation state of ions, and thus to derive that state from observations of atoms and ions, as well as to learn about the ionization state of a gas and thus to derive the level of ionization from observations.

3.2.1 Boltzmann statistics and excitation equation

The distribution of particles (atoms or ions) over the possible energetic states A and B is given by the Boltzmann equation

$$\frac{n_B}{n_A} = \frac{g_B}{g_A} e^{-\frac{\Delta E_{AB}}{kT}} \tag{3.18}$$

in which n_A is the number density of the particle in state A, g_A is the statistical weight of the state, ΔE_{AB} the energy difference between state A and B, and kT the product of Boltzmann constant k and temperature T.

In order to find the ratio of the number of particles in a given state to all particles of that kind one performs a sum of all ratios n_i/n_1 which, after some rearranging gives

$$n_{total} = \Sigma n_i = \frac{n_1}{g_1} \cdot \left(g_1 + g_2 e^{-\frac{\Delta E_{12}}{kT}} + g_3 e^{-\frac{\Delta E_{13}}{kT}} + \ldots \right) \equiv \frac{n_1}{g_1} Q(T) \tag{3.19}$$

in which $Q(T)$ is the **partition function** of that kind of particles. $Q(T)$ contains the sum over all exponential terms, and since (in practice) a particle does have only a finite number of excitable states, the sum is finite.

One can now use this expression to come to the general equation for the Boltzmann statistics

$$\frac{n_j}{n} = \frac{g_j}{Q(T)} e^{-\frac{\Delta E_{1j}}{kT}} \quad . \tag{3.20}$$

Using this expression, one can derive the relation between temperature and population state if the atomic constants are known (see for the application to stars Sect. 3.3.1).

3.2.2 Ionization and Saha equation

The statistics of the population of energetic states can also be extended to the case of ionization. One here has to deal with the statistics of three particles: the one of the lower ionization state (which will be called state a) with the population of all its levels, the higher ionization state (here called b) with all its levels, and the free electron which may occupy some energetic state given by the sum of the ionization energy (χ_{ab}) and the kinetic energy of the electron ($\frac{1}{2}mv^2$), so $E_{\text{el}} = \chi_{ab} + \frac{1}{2}mv^2$.

Consider the number of particles in the lowest excitation state of the two ionization states a and b. The number of particles in the lower ionization state is $n_{a,1}$ having statistical weight $g_{a,1}$. The statistical weight of the ion + electron system is $g_{\text{ion}} \cdot g_{\text{el}}$. The statistical weight of the ion in the ground state is $g_{b,1}$ and the number of particles in that state is $n_{b,1}(v)$ accounting for electrons in the velocity interval $[v,\ v+dv]$. The Boltzmann statistics now implies

$$\frac{n_{b,1}(v)}{n_{a,1}} = \frac{g_{b,1}}{g_{a,1}} \cdot g_{\text{el}}\, e^{-\left(\frac{\chi_{ab}+\frac{1}{2}mv^2}{kT}\right)} \tag{3.21}$$

where the first subscript always refers to the ionization state and the second to the excitation state. To find the statistical weight of the electron consider the total number of quantum states possible in the interval in space and in momentum

$$g_{\text{el}} = \frac{2\ dq_1\ dq_2\ dq_3\ dp_1\ dp_2\ dp_3}{h^3} \tag{3.22}$$

in which the factor 2 accounts for the two possible spin states ($s = \pm\frac{1}{2}$) of the electron. If this volume contains just one electron then $dq_1 dq_2 dq_3 = 1/n_e$ while $dp_1 dp_2 dp_3 = 4\pi\ p^2 dp = 4\pi\ m^3 v^2 dv$ (n_e being the number of electrons per unit volume). If one inserts these expressions in the Boltzmann equation for the statistics of ionization as

$$\frac{n_{b,1}(v)}{n_{a,1}} = \frac{4\pi\ m^3}{h^3} \frac{2}{n_e} \frac{g_{b,1}}{g_{a,1}} e^{-\left(\frac{\chi_{ab}+\frac{1}{2}mv^2}{kT}\right)} \cdot v^2\ dv \tag{3.23}$$

and subsequently integrates over velocity, then one obtains

$$\frac{n_{b,1}}{n_{a,1}} \cdot n_e = 2\frac{g_{b,1}}{g_{a,1}} \left(\frac{2\pi\ mkT}{h^2}\right)^{3/2} e^{-\frac{\chi_{ab}}{kT}} \tag{3.24}$$

which is the Boltzmann equation for the ionization.

For gas in a stellar atmosphere (in fact for gas anywhere) one has to account for the fact that each ionization state has its own excitation statistics. For each ionization state one has (Eq. 3.20)

$$\frac{n_{a,1}}{n_{a,\text{all}}} = \frac{g_{a,1}}{Q_a(T)} e^{-\frac{\chi_{a,1}}{kT}} \quad \text{and} \quad \frac{n_{b,1}}{n_{b,\text{all}}} = \frac{g_{b,1}}{Q_b(T)} e^{-\frac{\chi_{b,1}}{kT}}\ . \tag{3.25}$$

Inserting these in Eq. 3.24 one obtains

$$\frac{n_{b,\text{all}}}{n_{a,\text{all}}} \cdot n_e = 2\frac{Q_b(T)}{Q_a(T)} \cdot \left(\frac{2\pi mkT}{h^2}\right)^{3/2} e^{-\frac{\chi_{ab}}{kT}}\ . \tag{3.26}$$

One now may multiply that equation with kT and replace $n_e kT$ by the electron pressure P_e. Taking the logarithm leads to the **Saha equation** (Saha 1921):

$$\log\left(\frac{n_b}{n_a} P_e\right) = -\frac{5040}{T}\chi_{ab} + 2.5 \log T - 0.48 + \log 2\frac{Q_b(T)}{Q_a(T)} \tag{3.27}$$

in which the ionization potential χ_{ab} is expressed in units of eV. The ratio $5040/T$ is called Θ. All physical constants are collected into 0.48.

This equation shows that the ratio of the number of particles in two ionic states is governed by the ionization energy and the temperature. Since $Q_b(T)/Q_a(T)$ is not very temperature sensitive and thus mostly based on constants of atomic structure, while $\log T$ is only slowly varying with T, the dominant term for the ionization structure is $\Theta \cdot \chi_{ab}$ or $5040/T \cdot \chi_{ab}$.

The Saha equation can be used to calculate for any element the state of ionization in the gas. Note that, since the electron pressure plays a role, the content of, e.g., neutral hydrogen may vary for a given temperature a lot between stars. One therefore need not be surprised to find even in the very high density atmospheres of hot white dwarfs (WDs) traces of neutral hydrogen (see Fig. 17.1 for a WD at $T_{\text{eff}} = 59\,000$ K with hydrogen absorption lines).

3.3 Statistics and structure in stellar spectra

3.3.1 Excitation

The level of excitation within one ionic state can be recognized using the strength of absorption lines from the various excited states. If one would compare absorption lines with identical (or similar) $f\lambda$ but from different levels of excitation, the differences in strength of such lines must be due only to the difference in the number of particles in the levels n_l. Thus from such a sequence of n_l one can, with the help of the Boltzmann statistics (Eq. 3.20), find the temperature responsible for the observed distribution of particles over those levels. That temperature is also called an **excitation temperature** (T_{exc}) since it is derived from considerations of the excitation state.

In practice this procedure is not that simple. First, the strength of spectral lines depends in a complicated way on the conditions in the gas (Sect. 3.1.2 above). Furthermore, finding enough lines with similar $f\lambda$ in a spectrum, which in normal observations covers a limited range in λ and which shows numerous other spectral lines (many of them overlapping with the desired ones), is not easy. The method used to find T_{exc} is to do a full curve of growth analysis (see Sect. 3.3.7).

3.3.2 Ionization

The Saha equation tells which atoms and ions are to be expected in a stellar atmosphere depending on gas temperature and density. If absorption by, e.g., C IV is observed, one immediately knows that the temperature must be high enough for the ionization structure to produce C^{+3}.

This relation between temperature and ionization structure is the basis for the spectral classification schemes. Thus the sequence OBAFGKM is a temperature sequence based on the recognition of lines from the various ionization states. Further refinements ultimately led to the MK-classification (called after Morgan and Keenan working at the Yerkes Observatory in the 1950s-1960s). For the historic development of that classification and its value today, see Sect. 3.8.

If one can derive the ratio of the number of particles in two ionic states (full COG analysis; see Sect. 3.3.7), one can compare that ratio with the one from the Saha equation. In this manner one can derive the so called **ionization temperature** (T_{ion}). This does require, however, knowledge of the electron pressure. Note that the electrons are provided by all elements having ionization, so that to solve the Saha equation in fact needs solving a set of coupled equations.

3.3.3 Spectrophotometric methods

One can determine the temperature using **photometry**, utilizing the slope of the spectral energy distribution (the colour of the star) in relation with the slope of the Planck function. This is, however, not straightforward because all kinds of opacity effects change the simple Planck slope. Among these are the effects of temperature on hydrogen producing the Balmer jump, of the formation of H^- (Ch. 2.8.2) in cool atmospheres and of H_2 in very cool or very dense stellar atmospheres (Ch. 3.4).

3.3. STATISTICS AND STRUCTURE IN STELLAR SPECTRA

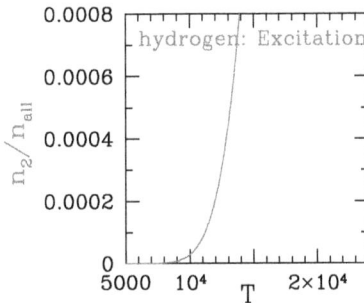
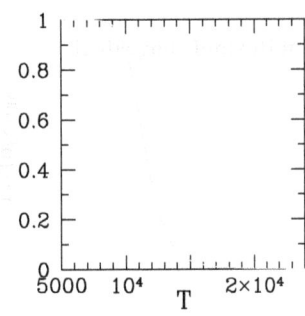
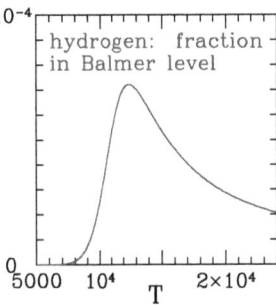

Figure 3.4: Behaviour of H with respect to excitation and ionization explaining the presence and **strength** of the **Balmer lines** and the size of the **Balmer jump** in relation with temperature. *Left*: The level of excitation to the 'Balmer level', n_2, given as $n_2/(n_1+n_2)$ using Boltzmann statistics. *Middle*: The fraction of neutral hydrogen, $n_I/(n_I+n_{II})$, versus temperature using the Saha equation. Note: choosing a larger P_e makes the ionization level become less at the same temperature (see Eq. 3.27). *Right*: The fraction of hydrogen in the Balmer level, n_2, with respect to the total amount of hydrogen, $n_I + n_{II}$, versus temperature (Boltzmann plus Saha).

Table 3.1: Nomenclature and wavelengths for hydrogen absorption

Lyman	$n_l \to n_u$	λ	Balmer	$n_l \to n_u$	λ	Paschen	$n_l \to n_u$	λ
Lyα	$1 \to 2$	1215.7	Hα	$2 \to 3$	6562.8	Pα	$3 \to 4$	18751
Lyβ	$1 \to 3$	1025.7	Hβ	$2 \to 4$	4861.3	Pβ	$3 \to 5$	12818
Lyγ	$1 \to 4$	972.54	Hγ	$2 \to 5$	4340.5	Pγ	$3 \to 6$	10938
continuum	$1 \to \infty$	912	contin.	$2 \to \infty$	3646	contin.	$3 \to \infty$	8203

3.3.4 Balmer jump and Balmer Series

A powerful method to determine the temperature of a star utilizes the *effects of temperature on hydrogen*. Hydrogen (with an ionization potential of 13.6 eV) plays a major role in the shaping of the spectral energy distribution and in the structure of the stellar atmosphere anyway.

Depending on temperature, hydrogen atoms get excited and populate the first excited level ($n = 2$) at 10.4 eV, the **Balmer level** (see Fig 3.4, left panel). Once the Balmer level is populated, absorption lines from that level, the **Balmer series** (see Table 3.1), become visible. One has, therefore, a temperature criterion from the Boltzmann statistics for excitation.

Ionization of atoms already excited to the Balmer level becomes also possible leading to absorption due to ionization from that level. For hydrogen this happens for $\lambda < 3646$ Å, equivalent to photons with $E_{\rm ph} > 3.2$ eV. This leads to a stronger continuum absorption shortward of the ionization edge at 3646 Å (see Fig. 2.5), a spectral structure called **Balmer jump**. For the Balmer jump in a range of spectral types see Fig. 1.4, for its effect on $U - B$ see Fig. 2.10. Since the photoionization cross section for hydrogen like atoms depends on frequency as $(\frac{\nu}{\nu_0})^{-3}$, with ν_0 the ionization edge frequency, the continuum depression quickly becomes shallower toward shorter wavelengths. Similar jumps exist for the other excited levels (see Fig. 2.5 for the Paschen and Pfund edges). The level of ionization of hydrogen in relation with temperature is shown in Fig. 3.4, middle panel.

At such levels of excitation, the temperature is high enough for hydrogen to be partially ionized (Saha equation). One now can calculate what fraction of hydrogen is neutral, and through Boltzmann statistics the population of the Balmer level. This behaviour is shown in Fig. 3.4, right panel. For high temperatures no neutral hydrogen is left and no Balmer absorption is visible.

Summarizing the temperature - Balmer level effects: starting at low temperatures, the Balmer level is not excited and no Balmer absorption lines are visible. With increasing temperature H

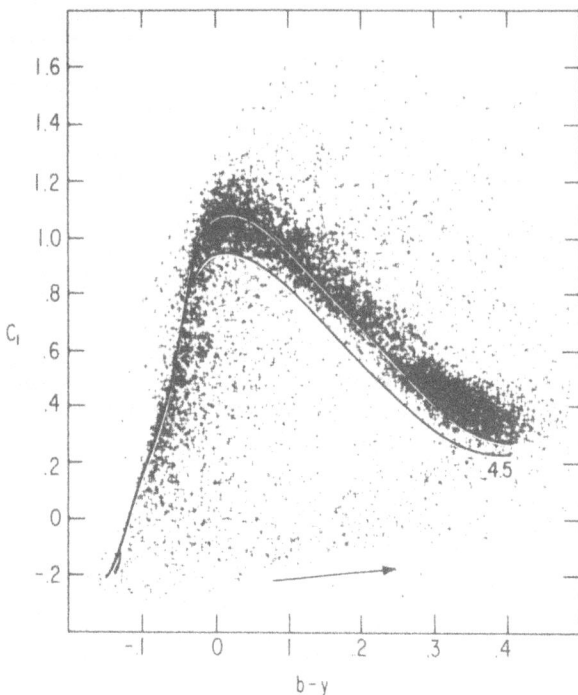

Figure 3.5: The c_1 versus $b-y$ diagram of Strömgren photometry shows how the slopes in the spectral energy distribution change with temperature. Overlain are models for solar metallicity with $\log g = 4.5$ (lower curve; black) and $\log g = 4.0$ (upper curve; 'white').
With ever bluer $b-y$ (progressively higher temperature), the Balmer jump index $c_1 = (u-v)-(v-b)$ first becomes 'redder' with a maximum for stars with $b-y \simeq 0$ equivalent to $T \simeq 10^4$ K (the temperature at which hydrogen has the maximum excitation to the 10.4 eV Balmer level; Fig. 3.4, right panel), followed by a 'blueing' of c_1 signifying a decrease in the strength of the Balmer jump (due to progressing ionization of hydrogen; Fig. 3.4 middle panel). The arrow shows the direction of the vector of interstellar reddening. Figure from Relyea & Kurucz (1978).

gets excited and the Balmer lines as well as the Balmer jump become visible. At yet higher temperatures, the hydrogen will be more and more ionized. Ever smaller amounts of neutral hydrogen are left so that the Balmer lines become weaker and the Balmer jump vanishes.

The **photometric method to find the temperature** makes use of the possibilities to measure with the Strömgren filter system the above described effects hydrogen has on the stellar spectral energy distribution. The Strömgren system (see Table 1.1) has y around 5470 Å and b near 4670 Å so that $b-y$ provides a spectral slope or temperature related colour (like $B-V$). Strömgren u is at 3500 Å and v around 4110 Å, so that $v-b$ is on the long side of the Balmer jump while $u-v$ bridges the Balmer jump (at 3642 Å). In the Strömgren system, the three shorter wavelength bands are combined into an index $c_1 = (u-v)-(v-b)$, being the difference in the spectral slopes of $u-v$ and $v-b$. Combining $b-y$ and c_1 into a diagram (see Fig. 3.5) makes the very characteristic run with temperature visible.

3.3.5 T_{eff} and $\log g$ from Strömgren photometry

In a further application of the Strömgren system one can determine T_{eff} and $\log g$. This is possible since gravity influences spectral line width (see Sect. 3.1.1.2 and Fig. 3.1) as well as the excitation conditions through the gas density (see Sect. 3.2.1), while temperature influences line width through Doppler broadening (see Sect. 3.1.1.3) as well as the line level population through the Boltzmann statistics (see Sect. 3.2.1). The Strömgren line strength index β represents the strength of the Hβ line, measured by comparing the signal strength in a wide filter with that in a narrow filter, both centered on the Hβ line (see Table 1.1). The c_1 index measures the Balmer jump (see Sect. 3.3.4). So combining β with c_1 allows to derive T_{eff} and $\log g$.

There are three approximate (temperature) regimes. In **cool stars** ($T_{\text{eff}} < 8500$ K) the line width (with β as indicator) is mostly due to temperature while the Balmer jump (excitation level) is still mostly formed through the gas density, so that c_1 shows mostly gravity. In **hot stars** ($T_{\text{eff}} > 11000$ K) the role is reversed: the line width (measured by β) is now dominated by gravity and the size of the Balmer jump (measured with the c_1 index) is mostly due to temperature. In **intermediate stars** (8500 K $< T_{\text{eff}} < 11000$ K) both gravity and temperature have similar sized influence on β and c_1. Note that the change of the size of the Balmer Jump with temperature

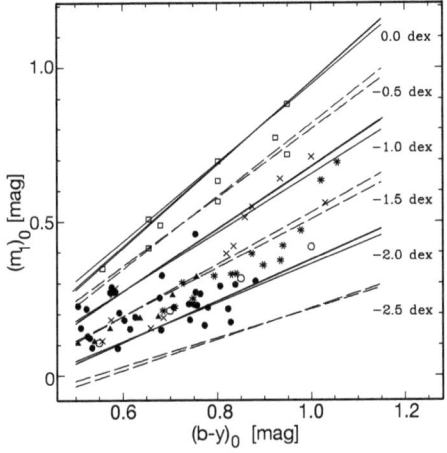

Figure 3.6: The m_1 vs. $b-y$ diagram (reddening correction applied) contains information about metal content (see the location of the v and b band in Fig. 3.8), so that metallicity values can be obtained photometrically. This diagram shows the calibration of metallicity in red giant stars. Different symbols are for stars in different globular clusters, whose metallicity was determined spectroscopically. The heavy lines (full lines for factors of 10 in metallicity, dashed lines for the intermediate $\sqrt{10}$) give an analytic fit with 4 coefficients, the thin lines with 5. Figure from Hilker (2000).

can be surmized from Fig. 3.4 and is visible in Fig. 3.5. More on these aspects can be found in Napiwotzki et al. (1993).

3.3.6 Metallicity from Strömgren photometry

The Strömgren system can be used to determine metal abundance. The location of the v-filter was choosen by Strömgren to fall right on a region with enhanced metal absorption (see Fig. 3.8). The metallicity index $m_1 = (v-b) - (b-y)$ is thus sensitive for metal content through $(v-b)$ and for temperature through $(b-y)$. [M/H] information can in this way be obtained, albeit only crudely since there is no information on individual elements. Remember that temperature and gravity both influence the excitation statistics of the absorbing species. One tries, in fact, to derive three parameters T, $\log g$, and [M/H] from four measured Strömgren bands, u, v, b, and y, using information from diagrams like Fig. 3.5 and 3.6.

A special case is the determination of the metal content in *red giant stars*. Considerable effort has gone into the calibration of the m_1 vs. $b-y$ diagram, using globular cluster red giant stars. The metal content of those can be determined spectroscopically, and one so can define lines of constant metallicity in the m_1 vs. $b-y$ diagram (see Fig. 3.6 for the related calibration).

3.3.7 Spectroscopy and the curve of growth

If the spectral resolution of the measurements is good enough to study lots of individual spectral lines, a detailed analysis of temperature and ionization structure is possible. Also the abundance of the elements can be studied this way. Classically, the method used is that of the COG. With powerful models and computers an overall fit to a fully synthesised spectrum may be attempted.

The COG is the relation between the strength of lines, W_λ/λ (Sect. 3.1.2), and the optical depth defined through $Nf\lambda$ (see Figs. 3.3 and 3.7). Due to saturation effects and aspects of line shape (damping wings) the relation is curved.

3.3.7.1 Excitation

If one selects lines from levels with approximately the same excitation energy, one may assume that (in LTE) these levels are equally populated. Thus N would be the same for all absorption from each of the levels in this narrow excitation energy range. One now makes a COG using the lines from levels at that selected small excitation energy range, which can be achieved by plotting $\log W_\lambda/\lambda$ against $\log f\lambda$ for all these lines. A (small) range in excitation energy is needed in order to find a sufficiently large number of lines for the COG! (Note that it is implicitly assumed also that the gas is uniform in density and temperature. But see Sect. 3.3.7.3)

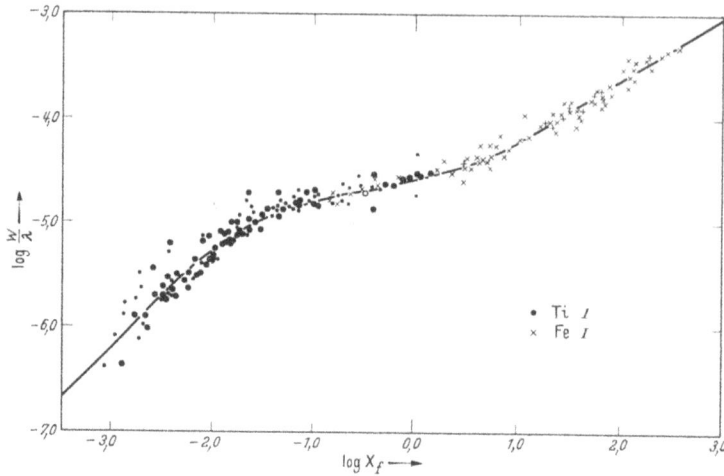

Figure 3.7: The COG for the Sun as determined from Fe I and Ti I lines. The CoG parameter $\alpha \simeq 0.03$ (see Fig. 3.3). Using lines coming from two well defined small excitation level ranges (see Sect. 3.3.7) the T_{exc} of, e.g., Fe I was determined. With this T all other lines of Fe I can be included and the value of T_{exc} improved ($T_{\text{exc}} = 4550$ K; the best value for the best defined COG). The same can be achieved for the lines of Ti I (having $T_{\text{exc}} = 4850$ K). Shifting the Fe I COG to that of Ti I leads to the value of $\log N(\text{Fe I}) - \log N(\text{Ti I})$, and thus to information about metal abundace. The horizontal scale is $\log f\lambda$ of the Fe I lines. Figure from Wright (1948) as given in Unsöld (1955).

A COG can be made for each of a selected different small range in excitation energy. These curves should show the same shape (being from atoms/ions in the same gas) but having a different location in the COG plot (being shifted in $Nf\lambda$), because for each of these energy level ranges (e.g., ranges r and s) the value of N is different according to the temperature based level of excitation. Thus from intercomparison of these curves of growth one can simply find the ratio of N_r to N_s (or graphically, $\log N_r - \log N_s$) from the COG of each of the selected energy level ranges.

Having established a set of ratios N_r/N_s one can now compare these with what would be expected from Boltzmann statistics. In this manner one finds the **excitation temperature** of the stellar atmosphere gas, T_{exc}.

Once the excitation temperature is found, all individual lines from all levels can be combined into one single COG, since the Boltzmann distribution can be accounted for by applying the appropriate N_j/N_{all} factors.

3.3.7.2 Ionization

Making a COG analysis (as just described) for different ions can be used to find the column density ratio N_b/N_a (or graphically, by shifting the plotted COG, $\log N_b - \log N_a$) of two adjacent ion states. Since the temperature is known (as well as the atomic constants), using the Saha equation one thus can calculate the electron pressure P_e. This procedure is, strictly speaking, not correct (see Sect. 3.3.7.3).

3.3.7.3 Depth structure of atmosphere

The procedures described above treat reality in a sloppy way. First, the COG analysis would only be valid for isothermal slabs of constant density. Second, although two adjacent ionization states clearly can exist at one temperature, the higher ions are in general found somewhat deeper in the stellar atmosphere. This follows from the fact that the excitation temperatures of adjacent ions are found to be different indeed, with T_{exc} of the higher ion being the larger of the two. In fact, two adjacent ions exist in overlaping layers so that N_a/N_b compares non-identical layers.

3.4. SPECIAL FEATURES

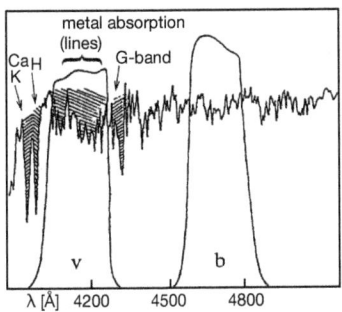

Figure 3.8: In the spectral energy distribution of cooler stars a prominent feature shows up. It is called the G-band (G = one of the original Fraunhofer letters) and is due to absorption by CH. To the left is a region with a strong depression due to metal line absorption.

In this figure, the response curves of the v and b filters of the Strömgren system are indicated. The $v - b$ index is used as a metallicity indicator (Sect. 3.3.6). The actual spectrum is that of a red giant. Figure adapted from a figure by Grebel (1991).

A stellar atmosphere has a temperature and a density gradient and each of the temperatures derived is based on line formation in a different layer. Nevertheless, the so derived excitation and ionization temperatures and the electron pressure collectively lead to insight in the depth structure of the stellar atmosphere. In particular a relation between T and n_e can be found. Finally, the translation into depth in km is made with the help of a full radiation transport analysis.

A tabular representation of the depth structure of the solar atmosphere is given in Table 2.2. Modern analyses treat the full radiative transfer including spectral line formation for a comparison with an observed spectrum.

3.3.7.4 Abundance of elements

Having derived the excitation temperature as well as the ionization structure with the methods described above, one may now proceed to compare the absorption from different elements. If two elements have an ionic state within the same ionization energy regime, these states may be compared directly with each other. Here one makes for each element separately the COG analysis of excitation (as done with different excitation levels; Sect. 3.3.7.1). Then the ratio of $N_{\text{element X}}/N_{\text{element Y}}$ follows from a comparison of the two COG. As an example the COG of Ti I and Fe I for the Sun is shown in Fig. 3.7. For ionization states with dissimilar ionization energy regimes some interpolation in the entire atmosphere structure has to be made.

In this manner it became possible to determine the abundance of the elements in stellar atmospheres. Because Fe is present in all stars, is relatively easy to detect in the various ionization stages over a large range of temperatures, as well as is the element with the largest nuclear binding energy (see Ch. 5), Fe became the reference element for abundance studies. Thus abundance values are mostly given as $[\text{X}/\text{Fe}] \equiv \log N(\text{X})/N(\text{Fe}) - (\log N(\text{X})/N(\text{Fe}))_\odot$.

Presently, computer codes exist which model the atmosphere with its entire spectrum including the possible spectral lines from all ionic states. A numerical fit to a spectrum is then achieved through iteration, in which excitation and ionization temperatures as well as the abundance ratios are allowed to vary until a (nearly) perfect fit is achieved.

3.4 Special features

3.4.1 The G-Band

For stars in the lower temperature range a depression in the spectral energy distribution is present near 4300 Å. It is called the G-band (one of the Fraunhofer lines!).

The depression is caused by the molecule CH. Further molecular bands as well as stronger absorption by metals is present in the vicinity (see Fig. 3.8).

The G-band is useful also for studies of the abundance of C and N (see Sect. 3.4.3).

3.4.2 Quasi-molecular absorption: H_2 and H_2^+

In dense stellar atmosphere gases, atoms and protons are so densely packed, that they may build molecules, which normally quickly dissolve. One can describe this process also by saying, that

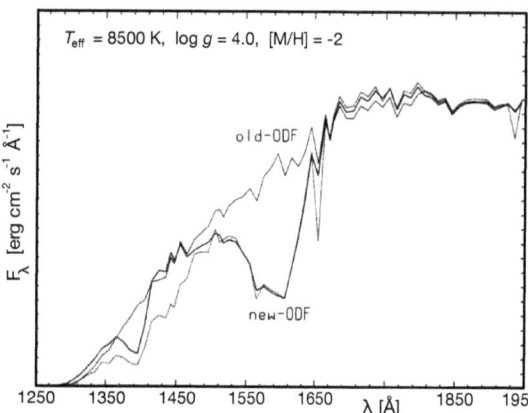

Figure 3.9: The spectrum (heavy line) of the star λ Boo (see Ch. 10.8.2) shows near 1600 Å and 1400 Å depressions due to 'quasi-molecular' absorption by H_2 and H_2^+, respectively. These molecules exist only fleetingly in the stellar atmosphere. Two models are shown. The 'old-ODF' (thin line) is the (pre 1995) optical depth function with the wide depression of the H I Ly α line at $\lambda_0 = 1216$ Å. The 'new-ODF' (dotted line) includes the contributions to the absorption at 1400 and 1600 å due to the quasi-molecules. Figure adapted from Castelli & Kurucz (2001).

during near passages of such particles in the gaseous atmosphere molecules exist fleetingly, building so-called '*quasi*'-*molecules*. Right during such events, the quasi-molecule may absorb photons with energy fitting to the molecular structure of that moment. Thus H_2 or H_2^+ and even He_2 may form

In the stable atmospheres of horizontal-branch (HB) stars, depressions were known to exist in the spectral energy distribution near 1600 Å and near 1400 Å, without being understood. Also some white dwarfs showed these spectral structures. Since several of these WDs contained only hydrogen in their atmosphere, the structures could be only due to special forms of H. The 1600 Å feature is due to H_2, the one near 1400 Å due to H_2^+. Similar structures exist near all the Lyman lines (see e.g., Wolff et al. 2001). Note that H_2^+ can be seen only in atmospheres where H is partially ionized (thus at $T > 8000$ K). In cooler atmospheres only H_2 can cause depressions ($T < 8000$ K).

The optical depth effects of the absorptions by H_2 and H_2^+ are always present (the opacity function of hydrogen should always include these effects), but they are only recognizable in stars with the 'right' atmospheric temperature and pressure. Furthermore, in stars with roughly solar metal abundance, numerous metal absorptions dominate the opacity so that the contribution by quasi-molecular absorption is proportionally small. Quasi-molecular absorption can normally only be seen in metal poor stars, such as HB stars, in white dwarfs where at high pressure H_2 can be present at higher temperatures for which there is also a bright UV continuum, and in a few special types, such as λ Boo stars (Ch. 10.8.2). Fig. 3.9 shows that the very wide hydrogen Lyman absorption profile without the quasi-molecular lines cannot fit such observations.

3.4.3 Molecular absorption in cool atmospheres

When the atmospheric gases are very cool, various molecules can come into existence (see Fig. 2.7). They all contribute to the continuous opacity (dissociation) as well as to the line opacity (molecular absorption bands). Since molecules have rather low dissociation energy and since the absorptions due to the molecules therefore lie at low energy (long wavelengths) many of the absorptions, often coming in bands, have been only discovered since instrumentation to observe spectra at $\lambda > 1$ μm became available.

As an example the spectrum of an 'asymptotic giant branch (AGB) star' (see Ch. 10.4.4) is given (Fig. 3.10), showing the modeled absorptions by CO, C_2 and CN, in relation with the observed spectrum. Other examples are given in Fig. 10.21 for cool subdwarf stars.

CN and CH have absorption bands between 3800 and 4300 Å, one of them being the G-band (see Fig. 3.8). The CH absorption of the G-band, in comparison with the feature of CN near 4216 Å and one with band-head at 3883 Å, shows whether an atmosphere is relatively rich in C or in N (CH in C-rich, CN in N-rich atmospheres).

The elements C, N, and O are, in more massive stars, in part products from nuclear fusion and they may, during later stages of stellar evolution, come to the stellar surface. Their abundance there tells about both the status of the fusion and about the convective properties of the star.

3.5. MAGNETIC FIELDS AND ZEEMAN EFFECT

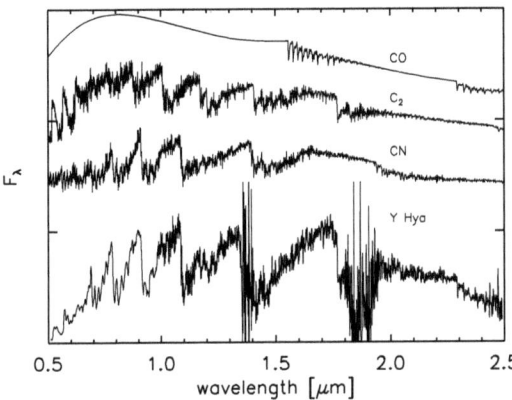

Figure 3.10: The atmosphere of a RG star cools with progress of the evolution up the GB. The temperature can become so low that molecules may form. The example is for a near solar metallicity AGB star, in which the fusion product carbon has been transported to the surface. In the C-rich atmosphere, in particular C-containing molecules are formed. The upper three curves are model calculations ($T_{\rm eff} = 3000$ K, $\log g = 0.0$), the lower curve shows the observed spectrum of the star γ Hya. Figure from Loidl et al. (2001).

Broad absorption due to H_2-dissociation in the near-IR may cause even blue $V - I$ values in the spectral energy distribution of cool white dwarfs (see Fig. 10.17).

H_2O absorption and absorption by yet larger molecules is present in atmospheres of very cool stars, such as, e.g., brown dwarfs (see Fig. 8.4). The atmospheres of Mira variables contain lots of molecules (e.g., AlH) as do those of cool subdwarf stars (see Fig. 10.21).

3.5 Magnetic fields and Zeeman effect

When the star has a strong magnetic field, spectral lines will split based on the interaction of magnetic field and electron spin. The wavelength *shift* is given by

$$\Delta \lambda = g \frac{e}{4\pi} \frac{\lambda^2}{mc^2} H \qquad (3.28)$$

in which H is the magnetic field strength [gauss]. The phenomenon is named after its discoverer, Zeeman (1897). An early astronomical detection was in the light from sunspots (by G. Hale).

When magnetic fields become very strong, the Balmer lines split and all magnetic components shift in wavelength (see Fig. 3.11). This has been observed in white dwarfs (see e.g. Fig. 3.12) which may have field strengths of the order of 10^5 gauss. For more on white dwarfs see Ch. 10.6.

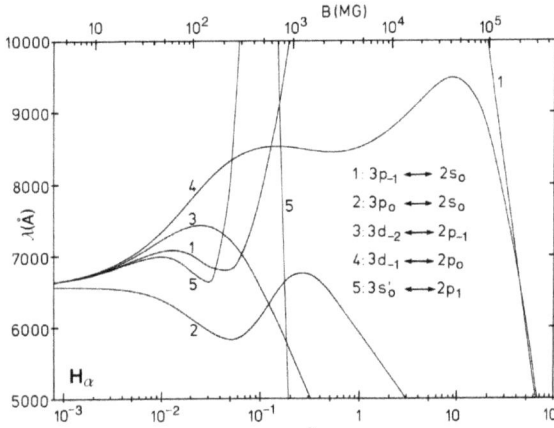

Figure 3.11: The effects of an atmospheric magnetic field on the wavelength of the Hα line is shown. Interaction of the magnetic field with the magnetic moment of the electron leads to splitting up of the lines and to further pronounced shifts related to the magnetic field strength (denoted with B, at the top). Figure from Wunner et al. (1985).

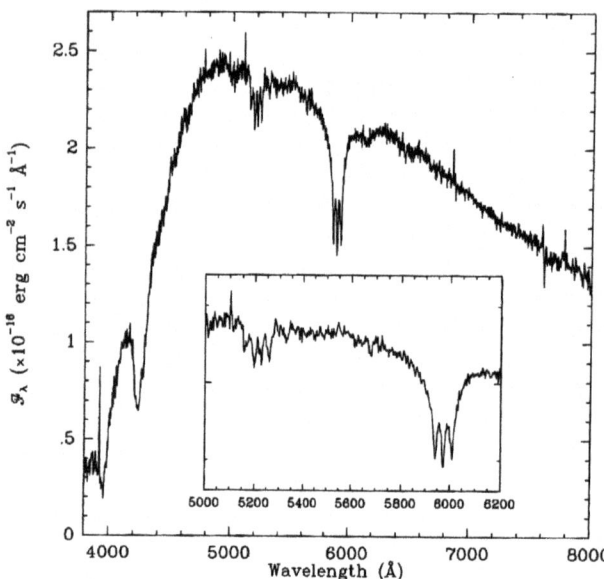

Figure 3.12: The severe compaction of WD stars leads to much enhanced magnetic fields at the stellar surface. Thus various spectral lines may exhibit Zeeman splitting. The example shows the spectrum of the white dwarf star LHS 2534. In the inset an enlarged part of the spectral range showing better the splitting of the Mg I and Na I D lines. Figure from Reid et al. (2001).

3.6 Gravitational settling and radiation levitation

In very stable atmospheres (no convection, no rotation, small luminosity) heavy elements are gravitationally pulled down preferentially. The heavy elements **settle down**. If such an atmosphere is stable for longer periods of time, the absorption lines of heavy elements may therefore become very weak. The star seems to contain only hydrogen and helium. In stars with large surface gravity also helium may become deficient even by a factor of 10.

On the other hand, radiation pressure may selectively push certain elements outward, leading to an atmospheric overabundance of such elements. This effect is called **levitation**. It works in particular when the maximum spectral intensity, $I_\lambda^{\max}$, is near strong resonance lines (efficient momentum transfer). It leads to (selective) metal enrichment of the atmosphere.

In fact, the balance between gravitational pull on heavy elements and the radiative momentum transferred onto elements with appropriate spectral lines will determine the balance. The process operates always, but the effects can be washed out in atmospheres even with the slightest amount of convection or turbulence. For more on the physics the reader is referred to Michaud (1980, 1982) and Michaud & Richer (1997).

The "peculiar" abundances are visible in two kinds of stars.

→ Ap stars (A peculiar stars) are stable MS stars with T_{eff} approximately between 20 000 and 8000 K. One sees unusually strong lines of, e.g., Si II, or Cr II and Eu II. Many of these stars show a strong magnetic field, some pulsate. The lower temperature limit is related with atmospheric convection; below $M = 1.15$ $\mathrm{M}_\odot$ (see Fig. 6.6) convection causes mixing washing out peculiarities. The crude upper temperature limit is related with the presence of stronger stellar winds blowing away gas with peculiar chemical effects. For more see, e.g., Vauclair & Vauclair (1982).

→ High surface gravity objects, such as white dwarfs (see Ch. 10.6) and horizontal branch sdB stars (see Ch. 10.4.2.2), being in long-lasting stable phases.

The consequence is, that the metal content of the atmosphere as derived from spectral lines can no longer be used as a (rough) indication for the age of a star[1].

Gravitational settling takes also place in the Sun. At the bottom of the outer convection zone (at $\simeq 0.7$ $\mathrm{R}_\odot$) He is dragged down, leading to a surface He abundance smaller than the cosmic

---

[1] Stars from the distant past of the Milky Way evolution formed in gas not yet enriched in metals, the results of nuclear fusion in the stellar centres and subsequent expulsion into the ISM (through stellar winds or supernova explosions). Stars having formed more recently were born in gas enriched in heavy metals. Hence there is a rough relation between age and metallicity (see Ch. 22). In this manner one attempts to get an indication for the age of a star if other methods (like determining the state of evolution and thus the original mass) fail.

3.7. STELLAR ROTATION

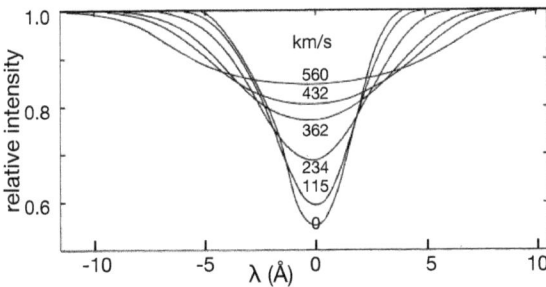

Figure 3.13: The effects of rotation on the shape of a spectral line (He I 4026 in a B3 star) is shown. The normalized profiles calculated *for the stellar equator* are labelled with values of $v_{\rm rot}$, the rotation velocity at the equator. Figure adapted from Slettebak (1949).

value. That this is so was uncovered by helioseismology (see Ch. 11.5). It occurs also in the atmospheres of brown dwarfs and planets.

3.7 Stellar rotation

When a star rotates, two effects are important when observing the star. One is the *change in the shape of the spectral lines*, the other is the *variation of surface temperature*. Both are related with the angle i at which the rotation axis is inclined toward us.

3.7.1 Rotation broadening of spectral lines

The light coming from one side of the surface (facing us) of a rotating star will be Doppler-shifted with respect to that from the other side. The spectral lines observed in the total spectrum will thus be a mix of blueward and redward shifted structures (as well as unshifted ones coming from the centre of the stellar disc).

One must integrate over the entire visible surface to get the overall effect of rotation on a spectral line in relation with rotation velocity. The maximum full width at half maximum b is given by

$$b = \frac{\lambda}{c} R \omega \sin i \quad \text{or} \quad \frac{b}{\lambda} = \frac{v_{\rm rot}}{c} \sin i \qquad (3.29)$$

with ω the angular speed of rotation, i the inclination angle of the rotation axis, $v_{\rm rot}$ the rotation speed at the equator. Note that rotating stars seen pole on do not show rotation broadening!

Fig. 3.13 shows how the shape of an absorption profile varies with rotation velocity. Clearly, for an observed star the inclination of the rotation axis is unknown. Values of the rotation derived from spectra are always given as $v_{\rm rot} \sin i$ (or $v \sin i$) with i unknown.

3.7.2 Rotation and average surface parameters T, M_V, $B - V$

A further effect of rotation has to do with the deformation of the star. At the equator the star will become more extended, while at the poles the distance from centre to surface will become smaller (thus higher surface gravity g). This has consequences for the temperature at the surface.

Consider a solid angle pointing outward from the centre of the star. The total flux F passing through that solid angle will reach the pole at smaller radius and will pass through a smaller surface area than at the equator. The same amount of energy will therefore lead at the poles to a higher surface temperature than at the equator. This was first noted by von Zeipel (1924) and $F \simeq g$ is known as "von Zeipel's theorem". This flux-gravity effect must also lead to differences in the ionization and excitation structure of those surface layers.

Since a rotating star will appear hotter seen pole on than seen at the equator the colour of a star will depend on both rotation speed and viewing angle (inclination). As an example, Fig. 3.14 shows the effect of rotation on the surface-integrated colour of main-sequence stars. The change of colour due to rotation has also implications for, e.g., the analysis of colour-magnitude diagrams (see Ch. 21.4).

Effects of rotation on the overall structure of a star will be addressed in Ch. 14.

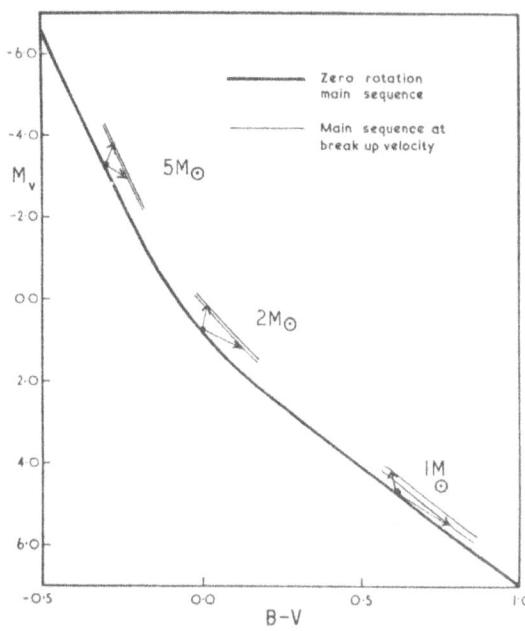

Figure 3.14: The effects of rotation on a star are manyfold. Shown here is the immediate effect on observed colour and M_V for main-sequence stars. At the equator a rotating star will have a larger radius and thus a lower surface temperature compared to that at the poles. Figure from Faulkner et al. (1968). (The figure caption in the original paper is flawed, which follows immediately from a comparison of the figure with their Table 2.)

3.8 Stellar classification: the MKK system and newer methods

3.8.1 Development of stellar classification towards the MKK system

The precise observation of stellar spectra using spectrographs with slits began at the beginning of the 19th century by Fraunhofer (1787-1826) but it took another 100 years until the formation of the spectral lines of the different elements was correctly interpreted. As in all areas of natural sciences, people tried to find similarities among the spectra (and stars) were looked for attempting to classify and sort them. Angelo Secchi (1819-1879) was the first who classified stars in five groups according to their spectra (and colours), ranging from hot stars like Vega to cool ones in the fourth class and special objects with emission lines in the last one. The next step towards a consistent spectral classification system was made by Edward Pickering (1846-1919) who divided Secchi's classes into several subclasses. These first spectral catalogs were published in the Henry Draper Memorial Catalogue (Henry Draper 1837-1882).

As the observing capabilities improved over the following years, more classification systems were introduced but also rejected. The most promising one was developed by Annie Cannon (1863-1941). She placed the former classes into a more logical physical order (temperature sequence) and added subclasses by the invention of a decimal division. This lead finally to the well known spectral sequence O,B,A,F,G,K,M with its typical letter-number combinations (numbers for the subtypes) as it is still in use. These and former observations and classifications were finally published in the 'Harvard Annals' with more than 250000 stars down to a visual magnitude $V = 9$ (the HD catalogue). Again, as observations improved and the number of observed stars steadily increased several refinements had to be made. Classes were combined or new spectral classes introduced, such as R and S for stars with chemical abnormalities in the lower temperature region. For a more detailed summary see Kaler (1989) or Hearnshaw (1996).

The work of Hertzsprung and Russell (Ejnar Hertzsprung 1873-1967, Henry Norris Russell 1877-1957) had shown that there is a clear distinction between giants (evolved stars) and dwarf stars (see also Ch. 1). This led to an inflation of the number of of indices and suffixes invented to extend the Harvard system.

A better differentiation was made by Morgan, Keenan and Kellman in 1943 (William Wilson Morgan 1906-1994; Philip C. Keenan 1908-2000; Edith Kellman 1911-2007) who defined so called luminosity classes reaching from supergiants with loose atmospheres to ultracompact dwarfs. In combination to the Harvard spectral system, a very useful two-dimensional classifiaction system was born; the MKK system (but in later years it has simply been called the MK system).

A star is classified in this system by certain line ratios seen in spectra having a spectral resolution of 1 to 3 Å, being observed in the wavelength range from 3500 to 5000 Å, the range being determined mainly by the prismatic dispersion and the sensitivity of photographic plates. These are compared to those of previously defined standard stars, which are assumed to best represent a given class of stars. There were several refinements to this system in the years after its introduction. In 1953 Morgan and Johnson introduced a revised system, called 'MK', which was superseded by the 'revised MK' system in 1973 by Morgan and Keenan. In 1984, Keenan introduced abundance indices for the cooler stars, which had to be added to the spectral type and luminosity class in case that certain features in the spectra indicated substantial differences to those of population I stars. Today the so called 'MK Process' tries to supplement the revised MK system with additional systems from other wavelength regions, as near infrared (NIR) and ultraviolet (UV).

Further and more recent refinements of the system foresee numerous indices, which represent indications for element abundances, extensions deal with new kinds of objects (see, e.g., Ch 8.3). Also certain peculiarities are attempted to be given by indices. With these, one approaches the field of detailed spectral analysis, which is, in fact, not the goal of the spectral classification system. Also, such details generally cannot be seen with spectra of the kind (wavelength range and resolution, mentioned above) as used for the first MK system.

3.8.2 Quality of the MK classification process

The MK system is an important tool in stellar astrophysics. It is a classification scheme based on the observed morphology of spectral structure[2]. From spectra one cannot, however, immediately derive the fundamental quantities of stars, the inital mass M_{init}, age t, and inital chemical composition. What can be derived are the most important *atmospheric parameters* like effective temperature T_{eff}, surface gravity $\log g$, and present chemical composition [M/H] (often represented by the easy to observe element Fe: [Fe/H]).

There are, unfortunately, numerous uncertainties with the MK system, both with the standard stars as well as with the system itself. E.g., the calibrations for the MK types change from year to year and from observer to observer but the classifications should (ideally) remain constant; in the high temperature regime the giants sometimes reside under the regime of the dwarfs; luminosity class IV can be spectroscopically easily confused with classes III and V; the luminosity *classes* are somewhat diffuse and do not lead to the desired parameter, the luminosity itself; and more such problems.

One important warning must be given about spectral types, too. After the relation between spectral type, temperature, and colour index $B-V$ became well established, these three parameters got 'hopelessly' interchanged. E.g., when $B-V$ was observed, papers often then also listed the spectral type belonging to that colour index. Later investigators, using such data, where then mislead to believe that a spectrum had been seen although such was not the case. Thus for many (in particular fainter) stars spectral types exist in the literature although no spectrum was ever taken for a proper classification in the MK system.

3.8.3 New classification methods

Since the goal of astrophysical studies is to derive the astrophysical parameters of the stars from the data, there is an advantage in getting there directly. Automated classification processes such as Neural Networks (NN) and Minimum Distance (MD) methods, which have already been included in the efforts to reach consensus on the MK classifiaction, may be much better suited to classify stars. This is in particular true for at least three reasons.

---

[2]The background lies, of course, in physics: the presence and strength of spectral lines is governed by the element abundance, the ionization state (Saha equation), the Boltzmann statistics and the gas density (collisional broadening).

1) An automatic approach can go directly from observational data (spectra, photometry, or any measurements available) to the astrophysically relevant parameters (T_{eff}, $\log g$, [M/H], v_{rot}); the accuracy reached will, of course, depend on the quality of the data.
2) An automatic process can handle the huge amounts of data becoming available from large surveys, such as the Sloan Digital Sky Survey, SDSS, being conducted, conceived for the astrometric satellite DIVA (see Willemsen et al. 2003) or for the planned astrometric satellite GAIA.
3) An automatic process perhaps leads to only crude values for the astrophysical parameters, but this allows to work with a well selected sample from a data base to study the kinds of objects of interest in detail.

It remains to be noted, that also for automatic processes a calibration is required. Thus the need to study as many kinds of stars in detail remains an important goal and requirement to further astrophysics.

This section (3.8) is adapted from the text on this subject by Willemsen (2001).

References

Böhm-Vitense, E. 1989, "Stellar Astrophysics, Vol 2: Stellar Atmospheres"; Cambridge Univ. Press
Castelli, F., & Kurucz, R.L. 2001 A&A 372, 260
Faulkner, J., Roxburgh, I.W., & Strittmatter, P.A. 1968, ApJ 151, 203
Grebel, E.K. 1991, Dipl. Thesis, Univ. Bonn
Hearnshaw, J.B. 1996, "The Measurement of Starlight", Cambrige University Press
Hilker, M. 2000, A&A 355, 994
Kaler, J.B. 1989, "Stars and their Spectra"; Cambridge Univ. Press
Loidl, R., Lancon, A., & Jørgensen, U.G. 2001, A&A 371, 1065
Michaud, G. 1980, AJ 85, 589
Michaud, G. 1982, ApJ 258, 349
Michaud, G., & Richer, J. 1997, Highlights of Astronomy 11, 667
Mihalas, D. 1970, "Stellar atmospheres"; Freeman & Co, San Francisco
Napiwotzki, R., Schönberner, D., & Wenske, V. 1993, A&A 268, 653
Reid, I.N., Liebert, J., & Schmidt, G.D. 2001, ApJ 550, L61
Relyea, L.J., & Kurucz, R. 1978, ApJS 37, 45
Saha, M.N. 1921, Proc. Roy. Soc. (A) 99, 135; Phil. Mag. 41, 267; Z. Physik 6, 40
Slettebak, A. 1949, ApJ 110, 498
Unsöld, A. 1955, "Physik der Sternatmosphären"; Springer, Heidelberg
Vauclair, S., & Vauclair, G. 1982, ARAA 20, 37; *Element Segregation in Stellar Outer Layers*
von Zeipel, H. 1924, MNRAS 84, 684
Wiese, W.L., Smith, M.W., & Glennon, B.M. 1966, "Atomic Transition Probabilities, Part I", NSRDS, NBS, Washington
Wiese, W.L., Smith, M.W., & Miles, B.M. 1969, "Atomic transition probabilities, Part II", NSRDS, NBS, Washington
Willemsen, P. 2001, "Characteristics of simulated dispersed images for the DIVA satellite", Dipl. Thesis Univ. Bonn
Willemsen, P.G., Bailer-Jones, C.A.L., Kaempf, T.A., & de Boer, K.S. 2003, A&A 401, 1203
Wolff, B., Kruk, J.W., Koester, D., Allard, N.F., Ferlet R., & Vidal-Madjar A. 2001, A&A 373, 674
Wright, K.O. 1948, Publ. Dominion Astroph. Obs. 8, 1
Wunner, G., Rösner, W., Herold, H., & Ruder H. 1985, A&A 149, 102
Zeeman, P. 1897, ApJ 5, 332

Chapter 4

Stellar structure: Basic equations

During the main-sequence phase stars release energy rather constantly for millions to billions of years. This is only possible if they are in a state of internal equilibrium. In the beginning of the chapter the four basic differential equations which express this equilibrium mathematically will be presented. After a short discussion of time scales relevant for stellar evolution the energy transport inside stars will be looked at in more detail. In a further section different equations of state and the opacity are discussed.

Since nuclear fusion is essential to stellar physics, it is presented in a chapter by itself (Ch. 5), including the role neutrinos play in the understanding of stars.

In the present considerations some simplifications will be made such as spherical symmetry. Therefore magnetic fields and rotation are neglected but they will be addressed in Chs. 12 and 14.

Other treatments of the topics of this chapter can be found in the textbooks of, e.g., Cox & Giuli (1968), Böhm-Vitense (1989) and Kippenhahn & Weigert (1990), as well as in the compendium by Lang (1980).

4.1 Four basic equations for the internal structure

4.1.1 Mass continuity

A spherical shell with thickness dr at some distance r around the centre of the star contains the mass $dM(r) = 4\pi r^2 \rho(r) dr$, where $\rho(r)$ is the (spherically symmetric) density of the gas. This immediately leads to the *mass continuity equation*

$$\frac{dM(r)}{dr} = 4\pi r^2 \rho(r) \quad . \tag{4.1}$$

4.1.2 Hydrostatic equilibrium

The gravitational force which holds the gaseous sphere (star) together is balanced by the internal pressure (with its various components) everywhere in the star.

To derive an expression for this we consider a volume element dV with a horizontally aligned base area dA and height dr in a distance r from the centre. The gravitational force on this volume element is

$$dF = -\frac{G M_r dM}{r^2} = -\frac{G M_r \rho(r)}{r^2} dA\, dr \tag{4.2}$$

with the mass dM is contained in dV and the mass M_r of the star within r.

The net force only vanishes if the gravitational force is compensated by a change in pressure $dP = \frac{dF}{dA}$. Substituting this we get the *equation of hydrostatic equilibrium*

$$\frac{dP(r)}{dr} = -\frac{G M_r \rho(r)}{r^2} \quad . \tag{4.3}$$

4.1.3 Energy conservation

As a further condition of internal equilibrium, energy has to be conserved. All energy produced in the interior is transported to the surface and radiated away. Consider spherical shells as in Sect. 4.1.1. Let $L(r)$ symbolize the total flux that enters the shell from the direction of the centre and $dL(r)$ the flux additionally produced within the shell. If $\epsilon(r)$ is the rate of energy production (see Sect. 4.4.3) per unit of mass we immediatelly get the *equation of energy conservation*

$$\frac{dL(r)}{dr} = 4\pi r^2 \rho(r)\, \epsilon(r) \quad . \tag{4.4}$$

Eq. 4.4 only holds if all of the energy produced is transported to the surface. In a non-equilibrium situation it is possible that an additional gain or loss of energy can occur, e.g. in the form of mechanical work during contraction or expansion. In this case Eq. 4.4 has to be modified to

$$\frac{dL(r)}{dr} = 4\pi r^2 \rho \left(\epsilon - \dot{Q}\right) \tag{4.5}$$

with $\dot{Q} = \frac{dQ}{dt}$ the change of energy per unit of mass and time.

In order to find a useful expression for $\dot{Q}$ we start with the *first law of thermodynamics* for an ideal gas

$$dQ = dU - dW \tag{4.6}$$

where dQ is the energy exchange per mass unit, dU the change of internal energy per mass unit, and

$$dW = -P\, dV \tag{4.7}$$

the work per mass unit performed by changing the *specific volume* $V = 1/\rho$ by dV. The change of internal energy dU for an ideal monoatomic gas under isobaric conditions is given by a change of temperature dT

$$dU = \frac{3}{2}\, n\, \Re\, dT \tag{4.8}$$

with n the number of moles per unit mass and $\Re = N_A k = 8.31 \cdot 10^7 \,\mathrm{erg\,mole^{-1}K^{-1}}$ ($N_A = 6.02 \cdot 10^{23}\,\mathrm{mole^{-1}}$ is *Avogadro's number* and $k = 1.38 \cdot 10^{-16}\,\mathrm{erg\,K^{-1}}$ *Boltzmann's constant*). Using the definition of the specific heat of a specific volume

$$C_V = \left.\frac{dQ}{dT}\right|_V \tag{4.9}$$

and inserting Eq. 4.7 and 4.8 into Eq. 4.6 leads to

$$dQ = C_V dT + P dV \quad . \tag{4.10}$$

From the *ideal gas law*

$$PV = n\Re T \tag{4.11}$$

we get

$$dV = n\Re \left(\frac{dT}{P} - \frac{T dP}{P^2}\right) \quad . \tag{4.12}$$

Inserting Eq. 4.12 into Eq. 4.10 yields

$$dQ = (C_V + n\Re)\, dT - n\,\Re\, T\frac{dP}{P} = C_P\, dT - V\, dP = C_P\, dT - \frac{dP}{\rho} \tag{4.13}$$

where we have used that the specific heat at constant pressure

$$C_P = \left.\frac{dQ}{dT}\right|_P \tag{4.14}$$

4.1. FOUR BASIC EQUATIONS FOR THE INTERNAL STRUCTURE

can be expressed as
$$C_P = C_V + n\Re \qquad (4.15)$$
which is always valid for ideal gas.

So finally we can rewrite the equation for energy conservation (Eq. 4.5) as
$$\frac{dL}{dr} = 4\pi r^2 \, \rho(r) \left(\epsilon - C_P \dot{T} + \frac{1}{\rho} \dot{P} \right) \quad . \qquad (4.16)$$

4.1.4 Temperature gradient

The fourth differential equation gives the temperature gradient $\frac{dT}{dr}$. It depends on the way the energy is transported. There are three mechanisms of energy transport: *radiation*, *convection* and *conduction*.

4.1.4.1 Radiative energy transport

To derive an equation for radiative energy transport in stars we have to start with the general equation of radiative transfer (Ch. 2.4)
$$\cos\theta \frac{dI_\nu}{dr} = -\kappa_\nu I_\nu + j_\nu \quad .$$

After multiplying this equation with $\cos\theta$ we integrate both over all directions and all frequencies
$$\int\int \cos^2\theta \, \frac{dI_\nu}{dr} \, d\omega d\nu = \int\int \cos\theta \, \kappa_\nu I_\nu \, d\omega d\nu + \int\int \cos\theta \, j_\nu \, d\omega d\nu \quad . \qquad (4.17)$$

On the *left* side the integration over all directions just gives a factor of $\frac{4\pi}{3}$ as only $\cos^2\theta$ depends on the direction. I_ν can be approximated with the Planck function B_ν, which after integration over all frequencies gives *Stefan-Boltzmann's law*
$$\int_0^\infty I_\nu d\nu = \int_0^\infty B_\nu d\nu = \frac{ac}{4\pi} T^4 \quad .$$

For the first term on the *right* side in Eq. 4.17 we can use that the flux density F_ν at a frequency ν is (with S_2 an integration over all directions) given as (Ch. 2.2)
$$F_\nu = \int_{S_2} I_\nu \cos\theta \, d\omega \quad .$$

The integral of the second term on the *right* in Eq. 4.17 vanished because j_ν does not depend on θ.

So far we obtained
$$\frac{4\pi}{3} \frac{d}{dr}\left(\frac{ac}{4\pi} T^4\right) = -\rho \int_0^\infty \kappa_\nu F_\nu d\nu \quad .$$

If we now simplify this integral by assuming $\int \kappa F d\nu = \int \kappa d\nu \int F d\nu = \overline{\kappa} \int F d\nu$, with $\overline{\kappa}$ the *Rosseland mean* (Ch. 2.6.4), use the relation
$$F = \frac{L(r)}{4\pi r^2}$$

for the total flux density $F = \int_0^\infty F_\nu d\nu$, and carry out the differentiation on the left hand side, we finally arrive at the *radiative temperature gradient*
$$\left.\frac{dT}{dr}\right|_{rad} = -\frac{3\overline{\kappa}\rho}{4acT^3} \frac{L(r)}{4\pi r^2} \quad . \qquad (4.18)$$

4.1.4.2 Convective energy transport

If the temperature gradient is steep, convection may set in which then dominates the outward transport of energy. Here hot gas from lower layers rises upwards into cooler layers, where it loses part of its energy and then sinks again. To find an expression for this *convective energy transport* we consider a bubble of gas at a distance r from the centre of the star. This bubble should be almost in thermal equilibrium with the surrounding material, but is slightly warmer. It will therefore expand and, due to the lower density, it will rise. As the bubble is almost in equilibrium with the surroundings we can approximate this process to be *adiabatic*, which means that there is no exchange of heat, so $dQ = 0$. Inserting this into Eq. 4.10 we get

$$dT = -\frac{P}{C_V} dV \quad . \tag{4.19}$$

Combining Eq. 4.19 with Eq. 4.12 and performing some basic manipulations we arrive at

$$\frac{C_V + n\,\Re}{C_V} \frac{dV}{V} = -\frac{dP}{P} \quad . \tag{4.20}$$

If we define the *adiabatic exponent*

$$\gamma = \frac{C_P}{C_V} = \frac{C_V + n\,\Re}{C_V} \tag{4.21}$$

and use $\rho = \frac{1}{V}$ we can rewrite (4.20) to obtain

$$\frac{dP}{dr} = \gamma \frac{P}{\rho} \frac{d\rho}{dr} \quad . \tag{4.22}$$

Because we need it later, we can already derive the adiabatic gas law by solving Eq. 4.22 to

$$P = \text{const}\, V^{-\gamma} = \text{const}\, \rho^{\gamma} \tag{4.23}$$

which has the form of a so-called *polytropic equation* (see Ch. 6.2.1).

Returning to our problem, we have to reformulate the *ideal gas law* (Eq. 4.11). Using the *molecular weight*

$$\mu \equiv \frac{\overline{m}}{m_H} \quad , \tag{4.24}$$

with $\overline{m}$ the average mass per particle (Eq. 4.65) and m_H the mass of the hydrogen atom, one gets

$$P = \frac{\rho}{\mu\, m_H} kT \quad . \tag{4.25}$$

If we take the gradient of Eq. 4.25

$$\frac{dP}{dr} = \frac{P}{\rho}\frac{d\rho}{dr} - \frac{P}{\mu}\frac{d\mu}{dr} + \frac{P}{T}\frac{dT}{dr} \tag{4.26}$$

and assume that μ is constant (at least at relevant scales) we can insert Eq. 4.22 into Eq. 4.26 to finally obtain the *adiabatic temperature gradient*

$$\left.\frac{dT}{dr}\right|_{\text{ad}} = \left(1 - \frac{1}{\gamma}\right)\frac{T}{P}\frac{dP}{dr} \quad \text{or} \quad \nabla_{\text{rad}} = \left(1 - \frac{1}{\gamma}\right) \quad , \tag{4.27}$$

the latter transition based on $\nabla_{\text{rad}} \equiv \left(\frac{\partial \ln T}{\partial \ln P}\right)_{\text{ad}} = \left(\frac{P}{T}\frac{dT}{dP}\right)_{\text{ad}}$ (see Eq. 4.51).

4.1.4.3 Conductive energy transport

A third mechanism of energy transport is *conduction*, where heat is transported via collisions between particles. Conduction can be of great importance in the dense stellar gas occuring in late stages of stellar evolution, notably in degenerate material (see, e.g., WDs and electron conduction; Ch. 17.1.2), but is usually negligible for the long and stable main-sequence phase. Therefore we will not discuss conduction in detail but add some remarks in Sect. 4.4.1.

4.2 Stability and time scales

4.2.1 Virial theorem

The virial theorem of statistical mechanics is of general importance. We start with multiplying the *equation of hydrostatic equilibrium* (Eq. 4.3) with $4\pi r^3$ and integrating from $r = 0$ to $r = R$

$$\int_0^R 4\pi r^3 \frac{dP}{dr} dr = -\int_0^R \frac{G M_r}{r} \rho(r) 4\pi r^2 dr \quad . \tag{4.28}$$

The term on the right hand side of Eq. 4.28 is equal to the total potential energy of the star E_{pot}. This can easily be seen if we estimate the potential energy of a spherical shell with radius r and thickness dr having a mass $dM(r) = \rho(r) dV = 4\pi \rho(r) r^2 dr$

$$dE_{\text{pot}} = -G \frac{M_r \, dM}{r} = -4\pi \, G \, M_r \, \rho(r) \, r \, dr \quad . \tag{4.29}$$

We can solve the term on the left hand side of Eq. 4.28 by partial integration

$$\int_0^R 4\pi r^3 \frac{dP}{dr} dr = 4\pi r^3 P \Big|_0^R - \int_0^R 3P \, 4\pi r^2 \, dr \quad . \tag{4.30}$$

The first term on the right hand side of Eq. 4.30 vanishes because $P(r = R) = 0$. We can rewrite the second term on the right hand side by inserting for P the pressure given by the ideal gas law (Eq. 4.25) so that

$$\int_0^R 4\pi r^3 \frac{dP}{dr} dr = -\int_0^R 3kT \frac{4\pi r^2 \rho}{\mu \, m_H} dr = -2E_{\text{th}} \quad . \tag{4.31}$$

That the right hand side equals the thermal energy $2E_{\text{th}}$ indeed is because for a monoatomic ideal gas $\frac{4\pi r^2 \rho}{\mu \, m_H} dr = N$ simply gives the total number of particles in a shell of dr with thickness dr.
Inserting this result into Eq. 4.28 we obtain the **virial theorem of statistical mechanics**

$$2E_{\text{th}} + E_{\text{pot}} = 0 \quad . \tag{4.32}$$

4.2.2 Kelvin-Helmholtz time scale

It is relevant to consider the time a star needs to collapse under its own gravity, e.g., from a very huge protostar with radius R_{init} to a certain radius R_{final}. For this the total potential energy of the star can in principle be calculated by integrating Eq. 4.29. Then $\rho(r)$ has to be known explicitly, which is normally not so. However, one can approximate E_{pot} assuming $\rho(r)$ to be constant and equal to its average value $\rho \approx \bar{\rho} = \frac{M}{\frac{4}{3}\pi R^3}$ (which also implies $M_r \approx \frac{4}{3}\pi r^3 \bar{\rho}$). One so obtains

$$E_{\text{pot}} = -4\pi G \int_0^R M_r \, \rho(r) \, r \, dr \approx -3G \frac{M^2}{R^6} \int_0^R r^4 dr = -\frac{3}{5} \frac{GM^2}{R} \quad . \tag{4.33}$$

With contraction one has to include that for moving particles $E_{\text{kin}} + E_{\text{pot}} = 0$ while the virial theorem has $2E_{\text{th}} + E_{\text{pot}} = 0$. Thus, if a star contracts with its gas roughly in TE, *half of its energy must be radiated away*, the second half remains as thermal energy E_{th}.

Using this and Eq. 4.33 one can calculate the time needed for the collapse from R_{init} to R_{final} if a constant luminosity L is assumed. This time is

$$t_{\text{KH}} = \frac{\Delta E}{L} = \frac{1}{2L}(E_{\text{pot}}(R_{\text{init}}) - E_{\text{pot}}(R_{\text{final}})) \approx -\frac{1}{2L} E_{\text{pot}}(R_{\text{final}}) = \frac{3}{10} \frac{GM^2}{RL} \quad , \tag{4.34}$$

called the *Kelvin-Helmholtz time scale*, which was found by these physicists in 1861 and 1854.

For the Sun the Kelvin-Helmholtz time scale is in the order of 10^7 years. Therefore gravitational energy cannot be the main source for the solar luminosity as it was still believed at the end of the 19th century, as described in Ch. 1.1.2. Thus the K-H time scale is an important aspect for theories about *stellar evolution*.

Note that in the literature the Kelvin-Helmholtz time scale is sometimes found without the factor $\frac{3}{10}$.

4.2.3 Nuclear time scale

The time it would take a star at luminosity L to consume all of its hydrogen supply in nuclear fusion is called *nuclear time scale*, t_{nuc}. The maximum amount of releasable energy is obtained if the final product of the fusion is ^{56}Fe. In this case the atomic mass difference between H and Fe is $\Delta m = 0.008\, m_{\text{H}}$. Therefore the maximum amount of energy a star with a hydrogen mass M can release is $E_{\text{nuc}} = 0.008\, M\, c^2$. The nuclear time scale is then given as

$$t_{\text{nuc}} = \frac{E_{\text{nuc}}}{L} = 0.008 \cdot c^2 \frac{M}{L} \quad . \tag{4.35}$$

However, real stars consume only a fraction of their hydrogen supply in nuclear fusion, because only the inner part of the star is hot enough for fusion. The fraction depends also on how efficient perhaps the material of the outer regions is mixed into the material of the fusion zone. The Sun will, e.g., only consume about 10 % of its hydrogen supply in nuclear fusion, which means a solar MS life time of approximately 10^{10} years.

Stars in the lower mass range are not able to continue the chain of fusion all the way up to ^{56}Fe. This results in a shorter (effective) nuclear time scale.

4.2.4 Dynamical time scale

The *dynamical time scale*, t_{dyn}, tells how long it would take a spherical object (e.g., a star) to collapse to a certain size if the inner pressure were removed suddenly (free fall). To derive an expression for the dynamical time scale one starts with the gravitational acceleration at a certain distance r from the centre

$$\frac{d^2 r}{dt^2} = -\, G\, M_r \frac{1}{r^2} \tag{4.36}$$

where M_r again stands for the total mass inside the sphere with radius r. While r decreases during the collapse M_r remains constant as the whole object collapses. After multiplying Eq. 4.36 with $\frac{dr}{dt}$ it can be integrated to

$$\frac{1}{2}\left(\frac{dr}{dt}\right)^2 = \frac{G\, M_r}{r} + C_1 \quad . \tag{4.37}$$

If one solves this equation for the object's surface ($r = R$) and requires that the velocity of the surface is zero at the beginning of the collapse (when $R = R_0$) one gets

$$\frac{dR}{dt} = -\left[2\, G\, M \left(\frac{1}{R} - \frac{1}{R_0}\right)\right]^{1/2} \quad . \tag{4.38}$$

Here the negative root is chosen because the cloud is collapsing. Substituting

$$S \equiv \frac{R}{R_0} \quad \text{and} \quad K \equiv \left(\frac{2\, G\, M}{R_0^3}\right)^{1/2}$$

into Eq. 4.38 leads to

$$\frac{dS}{dt} = -K \left(\frac{1}{S} - 1\right)^{1/2} \quad . \tag{4.39}$$

If one makes as further substitution $S \equiv (\cos \xi)^2$ then Eq. 4.39 becomes

$$(\cos \xi)^2 \frac{d\xi}{dt} = \frac{K}{2} \tag{4.40}$$

which one can integrate to become the equation of motion

$$\frac{\xi}{2} + \frac{1}{4} \sin 2\xi = \frac{K}{2} t + C_2 \quad . \tag{4.41}$$

4.3. CONVECTION VERSUS RADIATION

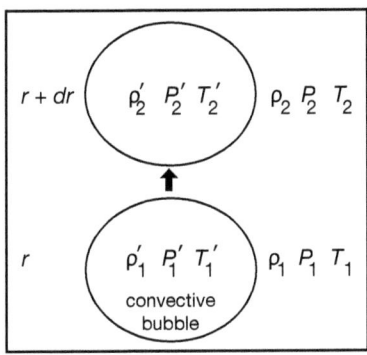

Figure 4.1: Sketch for the derivation of Schwarzschild's criterion: a convective bubble rises due to buoyancy. The buoyancy force is indicated by the black arrow. The force can be positive (convection) or negative (vertical motions are damped).

The constant C_2 vanishes if one requires $R = R_0$ for $t = 0$ and $\xi = 0$. One now can calculate the dynamical time scale, $t = t_{\mathrm{dyn}}$, which is the time for the collapsing sphere to reach radius zero ($S = 0$, $\xi = \pi/2$)

$$t_{\mathrm{dyn}} = \frac{\pi}{2\,K} = \left(\frac{\pi^2\,R_0^3}{8\,\mathrm{G}\,M}\right)^{1/2} = \left(\frac{3\,\pi}{32\,\mathrm{G}\,\rho_0}\right)^{1/2} \qquad (4.42)$$

where the initial mass M of the gas sphere is expressed in terms of the initial average density $\rho_0 = \frac{M}{4/3\,\pi\,R_0^3}$. Evidently, the dynamical time scale is independent of the radius of the initial sphere. For the Sun Eq. 4.42 gives $t_{\mathrm{dyn}\odot} \approx 30$ min.

The dynamical time scale is also called *free fall time scale*.

4.3 Convection versus radiation

4.3.1 Schwarzschild's criterion for convection

Whether convection or radiation is the dominant process of energy transport depends on the respective temperature gradients. **Schwarzschild** found for the occurence of convection the following criterion

$$\left|\frac{dT}{dr}\right|_{\mathrm{ad}} < \left|\frac{dT}{dr}\right|_{\mathrm{rad}} \quad .$$

To understand this, consider again the picture of the bubble from Ch. 4.1.4.2, which has a temperature T_1', density ρ_1' and pressure P_1', while the parameters of the surrounding material are T_1, ρ_1 and P_1 (see Fig. 4.1). The bubble has to be in pressure equilibrium with the surrounding material

$$P_1' = P_1 \qquad (4.43)$$

or it would expand until this equilibrium is reached. In most cases it will also be close to thermal equilibrium ($T_1' \simeq T_1$ and thus $\rho_1' \simeq \rho_1$), but if the bubble's density is only slightly lower than the density of the surrounding material ($\rho_1' < \rho_1$) it will start to rise due to a resulting buoyant force.

We use the index 2 for all parameters at distance $r + dr$ from the centre. Whether the bubble will continue to rise after traveling a distance dr (thus establishing convection) depends on the fulfillment of the buoyancy condition

$$\rho_2' < \rho_2 \quad . \qquad (4.44)$$

If one expands the densities in dependence of the radius to first order

$$\rho_2 \simeq \rho_1 + \frac{d\rho}{dr}\,dr \quad \text{and} \quad \rho_2' \simeq \rho_1' + \frac{d\rho'}{dr}\,dr$$

and inserts this into Eq. 4.44 using $\rho_1' \simeq \rho_1$ (at the onset of convective motion), the condition for convection becomes

$$\frac{d\rho'}{dr} < \frac{d\rho}{dr} \quad . \qquad (4.45)$$

We want to find a criterion which only depends on parameters of the surrounding material. Therefore it is useful to rewrite Eq. 4.45 in terms of pressure, and with Eq. 4.43 one has

$$\frac{dP'}{dr} = \frac{dP}{dr} \quad . \tag{4.46}$$

Using the differential form of Eq. 4.22 (adiabatic gas law) to rewrite the left–hand side of Eq. 4.45 and Eq. 4.26 to rewrite its right–hand side one obtains (assuming a homogeneous star, $\frac{d\mu}{dr} = 0$)

$$\frac{1}{\gamma}\frac{\rho'}{P'}\frac{dP'}{dr} < \frac{\rho}{P}\left(\frac{dP}{dr} - \frac{P}{T}\frac{dT}{dr}\right) \tag{4.47}$$

where the index 1 has been dropped. If one applies $P' = P$, $\rho' \simeq \rho$ and Eq. 4.46 one arrives at

$$\left(1 - \frac{1}{\gamma}\right)\frac{T}{P}\frac{dP}{dr} > \frac{dT}{dr} \quad . \tag{4.48}$$

Here the temperature gradient on the right–hand side is the actual temperature gradient of the surrounding gas, which is given by the radiative temperature gradient if convection does not occur. Comparing the left–hand side of Eq. 4.48 with Eq. 4.27 one sees that it is equal to the adiabatic temperature gradient. Taking into account that temperature gradients are negative inside stars one finally finds *Schwarzschild's criterion* for the occurence of convection

$$\left|\frac{dT}{dr}\right|_{ad} < \left|\frac{dT}{dr}\right|_{rad} \quad . \tag{4.49}$$

Eq. 4.48 can be rewritten to find another equivalent and very useful condition for convection. Since in stars naturally $\frac{dP}{dr} < 0$, Eq. 4.48 may be expressed as

$$\left(1 - \frac{1}{\gamma}\right) < \frac{P}{T}\frac{dT}{dr}\left(\frac{dP}{dr}\right)^{-1} \quad . \tag{4.50}$$

Defining the actual gradient as (see with Eq. 4.27)

$$\nabla \equiv \frac{\partial \ln T}{\partial \ln P} = \frac{P}{T}\frac{dT}{dP} \quad , \tag{4.51}$$

inserting this into Eq. 4.50 and taking ∇_{ad} from Eq. 4.27, the condition for convection becomes

$$\nabla_{ad} < \nabla = \nabla_{rad} \tag{4.52}$$

in which the actual gradient ∇ is set equal to ∇_{rad} (same argument as given below Eq. 4.48).

For an ideal monoatomic gas, which is either neutral or fully ionized, the specific heat at constant volume is given by $C_V = \frac{3}{2} n \Re$ (compare Eq. 4.8 and Eq. 4.9). Therefore, with Eq. 4.15 the adiabatic exponent becomes $\gamma = \frac{C_P}{C_V} = \frac{5}{3}$, which gives a value for ∇_{ad} of

$$\nabla_{ad} = \frac{2}{5} = 0.4 \quad . \tag{4.53}$$

This is, however, only an upper limit for ∇_{ad}. For a partially ionized gas ∇_{ad} depends on the fraction of ionization $x = \frac{N_i}{N}$ (with $x = 0$ for neutral and $x = 1$ for fully ionized gas) as

$$\nabla_{ad} = \left[\frac{5}{2} + \left(\frac{5}{2} + \frac{\chi_0}{kT}\right)\frac{T}{1+x}\frac{dx}{dT}\right]^{-1} \tag{4.54}$$

with χ_0 the ionization energy of the gas.

4.3. CONVECTION VERSUS RADIATION

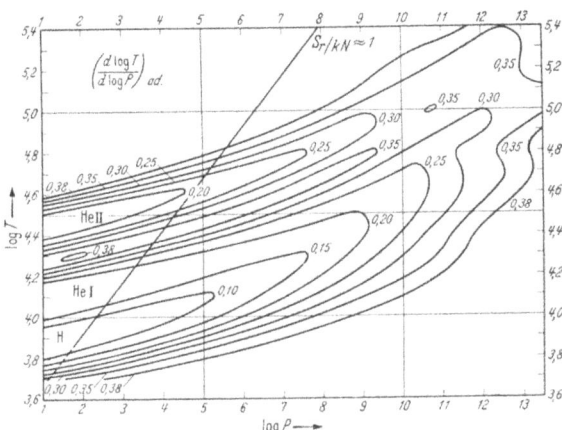

Figure 4.2: Contour lines ($\nabla_{\text{ad}} = \text{const}$) for the adiabatic temperature gradient ∇_{ad} in the $\log T - \log P$ diagram for stellar material of Population I.
The regions in which ionization takes place have $\nabla_{\text{ad}} < 0.4$. The regions are
H $\to$ H$^+$ (labelled with H),
He $\to$ He$^+$ (labelled with He I), and
He$^+$ $\to$ He^{2+} (labelled with He II).
Figure from Unsöld (1955).

4.3.2 Ledoux's criterion for convection

A refinement of the Schwarzschild criterion includes effects of chemical composition of the gas, so $\frac{d\mu}{dr} = 0$. If a volume element moves into a region with the same particle density but with a different mean molecular weight (e.g., convection in the core fusion zone), then this may either lead to a restoring force or to convection. The effect of different molecular weight is given as

$$\nabla_\mu \equiv \frac{d \ln \mu}{d \ln P} \tag{4.55}$$

and this gradient has to be added to the adiabatic one.

Thus there are, in fact, three aspects involved in the question of convection. If a volume of gas is radially shifted but is driven back, it may oscillate with the so-called *buoyancy frequency*

$$N^2 = \frac{g}{H_\text{P}}(\nabla_{\text{ad}} - \nabla + \nabla_\mu) \tag{4.56}$$

with ∇ from Eq. 4.51, ∇_{ad} from Eq. 4.54 and H_P the so-called pressure scale height defined as

$$H_\text{P} \equiv -\frac{dr}{d \ln P} = -P \frac{dr}{dP} \quad . \tag{4.57}$$

4.3.3 Estimates for $\nabla_{\text{ad}} < \nabla_{\text{rad}}$

4.3.3.1 Adiabatic gradient ∇_{ad}

In the regions where H and He become ionized an increase in inner energy mainly results in an increase of the fraction of ionization, but only a slight increase in temperature. Therefore C_V will be large. According to (4.21)

$$\gamma = \frac{C_P}{C_V} = 1 + \frac{n \mathfrak{R}}{C_V} \to 1$$

and so $\nabla_{\text{ad}} \to 0$.

Ionization of H normally sets in above $\simeq 6000$ K, whereas He becomes ionized as of 12000 K and doubly ionized around 25000 K. These species dominate the opacity in solar composition gas in this temperature range. In Fig. 4.2 contour lines of ∇_{ad} are plotted for stars of Population I in dependence of temperature and pressure.

4.3.3.2 Radiative gradient ∇_{rad}

∇_{rad} can be obtained from Eq. 4.3 and Eq. 4.18. Inserting the ideal gas law (Eq. 4.25) into Eq. 4.3 gives

$$\frac{d \ln P}{dr} = -\frac{GM}{r^2} \frac{\mu m_\text{H}}{kT} \equiv \frac{1}{H_\text{P}} \quad . \tag{4.58}$$

Dividing Eq. 4.18 by T and combining the result with Eq. 4.58 one obtains an expression for the radiative gradient

$$\nabla_{\mathrm{rad}} = \frac{3\,\overline{\kappa}\,\rho\,H_{\mathrm{P}}}{4ac\,T^4}\frac{L(r)}{4\pi r^2}. \qquad (4.59)$$

For convection to occur ∇_{rad} has to be large.

4.3.4 Absorption-driven or radiation-driven convection?

There are two parameters in Eq. 4.59 which govern the size of ∇_{rad}: $\overline{\kappa}$ and L (or F).

4.3.4.1 Large absorption coefficient κ

Convection due to a high value of $\overline{\kappa}$ (see Sect. 4.4.1 and Fig. 4.4) mostly occurs in the outer layers of cooler stars such as the Sun, stars lower on the main sequence, and red giants. They have absorption- or *opacity-driven convection* in the surface layers (see Fig. 6.6).

- κ (or $\overline{\kappa}$) is always large in layers where a new kind of ionization sets in (H or He, or some other element; see Fig. 4.4). This strengthens the effect of low ∇_{ad} and therefore often leads to convection.

- In layers at the boundary of hydrogen ionization at 6000 to 7000 K additionally two effects occur which lead to an even higher value of κ:

    - Due to the large number of free e^- the amount of H^-–ions increases which leads to an increase of $\kappa(H^-)$; see for H^- opacity also Ch. 2.8.2.
    - The energy is sufficiently large to excite a fraction of the H–atoms to the Balmer and Paschen levels. The now possible bound-free absorption from those levels leads to an increase of κ at $\lambda < 3650\,\text{Å}$ and $\lambda < 8200\,\text{Å}$ (see Ch. 3.3.4 and Fig. 2.5). Here ∇_{rad} can become really large.

- In the outer parts of very cool stars also molecules such as H_2 can exist. Additional absorption and thus an elevated κ exists where the photons may dissociate the molecules.

4.3.4.2 Large flux $F(r)$

In the central region of massive MS stars most of the energy is produced via the CNO–cycle (compare Sect. 5.1.5). As this cycle has a very strong temperature dependence ($\epsilon \propto T^{19.9}$) almost all of the energy is produced within the central part of the core ($r < 0.1 \cdot R$). This leads to a very high flux $F(r) = \frac{L(r)}{4\pi r^2}$, while the opacity is not very large (see Fig. 4.4).

Thus ∇_{rad} is large enough that the central regions of massive main-sequence stars are highly convective. This is, in fact, radiation- or *flux-driven convection* in the inner regions (see Fig. 6.6).

4.3.5 Convective overshoot

In the modeling of stellar structure an effect called convective overshoot became important for stellar evolution. Its inclusion reduced discrepancies with observed data (see Fig. 16.1).

For radii smaller than the outer limit of the convective zone, $r < r_0$, the radiative temperature gradient may be steeper than the adiabatic temperature gradient $|\frac{dT_{\mathrm{rad}}}{dr}| > |\frac{dT_{\mathrm{ad}}}{dr}|$. Therefore a convective cell (bubble) will be accelerated and start to rise. This leads to convection. However, convection continues beyond the radii where the Schwarzschild or Ledoux criteria are fulfilled (see Fig. 4.3 and its caption). So ultimately the bubble may have risen much higher than is actually allowed by these criteria. This behaviour is called convective overshoot. Thus convective energy transport also extends into regions outside the nominally convective zones as defined by the Schwarzschild and Ledoux criteria. Convection also results in mixing (Sect. 4.3.6).

4.4. MATERIAL FUNCTIONS

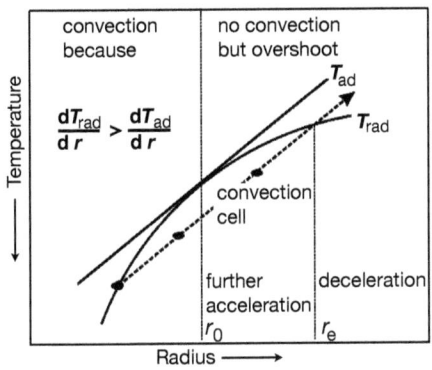

Figure 4.3: Gas units may overshoot the zone of convective conditions.
A convective bubble is, at r_0, still hotter than the surrounding material and it will continue its motion. Yet, as of r_0 the Schwarzschild criterion is no longer fulfilled. At r_e, where the temperature of the surrounding material and of the bubble are equal, the bubble will still have some (upward) momentum and the bubble may continue to rise while being decelerated. It then may be pushed back. All the time, exchange of gas with the surroundings (turbulence) will take place, leading to mixing.

4.3.6 Mixing length theory

Convection is a highly turbulent phenomenon. Therefore the Schwarzschild and Ledoux criteria are not sufficient to describe the conditions of convection. One has defined the so called **mixing length** l, the path length after which a convective cell dissolves. The mixing length is in the order of the pressure scale height $l \approx H_\mathrm{P}$. (A more detailed description can be found in Böhm-Vitense, 1989, Vol 2, Sect. 14.5.) For the modelling of stars one uses the parameter $\alpha = l/H_\mathrm{P}$.

In the Sun the mixing length is approximately $l \approx 1000$ km for its surface convection zone. The depth is governed by the ionization of H (the level of κ, see Fig. 4.4). The velocity of the rising cells is in the order of a few km s^{-1}.

4.4 Material functions

In Sect. 4.1 four differential equations have been presented:

$$\frac{dM(r)}{dr} = 4\pi r^2 \rho(r) \tag{4.1}$$

$$\frac{dP(r)}{dr} = -\frac{G\, M_r\, \rho(r)}{r^2} \tag{4.3}$$

$$\frac{dL(r)}{dr} = 4\pi r^2 \rho(r)\, \epsilon(r) \tag{4.4}$$

$$\frac{dT}{dr} = -\frac{3\bar{\kappa}\rho}{4acT^3} \frac{L(r)}{4\pi r^2} \quad \text{for} \quad \left|\frac{dT}{dr}\right|_{\mathrm{ad}} > \left|\frac{dT}{dr}\right|_{\mathrm{rad}} \tag{4.18}$$

$$\frac{dT}{dr} = \left(1 - \frac{1}{\gamma}\right) \frac{T}{P} \frac{dP}{dr} \quad \text{for} \quad \left|\frac{dT}{dr}\right|_{\mathrm{ad}} < \left|\frac{dT}{dr}\right|_{\mathrm{rad}} \tag{4.27}$$

which describe the gradients of $M(r)$, $P(r)$, $L(r)$ and $T(r)$ inside stars. Nevertheless these differential equations cannot be solved without knowledge of the material functions
   for the absorption coefficient or *opacity* $\kappa(P; T; X, Y, Z)$
   for density, or the *equation of state* $\rho(P; T; X, Y, Z)$
   for the rate of *energy production* $\epsilon(P; T; X, Y, Z)$
which all depend on the pressure P, the temperature T and the relative abundance of hydrogen X, helium Y and heavier elements Z (with $X + Y + Z = 1$). In the following subsections these material functions will be discussed.

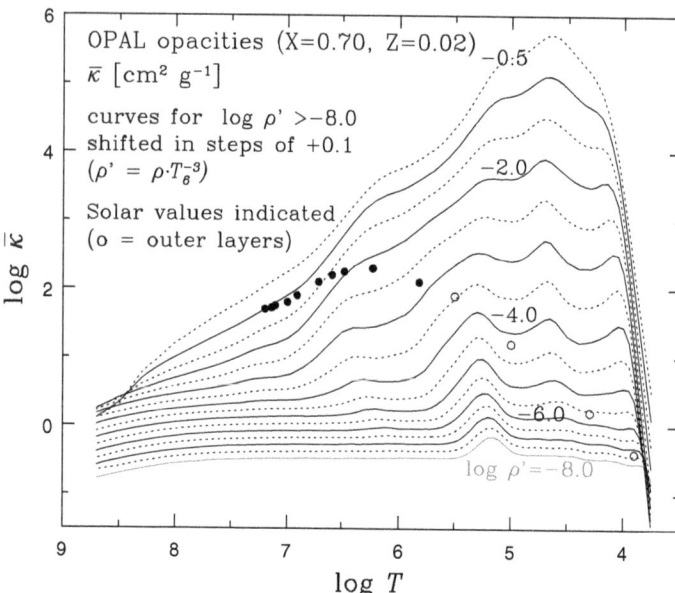

Figure 4.4: The Rosseland mean of the opacity $\bar{\kappa}$ (in cm^2 g^{-1}) as a function of T for a sequence of values of ρ (in g cm^{-3} T_6^{-3}; $T_6 = T/10^6$) for gas with $X = 0.70$, $Y = 0.28$ and $Z = 0.02$. For clarity the curves are shifted upward in steps of $\Delta \log \bar{\kappa} = 0.1$.
Note the sharp increase in $\bar{\kappa}$ at $\log T \simeq 3.8$ as well as the fact that $\bar{\kappa}$ is quite larger when ρ is larger. The braod maximum between $T \simeq 10^4$ and 10^5 K (density dependent) is due to H- and He-photoionization. The points indicate solar model values. The OPAL-opacities are available from Iglesias & Rogers (1996).

4.4.1 Opacity

Opacity has already partly been presented in Ch. 2.8. Knowledge about opacity is of fundamental importance for radiative energy transport (see Sect. 4.1.4.1). In the *radiative temperature gradient* (Eq. 4.18) an **opacity coefficient** $\bar{\kappa}$ is used, which is an effective absorption coefficient. One often uses the *Rosseland mean* (see Ch. 2.6.4). Unfortunately there is no simple equation which gives $\bar{\kappa}$ in dependence of temperature, density and chemical abundance because there are just too many different contributing elements and processes:

- **bound–free transitions** and **free–free transitions** which have been discussed in Ch. 2.8. Most of the bound–free opacity comes from the abundant species H and He. Once these elements are ionized, the opacity is smaller (see Fig. 4.4).

- **line absorption** (see Ch. 3) is very important in outer parts of stars. In inner parts ($T > 10^6$ K) most atoms are fully ionized and only heavy ions may still have some bound electrons. Therefore only few spectral lines occur with strong pressure broadening.

- **scattering** has been discussed in Ch. 2.8.5. If radiation pressure is dominant (high temperature and low pressure), **Thomson scattering** is of special importance.

- **heat conduction** by electrons occurs at higher densities and is especially strong in the case of stars like WDs (see Ch. 17.1.2) containing 'degenerate gas' (see Sect. 4.4.2.3). Conduction effectively reduces the opacity coefficient. With κ_{rad} is the Rosseland mean of the radiative opacity and κ_{cond} an effective opacity for conduction the total opacity coefficient becomes

$$\frac{1}{\kappa} = \frac{1}{\kappa_{\text{rad}}} + \frac{1}{\kappa_{\text{cond}}} \quad . \tag{4.60}$$

Table 4.1 gives an overview of typical values for κ_{cond}. A more detailed discussion of conduction can be found in, e.g., Meyer-Hofmeister (1982).

For the calculation of stellar models special opacity libraries are used. In Fig. 4.4 the Rosseland mean opacity is plotted in dependence of temperature and density.
Note that the run of opacity with T in Fig. 4.4 shows that opacity is (in the Sun) largest near $\log T \sim 4.5$ and is smaller both to lower T (the stellar atmosphere) as well as to higher T (the

4.4. MATERIAL FUNCTIONS

Table 4.1: Effective opacity coefficient for conduction $\log \kappa_{\text{con}}$ inside a red giant with $Y = 0.98$, $Z_N = 0.015$ and $Z_{Ne} = 0.005$ with κ_{cond} in $\text{cm}^2\,\text{g}^{-1}$ and ρ in $\text{g}\,\text{cm}^{-3}$.

$\log \rho$	-2.0	0	2	4	6
$T = 10^7\,\text{K}$	7.65	5.65	3.11	-0.10	-4.38
$T = 10^8\,\text{K}$			3.94	1.60	-1.94

stellar interior)! The lowest opacity inside a star occurs in highly ionized gas (little absorption possible) and is highest when H is nearly neutral.

The actual properties of the opacity govern the balance between radiative energy transport and convective energy transport, as discussed in Sect. 4.3.

The behaviour of opacity in the high temperature regime can be approximated by the Kramers opacity formula

$$\kappa \sim \kappa_0 \frac{\rho}{T^{3.5}} \quad \text{for } T > 10^5 \text{ K} . \tag{4.61}$$

4.4.2 Equation of state

The equation of state gives the pressure in dependence of temperature and density. The total pressure consists of the gas pressure P_{gas} and the radiation pressure P_{rad}

$$P = P_{\text{gas}} + P_{\text{rad}} \tag{4.62}$$

where the gas pressure is the sum of the partial pressures of the ions and electrons

$$P_{\text{gas}} = P_{\text{ion}} + P_e \tag{4.63}$$

(but see below for degenerate gas; Sect. 4.4.2.3).

4.4.2.1 Ideal gas law

In the inner parts of stars gas is nearly fully ionized and the size of particles is very small. Thus the ideal gas law can be used up to moderate densities (for $\rho < 10^4$ g cm^{-3}) in stellar interiors

$$P_{\text{gas}} = \frac{\rho}{\mu\, m_H} kT \quad (\sim nkT) \tag{4.25}$$

with the average molecular weight μ in units of hydrogen mass.

In case of fully ionized gas n and μ can easily be calculated from X, Y and Z if one takes into account the number of released electrons and the atomic weight of the components of the gas

$$n = (2X + \frac{3}{4}Y + \frac{1}{2}Z)\frac{\rho}{m_H} \tag{4.64}$$

$$\frac{1}{\mu} = 2X + \frac{3}{4}Y + \frac{1}{2}Z . \tag{4.65}$$

Here the term $\frac{1}{2}Z$ is the simplified expression for $(n_p + 1)/n_{\text{nucleons}}\, Z$. For solar metallicity ($X = 0.75$, $Y = 0.23$, $Z = 0.02$) the average molecular weight $\mu_\odot = 0.59$ for fully ionized gas.

4.4.2.2 Radiation pressure

The radiation pressure (see Ch. 2.2.1.3) depends on the actual intensity of the radiation. In local thermal equilibrium $I_\nu = B_\nu$ (the Planck spectrum) and the momentum transfer by the photons, $\int P_{\text{rad},\nu} d\nu$, follows from the integral over the Planck function (integrating over all photons available). In this case (LTE) the radiation pressure depends only on the temperature

$$P_{\text{rad}} = \int \frac{4\pi}{3c} I_\nu\, d\nu = \int \frac{4\pi}{3c} B_\nu d\nu = \frac{4\sigma}{3c}T^4 = \frac{a}{3}T^4 \tag{4.66}$$

with the radiation constant $a = \frac{4\sigma}{c} = \frac{8\pi^5 k^4}{15 c^3 h^3} = 7.566 \cdot 10^{-15}$ erg cm^{-3} K^{-4}.

Setting Eq. 4.25 (above) equal to Eq. 4.66 gives the temperature at which the gas pressure is equal to the radiation pressure

$$T^3 = \frac{3k}{a\mu m_{\rm H}} \rho \quad . \tag{4.67}$$

This boundary condition is given as the upper left dividing line in the "phase diagram" (Fig. 4.5). The consequences for a star in which the gas conditions surpass this limit are discussed in Ch. 6.4.

4.4.2.3 Degenerate gas

For higher densities (e.g., in the core of RGs and in WDs) gas 'degeneration' can occur.

Degenerate gas is gas, in which density and temperature are such, that the ideal conditions are no longer met. E.g., at high T, the fastest particles collide with neighbours that quickly that the high velocity tail of the maxwellian velocity distribution is quenched. At high ρ it becomes relevant that quantummechanics does not allow more than two electrons in one cell of six-dimensional phase space. At high densities, electrons are being forced into higher energy states requiring energy from the gas and the nuclei must move slower. Thus, the pressure is no longer dependent on T.

This contribution to the total pressure is the 'degeneration pressure', or 'Fermi pressure'. One differentiates between non-relativistic degeneration and relativistic degeneration.

The level of degeneracy in such gas is characterized by the **degeneracy parameter**

$$\psi \simeq 8 \cdot 10^{-6} \, T \, (\mu_e/\rho)^{2/3} \tag{4.68}$$

in which μ_e is the molecular weight per electron. It is beyond the scope of this text to *derive* the equations of state for degeneracy. A descriptions can be found in, e.g., Cox & Giuli (1968, their Ch. 24), Böhm-Vitense (1989, her Ch. 14.3-4) or Kippenhahn & Weigert (1990, their Ch. 15).

The pressure in a gas (omitting the radiation pressure) is thus composed of the contributions by the degeneration (or Fermi) pressure and the thermal pressure, the latter having contributions from the ion and the electron gas

$$P = P_{\rm Fermi} + P_{\rm iongas} + P_{\rm el.gas} \quad . \tag{4.69}$$

The individual contributions, the second two expressed in relation to $P_{\rm Fermi}$, are:

$$P_{\rm Fermi} = K_1 \left(\frac{\rho}{\mu_e}\right)^{5/3} \quad \text{with} \quad K_1 = \frac{1}{20}\left(\frac{3}{\pi}\right)^{2/3} \frac{h^2}{\mu_e} m_p^{-5/3} \tag{4.70}$$

$$P_{\rm iongas} = \frac{\mu_e}{\mu_a} \, \psi \, P_{\rm Fermi} \tag{4.71}$$

$$P_{\rm el.gas} = \mu^{2/3} \frac{\psi^2}{1+\psi} P_{\rm Fermi} \tag{4.72}$$

where $K_1 = 1.0036 \cdot 10^{13}$ [cgs] and μ_a the mean atomic weight (for non-relativistic gases $\psi \gg 1$).

The total pressure thus is

$$P = 10^{13} \, (\rho/\mu_e)^{5/3} \left(1 + \psi + \frac{\psi^2}{1+\psi}\right) \quad . \tag{4.73}$$

Depending on the value of ψ one finds a solution describing the particular gaseous object.

In the limiting case of a **completely degenerate non-relativistic** electron gas (Fermi–Dirac–degeneracy) the equation of state is given by

$$P_{\rm Fermi} = 10^{13} \, (\rho/\mu_e)^{5/3} \quad . \tag{4.74}$$

Recalling that $X + Y + Z = 1$, we are able to calculate μ_e for a fully ionized gas

$$\frac{1}{\mu_e} = X + \frac{1}{2}Y + \frac{1}{2}Z = \frac{1}{2}(1+X) \tag{4.75}$$

4.4. MATERIAL FUNCTIONS

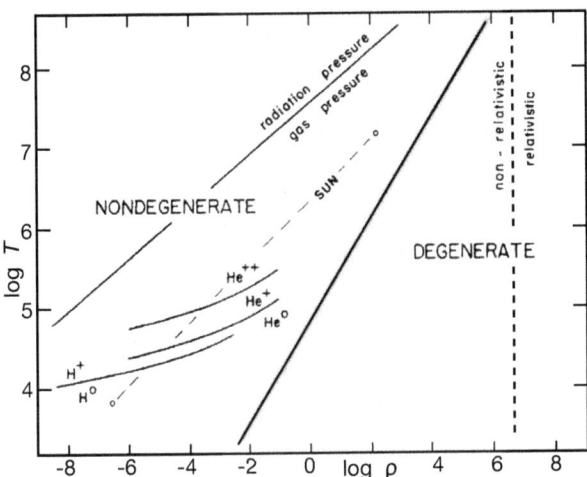

Figure 4.5: Phase diagram showing the conditions gas can have inside stars (T [K], ρ [g cm^{-3}]). The lines separate regions as indicated based on setting the pressure conditions on each side of each line equal (Eqs. 4.67, 4.70, 4.76). Thus, e.g., in the region at the upper left the radiation pressure is larger than the gas pressure. The long-dashed line indicates the approximate gas conditions in the Sun from the surface (lower left) to the centre. For the T_c and ρ_c in other stars see Fig. 6.5. Also the ionization limits for H and He are indicated (compare with Fig. 4.2). Diagram after Schwarzschild (1958).

which gives $\mu_e = 1$ for hydrogen gas and $\mu_e = 2$ for all other kinds of gas ($Y + Z = 1$). Such conditions exist in white dwarfs where the pressure of the degenerate electron gas is very large and much larger than any other pressure contribution.

For the **extreme relativistic case** of a **completely degenerate electron gas** the equation of state reads

$$P_e = K_2 \left(\frac{\rho}{\mu_e}\right)^{4/3} \quad \text{with} \quad K_2 = \left(\frac{3}{\pi}\right)^{1/3} \frac{hc}{8 \, m_u^{4/3}} \tag{4.76}$$

where $K_2 = 1.2435 \cdot 10^{15}$ in cgs units.

Limits between regimes.
The limit between the nondegenerate and the degenerate case can be estimated by setting the pressure for ideal electron gas as given by the ideal gas law (Eq. 4.25) equal to the pressure for completely degenerate non-relativistic electron gas (Eq. 4.70)

$$\frac{\rho}{\mu_e \, m_e} kT = K_1 \left(\frac{\rho}{\mu_e}\right)^{5/3} . \tag{4.77}$$

This condition is plotted in Fig. 4.5 (as is the limit between gas pressure and radiation pressure).

Finally the limit between nonrelativistic and relativistic degenerate gas can be found by equating the corresponding equations of state (Eq. 4.70 and Eq. 4.76). This leads to the limiting density

$$\rho = \mu_e \left(\frac{K_2}{K_1}\right)^3 \tag{4.78}$$

which is independent of temperature and is given by the vertical line in Fig. 4.5. The consequences for a star when the gas conditions surpass this line are discussed in Ch. 17.

4.4.3 Energy production functions: nuclear fusion and gravity

The energy radiated by stars is essentially produced through nuclear fusion. In general, the energy production rate, ϵ, can be approximated by a formula of the form

$$\epsilon_{\text{nuc process}} \simeq \epsilon_0 \, \rho^m \, X^n \, T^\theta \tag{4.79}$$

in which ϵ_0 is a normalization value, m, n and θ are exponents with m depending on the reaction, n = the number of participating particles in a fusion reaction, and θ some large exponent ($\theta \gg 4$) signifying the steep sensitivity with temperature. The various processes leading to this function are discussed in detail in Ch. 5.

A further energy source is based on gravity. When a star contracts, energy is liberated (Sect. 4.2.2); if a star expands, energy is needed. This energy is given by ϵ_{grav}. A first instance will be met in Ch. 7.5, another in Ch. 9.4.

4.5 Stellar winds

Almost all stars lose mass to interstellar space. Part of this mass loss may be due to the kinetic energy of the gas atoms of the stellar surface. E.g., gas atoms with mass m have in LTE a mean velocity $\frac{1}{2}mv^2 = \frac{3}{2}kT$. If the vertical component of the velocity would be large enough, and if no further collisions would occur, then the atom would leave the star (mass M, radius R). The minimum velocity for that is the **escape velocity**, v_{esc}

$$\frac{1}{2} m\, v_{\text{esc}}^2 = G\, \frac{M\, m}{R}\ . \tag{4.80}$$

However, there are various other mechanisms through which the gas at the surface can be accelerated (see below). This normally leads to outward velocities positively correlated with distance from the surface.

In general, the wind will exhibit a radial velocty profile. It can be given by

$$v(r) = v_0 + (v_\infty - v_0)\left(1 - \frac{R_*}{r}\right)^\beta \tag{4.81}$$

which can be is simplified to

$$v(r) \simeq v_\infty \left(1 - \frac{r_0}{r}\right)^\beta \tag{4.82}$$

in which $r_0 = R_*(1 - (v_o/v_\infty)^{1/\beta})$. This formulation contains parameters immediately identifiable with parameters which can be derived from observations. In general, $v_0 \simeq 0$ km s^{-1}.

As noted, there are energetic processes which lead to much enhanced mass loss. One considers, e.g., line driven winds and winds driven by the radiation continuum (these are further explored below and in Ch. 13.3.1), winds driven by accelerated dust or winds driven by sound waves emerging at the stellar surface from convective cells. Also rotation will play a role (see Ch. 14). For an early review see Cassinelli (1979) and a comprehensive treatment is available from Lamers & Cassinelli (1999). The number of stars for which the mass loss rates are well determined is small, while mass loss from population II red giants has not been observed yet. Here a cursory discussion follows since special aspects are dealt with in relation with particular types of stars or with other phenomena.

4.5.1 Coronal models

Stars that have surface convection zones or some other source of acoustic or mechanical wave energy are expected to have coronal zones (see Ch. 12) due to wave dissipation. Processes would be the same as those for the solar wind.

However, theoretical formulations lead to mass loss rates much smaller than observed for most stars (albeit fitting that of the Sun). Therefore these models are often combined with those for radiation driven winds, which are much more viable.

4.5.2 Radiative winds

Gas having sufficient opacity will pick up momentum from the photons, resulting in acceleration leading to a "radiation driven wind". Mostly the momentum transfer takes place through strong resonance lines. Even at small wind speeds, the gas then can absorb photons at Doppler-shifted wavelengths so that momentum transfer continues until some limiting wind velocity is reached. Also radiative forces on dust (as it is formed in red-giant star atmospheres) can drive the dust (and with it the gas) outward.

4.5.2.1 Line driven winds

Photons of appropriate energy can be absorbed in line transitions by gas in the stellar atmosphere. If the line is a 'resonance line', i.e., a permitted transition from the ground state (thus with large f-value), the absorption in such a line will quickly saturate (see Ch. 3.1.2). The atom will, after absorption, 'immediately' fall back to the ground state thereby emitting an identical photon.

If in the mean time the atom has gained some small outward velocity $v(r)$, the emitted photon will be Doppler-shifted and can then be absorbed in the wing of the line profile by atoms further out in the atmosphere. Such atoms thereby gain momentum and get accelerated. The absorbing capacity for these photones thus increases (because the velocity shift shifts the centre of the line). Photons of the kind emitted by the decay of the resonantly excited but velocity shifted atoms of interior layers can now be absorbed much better. Thus momentum is transfered more efficiently, leading to more acceleration of the gas.

Following Cassinelli (1979), the maximum mass loss rate is equivalent to the final mass momentum flux $\dot{M}v_\infty$, in which $v_\infty =$ **maximum wind velocity** obtainable (or observed). This maximum mass loss can be found by equating it to the photon momentum that is transferred by scattering all the radiation available between ν_0 and $\nu_0 + \Delta\nu_{max}$, where $\Delta\nu_{max} = \nu_0 v_\infty/c$. Thus

$$\dot{M}v_\infty = \frac{L_\nu \Delta\nu_{max}}{c} = \frac{L_\nu \nu_0}{c}\frac{v_\infty}{c} \quad . \tag{4.83}$$

If the spectrum is covered by non-overlapping spectral lines, so that adjacent spectral lines are separated by at least v_∞, the mass loss rate due to singly scattered photons can be derived from the momentum flux of the entire stellar luminosity, $\dot{M}v_\infty = L/c$, giving

$$\dot{M}_{max} = \frac{L}{v_\infty c} \quad . \tag{4.84}$$

The final kinetic energy of the wind, also referred to as the **wind luminosity**, is

$$\frac{1}{2}\dot{M}v_\infty^2 = \frac{1}{2}\left(\frac{v_\infty}{c}\right) L \quad . \tag{4.85}$$

Empirically it is found (Reimers 1977) that

$$\dot{M} \simeq 1 \cdot 10^{-13} \frac{L}{gR} \quad [\text{M}_\odot\ \text{yr}^{-1}]. \tag{4.86}$$

A star with larger luminosity has larger mass loss, a star with smaller radius and the same luminosity (leading to a more intense surface radiation field) has larger mass loss, too. In fact, there is an intimate interplay between g and R, so that it is the product gR that matters.

Assuming $v_{esc}^2 = v_\infty^2$ in combining Eq. 4.85 with Eq. 4.80 and applying Eq. 1.17 one arrives at

$$\dot{M} = \frac{1}{2}\frac{v_\infty}{c}\frac{L}{gR} \tag{4.87}$$

which has a structural form similar to the empirical relation of Reimers (considering that $v_\infty/c = \Delta\lambda/\lambda$, and that for stars with strong winds v_∞/c as observed is of the order $10^{-2.5}$ to 10^{-3}).

4.5.2.2 Continuum-driven winds

Also the continuum may drive winds through momentum transfer to free electrons. This is, in fact, similar to general radiation pressure (see Ch. 2.2.1.3). More on this can be found with massive stars (see Ch. 13.3.1.1).

4.5.2.3 Dust-driven winds

If a stellar atmosphere contains dust (which is possible in very cool atmospheres, $T_{eff} <$ 2500 K) then the opacity of the dust implies also momentum transfer. This may strengthen a stellar wind. Dust is in particular important in the case of AGB stars (Ch. 10.5).

4.5.3 Bi-stability winds: fast and dilute or slow and dense

Depending on the level of momentum transfer, a wind can be fast or slow.

A fast wind will disperse quickly around the star. It occurs foremost for stars with large luminosity and rather 'hot' spectral energy distributions. In particular, when in the Lyman continuum (radiation with $\nu > 3 \cdot 10^{15}$ Hz) the optical depth $\tau_{\rm LyC} < 1$, the fast wind establishes itself.

A slow wind will come into existence with stars having a 'cool' spectral energy distribution, the cases when $\tau_{\rm LyC} > 3$. The wind material is accelerated but not much, it can cool and so become more optically thick.

The boundary between the two stellar wind cases lies near $T_{\rm eff} = 21000$ K. Depending on the actual circumstances (binarity, effects of evolution, effects of rotation) a star may switch between the two wind modes; this led to the word "bi-stability". For more see Lamers & Cassinelli (1999).

4.5.4 Winds enhanced due to stellar rotation

When a star rotates, the surface gas senses an extra outward force. This leads to enhanced mass loss as well as to the formation of mass loss disks. For more see Ch. 14.6.

Stars with rotation may exhibit the slow wind and the fast wind simultaneously. Because of the latitude dependent $T_{\rm eff}$ (and thus a latitude dependent spectral energy distribution; see Ch. 3.7) the fast wind may emerge at high latitudes, the slow wind in the equatorial zone.

4.5.5 Pulsation-driven winds

When a pressure wave runs through gas (as in pulsating stars; Ch. 11), its amplitude is proportional to the amount of mass in the wave. In a homogeneous spherical object, a wave propagating outward will thus have an ever smaller amplitude. The outward decrease of gas density in stars means that the wave may propagate quite far.

Especially in the extended envelopes of AGB stars the density drops radially outward strongly (see, e.g., Fig. 10.12) so that the wave amplitude may even increase (the force of the wave becomes larger than the local gravity). This heats up the gas, which subsequently cools radiatively, leading to a dense and outward moving shell. Thus pulsation may contribute considerably to mass loss.

For details of the physics see the review by Gustafsson & Höfner (2004).

References

Böhm-Vitense, E. 1989, "Stellar Astrophysics"; Cambridge Univ. Press

Cassinelli, J. 1979, ARAA 17, 275; *Stellar Winds*

Cox, J.P., & Giuli, R.T. 1968, "Principles of Stellar Structure"; Gordon & Breach, New York

Iglesias, C.A., & Rogers, F.J. 1996, ApJ 464, 943

Gustafsson, B., & Höfner, S. 2004, in "Asymptotic Giant Branch Stars", H.J. Habing & H. Olofsson (eds.); Springer, Heidelberg; p.149

Kippenhahn, R., & Weigert, A. 1990, "Stellar Structure and Evolution"; Springer, Heidelberg

Lamers, H.J.G.L.M., & Cassinelli, J.A. 1999, "Introduction to Stellar Winds"; Cambridge Univ. Press

Lang, K.R. 1980, "Astrophysical Formulae"; Springer, Heidelberg

Meyer-Hofmeister, E. 1982, in "Landolt–Börnstein", Group IV, Vol. 2b, p. 178; Springer, Heidelberg

Reimers, D. 1977, A&A 57, 395

Schwarzschild, M. 1958, "Structure and Evolution of the Stars"; Princeton Univ. Press; Dover Publ.

Unsöld, A. 1955, "Physik der Sternatmosphären"; Springer, Heidelberg

Chapter 5

Nuclear fusion in stars

The fusion of light atoms into heavier ones leads to the release of energy because the mass of the fused particle is less than the sum of the mass of the original particles. Or, the potential well of heavier nuclei is deeper (binding energy E_b is larger) than that of light elements. This is so for elements with mass up to the vicinity of the mass of Fe. The mass 'deficit', Δm, is transformed into energy: $E = \Delta m\, c^2$. However, for fusion the charge barrier has to be overcome.

Stars generate their energy through fusion, thereby creating elements heavier than hydrogen from the ubiquitous hydrogen. Through stellar winds and through supernova explosions some of this processed material is returned to interstellar space and can participate in the formation of further generations of stars (as well as of planets, with all that is possible on them). Clearly, nuclear fusion is an important aspect of stellar physics and warrants a chapter by itself.

5.1 Energy production: fusion of H and He

5.1.1 Binding energy of nuclei

During their main-sequence phase and posterior giant phases stars produce energy in fusion processes. In a fusion process the energy difference of the binding energies of the final nuclei and the initial nuclei is released. In Fig. 5.1 the binding energy per nucleon is plotted in dependence of the mass number A. Fusion processes are only exothermic if the final nucleus is not heavier than ^{56}Fe as this nucleus has the highest binding energy per nucleon. The duration of the energy production can be estimated with the nuclear time scale (see Sect. 4.2.3).

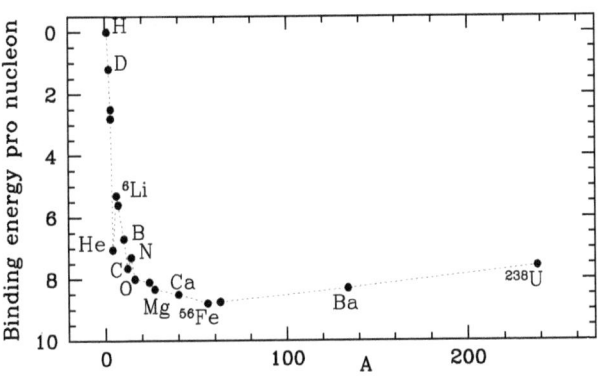

Figure 5.1: The binding energy per nucleon E_b/A as function of mass number A. Some elements are labelled. The energy scale is inverted to emphasize the depth of the potential well. If nuclei come to fusion they strive to the lowest total energy, which is to elements like Fe. The largest amount of energy freed comes from the fusion of H, much less energy per nucleon by fusion of He, C and N. Elements beyond Fe generally liberate energy only by falling apart (but see Sect. 5.2.4).

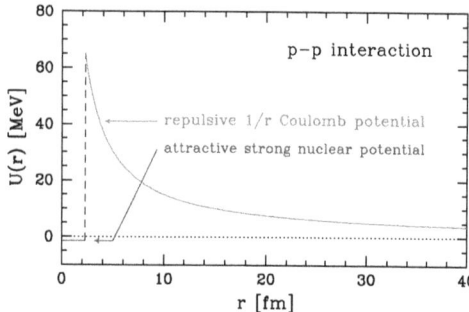

Figure 5.2: The potential energy in a proton–proton interaction is sketched as a function of the distance between the protons. 1 fm = 10^{-13} cm.

5.1.2 Estimates for the occurrence of hydrogen fusion

In Fig. 5.2 the potential energy in a *p–p* interaction is plotted in dependence of the distance between the two protons. For distances larger than approximately 2 fm the protons only experience their Coulomb repulsion. At smaller distances the attractive strong force quickly is much more important. In order to merge two protons into a deuterium nucleus they have to surmount the Coulomb barrier. One can classically (and in a general manner) calculate the required kinetic energy by setting it equal to the potential energy of the barrier

$$\frac{1}{2}\mu_p \overline{v^2} = \frac{3}{2} kT_{\text{classical}} = \frac{Z_1 Z_2 e^2}{r}$$

with Z the charge of each of the particles. Here $\mu_p = m_p/2$ is the reduced mass of the two protons. If we assume a range of the strong force of $r \approx 1$ fm and put $Z_1 = Z_2 = 1$ for two protons, we get as required temperature

$$T_{\text{classical}} = \frac{2 Z_1 Z_2 e^2}{3 k r} \approx 10^{10} \text{K} \quad . \tag{5.1}$$

This temperature is much higher than the temperature in the centre of the Sun, estimated to be $1.6 \cdot 10^7$ K. So fusion seems impossible and the Sun would not shine at all. However, nature does not behave according the classical description and protons have a good quantum mechanical chance to "tunnel" the repulsive barrier. We can estimate the temperature so required for fusion if we assume that for fusion to occur the distance of closest approach has to be in the order of a de Broglie wavelength $\lambda = \frac{h}{p}$

$$\frac{Z_1 Z_2 e^2}{\lambda} = \frac{p^2}{2\mu_p} = \frac{\left(\frac{h}{\lambda}\right)^2}{2\mu_p}$$

with $p = \mu_p v$. Solving for λ and inserting $r = \lambda$ into Eq. 5.1 gives

$$T_{\text{qm}} = \frac{4}{3} \frac{\mu_p Z_1^2 Z_2^2 e^4}{k h^2} \approx 10^7 \text{ K} \quad . \tag{5.2}$$

This fusion temperature requirement is in line with the central temperature in the shining Sun.

5.1.3 The Gamow peak

Only particles in the high energy end of the Maxwell–Boltzmann velocity distribution of particles will have a sufficiently high energy to surmount the potential barrier. But only very few protons in the high energy tail reach into the energy range where the Coulomb barrier becomes noticable. In brief, the high energy tail of the velocity distribution can be given by

$$N(E) \simeq e^{-E/kT} \tag{5.3}$$

where E is the actual kinetic energy of the particle. The Coulomb barrier can be given as a cross section or barrier penetration probability for a particle colliding with kinetic energy E by

$$\sigma(E) \simeq e^{-bE^{-1/2}} \tag{5.4}$$

5.1. ENERGY PRODUCTION: FUSION OF H AND HE

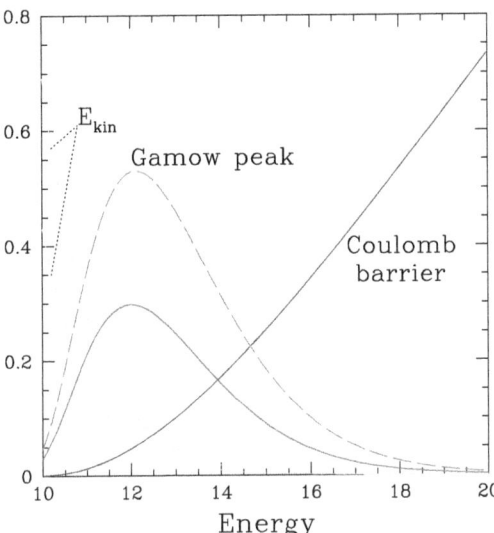

Figure 5.3: Fusion between nuclei occurs at temperatures well below the temperature equivalent of their Coulomb barrier because particles in the high-velocity tail of the Maxwell velocity distribution have sufficient probability to overcome the barrier.
Full curves: The curve at left (E_{kin}) shows the velocity distribution (Eq. 5.3), the curve at right the Coulomb barrier cross section (Eq. 5.4), the middle curve their product, leading to the "Gamow"-peak.
Dashed curves: For a **5%** higher temperature the E_{kin} curve shifts to the right and the Gamow peak is almost a factor **2** higher, explaining why fusion is so temperature sensitive (see also Fig. 5.6).
The Gamow Peak curves have been multiplied by 100 to make the shape and T-effect well visible. Data for hydrogen fusion (PP); relative values.

in which $b = 2^{3/2}\pi^2 \mu_m^{1/2} Z_1 Z_2 e^2/h$, a parameter dependent on the actual gas composition (μ_m is the mean molecular weight) and the charges Z_1 and Z_2 of the to be fused particles. The reaction rate follows by integrating the product of Eq. 5.3 and Eq. 5.4.

The shape of these functions as well as the value of their product G, which has a peak (Gamow 1928) as function of E is shown in Fig. 5.3. Since e-functions decrease steeply, the maximum of the product lies (depending on the normalisation of the two functions) somewhere near the middle. Clearly, when the temperature is increased *a little* the product increases *a lot*, thereby explaining the steep rise of the energy production functions of Fig. 5.6. If the temperature is so high that the peak of the Maxwell distribution approaches the formal threshold energy the steep increase in energy production levels off (see ϵ_{pp} in Fig. 5.6).

5.1.4 proton–proton chain

During the main-sequence phase stars produce energy in fusion reactions transforming four ^1_1H nuclei into one ^4_2He nucleus. The total amount of energy released is given by the difference in binding energy of 26.731 MeV, which corresponds to a mass defect of about 0.71 %.

The process which is dominant during the first phase of a stars' life is the **proton–proton** (PP) **chain**. It has three branches (PP I, PP II, PP III). In the PP chain the *first steps are the same for all branches*:

$$\begin{aligned} ^1_1\text{H} + ^1_1\text{H} &\longrightarrow\ ^2_1\text{H} + e^+ + \nu_e \\ ^2_1\text{H} + ^1_1\text{H} &\longrightarrow\ ^3_2\text{He} + \gamma \end{aligned} \qquad (5.5)$$

and ^3_2He is produced. Also an *electron neutrino* is generated which constitutes part of the Solar neutrino flux (Sect. 5.3.2). The reaction *rates* of these two steps in the PP chain differ considerably, and the rate of the whole chain is defined by the slowest step which is the first one. It is **slow** because it involves the weak decay of a proton $p^+ \longrightarrow n + e^+ + \nu_e$.

In the **PP I** branch ^3_2He then merges with another ^3_2He to become ^4_2He:

$$^3_2\text{He} + ^3_2\text{He} \longrightarrow\ ^4_2\text{He} + 2\,^1_1\text{H} \qquad (5.6)$$

In the **PP II** branch ^3_2He has the possibility to become ^7_4Be and then to proceed to ^4_2He:

$$\begin{aligned} ^3_2\text{He} + ^4_2\text{He} &\longrightarrow\ ^7_4\text{Be} + \gamma \\ ^7_4\text{Be} + e^- &\longrightarrow\ ^7_3\text{Li} + \nu_e \\ ^7_3\text{Li} + ^1_1\text{H} &\longrightarrow\ 2\,^4_2\text{He} \end{aligned} \qquad (5.7)$$

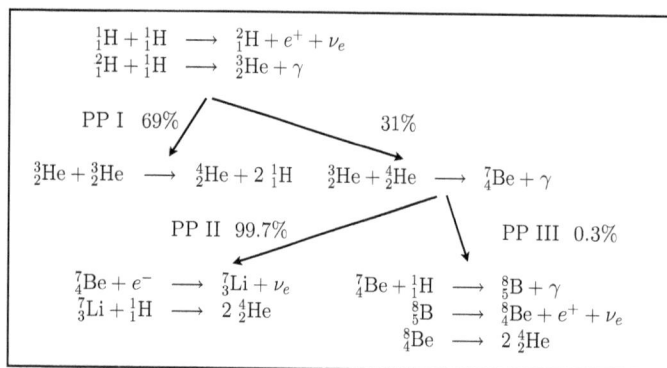

Figure 5.4: The three branches of the PP chain are shown. The indicated branching *ratios* are for the conditions in the core of the Sun.

There is a very small possibility that the beryllium-7 formed in the first reaction of the PP II branch captures a proton leading to the **PP III** branch:

$$\begin{aligned}
{}_2^3\text{He} + {}_2^4\text{He} &\longrightarrow {}_4^7\text{Be} + \gamma \\
{}_4^7\text{Be} + {}_1^1\text{H} &\longrightarrow {}_5^8\text{B} + \gamma \\
{}_5^8\text{B} &\longrightarrow {}_4^8\text{Be} + e^+ + \nu_e \\
{}_4^8\text{Be} &\longrightarrow 2\,{}_2^4\text{He}
\end{aligned} \quad (5.8)$$

All branches with their sequences are summarized in Fig. 5.4. The branching ratios, valid for the conditions in the Sun's core, are given there, too.

Following Schwarzschild (1958) the energy production rate ϵ_{pp} for the total PP chain is (and as shown in Fig. 5.6)

$$\epsilon_{\text{pp}} = 2.5 \cdot 10^6 \, \rho \, X^2 \, T_6^{-2/3} \, e^{\frac{-33.8}{T_6^{1/3}}} \quad \text{erg cm}^3 \text{ g}^{-2} \text{ s}^{-1} \quad (5.9)$$

with the temperature T_6 in 10^6 K and X the hydrogen fraction. Around $T = 1.4 \cdot 10^7$ K (thus locally) this equation can be approximated by a power law in temperature $\epsilon_{\text{pp}} \simeq \epsilon'_{0,\text{pp}} \rho X^2 T_6^{4.5}$ with $\epsilon'_{0,\text{pp}} = 1.07 \cdot 10^{-5}$ erg cm^3 g^{-2} s^{-1}. The PP chain fusion rate in the Sun has (with $T_c \simeq 1.4 \cdot 10^7$ K) a rather modest temperature dependence.

5.1.5 CNO cycle

Another hydrogen-fusion chain is the CNO cycle. Sometimes this chain is also called *Bethe–Weizsäcker cycle*. Here carbon, nitrogen and oxygen are used as catalysts:

$$\begin{aligned}
{}_6^{12}\text{C} + {}_1^1\text{H} &\longrightarrow {}_7^{13}\text{N} + \gamma \\
{}_7^{13}\text{N} &\longrightarrow {}_6^{13}\text{C} + e^+ + \nu_e \\
{}_6^{13}\text{C} + {}_1^1\text{H} &\longrightarrow {}_7^{14}\text{N} + \gamma \\
{}_7^{14}\text{N} + {}_1^1\text{H} &\longrightarrow {}_8^{15}\text{O} + \gamma \quad \text{(a slow reaction)} \\
{}_8^{15}\text{O} &\longrightarrow {}_7^{15}\text{N} + e^+ + \nu_e \\
{}_7^{15}\text{N} + {}_1^1\text{H} &\longrightarrow {}_6^{12}\text{C} + {}_2^4\text{He}
\end{aligned} \quad (5.10)$$

A further reaction of this cycle, or rather a second branch (with 0.04 % probability), exists:

$$\begin{aligned}
{}_7^{15}\text{N} + {}_1^1\text{H} &\longrightarrow {}_8^{16}\text{O} + \gamma \\
{}_8^{16}\text{O} + {}_1^1\text{H} &\longrightarrow {}_9^{17}\text{F} + \gamma \\
{}_9^{17}\text{F} &\longrightarrow {}_8^{17}\text{O} + e^+ + \nu_e \\
{}_8^{17}\text{O} + {}_1^1\text{H} &\longrightarrow {}_7^{14}\text{N} + {}_2^4\text{He}
\end{aligned} \quad (5.11)$$

These chains are schematically shown in Fig. 5.5.

5.1. ENERGY PRODUCTION: FUSION OF H AND HE

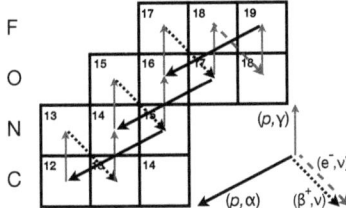

Figure 5.5: The reaction network of the CNO cycle is shown in a schematic manner. The sequence of reactions at the lower left are those of Eq. 5.10. The cycle has an extra loop given by Eq. 5.11, while a further loop, leading to ^{19}F, is also possible. Element names are given at left, the respective atomic numbers are given inside each box. At the lower right the arrows show the actual reaction types.

The energy production rate ϵ_{CNO} for the CNO cycle is

$$\epsilon_{\text{CNO}} = 9.5 \cdot 10^{28} \, \rho \, X \, X_{\text{CNO}} \, T_6^{-2/3} \, e^{\frac{-152}{T_6^{1/3}}} \quad \text{erg cm}^3 \, \text{g}^{-2} \, \text{s}^{-1} \tag{5.12}$$

which, around $T = 1.5 \cdot 10^7$ K (locally!) can be approximated by a power law in temperature $\epsilon_{\text{CNO}} \simeq \epsilon'_{0,\text{CNO}} \, \rho \, X \, X_{\text{CNO}} \, T_6^{19.9}$ (with $T_6 = T/10^6$), $\epsilon'_{0,\text{CNO}} = 8.24 \cdot 10^{-24}$ erg cm^3 g^{-2} s^{-1} and X_{CNO} the total mass fraction of C, N and O. Apparently the CNO cycle has at $T \simeq 10^7$ K a very strong temperature dependence. It requires, however, higher temperatures to noticeably contribute to the energy production (see Fig. 5.6). It therefore is unimportant in lower mass stars, in which, due to too low T_c, only the PP chain contributes to ϵ. For more massive main-sequence stars with sufficiently high central temperatures the CNO cycle is the dominant source of energy[1]. The strong dependence of ϵ on T induces in such stars central convection (Ch. 9.2.2).

An important aspect of the CNO cycle is that its fourth reaction $^{14}_{7}\text{N} + ^{1}_{1}\text{H} \rightarrow ^{15}_{8}\text{O} + \gamma$ (see Eq. 5.10) is a slow one. It leads in stars burning with the CNO cycle to a build up of N (*pile up of N*) at the cost of the C and O abundance in the burning zone. Ultimately, convection may bring this N to the surface leading to 'N-rich' stars. Or it leads to N-burning (see below).

5.1.6 Temperature dependence of H-fusion energy production

Fig. 5.6 shows the H-fusion energy production rate ϵ_H, i.e., ϵ_{pp} and ϵ_{CNO}, in dependence of the temperature for $\rho = 1$ g cm^{-3}, $X = 1$ and $X_{\text{CNO}} = 1$. (note that in the Sun $X_{\text{CNO}} \simeq 0.005 \cdot X$). The energy production is a strong function of T. Clearly, depending on T_c either the PP chain or the CNO cycle dominates energy production. In the Sun it is (still) the PP chain.

With all the fusion steps involved, the actual functions ϵ are more complicated than the simple two-particle fusion Gamow-peak analysis presented in Sect. 5.1.3.

5.1.7 He fusion: the triple alpha process

During the main-sequence phase hydrogen is transformed into helium. This leads to an increase of the mean molecular weight μ. If temperature and density stayed constant this would lead to a decrease in pressure according to the ideal gas law (Eq. 4.25). As a result the star is no longer in hydrostatic equilibrium and must contract. This contraction leads to an increase in density and temperature until a new equilibrium situation is reached (see Ch. 9).

If the temperature becomes high enough, helium burning will set in. The fusion toward $^{12}_{6}$C is a two-step process involving 3 α particles, thus known as the triple-alpha-process:

$$\begin{aligned} ^{4}_{2}\text{He} + ^{4}_{2}\text{He} &\longleftrightarrow ^{8}_{4}\text{Be} \\ ^{8}_{4}\text{Be} + ^{4}_{2}\text{He} &\longrightarrow ^{12}_{6}\text{C} \end{aligned} \tag{5.13}$$

In the first step an unstable beryllium nucleus is produced, which will decay back into two alpha particles within a few 10^{-16} s. However, if the density is sufficiently high, interactions are frequent

---

[1] For the CNO cycle to operate, C must be present. The first generation of stars in the universe did not contain C *ab initio*. Only once C was formed internally through the triple-α process (see Sect. 5.1.7) and convectively distributed in the star, could the CNO cycle operate. Matter returned to the ISM in the final supernova explosion of such massive stars provided further star generations with a certain amount of C. For more on metal enrichment see Ch. 22.

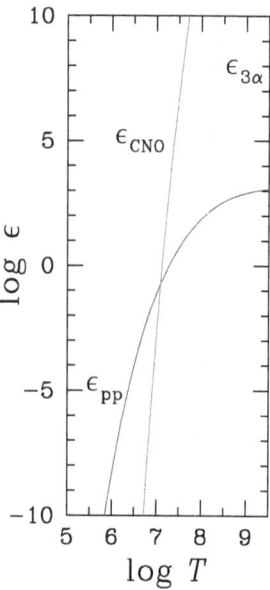

Figure 5.6: Energy production rate ϵ for burning of hydrogen and helium in $\mathrm{erg\,cm^3\,g^{-2}\,s^{-1}}$ in relation with the gas temperature T. For hydrogen the individual contributions of the PP chain and the CNO cycle, ϵ_{pp} and ϵ_{CNO}, are shown. For helium $\epsilon_{3\alpha}$ is plotted. The rates are those of Schwarzschild (1958) as given in Eqs. 5.9, 5.12, and 5.14 based on $\rho = 1\,\mathrm{g\,cm^{-3}}$, $X = 1$, $X_{\mathrm{CNO}} = 1$ and $Y = 1$.

The steepness of ϵ with temperature can be understood as follows. The larger the temperature, the further the velocity distribution of particles lies toward higher energy (see Fig. 5.3), allowing for more fusion.

The steepness of ϵ is part of the physical background of the mass–luminosity relation ($L \sim M^{\simeq 3}$) of main-sequence stars in that stars can in an easy/efficient manner extract energy from nuclear fusion to keep their shape (see Ch. 6.8.2.1).

When the energy maximum of the velocity distribution approaches the energy of the fusion potential barrier, the increase in ϵ naturally is considerably less pronounced (see the levelling off of ϵ_{pp} at T above 10^7 K). Similar curves can be created for further fusion reactions.

enough that the beryllium nucleus may react in time (before decay) with another alpha particle and forms stable $^{12}_{6}\mathrm{C}$. The total energy released in this reaction is 7.162 MeV.

One can estimate the temperature required for two helium–4 nuclei having the first reaction by inserting $\mu_p = \frac{m_{\mathrm{He}}}{2}$ and $Z_1 = Z_2 = 2$ into Eq. 5.2. This gives $T \approx 6 \cdot 10^8$ K. Hoyle suspected in 1952 that C should be formed already at lower temperatures and his prediction of a "resonance" at 7.65 MeV turned out to be correct. $^{8}_{4}\mathrm{Be}$ and $^{4}_{2}\mathrm{He}$ may meet forming a meta-stable bond which can decay to form $^{12}_{6}\mathrm{C}$ and the three reactions together can take place already at $T \approx 1 \cdot 10^8$ K (see Hoyle 1954 for the discovery story).

The existence of the said resonance is fortuitous. Since $^{8}_{4}\mathrm{Be}$ is unstable and decays quickly, its fleeting presence just functions as a stepping stone to form $^{12}_{6}\mathrm{C}$. No such resonance level exists for fusion of C with He to form O. Thus, the further build-up of heavier nuclei proceeds slowly instead of explosive. This allows time for a build-up of other elements (see Sect. 5.2.4.1).

The energy production rate $\epsilon_{3\alpha}$ for the triple alpha process is (Schwarzschild 1958)

$$\epsilon_{3\alpha} = 2 \cdot 10^{17}\, \rho^2\, Y^3\, T_6^{-3}\, e^{\frac{-4670}{T_6}} \quad \mathrm{erg\,cm^3\,g^{-2}\,s^{-1}} \tag{5.14}$$

which can, around $T = 1 \cdot 10^8$ K, be approximated by $\epsilon_{3\alpha} \simeq \epsilon'_{0,3\alpha}\, \rho^2\, Y^3\, T_8^{41.0}$ with the temperature T_8 in 10^8 K. The energy production rate increases even steeper than for the CNO cycle. The stronger dependence on Y and ρ is a result of the fact that the triple alpha process is, in essence, a three particle interaction.

5.2 Nucleosynthesis

Higher levels of fusion generate further energy and further heavy elements. However, the main effect and importance of higher level fusion is the nucleosynthesis of heavy elements. Since such high level fusion phases are only short in comparison with previous fusion phases (the PP and the further processes described in Sect. 5.1), they are not important for overall energy generation (also since H- or He-fusion still takes place in most of these stars as well).

5.2.1 Carbon and oxygen burning; α-capture

Nucleosynthesis after H and He fusion is the burning of C and O. Products from such fusion are generally called **the α-elements** because the nuclei formed can be thought of as $n \times \alpha$.

5.2. NUCLEOSYNTHESIS

C+α, O+α. Once sufficient $^{12}_{6}$C has been produced alpha particles can be captured, leading to the production of $^{16}_{8}$O, followed by further *alpha captures*. Such reactions are, e.g.,

$$\begin{aligned}
^{12}_{6}\text{C} + ^{4}_{2}\text{He} &\longrightarrow ^{16}_{8}\text{O} + \gamma \\
^{16}_{8}\text{O} + ^{4}_{2}\text{He} &\longrightarrow ^{20}_{10}\text{Ne} + \gamma \\
^{20}_{10}\text{Ne} + ^{4}_{2}\text{He} &\longrightarrow ^{24}_{12}\text{Mg} + \gamma \\
^{24}_{12}\text{Mg} + ^{4}_{2}\text{He} &\longrightarrow ^{28}_{14}\text{Si} + \gamma
\end{aligned} \quad (5.15)$$

where reactions going beyond $^{20}_{10}$Ne may take place in, e.g., AGB stars (Ch. 10.4.3) where He is convectively mixed into the outer shell of the C/O core of the star.

C+C. The lower temperature limit for pure carbon burning is $T > 5 \cdot 10^8$ K. Clearly, C-burning is only of relevance if sufficient C is present at all (or has been formed, as in high-mass stars). In C-burning an excited $^{24}_{12}$Mg nucleus is formed, which can decay in many different channels

$$\begin{aligned}
^{12}_{6}\text{C} + ^{12}_{6}\text{C} &\longrightarrow ^{24}_{12}\text{Mg} + \gamma, & +13.931\,\text{MeV} \\
&\longrightarrow ^{23}_{12}\text{Mg} + n, & -2.605\,\text{MeV} \\
&\longrightarrow ^{23}_{11}\text{Na} + p, & +2.238\,\text{MeV} \\
&\longrightarrow ^{20}_{10}\text{Ne} + ^{4}_{2}\text{He}, & +4.616\,\text{MeV} \\
&\longrightarrow ^{16}_{8}\text{O} + 2\,^{4}_{2}\text{He}, & -0.114\,\text{MeV}
\end{aligned} \quad (5.16)$$

and also the released energy (negative for endothermic processes) is given. The frequency of the processes depends on the temperature. The fastest processes are the third and fourth ones.

In white dwarfs C+C fusion may occur and is then explosive (Ch. 18.3.2).

O+O. The temperature limit for oxygen burning is $T > 1 \cdot 10^9$ K. The branches for O-burning are

$$\begin{aligned}
^{16}_{8}\text{O} + ^{16}_{8}\text{O} &\longrightarrow ^{32}_{16}\text{S} + \gamma, & +16.541\,\text{MeV} \\
&\longrightarrow ^{31}_{15}\text{P} + p, & +7.677\,\text{MeV} \\
&\longrightarrow ^{31}_{16}\text{S} + n, & +1.453\,\text{MeV} \\
&\longrightarrow ^{28}_{14}\text{Si} + ^{4}_{2}\text{He}, & +9.593\,\text{MeV} \\
&\longrightarrow ^{24}_{12}\text{Mg} + 2\,^{4}_{2}\text{He}, & -0.393\,\text{MeV}
\end{aligned} \quad (5.17)$$

The fastest process is the second one. Such O burning takes place in high-mass stars.

Production of free neutrons. The reactions of the C- and the O-burning produce (see above) free neutrons. These are of importance for reactions with neutron capture in the s- and r-processes, to be discussed below.

5.2.2 Nitrogen burning

One of the results of burning in the CNO-cycle (see Eq. 5.10) is a build-up of N, because the proton-capture by N in that cycle, $^{14}_{7}\text{N} + ^{1}_{1}\text{H} \rightarrow ^{15}_{8}\text{O} + \gamma$, is a slow reaction.

The 'overabundance' of N allows for another set of fusion reactions, starting with α-capture by N and then on to others. In particular the reaction sequence (in 'shorthand' description)

$$^{14}\text{N}(\alpha,\gamma)^{18}\text{F}(\beta^+)^{18}\text{O}(\alpha,\gamma)^{22}\text{Ne} \quad (5.18)$$

(see also Eq. 5.20) is effective. During the He-burning phase of stellar evolution this reaction chain can process lots of N into ^{22}Ne.

5.2.3 Fusion to heavier elements

In the later stages of evolution of higher mass stars fusion processes requiring also high temperatures (all based on surmounting the particular Coulomb-barrier) may occur, e.g., the burning of $^{20}_{10}$Ne

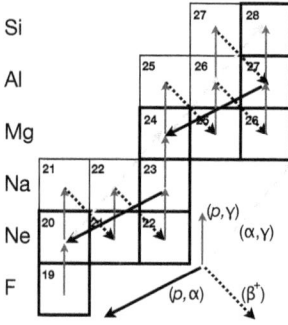

Figure 5.7: Proton capture reactions to come from F through Ne, Na, and Mg to Al and Si (Sect. 5.2.3) with repeated β decay. Arrows show the reactions. Element names are given at left, the atomic weight is given inside each box. Unstable isotopes are shown in thin-line boxes. Also two α-capture reactions are possible (see Eq. 5.15).
Note the possible decay roads of ^{23}Na and ^{27}Al each back to the starting nucleus but having produced He from $4p$. The possible reactions and their rates can be found in the NACRE compilation (Angulo et al. 1999). A similar figure can be found in Decressin et al. (2007).

to Mg. Such fusion involves more complicated networks of reactions. Shown in Fig. 5.7 is a build up like ^{20}Ne$(p,\gamma)^{21}$Na$(\beta^+)^{21}$Ne$(p,\gamma)^{22}$Na$(\beta^+)^{22}$Ne$(p,\alpha)^{23}$Na, which effectively is ^{20}Ne+$3p$ $\rightarrow 3\gamma + 2\beta^+ + {}^{23}$Na, with two possible exits: ^{23}Na$(p,\gamma)^{24}$Mg, with a stable Mg nucleus as result, or ^{23}Na$(p,\alpha)^{20}$Na, with the Na of the starting point of the chain but having produced an He nucleus (so effectively H-burning). These reactions may take place in, e.g., AGB stars (Ch. 10.4.3) where H is mixed convectively into deeper layers with products from core fusion reactions (Ch. 10.4).

In fact, any of the nuclei produced in the Eqs. 5.15 to 5.18 can be part of some further fusion reaction, depending on availability and temperature dependent fusion rate. To obtain fusion products havier than $^{56}_{26}$Fe reactions are endothermic as the binding energy per nucleon has a maximum for $^{56}_{26}$Fe (see Fig. 5.1).

Another (slow) road to build higher mass nuclei involves capture of neutrons.

5.2.4 General considerations (NSE); s-, r- and p-process

Heavy elements can also be made through **neutron capture**. This process works quietly, since there is no electric charge barrier to surmount. There are two types of neutron capture, the **slow** or **s-process** neutron capture and the **rapid** or **r-process** neutron capture. Furthermore, proton-rich nuclei exist, which were considered to be the product of **p-process** proton capture. The names r-, s-, and p-process were defined by Burbidge et al. (1957) in their seminal publication.

As Meyer (1994) notes in his review of all nuclear processes, *it is all a matter of nuclear statistical equilibrium* (NSE). Depending on the physical conditions one or another process will take place. In fact, if in a star the physical conditions change, then the structure will adjust itself such that such processes become active which will lead to a lower overall energetic state. Thus a star will, depending on its overall conditions, continue to extract energy from fusion (energy which is radiated away) to reach that lower energy state[2].

The so-called "s-process elements" form in an environment where free neutrons are present (see Scet. 5.2.4.1). Formed in a slow process, they are the stablest atoms known. The "r-process elements" form as a "freeze-out" from a high neutron density equilibrium state (however briefly that state may have existed). It occurs in supernovae or perhaps in disrupting neutron stars.

5.2.4.1 s-process

For neutron capture free neutrons are required. They can be released in several reactions such as the branches of burning of carbon (Eq. 5.16), oxygen (Eq. 5.17) and nitrogen (Eq. 5.18). Other important examples operating in the more massive stars (made possible by convective mixing) are

$$\begin{aligned} {}^{12}_{6}\text{C} + \text{H} &\longrightarrow {}^{13}_{7}\text{N} + \gamma \\ {}^{13}_{7}\text{N} + e^+ &\longrightarrow {}^{13}_{6}\text{C} + \overline{\nu_e} \\ {}^{13}_{6}\text{C} + {}^{4}_{2}\text{He} &\longrightarrow {}^{16}_{8}\text{O} + \text{n} \end{aligned} \quad (5.19)$$

---

[2]NSE applies also to the behaviour of matter right after the big bang. In the rapidly expanding and cooling universe elements can form according to the actual availability of building blocks and the reachable energetic state.

5.2. NUCLEOSYNTHESIS

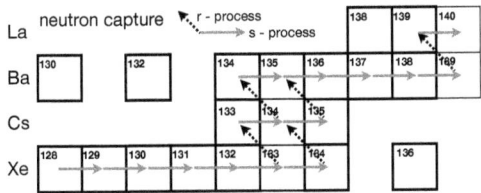

Figure 5.8: Sketch of a road for the build-up of heavy elements through slow neutron capture (s-process). Element names are given at left, the atomic weights inside the boxes, unstable isotopes in thin-lined boxes. These heavy nuclei have also roads for rapid neutron capture (r-process; e.g. Eq. 5.22).

and

$$\begin{aligned}
{}^{14}_{7}\text{N} + {}^{4}_{2}\text{He} &\longrightarrow {}^{18}_{9}\text{F} + \gamma \\
{}^{18}_{9}\text{F} + e^{+} &\longrightarrow {}^{18}_{8}\text{O} + \overline{\nu_e} \\
{}^{18}_{8}\text{O} + {}^{4}_{2}\text{He} &\longrightarrow {}^{22}_{10}\text{Ne} + \gamma \\
{}^{22}_{10}\text{Ne} + {}^{4}_{2}\text{He} &\longrightarrow {}^{25}_{12}\text{Mg} + \mathbf{n}
\end{aligned} \quad (5.20)$$

Then the s-process neutron capture becomes possible such as

$$\begin{aligned}
{}^{28}\text{Si} + n &\longrightarrow {}^{29}\text{Si} + \gamma \\
{}^{39}\text{Ca} + n &\longrightarrow {}^{40}\text{Ca} + \gamma \\
{}^{134}\text{Ba} + n &\longrightarrow {}^{135}\text{Ba} + \gamma
\end{aligned} \quad (5.21)$$

followed occasionally by a transformation to a higher element through an r-process like

$$\gamma + {}^{134}\text{Cs} + n \longrightarrow {}^{134}\text{Ba} + e^{-} \quad . \quad (5.22)$$

A sequence of such steps at the high mass end of the elements is displayed in Fig. 5.8. More can be found in Sneden & Cowan (2003) and Lattanzio & Wood (2004). The requirement is, that the elements formed are stable against β decay. Such stable elements build (in a graphic representation of all elements) the so-called "valley of β-stability".

The s-process takes place in all stars in which free neutrons are created, and leads to a slow build-up of heavy elements with their isotopic varieties. It thus is important in high-mass stars and in the AGB-phase of lower mass stars. In particular in AGB stars the evolution is slow enough for a "slow" process to come to a signifant building up of heavy elements (see Ch. 10.4.3).

5.2.4.2 r-process

Rapid neutron capture takes place when lots of free neutrons are available. One place is in well evolved stars, such as in the deep interior of those on the AGB (see Fig. 5.8).

The r-process occurs in supernovaexplosions, too, given the large numbers of photons and neutrons available there. Also endothermic reactions become possible in the first seconds after the core collapse of a massive star (see also Ch. 18). In those phases elements heavier than ${}^{56}_{26}\text{Fe}$ can be formed, all the elements heavier than Fe known on Earth.

Calculations concerning nucleosynthesis in supernovae are described in the reviews by, e.g., Arnett (1973) and by Hillebrandt et al. (1987).

5.2.4.3 p-process

Finally, the p-process is capture of protons by heavy elements. It leads to the so-called "p-process elements", or the proton-rich nuclei. However, proton capture is by and large unlikely for elements heavier than Fe. The p-process elements are rather the result of γ-decay, although various other mechanisms have been proposed. For details see Meyer (1994).

5.2.5 Nucleosynthesis and the Universe; Yields

According to the *big bang* model only $^1_1$H, $^2_1$H, $^4_2$He and a tiny fraction of $^7_3$Li were formed in primordial nucleosynthesis. General discussions of the processes of stellar nucleosynthesis can be found in the famous original paper by Burbidge et al. (1957, B²FH) and in, e.g., the review by Boesgaard & Steigman (1985).

Heavier elements (*all* those we observe and we ourselves are made of) have been and are produced in nuclear reactions in stellar interiors (see, e.g., the reviews by Trimble 1991 and by Sneden & Cowan 2003). The ultimate goal of theoretical nucleosynthesis is to calculate the yield (sometimes called Y) for the elements formed during stellar evolution and returned to interstellar space (see Ch. 22.3.2). An overview of nuclear fusion rates in massive stars is available from Rauscher et al. (2002), for massive metal poor stars from, e.g., Umeda & Nomoto (2002).

Decay of elements takes place, too. Nuclei heavier than Fe, such as those formed in supernova explosions, may break into smaller parts with a lower overall binding energy. This aspect has been used in nuclear chronometry, where the change of the ratio of the abundance of radio-nucleides having different half lives is used to determine of the age of stars (see Ch. 22.3.3).

5.2.6 The burning of Lithium

One of the few elements produced in the big bang is Li. It is destroyed in fusion (for the reaction see the PP II branch, Eq. 5.7) already at relatively low temperatures (e.g. in brown dwarfs, Ch. 8.2.2). In stars without convection, the surface abundance of Li should still be that *ab initio* so the oldest such stars in the universe should show the Big Bang Li abundance. Thus determining the abundance of lithium in very old stars has cosmological importance. For more on the Li abundance in old stars see the review by Spite & Spite (1985) and Ch. 15.5.

5.3 Neutrinos

During the fusion processes neutrinos are produced. These contribute in part to the cooling of the stellar interior. Only for the Sun such a neutrino flux can be observed (except the case of the neutrinos from a supernova explosion, like SN1987A). However, there was a problem with the neutrinos from the Sun and their properties. Therefore neutrinos are discussed in detail here.

5.3.1 Mean free path for neutrinos

The total cross section σ for a certain reaction is defined as the effective cross section, with which a particle interacts with other particles. This definition leads to the relation

$$n \sigma l = 1 \qquad (5.23)$$

where $n = \frac{\rho}{\mu m_p}$ is the number density of particles and l the mean free path. The cross section for neutrinos interacting with stellar material is approximately given by

$$\sigma_\nu = \left(\frac{E_\nu}{m_e c^2}\right)^2 \cdot 10^{-44} \quad \text{cm}^2 \qquad (5.24)$$

where E_ν is the neutrino energy. By solving Eq. 5.23 for l and inserting Eq. 5.24 we can calculate the mean free path for neutrinos in dependence of the matter density ρ

$$l_\nu = \frac{1}{n \sigma_\nu} = \frac{\mu m_p}{\rho \sigma_\nu} \approx \frac{1 \cdot 10^{20}}{\rho} \quad \text{cm} . \qquad (5.25)$$

where we have inserted in $E_\nu = 1$ MeV and $\mu = \mu_\odot = 0.59$ in order to get a rough estimate for l_ν.

For the central regions of **main-sequence stars** it is known from models that the density is in the order of 1 to 10^3 g cm^{-3} (see Fig. 6.5). Inserting $\rho = 10^3$ g cm^{-3} into Eq. 5.25 leads to

5.3. NEUTRINOS

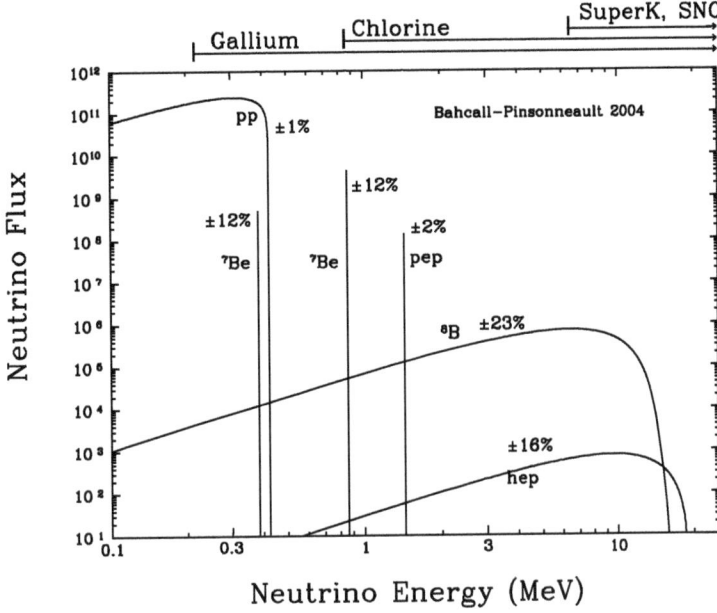

Figure 5.9: The neutrino spectrum of the Sun as predicted from the solar model. The various neutrino contributions are indicated. For all reactions see Sect. 5.3.2. Percentages give the uncertainty in flux from the model. Figure from Bahcall & Pinsonneault (2004).
pp marks the energy distribution curve of neutrinos from the first step in the PP chain burning, ^{8}B the curve for the PP III chain neutrinos. **hep** indicates neutrinos from the very rare reaction ^{3}He + p → ^{4}He + e$^+$ + ν. For these continuous spectra the scale is in neutrinos cm^{-2} s^{-1} MeV^{-1}.
^{7}Be marks wo spikes from the PP II chain, **pep** indicates neutrinos from the rare process of Eq. 5.27. For these "emission lines" the scale is in neutrinos cm^{-2} s^{-1}.
The detection thresholds for the gallium, the chlorine and the heavy-water (SuperKamiokande and Sudbury Neutrino Observatory) experiments are given along the top (see Sects. 5.3.3 and 5.3.6).

$l_\nu \approx 10^{17}$ cm $= 1.4 \cdot 10^6$ R$_\odot$. Therefore, neutrinos from fusion processes in the core region of stars should escape immediately. So solar neutrinos escape and allow to give insight in the processes taking place in the Sun's interior.

A fully different situation is that of a star near **stellar core collapse** leading to a supernova explosion and the formation of a neutron star in which $\rho \approx 10^{14}$ g cm^{-3}. Here Eq. 5.25 gives a mean free path of $l_\nu \approx 10$ km which is in the order of the radius of neutron stars (Ch. 17). To describe this collapse we would need a neutrino transport equation as interactions are now very important. Detailed models actually show that the inner part of the collapsing core becomes so dense, that the time the neutrinos would need to escape in a random walk exceeds the time of the collapse, as given by the dynamical time scale (see Eq. 4.42). The initial density in the beginning of the collapse is in the order of 10^{10} g cm^{-3}. The neutrinos inside this region are trapped, while an effective neutrino photosphere propagates outwards.

Currently it is believed that neutrinos play indeed a very important role in the mechanism of supernova explosions. The density of the collapsing stellar core exceeds about $8 \cdot 10^{14}$ g cm^{-3}, which is roughly three times the density of an atomic nucleus. According to the Pauli principle the neutrons suddenly experience a pressure due to degeneracy and the core rebounds creating a shock wave that moves outwards. Nevertheless this shock wave rapidly looses energy by photodisintegration mainly of $^{56}_{26}$Fe. Here the neutrino flux is believed to come into play depositing additional energy behind the shock which then is reinvigorated and blows the outer envelope away. The total energy liberated in the collapse by the neutrinos is approximately $3 \cdot 10^{53}$ ergs and therefore much larger than the total energy radiated away by photons ($\approx 10^{49}$ ergs).

5.3.2 Solar neutrinos

Neutrinos are produced in various nuclear reactions inside the Sun. They have been discussed in Sect. 5.1 (for the PP chain see e.g. Fig. 5.4) and can be summarized as

$$
\begin{aligned}
{}^1_1\text{H} + {}^1_1\text{H} &\longrightarrow {}^2_1\text{H} + e^+ + \nu_e & (\text{PP I, PP II, PP III}) & \quad \overline{E_\nu} = 0.263\,\text{MeV} \\
{}^7_4\text{Be} + e^- &\longrightarrow {}^7_3\text{Li} + \nu_e(+\gamma) & (\text{PP II}) & \quad E_\nu = 0.86\,\text{MeV}/0.38\,\text{MeV} \\
{}^7_4\text{Be} &\longrightarrow {}^8_4\text{Be} + e^+ + \nu_e + \gamma & (\text{PP III}) & \quad \overline{E_\nu} = 7.2\,\text{MeV} \\
{}^{13}_7\text{N} &\longrightarrow {}^{13}_6\text{C} + e^+ + \nu_e & (\text{CNO}) & \quad \overline{E_\nu} = 0.71\,\text{MeV} \\
{}^{15}_8\text{O} &\longrightarrow {}^{15}_7\text{N} + e^+ + \nu_e & (\text{CNO}) & \quad \overline{E_\nu} = 1.0\,\text{MeV}
\end{aligned} \quad (5.26)
$$

Additionally also some neutrinos are produced in the very rare reaction

$$
{}^1_1\text{H} + {}^1_1\text{H} + e^- \longrightarrow {}^2_1\text{H} + \nu_e \tag{5.27}
$$

In the last column of Eq. 5.26 the (average) neutrino energy is given. In reactions where in addition a positron is emitted, the neutrinos have a continuous energy spectrum as the total available kinetic energy is distributed over the two particles. In the other reactions the neutrino carries the total kinetic energy and therefore the decay is monoenergetic. Since ${}^7_4\text{Be}$ can either decay into the ground state or into just one of the excited states of ${}^7_3\text{Li}$, this decay has actually two spectral lines (Fig. 5.9). The total energy spectrum of the solar neutrinos predicted by the so-called *standard solar model* (the newer version) is plotted in Fig. 5.9. By comparing the predicted amount of solar neutrinos with measured numbers one can test the standard solar model (or the ideas we have about neutrinos).

5.3.3 Neutrino experiments

The first experiment measuring solar neutrinos was started by R. Davis in 1964. The experiment used a chlorine container in a goldmine in Homestake, South Dakota, to detect neutrinos in the reactions

$$
{}^{37}_{17}\text{Cl} + \nu_e \longrightarrow {}^{37}_{18}\text{Ar} + e^- \quad \text{and} \quad {}^{37}_{18}\text{Ar} \longrightarrow {}^{37}_{17}\text{Cl} + e^+ + \nu_e \quad . \tag{5.28}
$$

Here the Auger electron emitted when ${}^{37}_{18}\text{Ar}$ decays back to ${}^{37}_{17}\text{Cl}$ is counted ($t_{1/2} = 35$ d). This experiment is sensitive for neutrino energies above 0.814 MeV (compare Fig. 5.9). Therefore it mainly detects neutrinos from the decay of ${}^7_4\text{Be}$.

In the Gallium Experiment (GALLEX) and its successor Gallium Neutrino Observatory (GNO) the similar reactions

$$
{}^{71}_{31}\text{Ga} + \nu_e \longrightarrow {}^{71}_{32}\text{Ge} + e^- \quad \text{and} \quad {}^{71}_{32}\text{Ge} \longrightarrow {}^{71}_{31}\text{Ga} + e^+ + \nu_e \tag{5.29}
$$

are used, but with gallium instead of chlorine. These reactions have the advantage of a lower threshold energy (0.23 MeV). Therefore also a large fraction of the neutrinos from the initial reaction in the PP chain ${}^1_1\text{H} + {}^1_1\text{H} \longrightarrow {}^2_1\text{H}$ can be detected. For the experiment, which is located in the Gran Sasso tunnel (Italy) 30 tons of gallium are used. ${}^{71}_{32}\text{Ge}$ decays back with a half–life $t_{1/2} = 11.4$ d.

A third type of experiment uses huge containers filled with water. Here Cherenkov radiation from neutrinos scattering with electrons of the water is detected with photomultipliers. This type of detector has the advantage that the direction of the in-falling neutrino can be determined. However, it has the disadvantage of a high threshold energy ($E_\nu > 8$ MeV). One such experiment is the "Irvine-Michigan-Brokhaven" (IMB) detector. The **KamiokaNDE** (the Nucleon Decay Experiment in the Kamioka mine in Japan) was specifically built to investigate the solar neutrinos. The biggest detector of this type is the Super–Kamiokande which uses 50,000 tons of water as target. Neutrinos from SN 1987A were discovered both by IMB and Kamiokande.

5.3.4 The "solar neutrino problem"

The rate of neutrino captures is usually measured in *solar neutrino units* (SNU). One SNU corresponds to 10^{-36} captures per second and per target nucleus. From the measurements of the Homestake experiment between 1970 and 1981 a production rate of $1.3^{+0.7}_{-0.8}$ SNU was determined with ^{37}Ar, whereas the solar standard model predicts 7.6 ± 3.3 SNU. This discrepancy became known as the *solar neutrino problem*. Either the solar standard model or the standard model of particle physics has to be wrong somehow. Many physicists prefered the first explanation especially because, although the Homestake experiment mainly measured neutrinos from the decay of $^8_5$B in the PP III chain which only contributes to the energy production in 0.3 % of the reactions, its production depends on the temperature very strongly. If for example the helium content in the solar core were smaller than predicted by the standard solar model, the central temperature would be lower and therefore much fewer $^8_5$B neutrinos would be released, while the total luminosity of the Sun would still be the same. Already a decrease of the central temperature from $1.56 \cdot 10^7$ K to $1.47 \cdot 10^7$ K would be sufficient to explain this discrepancy. For an early review on the problem see Haxton (1995).

Whether a chnage in the helium content would solve the neutrino problem or not was tested with the new gallium experiments, which are also sensitive to neutrinos produced in the first fusion step of the PP chain. The combined results of GALLEX and GNO gave a mean value of $74.1^{+6.7}_{-6.8}$ SNU, whereas the value predicted from the standard Solar model is in the range of 120 to 140 SNU. Therefore the neutrino problem remained unsolved.

5.3.5 Neutrino oscillations

In the standard model of particle physics three generations of neutrinos exist: the electron neutrino ν_e, the muon neutrino ν_μ and the tau neutrino ν_τ, each belonging to one of the charged leptons e, μ and τ. In β^+ decays (and therefore in the fusion reactions in stars) only electron neutrinos ν_e are produced. If neutrinos are massless (as assumed in the standard model) they are stable and always stay what they are.

Theory predicts that in case of a non-vanishing mass, oscillations can occur that transform a neutrino of one type into a neutrino of another type and back. This transformation depends strongly on the difference of the square of the mass of the two neutrinos

$$\Delta m^2 = m_1^2 - m_2^2 \quad . \tag{5.30}$$

The oscillations happen on a length scale of

$$L = \frac{2.5\, E_\nu}{\Delta m^2} \quad \text{m} \tag{5.31}$$

where the neutrino energy E_ν has to be inserted in MeV and Δm^2 in $\frac{\text{eV}^2}{c^4}$. For example $\Delta m^2 = 10^{-11}\,\frac{\text{eV}^2}{c^4}$ gives $L \approx 1$ AU for $E_\nu = 1$ MeV.

Oscillations could solve the neutrino problem if indeed a large fraction of the produced electron neutrinos ν_e convert into muon neutrinos ν_μ somewhere between the Sun and Earth. These converted neutrinos cannot be measured with the detectors used.

Until 1998 the concept of neutrino oscillations had been only theoretical but then the necessity of having neutrino oscillations became evident from the measurements by the Super–Kamiokande. Since its detector has a directional sensitivity, the angular distribution of atmospheric muon neutrinos could be examined. These neutrinos are produced in the decay of pions, which have been created in interaction of cosmic ray particles with molecules in the upper atmosphere. In Super–Kamiokande more neutrinos are observed from the direction above the experiment than from below the experiment. This result is interpreted such that a fraction of the neutrinos travelling through Earth have transformed into tau neutrinos, the ones that cannot be detected in the experiment. In Fig. 5.10 the angular distribution of the detected electron and muon neutrinos is plotted demonstrating the size of the effect.

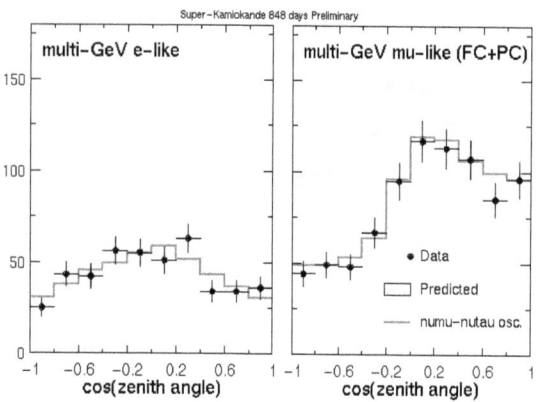

Figure 5.10: Angular distribution of intensity of *cosmic ray generated* electron neutrinos (left) and muon neutrinos (right) detected with the Super–Kamiokande ($\cos\theta=-1$: neutrinos from "below" having passed through the Earth; $\cos\theta=+1$: neutrinos directly from the CR event "above"). For the muon neutrinos the detected amounts differ significantly from predicted (shaded blocks in right panel). The histograms show fits to a model with oscillation of ν_μ to ν_τ. For the electron neutrinos no significant difference between measurements and expected values is observed. Figure from Fukuda et al. (1998).

The result that apparently oscillations from ν_μ to ν_τ exist, made the proposition that oscillations from ν_e to ν_μ might be the answer to the solar neutrino problem much more convincing. The neutrino oscillation concept together with the interior temperature structure of the Sun from standard Solar models and its confirmation from helioseismology (see Ch. 11.5) has resolved the dispute about solar neutrino production and the detection of these. Confirmation from neutrino detections came from the SNO.

5.3.6 The Sudbury Neutrino Observatory and solution of the problem

The *Sudbury Neutrino Observatory* (SNO) is located 2070 m below the surface in Inco's Creighton Mine near Sudbury, Ontario. Measurements of the flux of solar neutrinos (from ^{8}B) started in 1999. SNO is the first neutrino observatory which detects Cherenkov radiation from neutrino reactions in **heavy water**, D_2O. SNO uses 10^6 kg of that. For further information on SNO refer to http://www.sno.phy.queensu.ca.

5.3.6.1 Relevant neutrino reactions

The following interactions of neutrinos with deuterons and electrons occur in heavy water.

In **charged current reactions** an electron neutrino interacts with the neutron inside a deuteron transforming it into a proton:

$$\nu_e + D \longrightarrow p + p + e^- \quad . \tag{5.32}$$

The Cherenkov radiation of the emitted electron can be detected. This reaction is only possible for electron neutrinos.

In **neutral current reactions** the deuteron is broken up into a proton and a neutron:

$$\nu_x + D \longrightarrow \nu_x + p + n \quad . \tag{5.33}$$

The reaction can be observed because gamma rays are emitted when the neutron is captured by another nucleus. Since the capture by deuterons has only a low efficiency, 2000 kg of salt (NaCl) have been added in the second phase of SNO, providing a much higher efficiency through capture of neutrons on ^{35}Cl. This reaction is equally sensitive to all three neutrino types.

In **electron scattering** a neutrino scatters with an electron of the water:

$$\nu_x + e^- \longrightarrow \nu_x + e^- + \text{photon} \quad . \tag{5.34}$$

This reaction is the primary mechanism in normal light water detectors. It is sensitive to all types of neutrinos. However, the cross section for electron neutrinos is by a factor of six larger than for the other neutrino types.

5.3.6.2 Advantages of heavy water

Earlier experiments with normal water detectors mainly measured electron scattering (Eq. 5.34). They were therefore unable to distinguish between the different neutrino types. Additionally, if a significant number of ν_μ or ν_τ were coming from the Sun, they would have been almost unnoticed, since the cross section for these neutrinos is much lower.

SNO on the other hand is able to measure both the total flux of all types of neutrinos via the neutral current reaction (Eq. 5.33) in heavy water and the flux of electron neutrinos via the charged current reaction (Eq. 5.32).

5.3.6.3 The solution of the solar neutrino problem

After neutrino oscillations were suspected to occur for atmospheric neutrinos, many people felt this could also be the solution of the solar neutrino problem.

Nevertheless the excitement especially among particle physicists was large when the first SNO results were published on 18 June 2001. They clearly showed that only about $\frac{1}{3}$ of the solar ^{8}B neutrinos arriving at Earth are detected as ν_e and that the measured total flux of ^{8}B neutrinos is consistent with the solar standard model. Thus the particle physicists had to acknowledge that the astrophysicists had been right all along and that helioseismology (Ch. 11.5) is a viable diagnostic tool also for these problems.

Therefore it is clear that the ν_e emitted from the Sun transform partially into ν_μ and ν_τ on their way to Earth. The "solar neutrino problem" has been an "elementary particle theory problem".

5.4 Nobel prize 2002 for neutrino research

The Nobel prize 2002 was shared between Neutrino research and X-Ray astronomy.

Raymond **Davis** was honoured for his initiatives in neutrino research. He and his team detected neutrinos near nuclear reactors. He then went on to look for Solar neutrinos (Homestake), found them, and showed there were too few.

Masatoshi **Koshiba** initiated Kamiokande, detected Solar neutrinos, and was the one who with his team found the first evidence for the mass and oscillations of neutrinos using neutrinos produced by cosmic rays.

Neutrinos from SN 1987A were detected both by the Irvine-Michigan-Brookhaven (IMB) detector and by Kamiokande.

Giacconi was honoured with part of the 2002 Nobel prize for his work on X-rays (see Ch. 17.5).

References

Angulo, C., Arnould, M., Rayet, M., et al. 1999, Nuclear Physics A, 656, 3
Arnett, W.D. 1973, ARAA 11, 73; *Explosive Nucleosynthesis in Stars*
Bahcall, J.N., & Pinsonneault, M. 2004, http://www.sns.ias.edu/~jhb
Boesgaard, A.M., & Steigman, G. 1985, ARAA 23, 319; *Big Bang Nucleosynthesis; Theories and Observations*
Burbidge, E.M., Burbidge, G.R., Fowler, W.A., & Hoyle, F. 1957, Rev. Mod. Phys. 29. No. 4 (B^{2}FH)
Decressin, T., Meynet, G., Charbonnel, C., Prantzos, N., & Ekström, S. 2007, A&A 464, 1029
Fukuda, Y., et al., Super-Kamiokande Collaboration. 1998, Phys. Lett. B346, 33 (hep-ex/9805006)
Gamow, G. 1928, Z. Physik 51, 204
Haxton, W.C. 1995, ARAA 33, 459; *The Solar Neutrino Problem*
Hillebrandt, W., Kuhfuß, R., Müller, E., & Truran, J.W. 1987, in "Nuclear Astrophysics", Lecture Notes in Physics 287, Springer, Heidelberg
Hoyle, F. 1954, ApJS 1, 121
Lattanzio, J.C., & Wood, P.R. 2004, in "Asymptotic Giant Branch Stars", H.J. Habing & H. Olofsson (eds.); Springer, Heidelberg; p.23

Meyer, B.S. 1994, ARAA 32, 153; *The r-, s-, and p-Processes in Nucleosynthesis*
Rauscher, T., Heger, A., Hoffman, R.D., & Woosley, S.E. 2002, ApJ 576, 323
Schwarzschild, M. 1958, "Structure and Evolution of the Stars", Princeton Univ. Press; Dover Edition
Sneden, C., & Cowan, J.J. 2003, Science 299, 70
Spite, M. & Spite, F. 1985, ARAA 23, 255; *The Composition of Field Halo Stars and the Chemical Evolution of the Halo*
Trimble, V. 1991, A&ARev 3, 1; *The Origin and Abundances of the Chemical Elements Revisited*
Umeda, H. & Nomoto, K. 2002, ApJ 385, 404

Chapter 6

Stellar structure: Making star models

6.1 The equations of state and their complications

The differential equations describing the overall structure of a star cannot be solved analytically. The reason is that the functions for the opacity, κ, and the energy generation, ϵ, as well as the equation describing the behaviour of the gases, are (see Ch. 4) rather complicated functions of temperature, T, and chemical composition, X, Y, and Z. In fact, one must write $\kappa = \kappa(T, X, Y, Z)$, $\epsilon = \epsilon(T, X, Y, Z)$ and $P \sim \rho^{\alpha} T^{\beta}$, where $\rho = \rho(T, X, Y, Z)$, and α and β as yet undefined.

The *opacity* κ is a function of T, because the temperature governs the ionization structure while the opacity for a given T depends on which atoms and ions contribute to the opacity. So here the dependence on chemical composition (X, Y, Z) enters. The opacity κ is well tabulated (the latest version being the "Los Alamos" tables; Iglesias & Rogers 1996). For a description of the most important processes contributing to opacity see Ch. 2.8 and for values Fig. 4.4.

The *energy production rate* ϵ is a function of T, because the temperature governs the efficiency of the tunneling in the nuclear fusion processes. In turn, the energy production rate in a gas at a given temperature depends on which atoms and ions participate in the nuclear fusion. So also here the dependence on chemical composition (X, Y, Z) enters. The function ϵ is rather well established (see Ch. 5) and tabulated based on knowledge from nuclear physics (but new possibilities for astrophysically relevant fusion processes are sometimes discovered).

The *equation of state of the gas* is a function of P, ρ, and T, which depends on the composition of the gas (mean molecular weight) as well. For an ideal gas, $P \sim \rho T$. However, the relation can be quite different in gas of more extreme conditions (see Ch. 4.4.2). In gases of high temperature, e.g., the contribution to the pressure by the electrons has to be included, in other words *degeneration*, has to be taken into account. The consequences of such varying conditions are presented below.

For the gas equation one uses special functions, the so called *polytropes*. These are simplified equations characterised by the *polytrope index*.

6.2 Polytropes; Consequences of differing equations of state

The relation for the equation of state of the gas is normally

$$\rho = \rho(P, T) \ . \tag{6.1}$$

In pure adiabatic, hot $(T \gg 10^4$ K) hydrogen gas one simply would have $\rho = \rho(P)$. Under more extreme conditions, various further aspects have to be considered. These lead to a state of the gas more complicated than that for an ideal gas, the degeneracy (see Ch. 4.4.2.3) with equations of state having different functional forms. A good solution for the numerical problems in calculating stellar models is the use of polytropes (see, e.g., Ch. 19 in Kippenhahn & Weigert 1990).

6.2.1 The general polytropic equation

One defines the polytropic equation, a power law of the form

$$P = K \rho^\gamma \tag{6.2}$$

in which K = the polytropic constant, γ = the polytropic exponent; one also uses n = the polytropic index, with $\gamma = 1 + \frac{1}{n}$, or $n = 1/(\gamma - 1)$. With such a relation one can simplify the equations for stellar structure substantially, thereby shifting the complexity of the mathematical problem to the choice of values of the constant and the exponent (or index). In fact, if the gas behaves in one radial shell different from that in the next, one can accommodate the change by just changing the index and exponent, without requiring an otherwise new mathematical formulation.

For stars in hydrodynamic equilibrium (with Φ the gravitational potential) one has

$$\frac{dP}{dr} = \frac{d\Phi}{dr} \rho \tag{6.3}$$

and the spherically symmetric solution to the Poisson equation

$$\frac{d\Phi}{dr} = \frac{G\,M(r)}{r^2} \quad . \tag{6.4}$$

Assuming that a simple relation like a polytropic equation exists, the equations above give

$$\frac{d\Phi}{dr} = -\gamma\,K\,\rho^{\gamma-2}\,\frac{d\rho}{dr} \quad . \tag{6.5}$$

This equation can be integrated for $\gamma \neq 1$ ($\Phi = 0$ for $\rho = 0$) and one finds

$$\rho = \left(\frac{-\Phi}{(n+1)\,K}\right)^n \quad . \tag{6.6}$$

Choosing γ (or n) defines the exact dependence of the relation and the P, T-stratification can be described by the run of γ or n. (During integration of the stellar structure equations it has to be checked at every location that the choice of γ is still valid.)

6.2.2 Special polytropes

6.2.2.1 Polytrope for ideal gas

For ideal gases one has the well known relation $\rho = \mu P/\Re T$ or $P = \Re T \rho/\mu$ (with $\Re = k/m_p =$ 8.314 J mol^{-1} K^{-1}, the gas constant). Here, clearly, the polytropic exponent $\gamma = 1$, the index $n = \infty$ and $K = \Re T/\mu$. Actually, K rather is a free parameter because the conditions in the gas as given by T and μ are not defined beforehand.

Special case: an isothermal sphere. With an ideal gas one has

$$\frac{d\Phi}{dr} = -K\,\rho^{-1}\,\frac{d\rho}{dr} \tag{6.7}$$

which can be solved to

$$\Phi/K = \ln \rho - \ln \rho_c \quad \text{(with} \quad \Phi = 0 \text{ at } r = 0) \tag{6.8}$$

so that $\rho = \rho_c\,e^{-\Phi/K}$, which describes an infinitely large star with infinite mass. This structure has similarities with that of protostars which, initially, probably are isothermal. It also is similar to the central portion in evolved stars, which (if without central fusion) have isothermal cores.

Note: also interstellar clouds can be modelled with the help of polytropes. In that case the polytropic index is even negative: the temperature rising outward.

6.2. POLYTROPES; CONSEQUENCES OF DIFFERING EQUATIONS OF STATE

6.2.2.2 Completely convective stars

In a completely convective star one has the relation between temperature and pressure

$$(d\ln T/d\ln P)_{\text{ad}} = \nabla_{\text{ad}} = 2/5 \qquad (6.9)$$

so that $T \sim P^{2/5}$. Thus, if the gas is fully ionized and has negligible radiation pressure, then $P = K\rho^{5/3}$ with $K = \Re/\mu$. Thus $\gamma = 5/3$ and $n = 3/2$ allowing for a polytropic formulation. Note that K varies from gas to gas or star to star due to differences in the value of μ.

Examples of (almost) completely convective objects are pre main-sequence stars (Ch. 7.5), brown dwarfs (Ch. 8), stars of the mass extremes of the main sequence (Fig. 6.6), and stars approaching the end of their red giant phase (Chs. 10.2 & 13.4).

6.2.2.3 Non-relativistic degenerate electron gas

In a non-relativistic degenerate electron gas one has $P = K\rho^{5/3}$ so $\gamma = 5/3$, $n = 3/2$, and the constant $K = K_1$ (see Ch. 4.4.2.3) is

$$K_1 = \frac{1}{20}\left(\frac{3}{\pi}\right)^{2/3}\frac{h^2}{\mu_e}\frac{1}{m_p^{5/3}} \quad . \qquad (6.10)$$

This leads to the solution for the radius R of the object as function of central density ρ_c

$$R \sim \rho_c^{(1-n)/(2-n)} \quad . \qquad (6.11)$$

For $1 < n < 2$, R will be smaller for larger ρ_c! The relation for the mass $M \sim \rho_c R^3$ thus becomes

$$R \sim M^{(1-n)/(3-n)} \qquad (6.12)$$

which indicates that for $1 < n < 3$ one has $R \sim M^x$ with x being *negative*. In other words, for larger M the object has smaller R.

These conditions exist in white dwarf stars (Ch. 17), being the final stage of evolution for stars which started on the MS with up to a few $M_\odot$.

If the objects have gas with conditions where n increases, then the gas approaches degeneration, and $n \rightarrow 3$. Clearly, in stars the transition from $n = 3/2$ to higher values is smooth in the gas.

6.2.2.4 Relativistic completely degenerate electron gas

Choosing $n = 3$ so that $\gamma = 4/3$, one has the case for relativistic completely degenerate gas (see Ch. 4.4.2.3). Here one allows K to be a free parameter.

In the case of gas with $n = 3$ in the stellar core but with ideal gas in the shell, one arrives at the so called **Chandrasekhar limit**, which is the mass at which the gas of the entire star reaches the relativistically degenerate case. This leads to a natural limit with just physical constants

$$M_{\text{Ch}} \simeq \frac{5.8}{\mu_e^2} M_\odot \qquad (6.13)$$

which, for gas without hydrogen consisting of Helium (or C and N) for which $\mu_e = 2$, results in

$$M_{\text{Ch}} \simeq 1.4 \, M_\odot \quad . \qquad (6.14)$$

This is the maximum mass a white dwarf can have. Were a WD to become more massive than this limit, 'other things will happen': a phase transition will take place (see Ch. 17.1.4).

The case of relativistically degenerate gas but with $P \sim \rho \Re/\mu T$ (formally ideal gas) together with a constant temperature gradient $\nabla = 1/4$ ($T \sim P^{1/4}$) holds for the interior of the Sun. With these parameters, Eddington calculated in 1925 the first realistic models for the Sun.

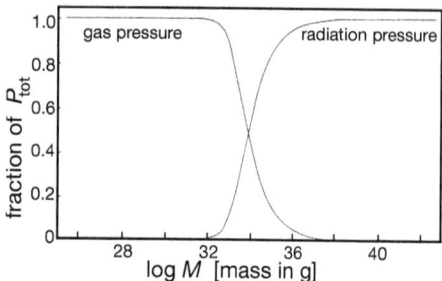

Figure 6.1: Balance between gas and radiation pressure in gas spheres of various masses (after Eddington 1926). Note that the minimum mass of an object called star is $0.08\,M_\odot$ (see Sect. 6.5), the maximum mass is $\simeq 100\,M_\odot$ ($M_\odot = 2 \cdot 10^{33}$ g). Apparently stars exist only in the middle regime where the balancing act makes things interesting!

6.3 Balance between internal pressure and gravitation

Stars are gaseous spheres, structures held stable by the delicate balance between inner pressure, tending to drive the gas outward, and self-gravitation, holding the gas together. (For discussion of what happens when there are deviations from that balance see Ch. 9.)

The **pressure** consists of three contributions, the normal gas pressure (thermal pressure), the pressure due to degeneration (electron gas) or Fermi pressure, and the radiation pressure, so

$$P = P_{\text{gas}} + P_{\text{Fermi}} + P_{\text{rad}} \tag{6.15}$$

in which the individual pressure terms are

$$P_{\text{gas}} = \frac{N_{\text{tot}}\, k\, T}{V} \tag{6.16}$$

$$P_{\text{Fermi}} = 2\,\frac{\hbar^2}{m_e}\left(\frac{N_e}{V}\right)^{5/3} \tag{6.17}$$

$$P_{\text{rad}} = \frac{4\pi}{3}\frac{I}{c} = \frac{a\,T^4}{3}\quad . \tag{4.66}$$

Eqs. 6.16 and 6.17 give the pressures as a mean for the entire star. Only in very massive stars (with large T_c) will the radiation pressure give an appreciable contribution to P. Therefore, simplifying the equations above, and noting that the total number of particles N_{tot} is roughly 2 times the number of electrons, the pressure due to atoms (P_{gas}) and electrons (P_{Fermi}) is

$$P = P_{\text{gas}} + P_{\text{Fermi}} = \frac{2k\,\rho\,T}{m_p} + 2\,\frac{\hbar^2}{\mu_e}\left(\frac{\rho}{m_p}\right)^{5/3}\quad . \tag{6.18}$$

The **gravitation** balances the internal gas pressure. Based on the virial theorem one thus has

$$-\int P\,dV = \frac{1}{3}\,U_G \tag{6.19}$$

and, simplifying by assuming that the actual gas density can be replaced by the average gas density so that $U_G = \frac{3}{5}\,G\,M^2/R$, one has the final form of the balance equation

$$-\int P\,dV = \frac{1}{3}\frac{3}{5}\frac{G\,M^2}{R}\quad . \tag{6.20}$$

6.4 The maximum mass of a normal star

A fundamental question in astrophysics is: what is the maximum mass a normal star can have? With a relatively simple argument one can derive an approximate limit for this mass (Eddington 1920s; see also, e.g., Celnikier 1989, Ch 5.1).

6.5. THE MINIMUM MASS OF A STAR

For that, consider stars of the main sequence, going from lower to higher mass. Stars are stable due to the balance of internal pressure and selfgravitation. As long as the pressure is set by the gas only, there is no problem. However, inside more and more massive stars the temperature may be very high, so that the contribution to the total pressure by the radiation pressure (when photons interact with matter, momentum is transferred, so pressure) can no longer be neglected.

At some mass (and thus central temperature) the radiation pressure becomes comparable in size to the gas pressure (Eqs. 6.16 & 4.66 above: $P_{gas} \sim T$ and $P_{rad} \sim T^4$). The radiation pressure threatens to dominate gravity and so would blow the star apart. Assume as limit for stability the case when gas pressure and radiation pressure are similar (upper-left boundary line in Fig. 4.5), so

$$P_{gas} \simeq P_{rad} \quad \text{and, with substitutions,} \quad \frac{2k\rho T}{m_p} \simeq \frac{a}{3} T^4 = \frac{1}{3} \frac{8\pi^5}{15} \frac{(kT)^4}{(hc)^3} \quad (6.21)$$

a relation which can be solved for kT giving

$$(kT)^3 = 2 \frac{\rho}{m_p} \frac{3 \cdot 15}{8\pi^5} (hc)^3 \quad . \quad (6.22)$$

The balancing act by gravitation on gas pressure was $1/3 \, U_G = -\int P \, dV$ (Eq. 6.19) so that

$$\frac{1}{3} \frac{3GM^2}{5R} = \frac{2k\rho T}{m_p} \frac{4\pi R^3}{3} \quad (6.23)$$

which is equivalent to

$$M^2 = \frac{2 \cdot 5}{G} R \, kT \, \frac{\rho}{m_p} \frac{4\pi}{3} R^3 = \frac{2 \cdot 5}{G} \frac{kT}{m_p} \left(\frac{3}{\rho 4\pi}\right)^{1/3} \left(\rho \frac{4\pi}{3} R^3\right)^{4/3} \quad . \quad (6.24)$$

With $(\rho(4\pi/3)R^3)^{4/3} = M^{4/3}$ and inserting kT from Eq. 6.22 (for $P_{gas} \simeq P_{rad}$) gives

$$M_{max} \simeq \left(\frac{2^2 3^5 5^4}{\pi^3}\right)^{1/2} \left(\frac{\hbar c}{G}\right)^{3/2} \frac{1}{m_p^2} \quad , \quad (6.25)$$

a relation based on natural constants only. After insertion of all values one obtains as mass limit

$$M_{max} \simeq 1.8 \cdot 10^{32} \text{ kg} \quad = \quad 90 \, M_\odot \quad . \quad (6.26)$$

Reality is more complex and the above is only a crude and simple analysis. Very massive stars have only in the core region a radiation pressure surpassing the gravitational balance leading to an expanded core but the hughe mass of the outer layers contains the star. Also pulsational instabilities may occur (see e.g., Appenzeller 1970, Baraffe et al. 2001). For more on massive stars with M up to 120 $M_\odot$ see Ch. 13, for stars up to 2000 $M_\odot$ see Ch.15.

The behaviour of gas and radiation pressure can be given in a graph (see Fig. 6.1), demonstrating the balancing act of stars. The derivation was first made by Eddington in 1926.

6.5 The minimum mass of a star

The lower limit to the mass of objects called star is the mass below which gaseous objects do not have sustainable nuclear fusion (hydrogen fusion).

The pressure in small and cool gas spheres is composed of the contributions by the degeneration pressure (see Ch. 4.4.2.3) and the thermal pressure, the latter having contributions from the ion and the electron pressure

$$P = P_{Fermi} + P_{iongas} + P_{el.gas} \quad . \quad (6.27)$$

The total pressure can be expressed as

$$P = 10^{13} \left(\rho/\mu_e\right)^{5/3} \left(1 + \psi + \frac{\psi^2}{1+\psi}\right) \quad (6.28)$$

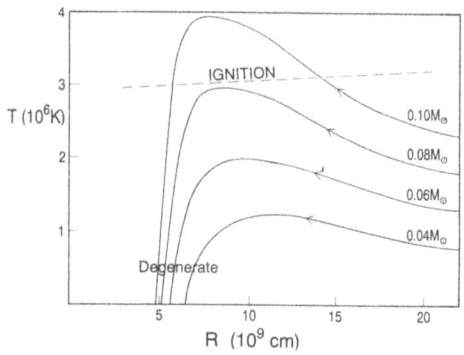

Figure 6.2: Central temperature as function of radius for $n = 3/2$ polytropes for a variety of masses. The line above which hydrogen ignites is indicated. The minimum mass of a gas sphere becoming star is on the curve just crossing the ignition line. Figure from Stevenson (1991).

in which
$$\psi \simeq 8 \cdot 10^{-6} \, T \, (\mu_e/\rho)^{2/3} \quad . \tag{6.29}$$
This describes a polytrope for which, depending on the value of the degeneration parameter ψ, one finds a solution for the particular cool gaseous object.

The radius of speres with such gas and surface temperature $T = 0$ and mass M (solar units) is
$$R_0 \equiv 2.8 \cdot 10^9 M^{-1/3} \mu_e^{-5/3} \tag{6.30}$$
so that
$$R = R_0 \left(1 + \psi + \frac{\psi^2}{1 + \psi}\right) \quad . \tag{6.31}$$
Since in polytropes with $n = 3/2$ the central density $\rho_c \simeq 6\, \overline{\rho}$, one arrives at
$$\psi = 1.9 \cdot 10^{-9} \, T_c \, M^{-4/3} \, (R/R_0)^2 \, \mu_e^{2/3} \quad . \tag{6.32}$$
This equation gives the conditions of gaseous spheres without an inner nuclear energy source. Initially $\psi > 1$ and the equation defines R and T_c as a function of decreasing ψ. To find the best conditions for nuclear fusion to occur one needs to find the mass for which T_c becomes large enough. Differentiating the expression for R above and setting $dT_c/dR = 0$ it is found that T_c exhibits an extremum as function of R at $\psi \simeq 0.55$ and $R/R_0 \simeq 1.9$ (see Stevenson 1991). This extremum clearly depends on R itself (see Fig. 6.2).

For $\mu_e = 1.15$ (which holds for normal gas of solar composition) one finds that
$$T_{c,\max} = 8.1 \cdot 10^7 M^{4/3} \quad [M \text{ in } M_\odot]. \tag{6.33}$$
Since fusion of hydrogen really can only be sustained above $3 \cdot 10^6$ K, the **minimum mass** of a gas sphere being just star is, following this simple analysis, $M_{\min} = 0.084$ $M_\odot$. This theoretical limit is confirmed by the lowest mass MS star seen in a globular cluster (Fig. 6.3). Deuterium may ignite before H (see Ch. 8.2.1 and Fig. 8.1) and thus will help to make the object to star.

Objects near and below this mass limit are called *brown dwarf* (BD). This name has been used for both the lowest mass stars as well as for objects just failing to be star. For more on the name giving as well as on BDs and their evolution see Ch. 8.

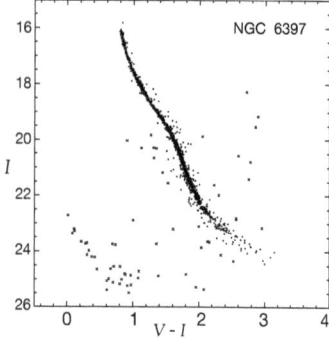

Figure 6.3: The presence of a lower limit for the mass of MS stars was first supported by the observations of the low end of the main sequence in the globular cluster NGC 6397.
The diagram shows the $I, V - I$ CMD as observed with HST. Stars in the MS band are plotted as •, other objects as ×. Note the faint white-dwarf stars at left, showing that the detection limit for stars is clearly fainter than the lower end of the main sequence. Figure adapted from King et al. (1998).

6.6 Methods for solving the differential equations

6.6.1 Numerical solutions

The differential equations can, because of the nature of κ, ϵ, and ρ, only be solved numerically. For that the parameters are renormalized so that the numerical values of each become comparable. Several strategies for calculation exist (see, e.g., Ch 9, 10, 11 in Kippenhahn & Weigert 1990).

a) *Start in the centre of the star.* Here the boundary conditions are $M = 0$ and $R = 0$ while T_c, ρ_c and P_c have to be choosen. One integrates outward and hopes to land at realistic surface parameters. Almost always the values at the surface for M or T turn out be quite different than the desired values. It needs a lot of trial and error to get close to a good solution.

b) *Start at the surface of the star.* Here one chooses M and R while $T \simeq 0$, $\rho \simeq 0$ and $P \simeq 0$. One integrates inward and hopes to land at $R = 0$ when also $M = 0$. Clearly, most integrations will end at $R = 0$ with $M \neq 0$!

c) *Start from both sides.* Halfway between surface and centre the change in the values of the various parameters is very small with respect to dM. Integrating the equations starting at the surface as well as in the centre, but only to halfway, leads to a stretch in dM with very smooth gradients for the variables. One declares a model to be complete if values and gradients approaching from each of the boundaries have become very similar.

d) *One does a relaxation calculation.* Both the inner boundary values and the outer boundary values are set and the intermediate values are estimated (or calculated) accordingly. Now the change of the parameters with dM is locally calculated and compared with what the material functions allow. The inner and outer boundary conditions now are adjusted to get ever better fits of the system.

e) *Start with three (L, T_{eff}) pairs* and make the inward integration to, e.g., the shell of mass fraction of 0.98 M_* resulting in for that mass shell three sets of values for R, ρ and T. These values define which model for the interior fits best and the ultimate values for (L, T_{eff}) are then found by interpolation. Or a different set of three (L, T_{eff}) pairs has to be choosen to better match an interior model. This is the method used by the Geneva group (Meynet, priv. comm.).

By the end of the first half of the 20th century, stellar structure was largely understood based on numerical solutions (by hand! no computers yet). Since the development of powerful computers, solving the equations numerically is not too difficult any more. Computers also allow the use of extensive tabular material for κ and ϵ.

6.6.2 Differential equations against mass shell

The equations of stellar structure given thusfar are in the so-called Eulerian description, structure as a function of radius (summarized at the beginning of Ch. 4.4). Since evolution changes stars rather against mass shell (becoming a red giant changes the radius but not the mass structure) it is for modelling more convenient to work with mass shells, dm. The equations have to be rewritten against mass, which is the so-called Lagrangian description.

The differential equations (without time dependence; only radiative energy transport) then are

$$\frac{dr}{dm} = \frac{1}{4\pi r^2 \ \rho(m)} \quad (6.34)$$

$$\frac{dP}{dm} = -\frac{G\ m}{4\pi r^4} \quad (6.35)$$

$$\frac{dL}{dm} = \epsilon(m) \quad (6.36)$$

$$\frac{dT}{dm} = -\frac{Gm}{4\pi r^4}\frac{T}{P}\nabla_{\text{ad}} = \frac{1}{4\pi r^4 \ \rho(m)} \frac{3\ \kappa(m)\ L}{16\pi\ ac\ T^3} \quad (6.37)$$

For the modelling, the star is subdivided into a set of **grid points** (shells dm). The number of grid points is large when variables change considerably over small dm and can be smaller when all variables change only little with dm.

6.6.3 Adding stellar evolution

A further step is the calculation of **stellar evolution**. Here one inserts in the equations functions describing how the parameters *change with time*. The simplest method is to calculate how in a time Δt the material changes, such as the composition in the interior due to fusion products or of other layers due to (convective) mixing, then calculate a new model for these changed conditions, and then see how the parameters in the other parts of the star have to be adjusted. A sequence of such models builds an "evolutionary sequence".

6.6.4 A model using gaussian functions

Since the differential equations describing a star are not very illustrative in terms of how the solutions for stellar structure look like, in particular since the material functions are rather complicated, an attempt has been made to make stellar structure more transparent by assuming that gaussian functions can be used to describe stellar structure. This assumption is, in fact, not far from reality (see, e.g., Fig. 6.4). Details of the method can be found in Celnikier (1989, Ch 6.6).

Briefly, one adopts functions of the kind

$$M = M_{\text{total}} \left(1 - \exp(-r^2/\lambda^2)\right) \tag{6.38}$$

$$L = L_{\text{total}} \left(1 - \exp(-r^2/\lambda_L^2)\right) \tag{6.39}$$

$$\rho = \rho_{\text{centre}} \exp(-r^2/\lambda_\rho^2) \tag{6.40}$$

$$T = T_{\text{centre}} \exp(-r^2/\lambda_T^2) \tag{6.41}$$

in which r is the running value of the radius and the λ's are the usual scaling constants of the gaussians. The shape of the functions differs according to the increase outward or inward of the variables. The normalisation parameter $\lambda = \lambda_M$ is taken as base for the other normalisations.

In addition simplified functions for the opacity κ (Eq. 4.61) and the energy production rate ϵ are adopted based on a numerical fit to known experimental and theoretical values. Including the ideal gas formula (with mean molecular weight μ) one has

$$\kappa = \kappa_0 \, \rho \, T^{-3.5} \quad \text{and} \quad \epsilon = \epsilon_0 \, \rho \, X^2 \left(\frac{T}{10^6}\right)^\nu \quad \text{and} \quad P = k \, \rho \, T \, / \, \mu \, m_p \; . \tag{6.42}$$

Inserting the gaussian functions into the differential equations, and trying solutions at particular locations (e.g. at $r/\lambda = 0.5$ and $r/\lambda = 2.0$) one finds the approximate proportionalities of the λ's:

$$(\lambda/\lambda_\rho)^2 = 1.37, \quad (\lambda/\lambda_T)^2 = 0.7, \quad (\lambda/\lambda_L)^2 = 2.34 + 0.7\nu \; , \tag{6.43}$$

values which can be inserted in the Gaussian functions. Now one requires consistency at some point in the star, e.g. at $M = 0.5 \, M_{\text{total}}$. This is at $(r/\lambda)^2 = 0.7$.

In the above manner, one obtains a set of 4 equations with 4 unknowns, the parameters λ, L_{total}, ρ_c, T_c. After extensive algebra one obtains the solution to the equations. Details can be found in Celnikier (1989, p. 164-167).

6.7 Vocabulary for stellar structure: definitions

In the study of stellar structure a certain vocabulary has developed with the following definitions.
◦ The word **core** means the inner region of a star having nuclear burning including the parts containing the ashes of the burning process. A star may contain, e.g., an inert He core.
◦ The word **shell** refers to a mass shell at some location in the star. *Shell burning* refers to a shell having nuclear fusion. If more than one such shells is present one has also an *intershell* region.
◦ The word **envelope** is used for the layers above the outermost region having nuclear fusion.
◦ The word **atmosphere** refers to the surface layers of the star, the layers in the vicinity of $\tau \simeq 1$, about which information can be derived from the spectrum.

6.8. ZERO-AGE-MAIN-SEQUENCE STAR PARAMETERS FROM MODELS

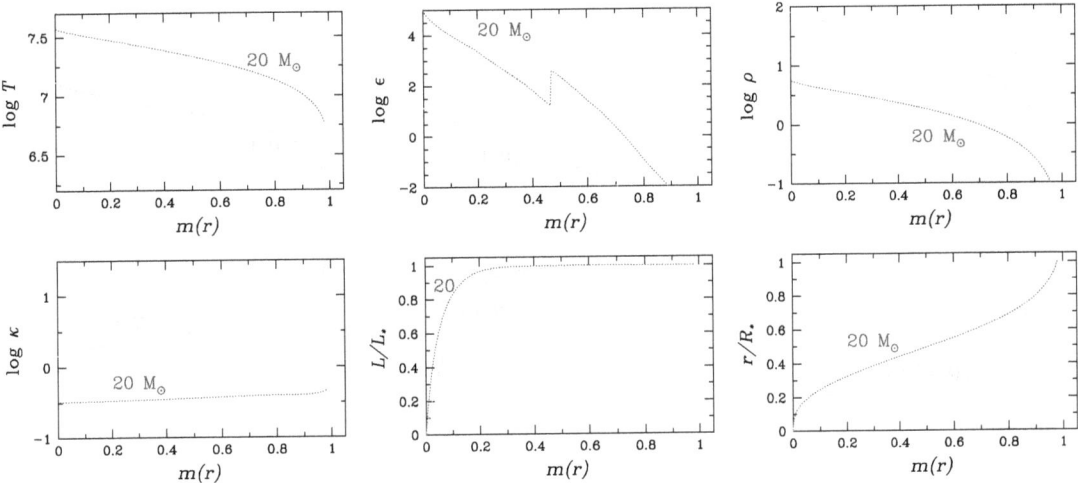

Figure 6.4: Diagram with the results from model calculations (Meynet 2008, priv. comm.) showing the value of parameters versus the fractional mass $m(r)$ for ZAMS stars of 1 $M_\odot$ and 20 $M_\odot$. The parameters are (in logarithmic fashion) temperature T [K], opacity κ [cm^2 g^{-1}], energy production ϵ [erg cm^3 g^{-2} s^{-1}], and gas density ρ [g cm^{-3}], and (in linear fashion) luminosity L/L_* and fractional radius r/R_*. Note the following:
The density in a 1 $M_\odot$ star is much larger than in a 20 $M_\odot$ star, with the temperature it is reverse! A 20 $M_\odot$ star produces 90% of its luminosity L within 11% of its mass, a 1 $M_\odot$ star produces 90% of L from $\simeq$ 30% of its mass. A 20 $M_\odot$ star has 1/2 its mass inside 1/2 its R, a 1 $M_\odot$ star has $\simeq$ 70% of its mass inside 1/2 R. The reason for the jump in the ϵ curve of the 20 $M_\odot$ star is given in Sect. 6.8.1. The behaviour of T and κ for the 1 $M_\odot$ star are almost mirrored (see Sect. 6.8.1).

Furthermore, a star normally is in overall **stellar thermal equilibrium** (see Ch. 1.2). This means that the amount of energy produced by nuclear fusion equals the amount of energy radiated from the surface. When a star is out of equilibrium, it means that the balance between internal pressure and gravity (see Ch. 6.3) is, perhaps only locally, disturbed. The nature of the balancing forces is that a star strives to regain equilibrium. Thus in case of an imbalance, structural changes will take place so as to reach a new equilibrium (Ch. 9.4). These changes may involve contraction and/or expansion, other/new nuclear processes, changes in opacity structure, etc. These processes make for the interesting aspects of stellar evolution.

6.8 Zero-age-main-sequence star parameters from models

Using the full equations for stellar structure and the full knowledge about the material functions, numerical recipies have resulted in extensive models for stars. These results can be best summarized in a set of figures for stars of initial structure, the so called **zero-age main-sequence (ZAMS)** stars. The ZAMS is the set of models of gaseous objects which just became star, i.e., these are the objects having sustainable nuclear fusion (for more on the ZAMS see Ch. 9.1.1.2, for the low-mass limit see Sect. 6.5). Note that the formation process takes time, as can be seen in Fig. 7.10.

6.8.1 ZAMS: structure as a function of mass shell

Fig. 6.4 shows the values of the parameters T, ϵ, ρ, κ, and L throughout two typical stars (1 and 20 $M_\odot$) plotted against mass fraction, $m(r)$, and of shell radius r/R_* against $m(r)$. The models are for ZAMS stars (Meynet, 2008; updated models compared to those of figures in later chapters).

As soon as nuclear fusion sets in the stellar structure changes (see Ch. 9.1.1) due to fusion. In making such models, one aims at finding a stable model, i.e., a model in which the parameters

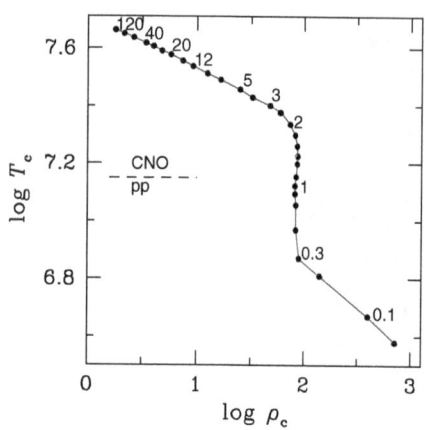

Figure 6.5: The values of the central temperature T_c [K] and central density ρ_c [gm cm^{-3}] for main-sequence stars (after $\simeq 2\%$ hydrogen is burnt) using models from Meynet (2008) for $M = 0.8$ M$_\odot$ and larger (for smaller M data are from Kippenhahn & Weigert 1990). Above the MS the stellar mass values are given. The dashed lines give the degeneration parameter ψ (see Ch. 4.4.2.3) of the gas. The full line is the approximate ZAMS.

Note two changes in the slope of the curve. Near $M = 0.3$ M$_\odot$, above which mass the outer convection zones (see Fig. 6.6) become shallow, the energy transport to the surface is slowing down therefore raising T_c. The change for $M > 1.3$ M$_\odot$ is due to the contribution of fusion in the CNO-cycle, raising the energy production (increased ϵ), leading to a steep inner temperature gradient and so the onset of inner convection (see Fig. 6.6). Compare the presented T_c and ρ_c values with the information of Fig. 4.5.

change little upon varying the starting values. The models given in Fig. 6.4 are then models for stars just beyond the true ZAMS (so after the "birth line", see Fig. 7.10), for a MS state in which $\simeq 2$ to 4% of the H has been burnt.

The result is that, e.g., the ϵ curve for the 20 M$_\odot$ star in Fig. 6.6 has a kink. This is an effect of the nuclear fusion because it leads to immediate changes in the chemical structure, while the ϵ functions depend strongly on T (see Ch. 5). This means that in stars with inner convection (see Fig. 6.6) the CNO-cycle fusion (see Ch. 5.1.5) quickly leads to a reduction of C and an increase of N, thus effectively slowing down CNO-cycle fusion to the rate of the slowest reaction in the cycle, being the N+H reaction (see Eq. 5.10). Since interior to the jump there is convection, which mixes efficiently the C nuclei (reduced in number) and the N nuclei (enhanced in number, N build-up), this leads to a reduced ϵ in the entire convection zone. That kind of structure is present in all stars with central convection.

The behaviour of κ (Fig. 6.4) follows from the physics presented in Ch. 4. The κ curve can be understood combining the run of T and ρ (as in Eq. 4.61). In the solar model κ rises toward the edge because there the temperature gets smaller with the concomittant strong increase in κ (Fig. 4.4).

The luminosity is produced in the core. Clearly, the strong T dependence of the ϵ functions explains why most of L is produced at the highest T, thus close to the stellar centre. This need no longer be so in slightly evolved stars once fusion has reduced the fuel supply near the centre (see, e.g., Ch. 9.2.1 and Fig. 10.4).

6.8.2 ZAMS: parameters along the ZAMS - a star as a leaky ball

Fig. 6.5 shows the conditions in the centre of ZAMS stars. Note that the actual central conditions do not follow the relation for an ideal gas! The reasons for the "kinks" in the curve are related with the aspects of internal structure as explained with Fig. 6.4.

The run of the parameters for the internal structure of main-sequence stars is summarized in Fig. 6.6. Note also the convective zones. Figs. 6.4, 6.5 and 6.6 give, in combination, a good overview of how MS stars are structured.

6.8.2.1 Similarity along the MS; homology; thermostat; luminosity and mass

Homology. MS stars are core H burners and, if of similar mass, have similar structure. To see this, one uses the concept of 'homology'. Take Eqs. 6.34 and 6.35 and formulate them for a similar star, using primed versions of the original variables, like r', etc. (for details see, e.g., Cox & Giuli 1968, their Ch. 22; or Kippenhahn & Weigert 1990, their Ch. 20). The same is done for the

6.8. ZERO-AGE-MAIN-SEQUENCE STAR PARAMETERS FROM MODELS

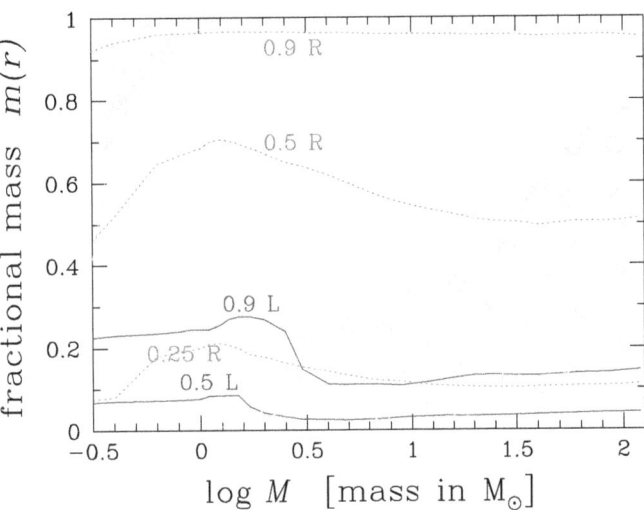

Figure 6.6: The internal structure is shown of main-sequence stars with $0.3 < M_{\text{ZAMS}} < 120$ M$_\odot$ (models from Meynet (2008) for masses of 0.8 M$_\odot$ and larger, data from Kippenhahn & Weigert (1990) for masses below 0.8 M$_\odot$).
The horizontal axis marks the MS mass ($\log M$), vertically some of the radial structure is given against fractional mass $m(r)$. For T_c and ρ_c see Fig. 6.5, for the run of several parameters against $m(r)$ see Fig. 6.4.
The dotted curves show at which $m(r)$ r is at 0.25, 0.5 or 0.9 of its full value. Especially the 0.5 R curve shows that stars around 1 M$_\odot$ are more compact than high mass stars. The luminosty curves show at which $m(r)$ 0.5 or 0.9 of L is reached. Note that for stars with $M > 3$ M$_\odot$ 90% of L is produced within $\simeq 12$% of the mass.
The curls mark regions having convection. Low mass stars have an outer convection zone driven by the opacity of the outer layers (Ch. 4.3.4); for lower stellar mass also the interior temperatures are lower thus leading to higher opacity (see Fig. 4.4) and so to deep convection zones. Higher mass stars have an inner convection zone because the energy production (CNO cycle, steep ϵ_{CNO}) leads to a steep temperature gradient and radiation driven convection (Ch. 4.3). The Sun has barely an outer convection zone (for its depth see Fig. 11.14) and no inner convection.

material functions κ, ρ and ϵ. Defining $\rho/\rho' = d$ (and similarly for the other variables) one arrives at 8 linear equations relating the two stars. These lead to solutions of which the first two are

$$\frac{\rho}{\rho'} = \frac{M/M'}{(R/R')^3} \quad \text{and} \quad \frac{P}{P'} = \frac{(M/M')^2}{(R/R')^4} \qquad (6.44)$$

showing that in homologous stars the parameters ρ and P scale from star to star by the particular ratios of the overall values of M and R. One can also eliminate R/R' and find that

$$\frac{P}{P'} = \left(\frac{\rho}{\rho'}\right)^{4/3} \left(\frac{M}{M'}\right)^{2/3} . \qquad (6.45)$$

The homology considerations show that for an entire star $\overline{\rho} \simeq \overline{M}/\overline{R}^3$ and $\overline{P} \simeq \overline{M}^2/\overline{R}^4$.

Assume one has an ideal gas so $P = (\Re/\mu) \rho T$. For a homologous star, inserting Eqs. 6.44 and 6.45 and rearranging leads to

$$\frac{T}{T'} = \frac{\mu}{\mu'} \frac{M}{M'} \left(\frac{R}{R'}\right)^{-1} \qquad (6.46)$$

of which the simplified version is

$$T(m) \simeq \mu M / R . \qquad (6.47)$$

This shows that when considering similar stars, at an internal location with given T one is at the same M/R. So if the similar star has larger M, then at the point with $T(m)$ also R is larger.

The homology relations are only valid for similar stars. This means that, in a strict sense, the relations can only be applied for small differences in the conditions, so small differences in M and, e.g., the behaviour of ρ, κ and ϵ. Any larger structural difference, like PP-chain fusion or CNO-cycle, convection or no convection, are not well handled by homology. Still, the crude general relations derived appear to hold over larger ranges of stellar mass.

Eq. 6.47 also indicates that when in a star μ becomes larger (due to ongoing fusion) T will rise and/or R will increase. This is the behaviour of ZAMS stars (see Sect. 6.8.2.2 and Ch. 9.1.1).

Another consequence of Eq. 6.47 is that for stars with the same mass but, for some reason, a higher T_c (see Sect. 6.8.2.2) its radius R must be smaller. Models show that at equal M a metal poorer MS star lies indeed at smaller R (see Sect. 6.9.2 and Fig. 6.7).

Thermostat. Eq. 6.47 shows in general that a MS star operates as if it has a thermostat. Consider a star of a given mass which, through some external agent, is compressed. *If the radius is reduced, the temperature must become larger, thus leading to more fusion (function ϵ), a higher internal energy and so to expansion and cooling. The star strives back to its original condition.*

Luminosity and mass. Assume the star has only radiative energy transport. Assume also that one deals with an ideal gas. Then, combining Eqs. 6.37 and 6.44 one obtains

$$\frac{L}{L'} = \left(\frac{\kappa}{\kappa'}\right)^{-1} \left(\frac{M}{M'}\right)^3 \left(\frac{\mu}{\mu'}\right)^4 \tag{6.48}$$

of which the simplified version (Eddington 1926) is

$$L = \kappa^{-1} M^3 \mu^4 \quad . \tag{6.49}$$

It shows that, simply speaking, for MS stars $L \simeq M^3$ (for real stars see Fig. 9.6).

6.8.2.2 A star as a leaky ball: general behaviour, effects of chemical composition

In general one can describe a star as a "leaky ball". Energy is generated inside the star and this energy leaks out as photons. The star is kept in balance by gravity which, after loss of stellar energy, tends to contract the star (overall thermal balance), raising its internal temperature, which spurns on nuclear fusion. This balancing act has been introduced in Ch. 1.2.

An important result of the homology considerations is that one can derive how the luminosity of a star depends on the material. Eq. 6.49 not only indicates $L \simeq M^3$ but shows also how L differs for different (or may change for changing) values of κ and μ.

$\to \kappa$. Stars with smaller κ (lower overall metal content) have larger L; κ is smaller in, e.g., metal poor stars, they more "less leaky" and gravity has to make up for that, raising T_c.

$\to \mu$. Stars with larger μ (larger mean molecular weight) have larger L; μ can be larger e.g. due to progressing core fusion; the stars are more compact leading to higher T_c.

A more extensive summary of these effects can be found in, e.g., Cox & Giuli (1968) or Mowlavi et al. (1998).

6.9 Internal structure and chemical composition

6.9.1 Consequences of nuclear enrichment for stellar structure

In the course of stellar evolution, the core region creates He from H and in more massive stars also creates further heavy elements. This implies that the mean molecular weight in the interior increases with time and the star is no longer homogeneous.

It is relatively easy to calculate what the consequences are of the change in the value of μ in the case of the Sun.

The general relation for L of a star is given in Eq. 6.49. For the Sun with 1 $M_\odot$, and assuming κ does not change, this relation reduces to $L \simeq \kappa^{-1}\mu^4$ [L in $L_\odot$], in which $\mu^{-1} = 2X + 0.75Y + 0.5Z$ (Eq. 4.65). The Sun is now at about half of its MS phase. In the central zone (about $1/3\ M$) X evolved to $\simeq 1/2\ X$ with the concomittant increase in Y. With these numbers, μ changed thus from 0.60 to 0.64. Opacity is normally dominated by hydrogen, so in the solar interior κ is reduced to about half its original value. Due to these changes, and according to Eq. 6.49, since its birth the luminosity of the Sun must have risen by $\simeq 60\%$ (models indicate it is $\simeq 40\%$).

The derivation above describes the change in a star like the Sun in the MS phase (see also Ch. 9.1.1.1). Solar evolution beyond the MS involves many other aspects (see Chs. 9 and 10).

6.9. INTERNAL STRUCTURE AND CHEMICAL COMPOSITION

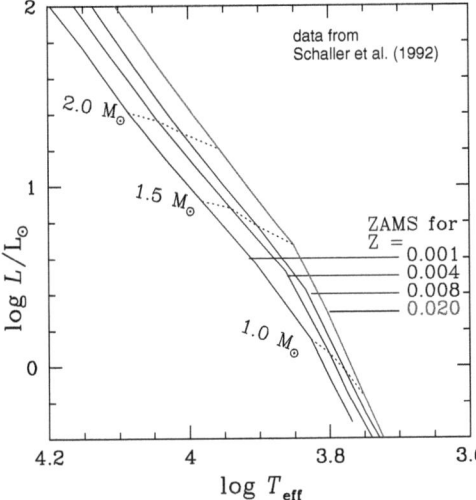

Figure 6.7: Location of the main sequence for stars with different chemical composition. Some points of equal mass are connected.
The elemental compositions by mass fractions are: $X = 0.70$, Z as indicated, and thus $Y = 1.0 - X - Z$. The rightmost ZAMS has $X = 0.70$, $Y = 0.28$ and $Z = 0.02$, a composition just a bit below solar metallicity. For the same mass, fewer metals means less opacity, more compact envelopes and so denser cores with higher T_c and thus larger L (Eq. 6.49). For the eye the ZAMS for lower metallicty is "below" the normal ZAMS, but it actually is to the left and at higher L than the normal ZAMS (see the equal mass points). The reason for the shifts is that such stars are 'more leaky' (see Sect. 6.8.2.1).

6.9.2 Non-hydrogen stars

One can, of course, calculate models for stars with a composition different from that of the Sun. Such models are of relevance for two reasons.

First, one can model stars which came into being in the early phases of the evolution of the Milky Way (in fact, of the Universe), where hardly any heavy elements existed ($Z = 0$ and so $\kappa < \kappa_\odot$), to see if evolution was different from that of present day stars. Such models lead to diagrams showing the location of the MS in relation with original chemical composition (Eq. 6.49). The model sequence of Fig. 6.7 is (from left to right), in fact a sequence following chemical evolution in the Universe (see also Ch. 22). More recently built stars formed in gas enriched with metals coming from the metal-enriched material shed by earlier generations of stars through stellar winds or in supernova explosions.

Second, one can model stars (or parts of stars) containing, e.g., just helium, as in the case of the bluest horizontal-branch (HB) stars (see Ch. 10.4.2). Also the interior of massive stars will turn into pure helium (Ch. 13.4), interiors in which He fusion will start. One therefore can regard these stars or their central parts as 'helium stars'. In such stars the mean molecular weight is larger than in 'hydrogen stars'. Their inner temperature is, therefore, higher than in hydrogen stars of the same mass. Thus, according to Eq. 6.49 their luminosity is higher. Fig. 6.8 (left panel) shows the theoretical surface parameters for H, He and C stars.

Models for the evolution of zero metallicity stars (those likely having been the first stars to be formed at all) are addressed in Ch. 15.

6.9.3 Central temperature and density of He and C stars

The values of the internal T_c and ρ_c for the models of H, He and C stars of Fig. 6.8 (left panel) are shown in the right panel of Fig. 6.8. This figure extends the information of Fig. 6.5, which displayed these values only for the hydrogen ZAMS. Calculating models with very different chemical composition is useful because such models show properties similar to those of the interior parts of evolved stars, like those which have after long H-fusion accumulated a core full of He.

Of great importance would be a diagram giving the change in the values of the parameters T_c and ρ_c in the course of evolution. Such diagrams may show in a very simple manner how the interior of a star evolves. Such diagrams will be presented in the chapters on the evolution of stars in the lower mass range (Ch. 10) and those in the higher mass range (Ch. 13).

That interior gas becomes degenerate in the course of evolution is indicated in, e.g., Fig. 10.2.

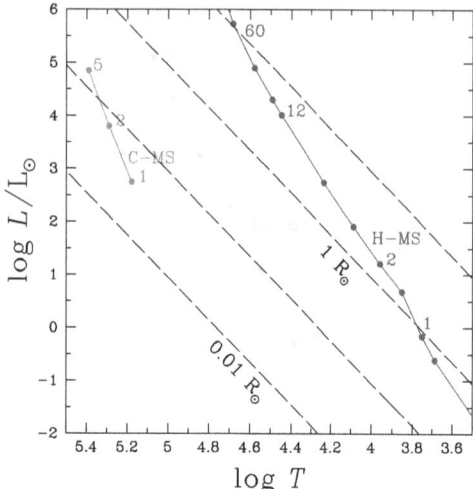 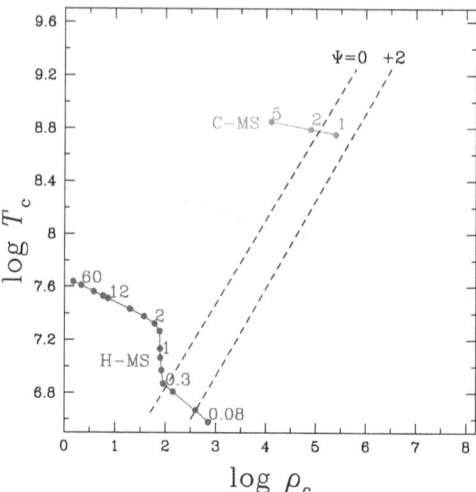

Figure 6.8: Parameters of models of stars with gas of different initial composition. H-MS: solar composition stars ($X = 0.70, Y = 0.28, Z = 0.02$) from Schaller et al. (1992). He-MS: helium stars ($X = 0, Y = 0.979$) and C-MS: stars consisting of just C+O ($X = 0, Y = 0, Z(C+O) = 0.994$), both after Kippenhahn & Weigert (1990).
Left: Surface parameters of the stars (the HRD; compare this diagram with Fig. 1.3). Note the difference in location of the H, He and C main sequences. Mass values and radii are indicated.
Right: Central values T_c and ρ_c for the same stars. This diagram extends Fig. 6.5. Mass values are indicated as well as two lines of the degeneracy parameter. Note that *there are no lines of evolution* in this diagram. Such lines will be presented in Chs. 10 and 13.

6.10 Summary

The gaseous spheres having (or having had) nuclear fusion are called stars. Their structure is determined by the total mass (Sect. 6.4, 6.5) and by the nature of the gas (equation of state; Ch. 4.4.2, Fig. 4.5, Sect. 6.2). The latter depends on the density of the gas, $\rho(T, X, Y, Z)$, in part set by the total mass M and total size R of the star. The conditions leading to $\rho(T, X, Y, Z)$ are intimately related with the level of energy production based on fusion, $\epsilon(T, X, Y, Z)$, and energy loss (the "leakyness of the ball") based on transparency of the gas, $\kappa(T, X, Y, Z)$. A star produces its energy because of gravity which forces the material to respond (Ch. 1.2 and Sec. 6.8.2).

The diagrams most relevant to understand the structure of (main-sequence) stars and the run of interior parameters along the MS are: Figs. 6.4, 6.5, 6.6, 6.7, and 6.8. Evolution is not shown in this chapter.

References

Appenzeller, I. 1970, A&A 5, 355 and 9, 216
Baraffe, I., Heger, A., & Woosley, S.E. 2001, ApJ 550, 890
Celnikier, L.M. 1989, "Basics of Cosmic Structures", Editions Frontières, Gif-sur-Yvette
Cox, J.P., & Giuli, R.T. 1968, 'Principles of Stellar Structure'; Gordon & Breach, New York
Eddington, A.S. 1926, "The Internal Constitution of the Stars", Cambridge Univ. Press
King, I.R., Anderson, J., Cool, A.M., & Piotto, G. 1998, ApJ 492, L 37
Kippenhahn, R., & Weigert A. 1990, "Stellar Structure and Evolution"; Springer, Heidelberg
Meynet, G. 2008, private communication
Mowlavi, N., Meynet, G., Maeder, A., Schaerer, D., & Charbonnel, C. 1998, A&A 335, 573
Schaller, G., Schaerer, D., Meynet, G., & Maeder, A. 1992, A&AS 96, 269
Stevenson, D.J. 1991, ARAA 29, 163; *The Search for Brown Dwarfs*

Chapter 7

Star formation, proto-stars, very young stars

Stars form regularly almost everywhere in our Milky Way galaxy. A sure sign of star formation is the presence of luminous hot stars. Such upper MS stars are, given their mass and large luminosity, only short lived and thus astronomically "young". Theories of star formation are concerned with cloud contraction, cloud fragmentation, formation of protostars, manifestations of protostars with accretion disks up to the onset of nuclear fusion and the emergence of stars as MS stars.

7.1 Evidence of star formation, populations, IMF

7.1.1 Signs of present star formation

- *Massive OB-type stars*

  Simple observations demonstrate present-time star formation, e.g. because of the existence of H II regions. These are proof of the presence of hot stars with strong ionizing UV flux: O- and early B-type stars (actually to about B2V). Only these produce H II regions (there is a relation between ionizing flux, gas density and nebular radius).

  The life time of OB stars is short compared to the time scale of Galactic evolution. An O5-type star spends only $\simeq 2 \cdot 10^5$ years on the MS (see Ch. 13), followed by very fast evolution via different stages up to supernova. O and B stars must thus have formed recently.

- *Less massive stars, T Tauri stars*

  Low luminosity stars with variable light and with emission lines of H, Ca II and Fe I are found in dark clouds and diffuse nebulae in Orion, Taurus, Monoceros, etc. T Tauri is the prototype of such variables. Such stars populate the pre-MS region in the lower part of the HRD. Theory shows they are young stars in the contracting phase evolving toward the MS.

- *Places of star formation*

  Another argument (*not* from theory of stellar evolution) comes from the vertical (*z*-direction) distribution of young OB stars in the Milky Way. Star counts show that, in general, the number density in *z*-direction can be approximated by

  $$n(z) = \text{const}\, e^{-z/h} \qquad (7.1)$$

  where z is the distance from the Galactic plane and h is the disk *scale height*. Following Bahcall & Soneira (1980), $h < 100$ pc for OB stars (see Bahcall 1986 for a review of star count investigations). In addition, the velocity dispersion of the OB stars perpendicular to the Galactic plane is small, $v_z \simeq 10$ km s^{-1}.

Observations show that the thickness of the disk of the much older stars of horizontal-branch type is of the order of 1 kpc[1]. Furthermore, a thin disk with OB stars would start to widen after a time h/v, the short period of about 10^7 years. Since the OB star disk is thin the reservoir of young stars is being replenished constantly.

- *Star forming material*

  OB-type stars are often connected with H II regions (see also Ch. 13). But the temperature in H II gas is too high and the density too low for cloud collapse to occur. The association with molecular gas showed that H II gas is the state the gas gets to right after star formation. The discovery of **molecular gas** near H II regions gave the breakthrough for our understanding of star formation.

These ideas led to the concept of *Populations*: **Pop. I** for relatively young disk stars, and Population II (**Pop. II**) for relatively old stars (in particular the halo stars), concepts considerably influenced by the nature of the CMD of such star groups. Later the name **Pop. III** stars was added for the very first stars, thus being very metal poor (the Milky Way was then not enriched yet with fusion products).

The process of the formation of the very first generation of stars (of Pop. III) is very speculative (see Ch. 15). A possible Pop. III candidate star is, e.g., CS 22885-96 (Molaro & Bonifacio 1990) with a metal abundance of [Fe/H] $= -4.21 \pm 0.20$ dex.

7.1.2 Star-formation processes and results of star formation

How the formation of stars in IS gas proceeds in detail is not well known yet. Two general categories of star-formation process can be given:

- *spontaneous star formation* (as gravity wins out over various cloud support mechanisms),
- *stimulated star formation* (triggered by some outer inducing conditions, e.g., spiral density wave shocks, cloud-cloud collisions, compression by expanding H II regions and/or winds from massive stars and/or supernovae).

The result of a theory of star formation should be that one can predict the statistics of stellar masses, i.e., the initial mass function (IMF). It normally is approximated by a power law

$$N(M) \propto M^{+\alpha} \quad \text{or} \quad dN(M) \propto M^\alpha \, dM \tag{7.2}$$

where $dN(M)$ is the number of stars in the mass interval M to $M + dM$. Evaluation by Kroupa (2002) of observational data indicates a tripartition of the IMF:

$$\alpha = \begin{cases} -0.3 & \text{for } 0.01 \leq M < 0.08 \quad [M_\odot] \\ -1.3 & \text{for } 0.08 \leq M < 0.50 \quad [M_\odot] \\ -2.3 & \text{for } 0.50 \leq M \quad\quad\quad\quad\ [M_\odot]. \end{cases} \tag{7.3}$$

The value of α for the stars in the solar neighbourhood is the famous "Salpeter value" of $\alpha = -2.35$ (Salpeter 1955). For more on the mass function and its diverse notations see Ch. 20.

7.2 Molecular clouds: places of star formation

7.2.1 Discovery and importance of interstellar molecules

The first discoveries of interstellar molecules were made between 1968 and 1970. The most abundant molecule in interstellar space is H_2. Carruthers (1970) discovered H_2 through UV line absorption present in the spectrum of the star ξ Persei (see Fig. 7.1).

---

[1]Of course, there are old and very old stars in the Galaxy: stars in globular clusters, other stars of the halo. A number of facts elucidate that star formation was a continuous process during the evolution of the Galaxy – not necessarily at all places, but at all times. There is no evidence for long interruptions in the global galactic star formation rate. Young stars are, however, only found in the disk.

7.2. MOLECULAR CLOUDS: PLACES OF STAR FORMATION

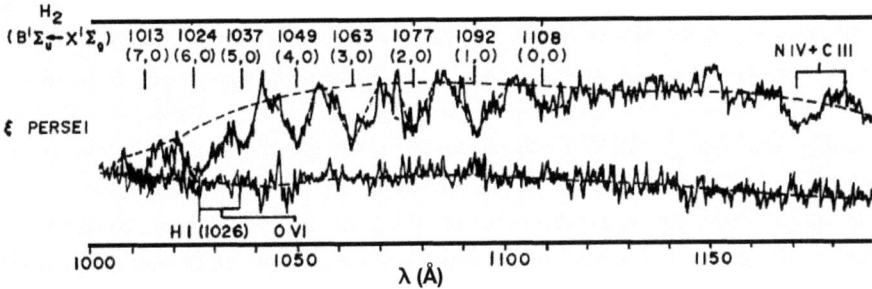

Figure 7.1: First detection of H$_2$ in the UV. The spectrum (top) is of ξ Persei (Sp.T. O7, $V=4.06$ mag), taken with a spectrograph on an Aerobee-150 rocket (1970 III 3), 12 s exposure. The H$_2$ absorption structures are marked with λ and (J_u, J_l). Fig. from Carruthers (1970).

Table 7.1: Known interstellar molecules (Herbst 1988); c- refers to a cyclic form

H$_2$	CH	CH$^+$	CN	CO	CS	OH	SiO
NS	SO	SiS	NO	CC	HCl	SO$^+$	PN
H$_2$	H$_2$S	HCN	HNC	HCO$^+$	HOC$^+$(?)	H$_2$D$^+$(?)	HN$_2^+$
HCO	CCH	HNO(?)	OCS	SO$_2$	SiC$_2$	HCS+	CCS
NH$_3$	H$_2$CO	H$_2$CS	HNCO	HNCS	CCCN	CCCH	CCCO
HOCO$^+$	HCNH$^+$	C$_2$H$_2$	c-C$_3$H	CCCS	H$_3$O$^+$(?)	H$_2$CNH	H$_2$NCN
HCOOH	HC$_3$N	C$_4$H	CH$_4$	SiH$_4$	CH$_2$CO	c-C$_3$H$_2$	CH$_2$CN
CH$_3$OH	CH$_3$SH	C$_2$H$_4$	HCONH$_2$	C$_5$H	CH$_3$CN	CH$_3$NC	CH$_3$NH$_2$
CH$_3$CCH		CH$_3$CHO		C$_2$H$_3$CN	HC$_5$N	C$_6$H	
HCOOCH$_3$		CH$_3$C$_3$N	HC$_7$N	CH$_3$OCH$_3$		CH$_3$CH$_2$OH	
CH$_3$CH$_2$CN		CH$_3$C$_4$H	CH$_3$C$_5$N(?)		CH$_3$COCH$_3$		HC$_9$N
HC$_{11}$N							

H$_2$ is by far the most abundant molecule. It is, however, difficult to detect because of unfavourable emission conditions: it has no dipole moment, only a quadrupole transition at 28μ. Thus H$_2$ can be seen in near-IR emission lines from radiatively and collisionally excited states such as present after recent star formation near hot stars. But H$_2$ is always associated with dust, which radiates strongly in the infra-red, too. H$_2$ can also be detected from cool gas with UV absorption lines (near 1000 and 1100 Å) but at those wavelengths the dust causes tremendous extinction so the stellar light is considerably dimmed.

The association of molecules with dust is quite typical. Dust has two functions:
– Dust screens the interstellar gas from the radiation field protecting molecules from dissociation.
– The surfaces of the dust grains give the possibility for catalytic molecule formation; the molecular binding energy can be carried over to the dust particles without the necessity of photon emission.

The first more complex IS molecules were discovered at radio wavelengths: NH$_3$ by Cheung et al. (1968) and H$_2$CO by Snyder et al. (1969). The second most abundant molecule, CO, was discovered by Wilson et al. (1970). More than 80 molecular species have been discovered up to 1988 (see Table 7.1), and several more since then.

The most important molecule to investigate star formation is CO (^{12}C^{16}O, and the isotopes ^{13}C^{16}O, ^{12}C^{18}O, etc.) because it is abundant and easy to observe at 2.6 mm (with mm-wavelength telescopes, e.g., the Swedish-ESO-Sub-mm-Telescope SEST at La Silla). It is believed to be a tracer of H$_2$. The conversion factor (the infamous and highly controversial "X factor") is

$$n(\text{H}_2)/n(\text{CO}) \approx 5 \cdot 10^5 \ . \tag{7.4}$$

Originally, the X-factor is the ratio between the H$_2$ column density derived from the far-IR dust emission and the velocity-integrated CO emission intensity (in units K km s^{-1}; see Heithausen & Mebold 1989). In spite of the uncertainty about X, CO is widely used for star-formation studies.

Table 7.2: Parameters of dwarf and giant molecular clouds (DMCs & GMCs)

Parameter		DMCs	GMCs
mass M	$M_\odot$	$10^3 \ldots 10^3$	$10^5 \ldots 3 \cdot 10^6$
particle density $n(H_2)$	cm^{-3}	$100 \ldots 1000$	10000
temperature T	K	10	10
diameter D	pc	$2 \ldots 5$	30 *)
position in Milky Way		in general disk	in spiral arms
structure		inhomogeneous •)	inhomogeneous •)

*) GMC complexes reach up to 100 pc.
•) Clumps in GMCs are similar to DMCs.
Cores in GMCs and DMCs have $n(H_2) > 10^4$ cm^{-3}, $d \approx 0.1$ pc and
frequently contain IR point sources.

7.2.2 Characteristics of molecular clouds

All phenomena of star formation are associated with molecular clouds. For conditions in these clouds, see, e.g., the reviews by Evans (1999) and van Dishoek & Blake (1998). Based on morphological considerations one defines two cloud classes (see summary in Table 7.2).

- *Dwarf Molecular Clouds (DMCs)*

  DMCs have masses and densities as indicated in the table. Temperatures determined from the thermal broadening of the CO lines and are in the order of 10 K. Optically, DMCs can be recognized by their association with dust (= dark clouds, often with sharp borders). Typical DMCs are found in Orion, Taurus, etc. (see, e.g., Fig. 7.2, right).
  In DMCs, often "cloud cores" are embedded: spots of very high density ($n(H_2) > 10^4$ cm^{-3}). DMCs are *not* necessarily bound to spiral arms, i.e., to OB-type stars and H II regions.

- *Giant Molecular Clouds (GMCs)*

  GMCs are the most massive objects of the Galaxy. They have masses between 10^5 and $3 \cdot 10^6$ $M_\odot$ and diameters of typically 30 pc (see, e.g., Fig. 7.2, left). They are often associated with massive stars and H II regions (in spiral arms). GMCs have extremely inhomogeneous structures with filaments, arcs, or lobes; embedded are
  → "clumps" (radii $\simeq 3$ to 5 pc, masses 10^3 to 10^4 $M_\odot$), which are similar to DMCs, and
  → "cores" (see DMCs, above).
  The smallest features show diameters down to 0.014 pc (Falgarone & Phillips 1996).

Several detailed analyses of masses and sizes of the clouds have been performed (see, e.g., Elmegreen & Falgarone 1996; Stutzki et al. 1998). The characteristics of the clouds can be best interpreted by the concept of fractal structure: size l ("diameter") and mass M show power-law distributions of the form

$$n(l)\, dl = l^{-\alpha_l} dl \quad \text{and} \quad n(M)\, dM = M^{-\alpha_M} dM \quad (7.5)$$

where $M \sim l^\kappa$ is also a power law. Actual values (Elmegreen & Falgarone 1996) are $\alpha_l = 1 + D$ and $\alpha_M = 1 + D/\kappa$ for the interstellar fractal dimension $D = 2.3 \pm 0.3$. The value of κ (from observations) is in the range of 2.4 to 3.7.

One might ask if the smallest structure elements in molecular clouds are unstable. Might they be the progenitors of the protostars? Since such structures are very dusty one cannot see what is going on inside. Their role in the star-formation scenario is not yet understood.

7.2.3 Observed phenomena in star forming regions

When observing star-forming regions one has to work with those species which show themselves by their radiative processes. Mentioning molecules such as CO or NH$_3$ only indicates that such species are easy to observe and/or give useful information.

7.2. MOLECULAR CLOUDS: PLACES OF STAR FORMATION

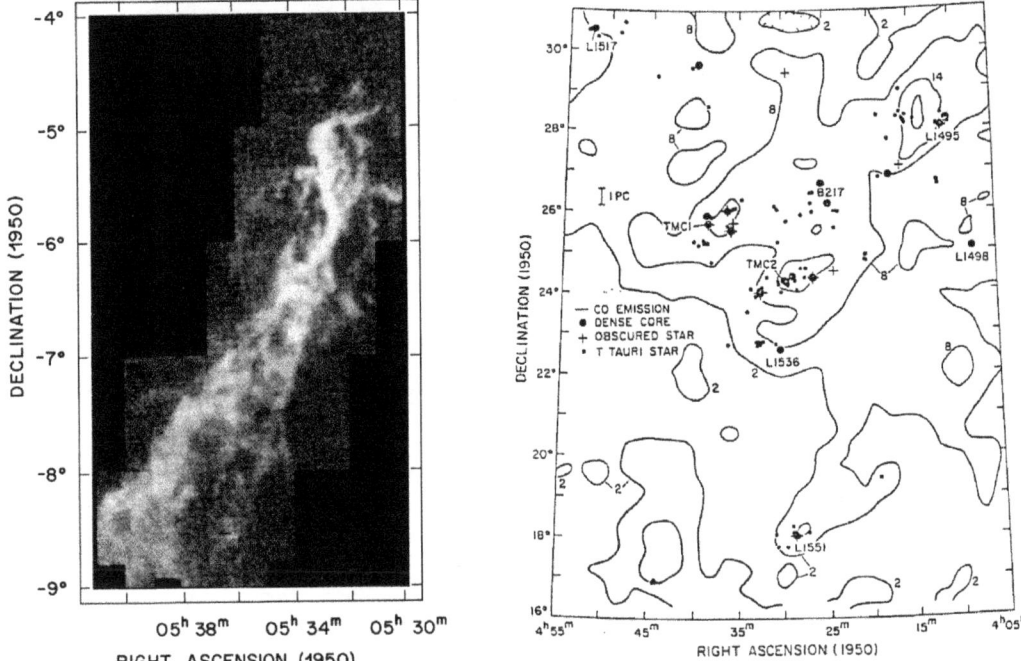

Figure 7.2: Two star forming regions are shown to demonstrate various observational features. *Left*: Southern portion of the Orion GMC mapped in the optically thin ^{13}CO emission at 2.7 mm (7m antenna at Holmdel). The filamentary emission is bright. Figure from Bally et al. (1987). *Right*: Phenomena associated with star formation in the constellation Taurus. Objects labelled (Lxxx, Byyy) are notable ones (emission lines, special colours) from various catalogues. Also marked is the TMC1 (= Taurus Molecular Cloud 1) as well as TMC2. Figure from Shu et al. (1987). Details are given in Sect. 7.2.3. For L 1551 see also Fig. 7.11.

One of the most interesting regions of the sky is the constellation Taurus. Optical images show a region relatively poor in stars, whereas radio and IR telescopes reveal a number of exciting features, among them some Taurus Molecular Clouds (TMCs). In Fig. 7.2 (right) we can distinguish:

1. Isophotes of CO emission (full lines).
    Some dense molecular clouds (TMC1, TMC2) can be distinguished. TMC1 is an example of a dense cloud without embedded IR sources (see below).
2. Dense cores of MCs (big filled circles).
    These have been detected in the strong 1.3 cm $(J, K) = (1,1)$ line of NH$_3$. Mean properties: diameter 0.1 pc, $\rho \sim 3 \cdot 10^4 \, \text{cm}^{-3}$, $M \sim 4 \, \text{M}_\odot$, gas $T_\text{exc} = 11$ K (Myres & Benson 1983).
3. Obscured stars (crosses).
    These are star-like IR sources. Not all MCs and dark clouds/nebulae are connected with IR sources. But sometimes the density of IR sources may be extremely large: see the nebula NGC 2024 in Orion (Fig. 7.3).
4. T Tauri stars (small dots).
    These stars can be detected through Hα emission in their spectra and light variability. They are interpreted as stars in the pre-main-sequence phase (more on preMS stars in Sect. 7.7).
5. Optically prominent dark clouds (prefixes L and B).
    Lynds (1962) made a comprehensive search for dark nebulae on Palomar Schmidt plates (objects with name Lnnn). An old list is from Barnard (1919, names Bnnn). It is apparent (and plausible) that the dark clouds are physically associated with molecular gas.
6. Bipolar molecular outflows (BMOs).
    Very notable and important phenomena are the *bipolar molecular outflows*. BMOs are an

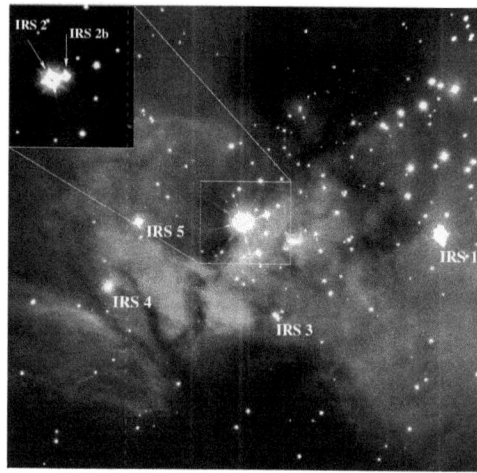

Figure 7.3: Densely packed infrared sources in the Flame Nebula (NGC 2024) in upper Orion.
Top: The early exploration K-band ($2.2\,\mu$) image is a mosaic of 64 $1' \times 1'$ fields (from Lada et al. 1991).
Left: Larger field as observed with SOFI on the NTT at La Silla (Bik et al. 2003). Image size: $5' \times 5' = 0.5 \times 0.5$ pc. Blue = Pγ, red = Brγ, green = H_2.

ubiquitous and spectacular part of the process of star formation (Bachiller & Gomez-Gonzales 1992, Staude & Elsässer 1993). For an example see Fig. 7.4.

The BMOs are often accompanied by *bipolar nebulae*, characterized by highly collimated bipolar jets (also optically), disks, and Herbig-Haro objects (tiny emitting knots in nebulae). For more see Sect. 7.6.

7.3 Instabilities in the interstellar gas

For star formation to occur it is clear that non-equilibrium processes play a major role. They transform the gas from low *interstellar* density ($\sim 10^{-23}\,\text{g cm}^{-3}$) to high *stellar* density ($\sim 10^2\,\text{g cm}^{-3}$). Instabilities in the IS gas must be responsible for this transformation.

In the physics of gases several instabilities are known. In IS gas three "instabilities" may be important: the *gravitational instability*, the *thermal instability*, and *ambipolar diffusion*. But various other aspects are relevant for the stabilization and support of the IS medium too, such as *gas pressure, rotation, turbulences, magnetic pressure*. Finding out what the balance is between supporting and contracting forces is tricky and the problem has not yet been solved.

7.3.1 Gravitational instability (Jeans instability)

The theory of gravitational instability starts with Sir James H. Jeans (1877–1946). Two forces act on a volume element (of 1 cm^3, say) in a spherically symmetric mass distribution (Jeans 1902). These are:
(1) The inwards directed gravitational force

$$F_g = \text{G} \cdot \rho \cdot M \cdot R^{-2} \tag{7.6}$$

where ρ is the density, G the constant of gravitation, and M the mass within the distance R from the centre.
(2) The outwards directed force due to the pressure gradient

$$F_p = \frac{dP}{dr} \tag{7.7}$$

or, using $P = \rho \cdot T \cdot \Re/\mu$ (with μ the mean molecular weight, $\Re$ the gas constant),

$$F_p = \frac{d}{dr}[T \cdot \rho \cdot \Re \cdot \mu^{-1}] = T \cdot \Re \cdot \mu^{-1} \cdot \frac{d\rho}{dr} \quad . \tag{7.8}$$

7.3. INSTABILITIES IN THE INTERSTELLAR GAS

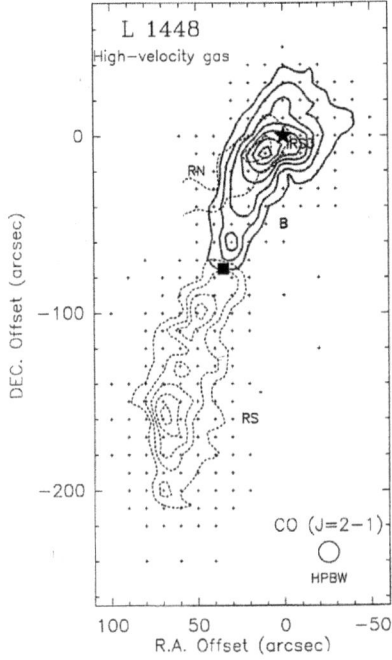
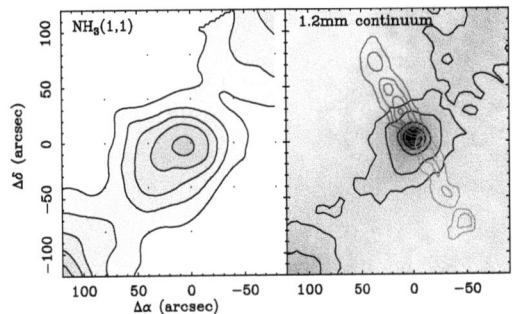

Figure 7.4: Bipolar outflows. *Top*: Bipolar outflow seen in CO emission from the IR source IRAS 04166+2706 in the core of a molecular cloud detected in NH$_3$. Contourlines are for gas with $-50 < |v| < -30$ km s^{-1}. Figure from Tafalla et al. (2004).
Left: L 1448 in Perseus (Bachiller & Gomez-Gonzales 1992, Bachiller et al. 1990) as determined from CO at 230 GHz (1.3 mm). The solid contours are for the emission with $-55 < v < 0$ km s^{-1} (B; blue, northern lobe), the dashed contours for velocities $10 < v < 65$ km s^{-1} (RS; red, southern lobe). A little receding gas is present in the North (RN; red, north lobe). The black square marks the central position, probably an unseen young stellar object. IRS 3 (*) is an IR source which excites a second, but minor outflow.

Both combine to the equation of motion

$$\frac{d^2 r}{dt^2} = -\frac{GM}{R^2} - \frac{\Re T}{\mu} \frac{1}{\rho} \frac{d\rho}{dr} \tag{7.9}$$

where the minus sign signifies inward motion.

The gas will start to contract as soon as the acceleration is negative, and this means

$$\frac{GM}{R^2} > \frac{\Re T}{\mu} \frac{1}{\rho} \left| \frac{d\rho}{dr} \right| . \tag{7.10}$$

We use the approximation

$$\frac{d\rho}{dr} = \frac{\rho}{R} \quad \text{and also} \quad R = \left(\frac{3}{4\pi} \frac{M}{\rho}\right)^{1/3},$$

and get finally

$$M_{\text{Jeans}} = \left(\frac{3}{4\pi}\right)^{1/2} \left(\frac{\Re T}{\mu G}\right)^{3/2} \left(\frac{1}{\rho}\right)^{1/2}, \tag{7.11}$$

the so-called "Jeans mass", i.e. the critical mass of instability for contraction (or the minimum mass of contraction for the ambient values of T, ρ, and μ).

An exact analysis does not give the Jeans mass but rather the "Jeans length", i.e. the wavelength of a superimposed sinusoidal density perturbation. From the Jeans length one derives the Jeans mass by multiplication with density and volume. Such a mathematical analysis results in a factor $\pi^{3/2}$ instead of $(3/4\pi)^{1/2}$.

Approximations are
 (1) $M_{\text{Jeans}} = 10^2 \, T^{3/2} \, n^{-1/2}$ (for H I gas)   and
 (2) $M_{\text{Jeans}} = 1/4 \, 10^2 \, T^{3/2} n^{-1/2}$ (for H$_2$ gas),
with M_{Jeans} in M$_\odot$, T in K and n in number of atoms resp. molecules in a cm^3.

For typical conditions in IS clouds this means:
 (1) for neutral hydrogen ($\mu = 1$, $n = 1$ cm^{-3}, and $T = 100$ K): $M_{\text{Jeans}} \approx 10^5$ M$_\odot$.
 (2) for dense molecular clouds ($\mu = 2$ [H$_2$!], $n = 1000$ cm^{-3}, $T = 10$ K): $M_{\text{Jeans}} \approx 25$ M$_\odot$.

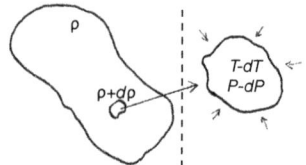

Figure 7.5: Sketch of a thermal instability. In gas with parameters temperature T, pressure P, density ρ a change of ρ to $\rho + d\rho$ in the cloud core (shown enlarged at right) may lead to $T - dT$ and $P - dP$.

7.3.2 Thermal instabilities

Thermal instabilities occur in IS gas under certain conditions. For general information on this topic see Spitzer (1978). Which aspects of ISM physics influence the occurence of instabilities?

7.3.2.1 Energy input and energy loss

Which are the processes of energy input and output of the ISM? One has to consider the processes of *heating and cooling* (see, e.g., Dalgarno & McCray 1972). Is there a balance?
  ○ IS gas can be heated by
  (1) UV sources (neighbouring young hot stars) and ionization,
  (2) X-ray sources (e.g. X-ray binary stars) and ionization,
  (3) cosmic radiation and ionization.
The total heating can be given by a heating function $H(u_\nu, (X, Y, Z))$.
  ○ Cooling mechanisms in the IS gas are (all radiative)
  (1) free-free radiation,
  (2) recombination,
  (3) collisional excitation of metal atoms and ions and radiative energy loss,
  (4) radiative energy loss from excited molecules and from dust grains.
The total cooling can be given by a cooling function $\Lambda(T, (X, Y, Z))$.
For dense molecular clouds, radiation by dust grains is the most important cooling process. These clouds may become optically thick for all radiation except IR radiation – exactly the domain where dust grains radiate.

7.3.2.2 Density fluctuations and their growth

Consider a volume element in a gas with constant pressure (as is assumed for the normal ISM) with density ρ. By random or superimposed fluctuations this volume element may get a density $\rho + d\rho$ (see Fig. 7.5).
• If cooling is effective the temperature T will decrease by an amount dT. Hence the pressure P decreases with respect to the surroundings and the volume element will be compressed. Cooling is effective when the energy gain from contraction is radiated fast enough for the density to increase further. In this manner an instability may develop.
• If energy loss through radiation from the element is slow (ineffective), then upon compression the temperature will increase again and, possibly, re-expansion will occur.

The efficiency of cooling can be described by the cooling time t_c during which a cloud at given temperature T would cool by a factor of e, so

$$\frac{d \ln T}{dt} = \frac{1}{t_c} \quad . \tag{7.12}$$

Correspondingly, one can define a heating time t_h. A thermal instability sets in when cooling is more efficient than heating, i.e., when $t_c < t_h$. The density fluctuations will then increase exponentially with the metal content dependent cooling time scale

$$t_c = \frac{3}{2} k \cdot T \cdot n^{-1} \cdot \Lambda(T, (X, Y, Z)) \tag{7.13}$$

where k is the Boltzmann constant, n the particle number [cm^{-3}] and Λ the cooling function for the particular IS gas, with X, Y, Z representing the cooling agents (chemical composition and ionization structure).

7.3.3 Stability and ambipolar diffusion in molecular clouds

7.3.3.1 Low efficiency of star formation

For molecular gas with $T = 10$ K and $n = 20$ molecules cm^{-3} the Jeans mass is about 170 M$_\odot$. Suppose all clouds are gravitationally unstable and collapse in a time like the free-fall time t_f which is (see Ch. 4.2.4)

$$t_f = \left(\frac{3\pi}{32\,G\,\rho}\right)^{1/2} = \frac{3.6 \cdot 10^7}{\sqrt{n}} \text{ yr} . \tag{7.14}$$

The total mass of molecular matter in the Galaxy is 1 to $3 \cdot 10^9$ M$_\odot$. Thus the galaxy should have a star forming rate in the order of 25 M$_\odot$ yr^{-1}. Since the observed rate is much lower, only about 5 M$_\odot$ yr^{-1}, one must conclude that most of the clouds will not become unstable!

7.3.3.2 Cloud support mechanisms

Three factors may prevent molecular clouds from collapsing. These are (1) rotation, (2) internal turbulence, (3) magnetic pressure.

(1) Rotation

Rotation may hamper star formation through three mechanisms.

- *Cloud angular momentum due to galactic rotation*
  Assume a spherical portion of the ISM in the plane of the Galaxy with density $\rho = 10^{-24}$ g cm^{-3}, mass $M = 1$ M$_\odot$, and radius $R = [(3/4\pi)\,M/\rho]^{1/3} = 2.6$ pc. Its angular momentum is $J = \theta \cdot \omega$ where the moment of inertia of the cloud is $\theta = (2/5)\,M\,R^2$ and where $\omega = 10$ km s^{-1} kpc^{-1} $= 3 \cdot 10^{-16}$ rad s^{-1}, the gradient of the rotational velocity around the centre of the Galaxy. One so obtains for the rotation (see Fig. 7.6) the angular momentum per unit mass

$$J/\text{M}_\odot \approx 10^{18} \text{ m s}^{-2} . \tag{7.15}$$

Contraction of the cloud to $R_1 = 1$ R$_\odot$ and conservation of angular momentum (i.e. $J = \theta \cdot \omega = \theta_1 \cdot \omega_1$) leads to an angular velocity of

$$\omega_1 = \omega\,(\theta/\theta_1) = \omega\,(R^2/R_1^2) \approx 4 \text{ rad s}^{-1} . \tag{7.16}$$

This corresponds to the completely unrealistic rotational velocity of $V_1 \approx 3 \cdot 10^6$ km s^{-1} at the surface of the star to be formed.
Actually, the rotation rates of cloud clumps/cores (Feigelson & Montmerle 1999) measured are in the order of $\omega \approx 1$ km s^{-1} pc$^{-1} = 3 \cdot 10^{-14}$ rad s^{-1}; the rotation rates of the NH$_3$ cores observed in the Taurus complex are always less than 5 km s^{-1} pc^{-1}.
Concluding, the angular momentum of the clouds must have been lost by a certain mechanism and rotation cannot be important in supporting cloud cores.

- *Magnetic braking*
  The process of magnetic breaking seems to be most important to produce low-angular momentum material. The magnetic field in our Galaxy (which can be deduced from the observed degree of polarisation of stellar radiation) is in the order of

$$B \approx 10^{-6} \text{ Gauss} = 10^{-10} \text{ Tesla} . \tag{7.17}$$

Note that the magnetic energy per unit of mass ($P_m = B^2/8\pi$; also called magnetic pressure) is in the order of 10^4 J g^{-1} (huge in comparison with the solar value of $< 10^{-7}$ J g^{-1}).

The magnetic field follows roughly the spiral arms. A contracting cloud will rotate with increasing angular velocity and will wind up the frozen-in field (see Fig. 7.6, left). The field lines will be stretched and produce an opposite angular momentum (like an elastic spring), leading to "magnetic braking". Quantitative estimates show that angular velocities of such clouds remain nearly constant (in spite of contraction).

- *Rotation and development of accretion disks*

  In very dense contracting clouds one can assume there is *solid body rotation* due to friction at high viscosity. Then the angular velocity is constant with distance R from the centre.

  In the equatorial plane we obtain for the centrifugal force F_z and the graviational force F_G

  $$F_z = m\omega^2 R \quad \text{and} \quad F_G = GmM/R^2 \;. \tag{7.18}$$

  It follows that within a critical radius R_c one has $F_z = F_G$, thus

  $$R_c^3 = \frac{GM}{\omega^2} \;, \tag{7.19}$$

  and the collapse can not be prevented by rotation. Outside this radius the collapse can only proceed perpendicular to the vector of angular momentum. The result is: *formation of a central object surrounded by a disk* (see Sect. 7.6).

  For more on the problem of the evolution of angular momentum see, e.g., Bodenheimer (1995).

(2) Turbulence

The influence of turbulence is not very well understood (see Shu et al. 1987, their Ch. 2.3.1). Typical line widths Δv (FWHM) of CO in molecular clouds are in the order of 1 km s^{-1} for small clouds and 10 km s^{-1} for giant cloud complexes.

When interpreted as turbulence these widths correspond to the virial values

$$v \approx \left(\frac{2\,GM}{R}\right)^{1/2} \;, \tag{7.20}$$

with R being the radius of the cloud. The *line widths* and observed *cloud sizes* correlate through a power-law: $\Delta v \sim R^a$, with $0.3 < a < 0.6$ from observations (Larson 1981, Myers 1987).

The problem is that theory has it that the turbulence should be highly *dissipative*. Furthermore, observations (maps of polarization) show that the magnetic fields are well ordered over the dimensions of the clouds. Yet, theoretical work seems to indicate that turbulence can dominate over the magnetic fields (Mac Low & Klessen 2004).

(3) Magnetic pressure

The pressure contributed by magnetic fields "supports" molecular clouds. This involves two aspects. One is that the presence of a magnetic field prevents a cloud to obtain high angular momentum through the process of magnetic braking (see above at 1). The other is that magnetic fields are a major agent for the support of the molecular clouds by magnetic pressure. Surely, the clouds are partially ionized (though the degree of ionization can be very small). The gas is then a very good conductor and *the magnetic field is frozen-in*. When the cloud contracts, the magnetic lines of force also contract and the field strength is inversely proportional to the radius squared, i.e., $B \sim R^{-2}$ (note that the same R^{-2} dependence holds for gravitation!).

Measurements of the Zeeman splitting of OH lines give typical values of the magnetic field strength B (components in the line of sight only). These are $B = 15$ to $30~\mu$G; some very large observed values are $B = 130\,\mu$G (S 106 in Cygnus), $125\,\mu$G (Orion A cloud).

Theoretical analysis allows to derive a value for the critical mass M_{crit} of a molecular cloud in the presence of a magnetic field (see the refs. in Shu et al. 1987, their Ch. 2.3.2). This is approximately

$$M_{\text{crit}} \approx 10^3 \left(\frac{B}{30\,\mu\text{G}}\right)\left(\frac{R}{2\,\text{pc}}\right) \; \text{M}_\odot \;, \tag{7.21}$$

7.4. THEORETICAL SCENARIO OF STAR FORMATION

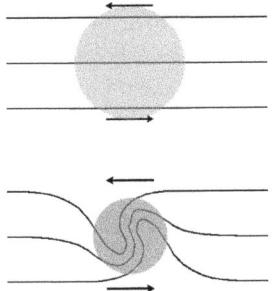
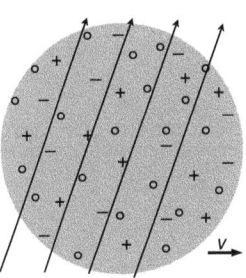

Figure 7.6: Effects of a magnetic field.
Left: When a rotating IS cloud contracts due to gravitation the magnetic field will be wound up. The rotation may originate in shear-effects of galactic rotation.
Right: The charged plasma particles interact with the magnetic field, e.g., when the gas moves (velocity v) with respect to the magnetic field, or when the gas condenses/contracts.

leading to values much larger than the classical Jeans mass. For a typical GMC with $T = 10\,\text{K}$, $n = 10\,\text{cm}^{-3}$, $B = 30\,\mu\text{G}$, and $R = 30\,\text{pc}$ it leads to

$$M_{\text{crit magn.field}} > 2 \cdot 10^5 \, M_\odot \quad \text{vs.} \quad M_{\text{Jeans}} < 10^2 \, M_\odot \quad . \tag{7.22}$$

7.3.3.3 Ambipolar diffusion

Stars have a magnetic energy per unit mass which is some orders of magnitudes smaller than in the ISM (cf. above). The question is: how can the proto-stellar cloud get rid of the magnetic field? A commonly accepted process is the *"ambipolar diffusion"*.

Fig. 7.6 (right) illustrates the situation in a molecular cloud, in which ions, electrons and neutral particles co-exist. A typical value of the ratio of charged to neutral particles is about 10^{-7} (at a density of $10^4\,\text{cm}^{-3}$). The cloud may move with a velocity v. The magnetic field is coupled to the charged particles by electromagnetic forces. The charged particles, in turn, are coupled to the neutral component by viscous forces. Now, if a molecular cloud is moving relative to the magnetic field, then positive and negative charged particles move in opposite directions ("ion slip") and a deplenishing of charged particles follows. Thus the magnetic field decouples from the neutral gas and the neutal gas may form a cloud core essentially free from magnetic forces.

Two aspects of this scenario are noteworthy.
• The degree of ionization (thus the density of charged particles) and its dependence from the density within the cloud is poorly known (ionization is due mainly to cosmic rays).
• Magnetic forces support the clouds only *perpendicular* to the magnetic lines. *Along* the magnetic lines the plasma can of course contract. Such a contraction leads to a disk. Since cooling is more efficient in a thin disk (photon escape) than in a cloud extending in 3 dimensions it will cause further contraction combined with a decreasing ionization fraction.

7.4 Theoretical scenario of star formation

The basic difficulty with the investigation of the formation of stars is that it occurs in the densest parts of molecular clouds. The associated dust causes extremely high extinction so that star formation is, observationally, not directly accessible. The fundamental scenario is a theoretical one.

A cloud in which gravity dominates the forces of thermal and magnetic pressure starts to contract (and eventually tends to break into denser parts in the course of the collapse). The density will rise towards the centre. The free-fall collapse time is roughly given by $(G\rho)^{-1/2}$ (see Sect. 7.3.3) and is therefore smaller in the interior. The result is the **inside-out-collapse**: the deep interior begins to collapse *before* the outer regions. Observations show dense cores containing CO and, at even higher densities, also NH_3.

Theory predicts that the collapse ends in a stable body, called **"protostar"**, which is (Shu 1977) an *isothermal sphere in hydrostatic equilibrium* (see Ch. 6.2.2.2). The protostar pulls in cloud material gravitationally (see Fig. 7.7 and its caption for various details).

A few further comments on fragmentation and the actual observations are needed.

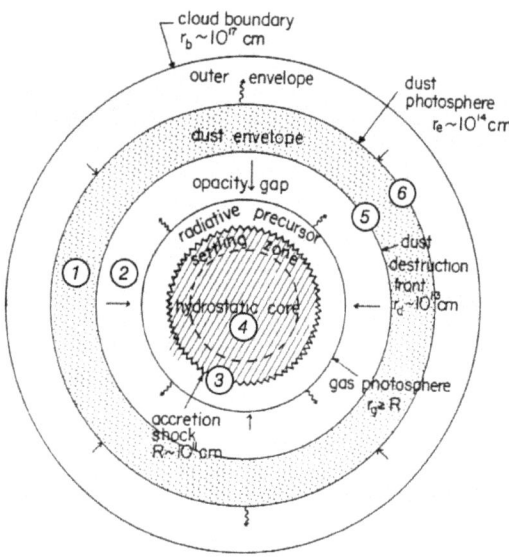

Figure 7.7: Structure of a protostar (Stahler 1994). The various layers are explained as follows.

○ The IS gas and dust from the dense birth cloud first has to pass through an optically thick dust envelope (1). The dust is then vaporized in the opacity gap (2).

○ The optically thin gas then hits the surface (3) of the hydrostatic core (4), giving rise to an accretion shock front (3) at $R \simeq 10^{11}$ cm ($\simeq 1\,R_\odot$). The intense radiation from this shock vaporizes the IS grains. Since the grains normally provide most of the opacity, the vaporization results in a nearly transparent region (2) around the protostar (4).

○ The photons flow outward freely through the opacity gap (2) before entering the thick dust envelope (1). The inner edge of the envelope is the dust destruction front (5) at $R \simeq 10^{13}$ ($\simeq 10\,R_\odot$). The temperature of both gas and dust is about 1500 K.

○ Due to the dust envelope (1) the spectrum of the proto-star becomes extremely red (multiple absorption and re-emission). The outer edge (6) is the dust photosphere where photons can escape.

Fragmentation.

The process of fragmentation of massive GMCs is not very well understood. Judging from the equation for the Jeans mass M_{Jeans}, the actual values in DMCs and GMCs (Table 7.2), and the fractal analysis, even smaller cloud masses can become unstable. Such clouds give the clues that parts of the first massive clouds fragment, initiated by spatial density fluctuations.

Apparently the dense cores seen in CO and, at higher densities, seen in NH_3, are the first signs of stars forming. The "dust photosphere" produces star-like infrared sources, the IRSs (see Fig. 7.2 right, and Fig. 7.3).

Observations of protostars.

In H II regions, in the neighbourhood of O- and B-type stars, one observes IRSs of high luminosity as is expected from very massive protostars. Famous examples are

- the Becklin-Neugebauer object in Orion ($L = 2 \cdot 10^3\,L_\odot$, $T = 500$ K),
- the Kleinmann-Low object in Orion ($L = 2 \cdot 10^3\,L_\odot$, $T = 150$ K),
- IRS5 in the H II region W3 ($> 3 \cdot 10^4\,L_\odot$) and
- IRS2 in the reflection nebula Mon R2.

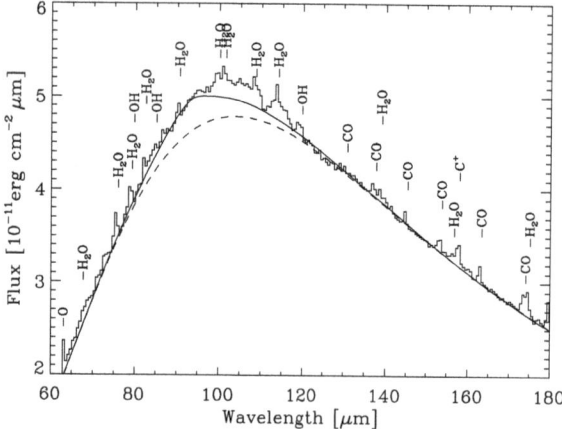

Figure 7.8: Mid- to far-IR spectrum of the solar-type protostar NGC 1333-IRAS 4 as observed with ISO. The solid line shows the spectrum with many spectral features identified. The O and C^+ emissions are well known interstellar cooling lines. CO, OH, and H_2O may come from the photon-dominated proto-stellar cloud surface, the collapsing envelope of the star-forming cloud, or from jet-like outflows. The dashed line is the best fit with a "greybody" procedure, the full line the fit including emission by calcite dust. From Ceccarelli et al. (2002) & Maret et al. (2002).

7.5. PRE-MAIN-SEQUENCE EVOLUTION (PMS EVOLUTION)

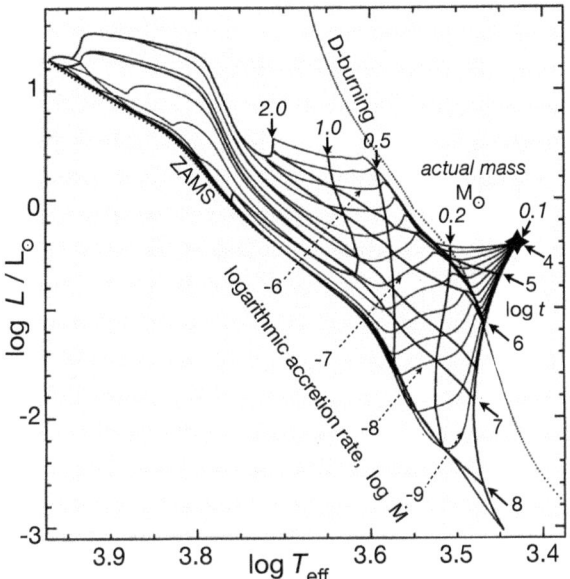

Figure 7.9: HRD showing the evolution of an *accreting* PMS star (solar metallicity, $Z = 0.02$) with $M_{\text{init}} = 0.1$ M$_\odot$ ($\star$) and thus an initial radius of 3 R$_\odot$ (for radii in the later stages use $L = 4\pi R^2 \cdot \sigma T^4$).
There are three sets of lines in the diagram. One set shows the evolution of the PMS for accretion rates between $\dot{M} = 10^{-5.5}$ and 10^{-9} M$_\odot$ yr^{-1} (steps of $10^{0.25}$ except below 10^{-8} where the step is $10^{0.5}$). A second set of lines gives the actual mass of the accreting PMS (from $M_{\text{init}} = 0.1$ M$_\odot$ to 2.0 M$_\odot$). The third set represents the isochrones ($\log t$, with $t = 10^4$ to 10^8 yr).
Further indicated are: at left the ZAMS and at right the deuterium-burning sequence. Note that star-forming material also has a small amount of D with $[D/H] \simeq 5 \cdot 10^{-5}$; D-burning through $D(p,\gamma)^3$He already starts at $T \simeq 10^6$ K (see Ch. 8, Ch. 6.5). Its ignition slows the contraction in the vicinity the D-burning sequence. Figure adapted from Tout et al. (1999).
For pre-MS evolution of higher mass stars see Fig. 7.10, for evolution of very low mass stars Fig. 8.3.

In the vicinity of high-luminosity IRSs one observes frequently microwave amplified stimulated emission radiation (MASER) sources, e.g. from OH (at 18 cm) or from H$_2$O (at 1.35 cm). Probably the maser source is in the envelope of massive protostars. The overpopulation of the higher energy levels of the molecules can be explained by absorption of the IR radiation.

The IR spectrum of a protostar is shown in Fig. 7.8. Note the broad continuum emission while emission lines by various molecules present in the birth cloud are superimposed.

7.5 Pre-main-sequence evolution (PMS evolution)

7.5.1 Energy source of PMS stars

In the pre-main-sequence phase the central temperature of the pre-main-sequence star (PMS), or protostar, is too low to ignite hydrogen burning. The protostar is not yet in, but close to, stationary equilibrium, it still is slowly contracting and is in "quasi-hydrostatic equilibrium".

The star's energy source is gravitational energy. Thus if all gravitational energy of infalling (or rather, contracting) material is being radiated the luminosity of a protostar of mass M_* and radius R_* is

$$L_{\text{proto}} = \frac{G M_* \dot{M}}{R_*} \tag{7.23}$$

where $\dot{M}$ is the accretion rate, i.e., the mass falling onto the star per unit of time. The star evolves in the Kelvin-Helmholtz time-scale (see Ch. 4.2.2)

$$t_{\text{KH}} = \frac{1}{2} \frac{G M_*^2}{R_* L_{\text{proto}}} \tag{7.24}$$

which in the case of a 1 M$_\odot$ star is about 10^7 years. We know, however (Ch. 4.2.2), that half the gravitational energy goes into heating of the gas (notably the core). Contraction can continue efficiently as long as the central regions of the cloud can cool by radiating away excess energy. This requires low opacity of the gas.

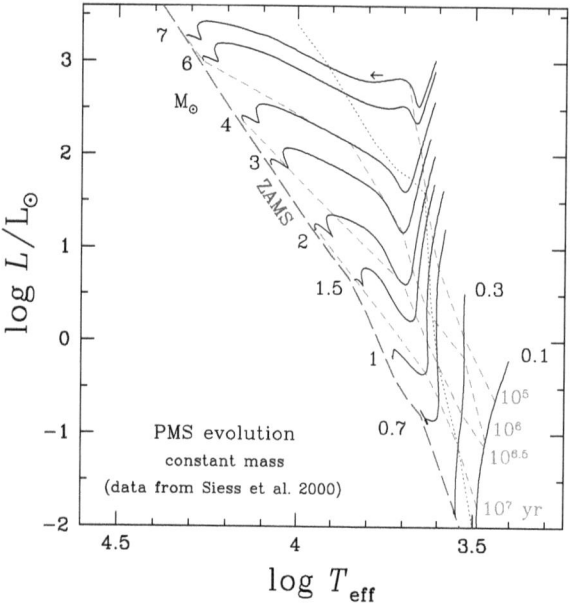

Figure 7.10: Evolution of pre main-sequence stars using models from Siess et al. (2000). The PMS objects contract, start H fusion and ultimately land on the zero age main sequence (ZAMS). During this PMS period internal energy transport is initially mostly by convection, later partly by radiation (for the structure of ZAMS stars see Fig. 6.6). The models are *without continuing accretion* (compare with Fig. 7.9). The numbers along the ZAMS indicate the mass of the star. Isochrones (- - - -) are labelled with time. The approximate location of the onset of sustainble H fusion is shown by the birth line (.....), taken from Behrend & Maeder (2001) and from the combination of Chabrier & Baraffe (2001) with Siess et al. (2000).

7.5.2 Theory of pre main-sequence stars

Theory aims at calculating stellar models with gravity as energy source. This leads to evolutionary tracks in the HRD and – for the observers – a "translation" into tracks in the CMD.

There are two main areas of uncertainty in the modelling. The first one is that the nature and condition of the gas in the contracting object is not well known (see Ch. 6.2 for various possibilities for the equations of the state of the gas). Another is that the accretion rate $\dot{M}$ is unknown.

7.5.2.1 Contraction along the Hayashi line in the earliest phase

In the simplest case one assumes an adiabatically contracting gas sphere. The relation between temperature T and density ρ is then $T \sim \rho^{\gamma-1}$ where $\gamma = C_P/C_V$ is the ratio of specific heats.

For a Jeans mass $M_{\text{Jeans}} \sim T^{3/2}\rho^{-1/2}$, thus $T \sim \rho^{1/3}$ (for $M_{\text{Jeans}} = \text{const}$). This means, a Jeans mass contracts only if in the course of contraction $\gamma - 1$ becomes greater than $1/3$; otherwise the mass is unstable. In reality the contraction is not adiabatic because the PMS radiates. As long as the temperature is too low for H and He to be ionized and still low enough to be in a low opacity regime (see Fig. 4.4) no equilibrium can be achieved. With increasing temperature, and thus ionization of H and He, γ grows and surpasses $4/3$ (for a completely ionized gas we have $\gamma = 5/3$, see Ch. 6.2). Temperature rises rapidly and the equilibrium radius R_{eq} is reached. If we assume for a moment that the total (gravitational) energy is used for ionization then we can give an approximation for R_{eq} (with E_{ion} the ionization energy)

$$\frac{M_*^2}{R_{\text{eq}}} = E_{\text{ion}} \quad . \tag{7.25}$$

In this phase the PMS star is fully convective. This means that between core and surface the temperature difference can only have a certain maximum value, otherwise the condition for convection $\nabla_{\text{ad}} < \nabla_{\text{rad}}$ would be violated. In that case, convection would stop and the star would again become unstable. The evolutionary track in the HRD, therefore, first runs perpendicular downwards – as shown in particular by the tracks for low mass stars in Figs. 7.9 and 7.10. Such a line for a given initial mass is called *"Hayashi line"* or *Hayashi track* after the first modern work on PMS evolution by Hayashi (1961).

Calculations by Wuchterl & Klessen (2001) show that solar mass stars are, during their descent in the Hayashi phase, radiative in the core with an outer convective shell.

After a sufficient amount of contraction the central density and temperature become high enough for H-fusion to become possible. The proto-star passes its "birth line" (see Fig. 7.10) and then proceeds with further structural adjustment to land on the "zero-age main sequence".

7.5.2.2 The accretion rate $\dot{M}$

If a PMS continues to accrete matter with some accretion rate $\dot{M}$ (see Fig. 7.9) the evolution is quite different from the constant M_{PMS} case (see Fig. 7.10). Model calculations unveil this dependence (see Tout et al. 1999). Fig. 7.9 shows the large range of final masses possible after an initial $M = 0.1$ M$_\odot$.

7.6 Bipolar outflows, jets, Herbig-Haro objects, disks

7.6.1 Definition of bipolar outflows and Herbig-Haro objects

A variety of phenomena can be observed which are intimately related with young stellar objects (YSOs), such as IRSs, protostars and TTSs. The basic observational aspect is the *bipolar nebula*, due to a *bipolar outflow*[2] coming from the YSO. For other clear observational cases see, e.g., Fig. 7.4 and 7.12. For reviews of the topic see Bachiller (1996), Reipurth & Bertout (1997) and Reipurth & Bally (2001).

For the definition of what a bipolar nebula is we follow Staude & Elsässer (1993, p. 166):

"Bipolar nebulae consist of two bright lobes of parabolic shape, which are located on opposite sides of the central star, thereby defining a polar axis of the system. Within the lobes,
 - dust grains reflect the light of the star, and
 - gas might be excited by the stellar radiation or by shocks.

An extended, optically thick and massive toroidal distribution of gas and dust (i.e. a *thick disk*), which surrounds the central star and seperates the bright polar lobes, defines the equatorial plane of the bipolar nebulae. Not in every individual case are these structure elements equally evident. Unfavourable orientation with respect to the observer or heavy extinction varying across the nebulae may alter their appearance."

The central YSO is often invisible (or only visible in the IR). But there are always indications of mass loss, such as:

− P Cygni spectral lines of the stellar object (i.e. violent stellar winds are present),

− low and (and sometimes high) outflow velocities as determined from emission lines of the excited gas in the lobes,

− large proper motions of embedded, clumps or "knots", called **Herbig-Haro (HH) objects**; these are probably consist of shock-excited gas,

− Doppler shifts of the emission lines.

The HH objects are always part of jets, called Herbig-Haro jets. Normally, the jets have knotty structure, sometimes there are several knots in a row (e.g. in the jet of the HH complex HH 34, see in Sect. 7.6.4), sometimes only a few single knots.

Herbig-Haro objects show a variety of emission lines. In the UV domain one sees in some cases higher energy emission in semi-forbidden lines (like 1909 C III] and 1663 O III]), in other cases low energy emission lines of 2968 C I] and bands of H_2 emission (like 1431, 1446 Å, etc.).

An object showing the entire morphology is the IR source L1551–IRS5 in Taurus (Fig. 7.11).

---

[2]The first observations of bipolar outflows and the recognition of the observed structures as due to outflow were made in the early seventies. A first explanation as due to rotation and anisotropic mass loss from YSOs was proposed by Herbig (1975).

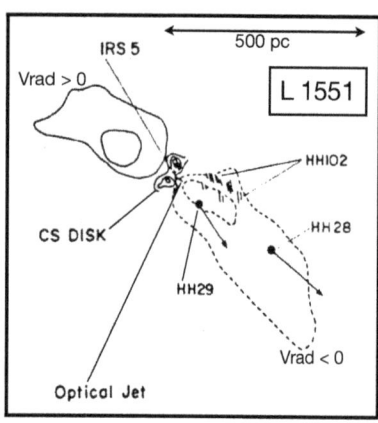

Figure 7.11: The outflow activity around L1551–IRS 5 as seen in the map of CO emission (figure adapted from Lada 1985). IRS 5 lies at the centre of the lobes.
This YSO is at a distance $d \simeq 140\,\mathrm{pc}$ and has $L \simeq 30\,\mathrm{L}_\odot$ and $A_V > 25$ mag. The figure indicates the velocity of the two lobes (full and dashed contours, respectively) of $^{12}\mathrm{CO}$ emission. In the approaching lobe one sees the HH objects HH 28 and HH 29. Their proper motion vectors (21 and 23 arcsec per 100 yr) are indicated. The lateral velocity is about $160\,\mathrm{km\,s^{-1}}$, and the arrow lengths correspond to a distance covered in 1000 yr.
The nebulosity HH 102 is essentially a reflection nebula. IRS 5 is also surrounded by a disk or torus, which has been observed in CS ($J = 1 \to 0$) at 49 GHz.

7.6.2 Some physical characteristics of bipolar flows

The characteristics of the reults of bipolar outflows can be divided in velocity cases leading to: slow ($\to$ nebulae), fast ($\to$ outflows), very fast ($\to$ jets).

1) *Bipolar nebulae* are characterized by
   ○ measured flow extent of 0.1 pc to about 1 pc,
   ○ the collimation (i.e. the ratio of outflow length to its width) is about 2 to 5,
   ○ velocities (from low-mass stars) with average $\simeq 5\,\mathrm{km\,s^{-1}}$,
   ○ estimated mass-loss rates are $\simeq 2\cdot 10^{-8}$ to $2\cdot 10^{-7}\,\mathrm{M}_\odot\,\mathrm{yr}^{-1}$,
   ○ short duration of the optically detectable outflow (of) phase, t_{of} (Staude & Elsässer 1993).
E.g., from the known number of outflow sources in the Taurus-Auriga cloud, N_{of}, and the number of low-mass PMS stars in that region N_{PMS}, whose average phase life is $t_{\mathrm{PMS}} \approx 3\cdot 10^5$ yr, follows that $t_{\mathrm{of}} = t_{\mathrm{PMS}} N_{\mathrm{of}}/N_{\mathrm{PMS}} \approx 2\cdot 10^4$ yr or $N_{\mathrm{of}}/N_{\mathrm{PMS}} \approx 1/15$.

2) *High velocity outflows* are characterized by
   ○ observed maximum velocity of a few $\mathrm{km\,s^{-1}}$ to about $100\,\mathrm{km\,s^{-1}}$,
   ○ measured flow extent of 0.1 pc to about 5 pc,
   ○ collimation of 2 to 5,
   ○ kinematical time scale (flow length divided by the maximum velocity) of $\simeq 10^3$ to $\simeq 10^5$ yr,
   ○ estimated total molecular mass from a few $0.01\,\mathrm{M}_\odot$ to about $100\,\mathrm{M}_\odot$,
   ○ sometimes detection of high-velocity components.

3) *HH jets* are characterized by
   ○ their length being shorter than that of the outflows, namely 0.03 to 0.1 pc,
   ○ typical opening angles range from 3° to 10° (collimation 6 to 20),
   ○ the speed is typically $300\,\mathrm{km\,s^{-1}}$ (200 to $400\,\mathrm{km\,s^{-1}}$),
   ○ estimated mass flow rates of about 0.05 to $2\cdot 10^{-8}\,\mathrm{M}_\odot\,\mathrm{yr}^{-1}$,
   ○ the gas is shock-heated to about 10^4 K.
Examples of jets can be found in Fig. 7.12 and Fig. 7.13.

7.6.3 Circumstellar disks

A circumstellar disk was first detected for the case of HL Tauri (see Fig. 7.13) in CO by Sargent & Beckwith (1987). HL Tauri became thus the prototype.
   Typical aspects of disks are:
   ○ an elongated disk-like structure with a long semi-axis; for HL Tau the extent is 2000 AU ($\simeq 50\times$ the semi-major axis of Pluto's orbit) but lengths down to 100–200 AU have been found.
   ○ the mass is in the order of $0.1\,\mathrm{M}_\odot$ (but also values of $10^{-7}\,\mathrm{M}_\odot$ can be found in the literature!).

7.6. BIPOLAR OUTFLOWS, JETS, HERBIG-HARO OBJECTS, DISKS

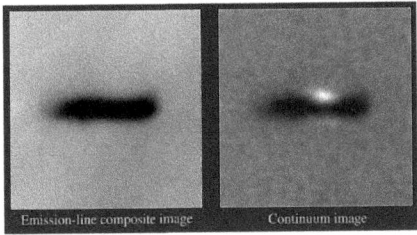 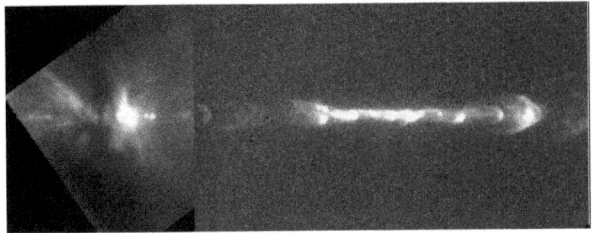

Figure 7.12: The disk around the YSO Orion 114 and the jet of YSO HH 111 in Orion B.
Left: CCD images of the object Orion 114 observed in the light of emission lines (left image, sum of O III, H α and N II) and in the continnum (right image). In emission lines, the diffuse gas is bright but the central object (the star) is not bright. The circumstellar disk absorbs all light from behind, both of the star and the diffuse gas. In the image taken in continuum light, the star is visible except for the heavy obscuration in the disk. Images from McCaughrean & O'Dell (1996).
Right: HST observations of the object HH 111 in Orion B including its 4 pc long jet. The left part of the image was obtained in the infrared, the right part in the visual (Hα and N II). At left, there is a dark circumstellar disk (vertical) around the YSO. The straightness of the jet is well visible as are the "knots" (bright regions) where the jet interacts with the local gas in bow-shock-like structures. Image from STScI News Release 2000-05. See also Reipurth & Bally (2001).

∘ high velocity resolution data demonstrate that this disk gas is bound to (and in Keplerian rotation about) the star.

The object L1551–IRS5 is also surrounded by a disk or torus, which has been observed in CS ($J = 1 \to 0$) at 49 GHz (see Fig. 7.11).

More than 50% of all YSOs in Taurus show circumstellar disks. The IR excess of the spectra of all TTSs is an additional proof that disks are a natural phenomenon for YSOs; the double-peaked emission lines support this. The disks can be considered to be proto-planetary clouds.

Fig. 7.18 tries to summarize all phenomena in a uniform picture.

7.6.4 Origin of outflows

Several origins for the outflows have been suggested.

Decay of magnetic fields or release of potential energy are discussed. The relation between the magnetic structure of the rapidly rotating central object with the outflow activity has been discussed in more detail by Camenzind (1990).

The jet of YSOs with bipolar nebulae is perhaps driven by some "engine" while the molecular outflow is powered by the infalling matter through a quadrupolar circulation pattern around the central object (Lery et al. 2002). Part of the material falling in through a disk may get redirected by the radiative force of the surface of the YSO to be driven out at the poles of the star. The increase in L together with the presence of the still active accretion disk means that a wind can only escape in the direction of the poles (see also Fig. 14.6).

Impressive examples of YSOs and their surroundings are:

∗ HL Tauri and HH 30: A large number of flows, jets, HH objects, and disks can be seen in this region in Taurus (see Fig. 7.13). The figures show the extremely complex structure as is normally present in star forming regions.

∗ HH 34: A most exciting region in the southern part of Orion around the giant complex HH 34. Wonderfull colour pictures have been taken with the ESO Very Large Telescope VLT2 "Kueyen" equipped with the camera FORS (see ESO Press release 17/99 from Nov. 17, 1999). Numerous protostars can be seen. A detailed analysis of the region is given by Devine et al. (1997). The jet length is 30″ (or about 0.03 pc), and the velocity 300 km s^{-1}. The "waterfall" at the lefthand side in that ESO press release picture is an enigma.

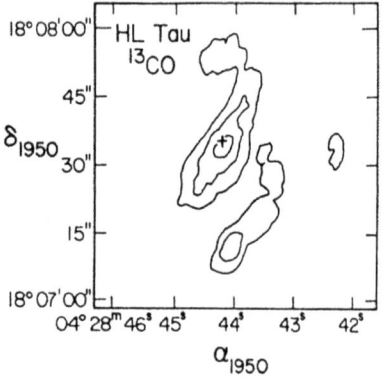
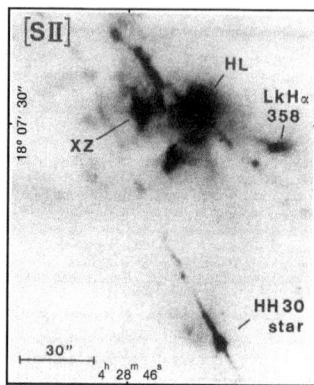

Figure 7.13: The region of the Herbig-Haro objects HL Tau, XZ Tau and the jet with HH 30.
Left: The discovery image of the circumstellar disk around HL Tauri detected in ^{13}CO emission (from Sargent & Beckwith 1987).
Right: CCD image of the region in the light of the forbidden sulfur lines [S II] 6716/6713 Å (from Appenzeller & Mundt 1989).

7.7 Very young stars

Very young stars show a variety of phenomena which characterize them as very young indeed. These vary depending on the mass of the star. Stars like T Tauri were the first to be recognized as young, followed much later by the Ae/Be stars.

7.7.1 General characteristics of T Tauri stars

The objects which can readily be related to the theoretical considerations are the so called T Tauri stars (T Tau stars, **TTSs**). They have been defined in a famous paper by Joy (1945) on the basis of photometric and spectroscopic criteria (and are sometimes called "objects of Joy").

- *Photometry, variability*

As the name indicates, TTSs are a special type of variable stars.
– Normally they are variable in the time scale of days. But variability down to hours or even minutes can be observed. Sometimes a weak periodicity might be present; sometimes one observes flare-like outbursts and irregular variations.
– The light amplitudes are in the order of 1.5 to 4 mag.
– TTSs are variable in all wavelength regions *except* in the far IR and sub-mm regime. Note that the classification of young variable stars is a bit sophisticated (see Duerbeck & Seitter 1982).

- *Spectral type*

Spectral types in the range F to M (Fe to Me, with "e" for emission). There are TTSs known which have earlier spectral types. These are called *Herbig Ae and Be stars* (see the study by Herbig (1960) and Sect. 7.7.4). Prominent examples: Z CMa (B9pe), R Mon (A3e), T Ori (B8–A3ep).

- *Emission lines*

Emission lines are characteristic for TTSs (see Fig. 7.14 top). Strongest lines are those of the Balmer series but also Ca II H and K are strong. Many other elements appear in emission. The strength of the emission leads to a classification of the TTSs:

(1) classical T Tauri stars (*CTTS*), stars with strong emission,
(2) weak line T Tauri stars (*WTTS*), stars with weak emission or without.

For spectra of each type (CTTS and WTTS) see Fig. 7.15, left panel.

- *Forbidden emission lines*

A number of TTSs shows forbidden lines. Strongest line is [O I] at 6300 Å; other lines are [S II]

7.7. VERY YOUNG STARS

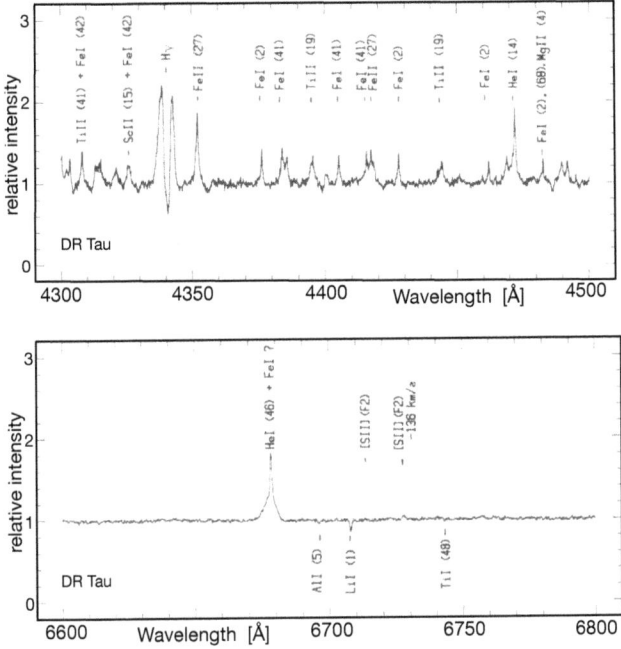

Figure 7.14: Spectra of the classical T Tau star DR Tau. The lines are identified by the ion name and the spectral multiplet number.
Top: Section of the blue spectrum. The double peaked Hγ emission probably comes from a rotating disk. One further sees lines of neutral and singly ionized metals.
Bottom: Section of the red spectrum. A few marginally detected lines are present. The emission of He I is strong (as is the line from He I in the blue spectrum). Note the Li absorption line.
Figures adapted from Appenzeller & Mundt (1989).

at 6716/6731 Å (see Fig. 7.14 bottom), or [O II], [Fe II], etc.

- *Shape of the emission lines*

The two-peak emission of the Hγ line (e..g., DR Tau: Fig. 7.14 top) can be explained by emission from a disk-like structure around the star. In several cases violet-displaced absorption edges of the emission lines (*P Cygni profiles*) have been observed They indicate the presence of *stellar winds*, thus expanding atmospheres. Crude mass-loss estimates lead to values of $4 \cdot 10^{-8}$ to $3 \cdot 10^{-7}$ $M_\odot$ yr^{-1}.

Some stars (e.g. XY Orionis) show red-displaced absorption edges (inverse P Cygni profiles) which point to the presence of infalling material. This group of stars is called *XY Ori stars*.

All emission peaks are blueshifted, displaying normally two different velocity components, one at about -100 to -200 km s^{-1}, the other one at -10 to -30 km s^{-1} (thus close to the stellar velocity). Their interpretation is still unclear. The faster component may originate in the jet, because the gas in jets and Herbig-Haro objects (see Sect. 7.6) has similar velocities.

- *Continuum*

The continuum of the TTSs displays a strong IR excess (see Fig. 7.15, right panel). The three CTTSs show much stronger excesses than the WTTS TAP 57. This can be explained by larger accretion disks around the CTTSs.

- *FUORs*

Some TTSs show light outbursts of up to 6 mag within the time scale of months, e.g., the star **FU OR**ionis. This group is called *FUORs*. More on these stars in the review by Hartmann & Kenyon (1996).

7.7.2 T Tau stars and X-ray emission

The Taurus molecular clouds (TMCs) shown in Fig. 7.2 have been investigated intensively in many wavelength ranges. After the detection of gas complexes in CO, the near-IR revealed many point sources and Herbig-Haro objects.

X-ray data from the XMM-Newton X-ray satellite showed that many pre-MS and T Tau stars produce X-rays (see, e.g., Güdel et al. 2007). This is understood as due to magnetic activity at the star's surface. The YSOs have a strong magnetic field from the star-formation process, which

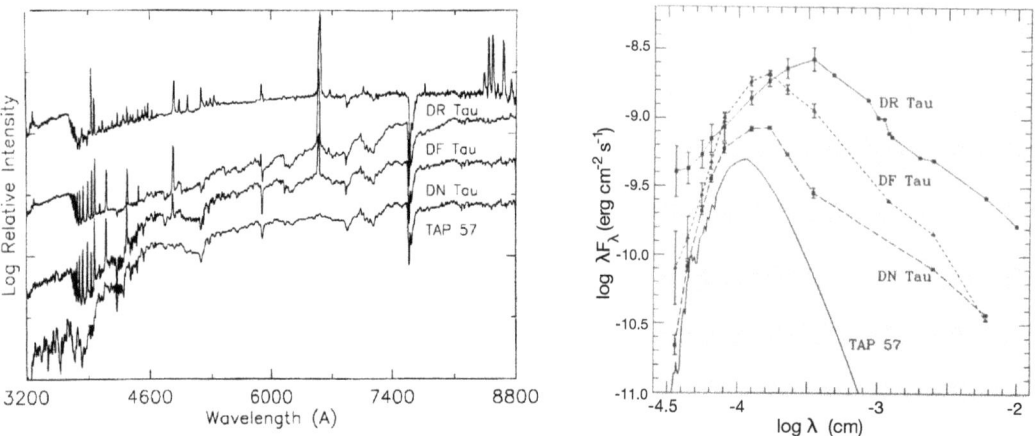

Figure 7.15: *Left:* Medium resolution spectra of four late-K early-M T Tau stars. The upper three stars are classical TTs, TAP 57 is a weak-line TTS. Note the Balmer series emission in the upper three spectra, including the emission in the Balmer continuum, signifying cooling. *Right:* Observed spectral energy distribution of TTSs from 3600 Å to 100 μ. Figures from Bertout (1989).

leads to strong effects in the surface layers (stellar coronae; see Ch. 12) of these objects. There are three trends:
- The more massive stars rotate faster; this clearly is the result of the larger angular momentum a contracting more massive object retains.
- The more massive stars have a stronger X-ray flux; a more massive star has included more of the ambient magnetic field leading to a more active corona.
- The less young stars have a less strong X-ray flux; this can be understood as due to the evolution of the stellar rotation due to magnetic braking (for the mechanism see Ch. 14), so that the coronal activity gets reduced.

7.7.3 T Tauri stars as young objects

For many TTSs it has been posssible to determine values for L and T_{eff}. A plot of these objects in an HRD supports the notion they are pre-main-sequence stars indeed (Fig. 7.16).

An additional hint for TTSs being young objects is the presence of *lithium* lines in their spectra (see the 'strong' Li 6707 Å doublet in the spectrum of DR Tau, Fig. 7.14 bottom). Lithium is not generated in stellar fusion processes but is consumed (see Chs. 5 and 8.2.2). At temperatures $T > 2 \cdot 10^6$ K, already occurring not far below the surface of late-type stars, Li captures protons and is subsequently changed into helium by α decay:

$$^7\text{Li}\,(p,\alpha)\,^4\text{He} \quad \text{or} \quad ^7\text{Li} + p \longrightarrow \,^4\text{He} + \,^4\text{He} \ . \tag{7.26}$$

Also, convection transports (originally) lithium-rich material to deeper layers where it is destroyed. The detection of rather strong Li lines in the spectra of TTSs (the Li line in DR Tau is really strong compared to the other lines in that spectral region!) indicates that this process can not have been active over a long time, i.e., the TTSs are very young stars (see also BDs, Ch. 8).

Summarizing, TTSs represent early phases of stellar evolution, because
- they are associated with regions of active star formation (dark clouds, molecular clouds),
- they show accretion disks and have often molecular outflows (see Sect. 7.6),
- their position in the HRD is in the region of pre-MS stars, and
- lithium has been detected in T Tau star spectra.

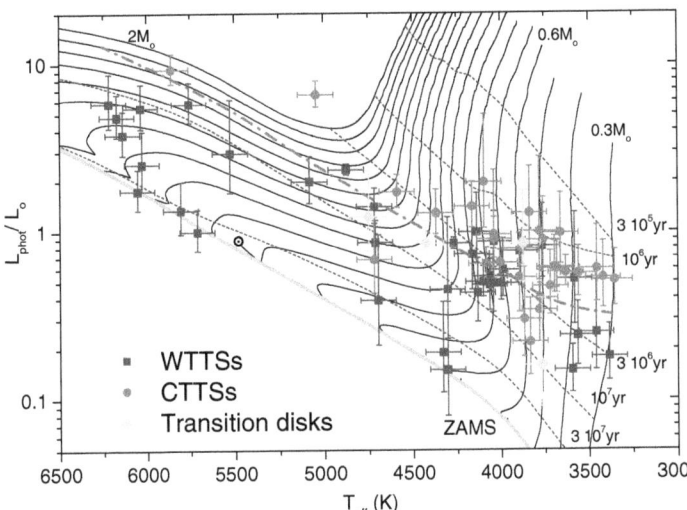

Figure 7.16: HRD of objects detected in the Taurus-Auriga T association. The symbols indicate the object type (see the explanation inside the figure.)
The evolution tracks are from Siess et al. (2000), are based on $Y = 0.277$, $Z = 0.02$ (not quite solar), and are labelled with the mass (steps of 0.1 $M_\odot$). There is no mass gain in this eveolution. The tracks support that the objects are pre-MS stars. Iso-age lines are indicated (compare with Fig. 7.10).
Figure from Bertout et al. (2007).

7.7.4 Herbig Ae and Be stars

Once stars have formed some signature from the star formation material remains as in the Herbig Ae and Be stars. These objects are the *massive versions* of spectral type A or B (with emission index e) of the lower mass versions known as T Tau stars. They are embedded in dusty material, too, but the luminosity of the star (which is already on the main sequence) has resulted in partial dilution of the birth cloud and/or of the left-over accreting matter.

Ae and Be stars are marked in particular by their Hα emission. That emission may be variable, sometimes is temporarily absent.

Ae and Be stars can be easily discovered by photometrically imaging a star field in at least two photometric bands (B and V) and in addition in the Hα filter (method first explored by Grebel et al. 1992). Combining the information from the CMD and a two colour diagram with $H\alpha - V$ versus $B - V$ allows to identify the Be and Ae stars immediately because the Be stars are brighter in Hα. See Fig. 7.17 for the first application using Strömgren filters.

An alternative explanation for Ae/Be stars is that the emission lines come from a shell shed *at the end of the MS evolution*. This mass would be lost by stars with still relatively fast rotation. For that see Ch. 14.8.

More on Ae/Be stars can be found in the review by Waelkens & Waters (1998). A few remarks on Oe and massive Be stars are made in Ch. 13.2.1.3.

7.8 Summary

Star formation takes place in dense interstellar gas clouds which become compact through cooling, molecule formation and gravitational contraction. Most of what we know comes from the formation of low-mass stars. Formation of massive stars must be very much obscured and is rather fast.
- As soon as a proto-stellar cloud has formed dense condensations, gravitation will proceed to make these more compact.
- Condensations can only become denser if the interior can sufficiently radiate away the energy generated by contraction.
- Details of star-formation are poorly known (effects of magnetic fields, formation of jets, etc.).
- Very dense clumps will ultimately heat up in the core (isothermal sphere with ideal gas).
- If the central temperature becomes high enough, some fusion will start (D $\rightarrow$ H). The object may proceed to H-fusion to become star.

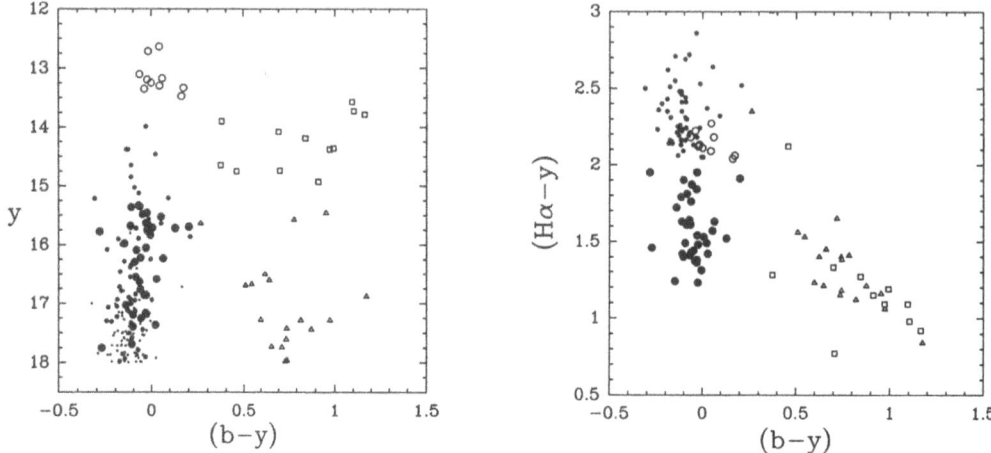

Figure 7.17: Finding Be and Ae stars in a cluster works well combining a CMD (at left) with the two colour diagram, $H\alpha - y$ versus $b - y$ (at right). Symbols: • = MS stars, ○ = blue supergiants, △ = red giants, □ = red supergiants. Here $H\alpha - y$ is defined as $2.5 \log I(H\alpha)/I(y)$ (thus without − sign!) so that being bright in $H\alpha$ means a small colour index. Stars *with* emission in $H\alpha$ deviate in the two colour diagram from the location of normal stars, the latter being expected near the diagonal through the two-colour diagram. Be and Ae stars are massive *and* young. Data for the young SMC cluster NGC 330, thus $y = 18.8$ mag is $M_V = 0$ mag. Figure from Grebel et al. (1992).

• The energy released may hamper further accretion; the final result of the formation depends on the total mass available around the protostellar core and the further accretion possible.
• Young stars show a multitude of phenomena. The most important ones are:
∗ bright and variable emission lines due to energy dumped into the remains of the birth cloud,
∗ polar outflow phenomena (see Fig. 7.18).
• Finally, the young star becomes visible unhampered by gas and dusty material of the birth cloud.

References

Appenzeller, I., & Mundt, R. 1989, A&A Rev. 1, 291; *T Tauri Stars*
Bachiller, R. 1996, ARAA 34, 111; *Bipolar Molecular Outflows from Young Stars and Protostars*
Bachiller, R., & Gomez-Gonzales, J. 1992, A&A Rev. 3, 257; *Bipolar Molecular Outflows*
Bachiller, R., Martin-Pintado, J., Tafalla, M., Cernicharo, J., & Lazareff, B. 1990, A&A 231, 174
Bahcall, J.N. 1986, ARAA 24, 577; *Star Counts and Galactic Structure*
Bahcall, J.N., & Soneira, R.M. 1980, ApJS 44, 73
Bally, J., Langer, W.D., Stark, A.A., & Wilson, R.W. 1987, ApJ 312, L 45
Barnard, E.E. 1919, ApJ 49, 1
Behrend, R., & Maeder, A. 2001, A&A 373, 190
Bertout, C. 1989, ARAA 27, 351; *T Tauri Stars: Wild as Dust*
Bertout, C., Siess, L., & Cabrit, S. 2007, A&A 473, L21
Bik, A., Lenorzer, A., Kaper, L., Comerón, F., Waters, L.B.F.M., de Koter, A., & Hanson, M.M. 2003, A&A 404, 249
Bodenheimer, P. 1995, ARAA 33, 199; *Angular Momentum Evolution of Young Stars and Disks*
Camenzind, M. 1990, Rev. Modern Astronomy 3, 234; *Magnetized Disk-Winds and the Origin of Bipolar Outflows*
Carruthers, G. 1970, ApJ 161, L 81
Ceccarelli, C., Caux, E., Thielens, A.G.G.M., Kemper, F., Waters, L.B.F.M., & Phillips, T. 2002, A&A 395, L29
Chabrier, G., & Baraffe, I. 2000, ARAA 38, 337; *Theory of Low-Mass Stars and Substellar Objects*

7.8. SUMMARY

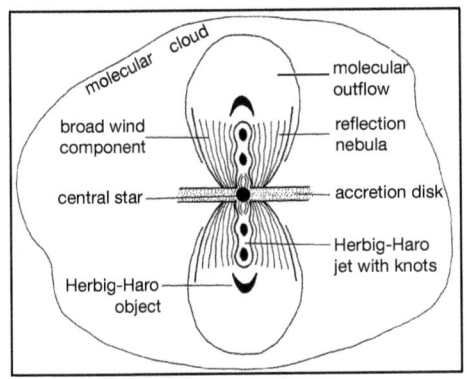

Figure 7.18: Schematic view of a bipolar nebula and its structure elements. These are: the young star, the jet-like outflow (with bright "knots"), the tip-of-the-jet HH object, the reflection nebula also exhibiting broader outflow (wind) and the molecular gas flowing out which interacts with the original birth (molecular) cloud. Compare with the observed features as shown in previous images. Figure after Staude & Elsässer (1993).

Cheung, A.C., Rank, D.M., Townes, C.H., Thornton, D.D., & Welch, W.J. 1968, Phys. Rev. Lett. 21, 1701
Dalgarno, A., & McCray, R.M. 1972, ARAA 10, 375; *Heating and Ionization of H I Regions*
Devine, D., Bally, J., Reipurt, B., & Heathcote, S. 1997, AJ 114, 2095
Duerbeck, H., & Seitter, W. 1982, in: "Landolt-Börnstein", Group VI, Vol. 2b, p. 258
Elmegreen, B.G., & Falagrone, E. 1996, ApJ 471, 816
Evans, N.J. 1999, ARAA 37, 311; *Physical Conditions in Regions of Star Formation*
Falgarone, E., & Phillips, T.G. 1996, ApJ 472, 191
Feigelson, E.D., & Montmerle, T. 1999, ARAA 37, 363; *High energy processes in Young Stellar Objects*
Grebel, E.K., Richtler, T., & de Boer, K.S. 1992, A&A 254, L 5
Güdel, M., Briggs, K.R., Arzner, K., et 18 al. 2007, A&A 468, 353
Hartmann, L., & Kenyon, S.J. 1996, ARAA 34, 205; *The FU Orionis Phenomenon*
Hayashi, C. 1961, PASJ 13, 450.
Heithausen, A., & Mebold, U. 1989, A&A 214, 347
Herbig, G.H. 1960, ApJS 4, 337
Herbig, G.H. 1975, ApJ 200, 1
Herbst, E. 1988, Rev. Modern Astronomy 1, 114; *Interstellar Molecular Formation Processes*
Jeans, J.H. 1902, Proc. Royal Soc., London, Vol. 71, 136
Joy, A.H. 1945, ApJ 102, 168
Kroupa, P. 2002, Science 295, 82
Lada, C.J. 1985, ARAA 23, 267; *Cold Outflows, Energetic Winds, and Enigmatic Jets*
Lada, E.A., Evans, N.J., Depoy D.L., & Gatley I. 1991, ApJ 371, 171
Larson, R.B. 1981, MNRAS 194, 809
Lery, T., Henriksen, R.N., Fiege, J.D., Ray, T.P., Frank, A., & Bacciotti, F. 2002, A&A 387, 187
Lynds, B.T. 1962, ApJS 7, 1
Mac Low, M.-M., & Klessen, R.S. 2004, Rev. Mod. Phys. 76, 125
Maret, S., Ceccarelli, C., Caux, E., Thielens, A.G.G.M., & Castets, A. 2002, A&A 395, 573
McCaugrean, M.J., & O'Dell, C.R. 1996, AJ 111, 1977
Myers, P.C. 1987, in "Interstellar Processes", D. Hollenbach & H. Thronson (eds.); Reidel; p. 71
Myers, P.C., & Benson, P.J. 1983, ApJ 266, 309
Molaro, P., & Bonifacio, P. 1990, A&A 236, L 5
Reipurth, B., & Bally, J. 2001, ARAA 39, 403; *Herbig-Haro Flows: Probes of Early Stellar Evolution*
Reipurth, B., & Bertout, C., (eds.). 1997, IAU Symp. 182, "Herbig-Haro Flows and the Birth of Low Mass Stars"; Kluwer
Salpeter, E.E. 1955, ApJ 121, 161
Sargent, A.I., & Beckwith, S.W.B. 1987, ApJ 323, 294
Siess, L., Dufour, E., & Forestini, M. 2000, A&A 358, 593

Shu, F.H. 1977, ApJ 214, 488
Shu, F.H. 1991, in: "Frontiers of Stellar Evolution", D.L. Lambert (ed.). ASP Conf. Ser. 20, p. 23; *Star Formation: A Theoretician's View*
Shu, F.H., Adams, F.C., & Lizano, S. 1987, ARAA 25, 23; *Star Formation in Molecular Clouds*
Snyder, L.E., Buhl, D., Zuckerman, B, Palmer, P. 1969, Phys. Rev. Lett. 22, 679
Spitzer, L. 1978, "Physical Processes in the Interstellar Medium", Wiley-Intersc. Publ., New York
Stahler, S.W. 1994, PASP 106, 337
Staude, H.J., & Elsässer, H. 1993, A&A Rev. 5, 165; *Young Bipolar Nebulae*
Stutzki, J., Bensch, F., Heithausen, A., Ossenkopf, V., & Zielinsky, M. 1998, A&A 336, 697
Tafalla, M., Santiago, J., Johnstone, D., & Bachiller, R. 2004, A&A 423, L21
Tout, C.A., Livio, M., & Bonnell, I.A. 1999, MNRAS 310, 360
van Dishoeck, E.F., & Blake G.A. 1998, ARAA 36, 317; *Chemical Evolution of Star-Forming Regions*
Waelkens, C., & Waters, L.B.F.M. 1998, ARAA 36, 369; *Herbig Ae/Be stars*
Wilson, R.W., Barrett, A.H., & Moran, J.M. 1970, ApJ 161, L 43
Wuchterl, G., & Klessen, R.S. 2001, ApJ 560, L 185

Chapter 8

The almost stars: Brown Dwarfs

8.1 Introduction and naming problems

The lowest mass objects that can be called "star" are the brown dwarfs (BDs) since they have energy production through fusion[1]. The surface of these objects is so cool ($T < 3000\,\mathrm{K}$) that their colour was redder than red and they thus were called brown (Tarter 1975).

In the early research of BDs there has been some confusion about the name giving. Sometimes a BD was taken to be a star at the lower end of the main sequence, in other cases objects below the MS low mass end were called BDs. Furthermore, the mass of such low mass objects may extend into the mass of planet like objects. Here the planet Jupiter is the example and often masses of BDs are given in units of Jupiter mass, M_{Jup} (1 $M_{\mathrm{Jup}} \simeq 0.001\,M_\odot$).

In the course of time the name giving has been standarized.
→ **MS stars** are those objects, which eventually come to sustainable H fusion.
→ **BDs** are objects, which for a while have deuterium fusion, but then turn into passive objects.
→ **Planets** are those objects that never came to fusion in their centres.

Another division in object kinds is also used: BDs are taken to be objects which formed in the same way as stars (see Ch. 7), while planets are those objects having formed in a **proto-planetary disk** around a young star (see Ch. 7.6.3). Since there is no information (yet) in observations about the formation history of the object seen (was it formed like star formation or in a proto-planetary disk?), this separation is at the moment only of theoretical value.

The first brown dwarf, Gl 229B, was discovered by Nakajima et al. (1995). Since then numerous BDs have been found, in particular from near-IR photometric surveys, and the namegiving as described above was settled.

8.2 Nuclear fusion in brown dwarfs

8.2.1 Deuterium burning

It is possible that objects with mass less than $M_{\mathrm{min\,MS}} \simeq 0.08\,M_\odot$ as derived in Ch. 6.5 become star. This has to do with the fact that deuterium reaches fusion conditions already at temperatures substantially less than needed for hydrogen fusion. If during star formation the central temperature of the young stellar object surpasses $T_c \simeq 0.5 \cdot 10^6\,\mathrm{K}$ then deuterium will start burning as

$$\mathrm{D} + \mathrm{H} \to {}^3_2\mathrm{He} + \gamma \quad . \tag{8.1}$$

This reaction is the same as step two in the PP I-chain (see Ch. 5.1.4), a chain which in its entirety operates only for stars with $M > M_{\mathrm{min\,MS}}$. The low mass limit for D-burning is near 0.012 $M_\odot$.

---

[1] Normally, only objects having fusion of hydrogen (or of heavier elements) are called "star". But note that also the objects remaining after the supernova explosion of higher mass stars, the neutron stars, are called stars, although they do not have fusion any more; and white dwarfs are often called white-dwarf stars.

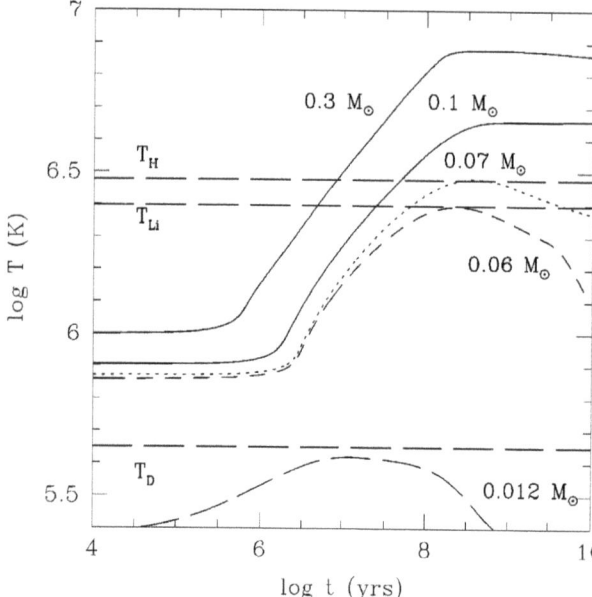

Figure 8.1: The evolution of proto-stellar objects shows that only when $M > 0.08\,M_\odot$ the central temperature rises far enough to sustain nuclear fusion. Objects with less mass will have a fairly high T_c for a while, but fail to become star.
Note the horizontal lines being the *temperature limits for fusion* by H, Li, and D. Note also that the time is given as **log** years thus showing the *slowness* of the evolution, the time it takes a low mass star to come to stable H-fusion but also the rather long time such objects may have very low-level D fusion. Figure from Chabrier & Baraffe (2000).

The abundance of deuterium in the Milky Way today (as found from stellar as well as interstellar spectroscopy) is $[D/H] \simeq -5$. Because the amount of D in a young star-like object is small, the fuell is, at some point, fully consumed in BDs and the fusion then stops completely.

In stars with $M_{\rm init}$ near $0.08\,M_\odot$ the D-burning will start first, raising T_c and thus helping the object to reach the H-burning state.

The considerations for the conditions for fusion form the basis for the name giving described above: the stars with $M_{\rm init} > 0.08\,M_\odot$ will become true MS stars, while those below that limit (but not too far) will burn only D for a while and are called BD. This division and behaviour can be seen in Fig. 8.1, where the evolution of the central temperature is shown in relation to the total mass. Clearly only objects with $M > 0.08\,M_\odot$ will reach the "star state" and will have T_c high enough for H-fusion over a long period of time. Only those lower mass objects with T_c just above $10^{5.65}$ K have a certain period of D-burning.

BDs are fully convective objects (Ch. 6.2.2.2), so that all D available in the star can be used for the fusion. One consequence is that the abundance of D in the atmosphere should decrease with time. The evolution of the D content is shown in Fig. 8.2.

8.2.2 Lithium burning

The element Li, an element formed in and remaining after the big bang, can also come to fusion at temperatures below that of H-fusion. Li is present in two isotopic forms, ^{7}Li, and ^{6}Li.

^{7}Li burns at $T > 2.5 \cdot 10^6$ K (see Fig. 8.1), a temperature reached in BDs of $M_{\rm BD} > 0.06\,M_\odot$. ^{6}Li burns even at $T > 1.5 \cdot 10^6$ K, thus only requiring $M_{\rm BD} > 0.035\,M_\odot$ (or 35 $M_{\rm Jup}$!).

The normal burning of ^{7}Li is

$$^7{\rm Li} + p \rightarrow 2\ {\rm He} \qquad (8.2)$$

equal to step three in the PP II-chain (see Ch. 5.1.4). If a star has an age larger than 10^8 yr (e.g. derived using the age of the parent star cluster) and is found containing ^{7}Li then it is a bona-fide BD (Rebolo et al. 1992). This parameter combination signifies an object of low mass whose T_c never came above the Li burning limit of $2.5 \cdot 10^6$ K.

Attempts have been made to prove the occurance of ^{6}Li burning by looking at the abundance of the Li isotopes in the spectra of stars and BDs. It requires distinghuishing the very weak ^{6}Li absorption in the wing of the weak ^{7}Li line near 6708 Å. The surface Li abundance should change with time (see next section).

8.3. EVOLUTION AND SURFACE PARAMETERS OF BDS

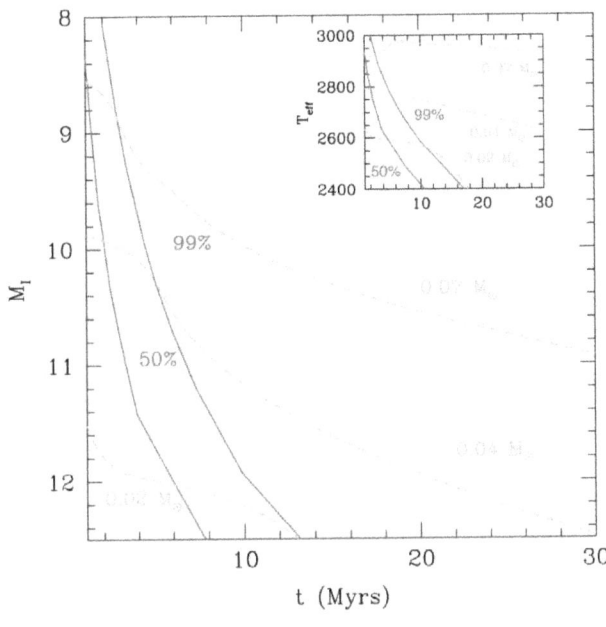

Figure 8.2: The evolution of the content of deuterium in the fully convective BDs is shown with the absolute brightness M_I against time. The dashed lines give the mass of 3 BDs and the change in M_I, the solid lines at which point in time 50% or 90% of D is consumed. The inset shows the same but with T_{eff} against time. Figure from Chabrier & Baraffe (2000).

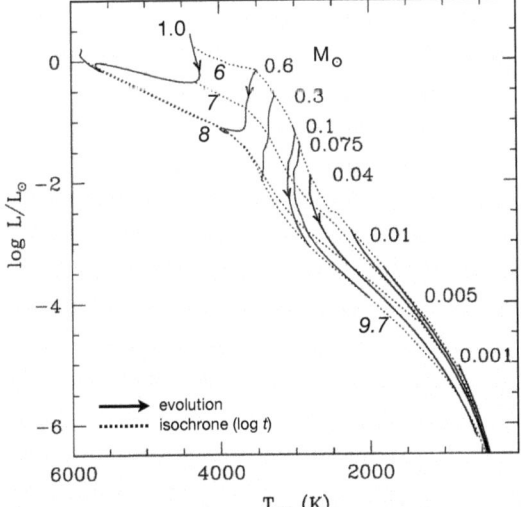

Figure 8.3: Evolutionary tracks for objects becoming brown dwarf or becoming star on the lower MS are shown along with isochrones. The tracks (———) are labelled with their M_{init}. Note that $M_{\text{init}} < 0.075\,M_\odot$ objects will decrease in luminosity continuously (arrow to lower right), above this mass limit they will stay on the MS for considerable time. The isochrones ($\cdots$) are (from top to bottom) for 10^6, 10^7, 10^8 and $10^{9.7}$ yr.
This figure is similar to Fig. 7.16 but shows in particular the tracks for the BDs. The models do not involve continuing accretion of matter. Figure from Chabrier & Baraffe (2000).

8.3 Evolution and surface parameters of BDs

Initially, BDs radiate due to the gravitational energy released in contraction. BDs then evolve rather slowly having slow D-burning in the interior with degenerate gas (see Ch. 6.5). Depending on M_{BD}, the evolution is either just slow (higher mass range, higher T_c) or very slow (lower mass range, lower T_c). During evolution, T_{eff} may reach a (cool) maximum of $T_{\text{eff}} \simeq 3000$ K (see Fig. 8.3) but the luminosity decreases continuously.

Since BDs are fully convective, the content in D in the atmosphere must decrease with time. If also Li is consumed, its abundance should change, too. Depending on the duration of the convection phase, much of the Li stays unaffected and can be used to test nucleosynthesis in big bang models (see Chs. 5.2.5 & 22.3).

Photometry shows that BDs have really cool atmospheres, $800 < T_{\text{eff}} < 2500$ K. In such conditions numerous molecules are present, such as H_2, CO, CH_4, etc. (see Figs. 2.7 & 8.4), as well as isotopic forms of these. TiO (also present in M-stars of the lower main sequence) is absent which in fact led to the definition of the spectral type L (very cool stars having no TiO). Most likely, dust is present too. If dust is present in large amounts, it may gravitationally settle and

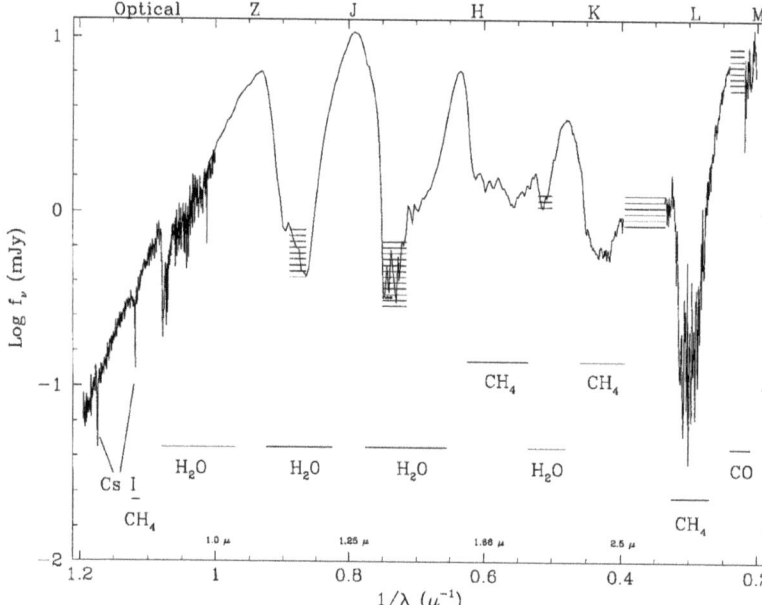

Figure 8.4: Near-infrared spectrum of the "first" brown dwarf Gl 229B (T_{eff} = 900 K, spectral type T6), shows many strong molecular absorption structures. Note that the spectrum is plotted against $1/\lambda$ [μm^{-1}], so that the figure covers the wavelengths of 8300 Å (at left) to 5 μm (at right). Photometric band names are given at the top. The near-IR photometric band I is centered near 8000 Å. Clearly $I - J$ is very red but $J - K$ is quite blue. Figure from Basri (2000).

so lead to reduced surface temperatures, down to $\simeq$ 800 K. Such BDs have the spectral type T, also signifying the presence of NH$_3$ (see the review by Kirkpatrick 2005). The molecular layer has a depth depending on run of the T, ρ combination (radiative or collisional dissociation). The atmospheres of warmer BDs are not too dissimilar from those of the coolest M dwarfs (see review by Allard et al. 1997) or M subdwarfs (Fig. 10.21).

Opacity in the visual comes in part from H$^-$. In the near IR there are numerous strong absorption features. In the photometric H and K bands there is strong methane (CH$_4$) absorption, such that it makes $J - K < 0$ (i.e., rather blue) and the index from the visual IR (I) to the IR J-band at 1.25 μm, $I - J > 5$ mag, very red (inspect Fig. 8.4).

Once fusion stops, the objects cool out (see Fig. 8.3).

8.4 How ubiquitous are BDs?

Numerous searches for BDs have been conducted. One performs photometry in the visual IR, mostly the I band near 8000 Å, as well as photometry in the near IR, like in the J, H, and K bands (see Fig. 8.4). A CMD showing the location of MS stars and BDs observed in the Pleiades cluster is given in Fig. 8.5.

BDs have been discovered in large numbers for the first time in the Pleiades. With newer near-IR techniques many star-forming regions have been investigated and numerous BDs have been discovered (Orion, Taurus, etc.).

From these observations the mass function (see Ch. 20) could be derived. If that function ($dN/dM \sim M^{-0.8(\pm 0.25)}$) holds for the entire Milky Way, then BDs contribute with their mass about 10% to the amount of stellar mass present in our galaxy.

8.5 Deuterium, litium and cosmology

According to the big bang model, deuterium and lithium are two of the species formed in the early phases of cooling of the Universe. Protons formed first and subsequently also D and He as well as some Li. The abundance of D from big bang models is D/H $\simeq 10^{-5}$.

The current abundance of D and Li need not be the same as that right after the big bang since D and Li are consumed in fusion in stars. In stellar cores D and Li are completely burned. In stars with extended convection zones also the D and Li from the outer layers is (after transport

8.6. THE LIMIT TO GIANT PLANETS

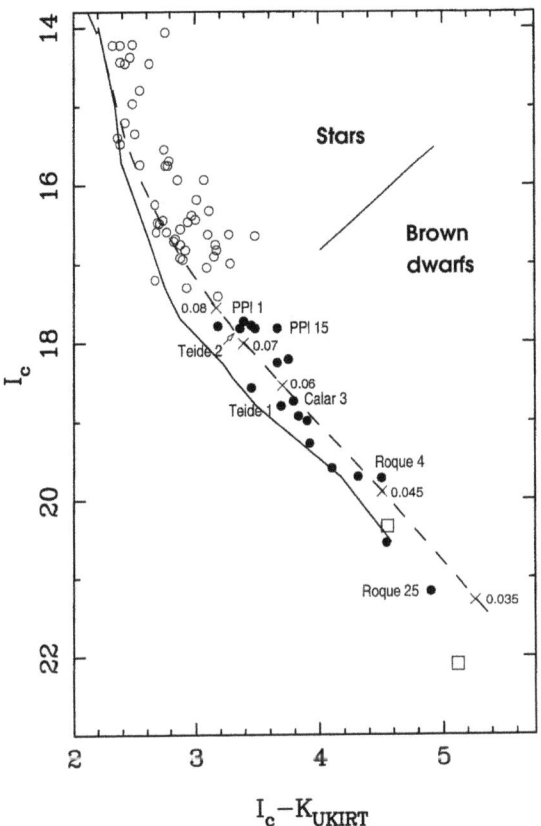

Figure 8.5: Near-IR CMD (I, I−K) of the Pleiades showing the lower end of the MS (stars are ∘) down to the brown dwarf (BDs are •) domain. The photometric band K_{UKIRT} is a special one used at the UK-IR-Telescope on Hawai. Several individual BDs are named. The solid line is the main sequence, the dashed line the 120 Myr isochrone (from Chabrier & Baraffe 2000) with mass values indicated. The two open squares are BDs with known parallax, shifted to the Pleiades distance. Figure from Basri (2000).

into the interior) consumed in the fusion. When at the end of the life of such stars stellar material is returned to the ISM it clearly does not need to have the D and Li abundance as when the star formed (see Sect. 8.3). The final important consequence is, that younger generations of stars may have somewhat less D and Li to start with. Note that in larger mass MS stars D and H will be consumed such (depending on the depth of convection zones and on mixing) that their relative abundance in material which did not partake in fusion is essentially unaltered.

Determination of the current abundance of D is not easy at all. The abundance of elements is found through analysis of absorption lines (see Ch. 3) of D, which have wavelengths just a little bit different from those of H. Only when lines of H are not wide and not too saturated can one expect to recognize the lines of D. This is all but impossible with stellar H and D absorption lines. The only chance to determine the abundance of D is through interstellar lines, again only when the H lines are narrow and the adjacent D lines strong enough to be recognizable. Perhaps the easiest way is to use the lines of H_2 and HD, which are among the narrowest IS absorption lines (very cold gas). For an example of an interstellar D determination using the molecule HD see, e.g., Bluhm & de Boer (2001). The disadvantage is, that the ratio of H_2 to HD has to be known, a ratio which depends on molecule formation physics.

The abundance of D found today is very similar to the one predicted by big bang models. However, the accuracies of determinations as well as predictions are not good enough to conclude more than that big bang models and present observations are not in conflict.

8.6 The limit to giant planets

At $M_{\mathrm{init}} < 0.12\,\mathrm{M}_\odot$ one enters the regime of non-stars, objects without fusion, the (giant) planets. The relation between R and M of such low-mass objects is shown in Fig. 8.6. Note that the gas planets of the solar system (with $M_{\mathrm{Planet}} \simeq 0.001\,\mathrm{M}_\odot$) have the same size as the smallest BDs (with $M \simeq 0.1\,\mathrm{M}_\odot$) because the gas in the latter is degenerate (see also Fig. 17.5).

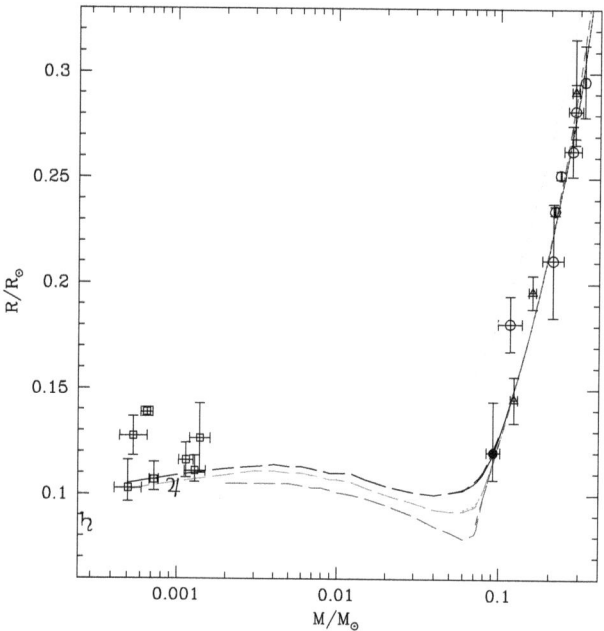

Figure 8.6: The relation of mass and radius for planets, brown dwarfs and low-mass stars is shown. Note the symbol for Jupiter (at 0.001 M$_\odot$), just below the BD detections as well as the one for Saturn, at the left edge. At right, the vertical lines are isochrones for low mass stars, the solid line for 5 Gyr, the dash-dot line for 0.1 Gyr. The dashed lines are isochrones for BDs from Baraffe et al. (2003), showing the relation for R and M of planets having formed 0.5, 1.0, and 5 Gyr ago. Figure from Pont et al. (2005).

8.7 Summary

Objects with a mass not sufficient for H-burning but which have D-burning for a limited period of time are called brown dwarfs.

- $M > 0.08$ M$_\odot$: → normal MS star.
- $0.08 > M > 0.012$ M$_\odot$: → BD. Atmosphere mostly molecular gas.
- $0.08 > M > 0.06$ M$_\odot$: → BD with also Li burning.
- Duration of significant D-burning: $\simeq 2 \cdot 10^6$ to $\simeq 2 \cdot 10^7$ yr.

References

Allard, F., Hauschildt, P.H., Alexander, D.R., & Starrfield, S. 1997, ARAA 35, 137; *Model Atmospheres of Very Low Mass Stars and Brown Dwarfs*

Baraffe, I., Chabrier, G, Barman, T.S., Allard, F., & Hauschildt, P.H. 2003, A&A 402, 701

Basri, G. 2000, ARAA 38, 485; *Observations of Brown Dwarfs*

Bluhm, H., & de Boer, K.S. 2001, A&A 379, 82

Chabrier, G., & Baraffe, I. 2000, ARAA 38, 337; *Theory of Low-Mass Stars and Substellar Objects*

Kirkpatrick, J.D. 2005, ARAA 43, 195; *New Spectral Types L and T*

Nakajima, T., Oppenheimer, B.R., Kulkarni, S.R., Golimowski, D.A, Matthews, K., et al. 1995, Nature 378, 463

Pont, F., Melo, C.H.F., Bouchy, F., Udry, S., Queloz, D., Mayor, M., & Santos, N.C. 2005, A&A 433, L21

Rebolo, R., Martín, E.L., Magazzù, A. 1992, ApJ 389, L 83

Tarter, J. 1975, PhD Thesis, Univ. Calif. Berkeley

Chapter 9

Stars out of balance: from MS star to red giant

There are three aspects of stellar evolution of general importance, which have to be discussed to understand the *basics of the evolution* of MS stars. These have to do with **changes in composition due to nuclear fusion**, with the **effects of inner convection**, and with the **transition of central fusion to fusion in a shell**. These changes lead to changes in the overall structure of a star, in particular to changes of its radius. Why does a star become a red giant? The discussion of these topics deserves a separate chapter.

9.1 Main-sequence stars

The formation of stars is regarded complete once nuclear fusion sets in at a sustainable level. Criteria for, e.g., the minimum mass required, were discussed in Ch. 6.5.

Once a star is born (fusion sets in, the "birth line", see Fig. 7.10) it develops further by further contraction and restructuring. This phase may take up to several times 10^5 years. During this phase energy transport to the surface is partly by convection and partly radiative (as already discussed in Ch. 7.5). When the star reaches stable conditions (in fact, reaches thermal equilibrium) it is in the state of a **main sequence (MS)** star on the **zero age main sequence (ZAMS)**.

Mass rich stars will produce quickly large amounts of energy and will thus quickly radiate away the rest of the birth cloud (hot stars also ionize it away). The clearing away of the birth cloud will take more time for stars of lower mass, since the energy produced is much smaller. They will show signatures of the surrounding birth cloud or of the accretion disk (through brightness variations and emission lines in the spectrum) over a longer period of time (T Tau stars, Ch. 7.7.1; Ae/Be stars, Ch. 7.7.4). Objects which ultimately have too little mass to become stars end up as brown dwarfs (see Ch. 6.5 and Ch. 8).

From observations it is long known that the main sequence in HRDs and CMDs has a certain width, which is larger than can be explained by observational noise. Stars apparently change during the MS phase.

9.1.1 Changes in the main-sequence phase

9.1.1.1 Evolution due to the changing composition of the interior

The MS phase of stars is a very stable phase. In it, nuclear fusion steadily turns hydrogen into helium. It is this change which leads to a *very slow change in the structure of MS stars*.

The fusion transforms 4 H nuclei into 1 He nucleus plus 2 ν_e, 2 γ and 2 e^+. The 2 e^+ will combine with the 2 excess e^- and turn into γs, too. The ν_e will hardly interact with matter and flow out of the star (Ch. 5.3). The sum of the PP I chain (Fig. 5.4) is, in fact,

$$4\,\mathrm{H}^+ + 4\,e^- \to 1\,\mathrm{He}^{2+} + 2\,e^- + 4\,\gamma + 2\,\nu_e \quad . \tag{9.1}$$

The number of particles reduces (8 → 3) so the mean molecular weight μ increases. With fewer particles the core[1] will shrink. When gas contracts, it will heat up (Ch. 4.2.2). Thus the continuous fusion leads to gradual contraction and heating up of the core. Since at higher T the fusion will be faster (Eq. 5.9), it leads to a larger luminosity. This general behaviour was presented in Ch. 6.8.2.1.

If T increases in the core, this rise will also propagate outward. It means that nuclear fusion may also become possible in the layers just outside the core[2]. This means that the amount of mass involved in the fusion increases and the core.

Summarizing: *MS stars increase their energy production slowly because the change in interior composition and increased mean molecular weight lead, based on the work of gravity (Ch. 6.8.2.1) to compaction, thus to higher T and thus to higher L.*

The increase in luminosity can be seen in the evolutionary tracks of all MS stars. With the increase in luminosity the star has to adjust its outer structure ($L = 4\pi R^2 \sigma T_{\text{eff}}^4$) a little as well to accommodate the increased energy flow through the surface. A new balance is established with a slightly larger radius and a slight decrease in surface temperature (see Fig. 9.1, combined with Fig. 1.3 for R). Or, the changes can be understood as due to a deviation from stellar thermal equilibrium (see Sect. 9.4) which the star redresses by simultaneously shrinking the core and increasing the outer radius (see also Ch. 1.2). In fact, most of the further evolution (the more advanced stages) is caused by similar changes in the material and the overall structure.

9.1.1.2 The end of the main-sequence phase

As long as the star is burning hydrogen in its core it is a MS star. Thus **MS stars lie in a strip in the HR-diagram** (see Fig. 9.1), of which the lower-left envelope is the **zero age main sequence (ZAMS)**. The band contains to the upper right stars being in the process of MS evolution. The limit to the right of this MS band is reached when central hydrogen burning stops. In the HRD this limit is called the **terminal age main sequence (TAMS)**. It is sharp in stars with interior convection (see Sect. 9.2.2) but diffuse in stars without interior convection (see Sect. 9.2.1).

9.2 Effects of convection on the MS phase

The structure of MS stars is also determined by the effects of convection in the interior. Low mass stars have no convection in their interior zones, more massive stars do have convection in the interior. The boundary between the two lies near 1.15 M$_\odot$ (see Fig. 6.6).

Depending on the presence or absence of convection the evolution in the MS phase differs.

9.2.1 Stars without inner convection ($M_{\text{init}} < 1.15$ M$_\odot$)

Fusion (H → He) is through the PP chain (Ch. 5.1.4). Without convection in the interior, the central fusion leads, simply put, 'immediately' to a central small sphere containing just He, the He core. Inside this core no further fusion reactions take place, since all H has been used up. This central sphere is almost isothermal (for lack of an energy source).

In practice the structure is, obviously, more smooth. Due to the high temperature, the gas particles move about quickly, so that mixing takes place. But the He will, due to its larger molecular weight, tend to gravitationally settle to the centre.

Nuclear burning in the core of such stars will last a long time. The rise of T_c will slowly lead to an increase of T at the outer edge of the H-burning core, so that the core mass (the mass partaking

---

[1] For the definition of "core", "shell", "envelope", and "atmosphere" see Ch. 6.7.
[2] The boundary between the core and the envelope is in reality not sharp. The structure of a star is normally continuous, with spatially continuous changes, of which some are drastic, requiring in the modelling a dense set of grid points (see Ch. 6.6), others are more smooth.

9.2. EFFECTS OF CONVECTION ON THE MS PHASE

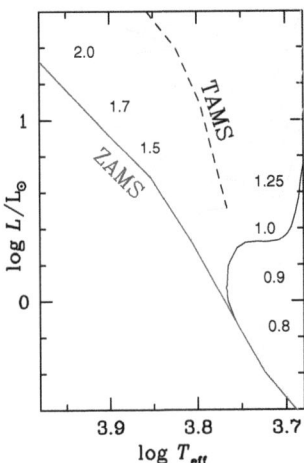

Figure 9.1: Sketch of the location of the Zero-Age Main Sequence (ZAMS) and the evolution of stars of mass as indicated in $M_\odot$. Due to H-fusion the mean molecular weight of the gas in the stellar core increases, leading to higher T_c, thus to more fusion (see Sect. 9.1.1.1). The ensuing larger L [$L_\odot$] leads to adjustment of the atmosphere with larger R and smaller T_{eff} [K] (evolution up and to the right of the MS). The Terminal-Age Main Sequence (TAMS) marks the end of central H fusion in stars with inner convection (Sect. 9.2.2). Stars with $M < 1.15$ $M_\odot$ have no inner convection and therefore evolve smoothly from the MS stage into the RG stage (Sect. 9.2.1). In their cores, the zone with the largest energy production slowly evolves away from the centre (see, e.g., the Sun: Fig. 10.4). For evolution of other mass ranges, see Figs. 10.1 and 13.9. Data from Schaller et al. (1992).

in fusion) will grow. Only when the fraction of He in the core has grown very large will the real centre be without fusion. Thus the star will smoothly change from one having core burning into one having fusion in the shell around the He core. As an example see the evolution of the Sun, for which the evolution of the interior structure is given in Fig. 10.4. The shell in which H fusion takes place will with time increase in radius and the mass of that fusion zone will grow too, ultimately leading to an appreciable increase in luminosity (see Fig. 9.1 and Sect. 9.6).

The He produced in the H burning shell will be added to the He core. In such stars the He core mass will continue to grow. Without energy source, the He core will be isothermal.

9.2.2 Stars with inner convection ($M_{\text{init}} > 1.15$ $M_\odot$)

In stars with $M_{\text{init}} > 1.15$ $M_\odot$ the central temperature is high enough for CNO cycle H burning (Eq. 5.10) in the centre. (Note that outside the centre, at lower T, H fusion is through the PP chain.) Since ϵ_{CNO} is a steep function of T (Eq. 5.12) the central temperature gradient is very steep leading to convection in the core (radiation driven convection, see Ch. 4.3.4). This will result in mixing of core material with material from layers closer to the surface (the envelope). The convective motions will therefore transport H-rich material into the He-enriched core, leading to replenishment of the nuclear fuel, and will move fusion products into the envelope. The burning in the core will therefore be able to continue (much) longer than in the no convection case.

How far convection reaches outward (Ch. 4.3) is still investigated. Modelling convection is difficult because it is difficult to assess how quickly convective motion will come to a halt (convective overshoot; Ch. 4.3.5). Effects of convective overshoot in the modelling of evolution will be addressed in Ch. 16. Furthermore, the extent of the convective core will decrease with evolution.

Convection will lead to a homogenization of the composition of the gases inside the star, out to the maxium radius of the convection. Even so, the mean molecular weight increases and as a consequence also T_c and P_c. Thus the luminosity increases (as discussed in Sect. 9.1.1.1). When in the core the amount of available H has decreased to a critically low level, the H-burning will stop rather suddenly. This happens when the core has reached a total mass of $\simeq 0.5$ $M_\odot$ in He.

As soon as the burning stops, T_c will decrease. This leads to core contraction and (Ch. 4.2.2) about half of the energy will emerge as radiation (making up for the absence of fusion), the other half will heat up the core (increase in T_c), with in total a slight overall contraction of the star.

This point in the evolution can be recognized in evolutionary tracks (see Fig. 9.1). The star has reached the **terminal age main sequence** (TAMS). Due to the contraction, the layers higher up will be drawn in and become denser, thereby increasing T and P there. In this process, T at the outer edge of the He-core will increase so much, that it rises above the lower temperature limit for H-fusion. Now H gets ignited in this mass shell and the star has an energy source again.

The 'sudden' stop of H-fusion with the following contraction shows up as the left going part of the evolutionary track after the TAMS as shown in Fig. 9.1 and in many figures in Chs. 10 and 13.

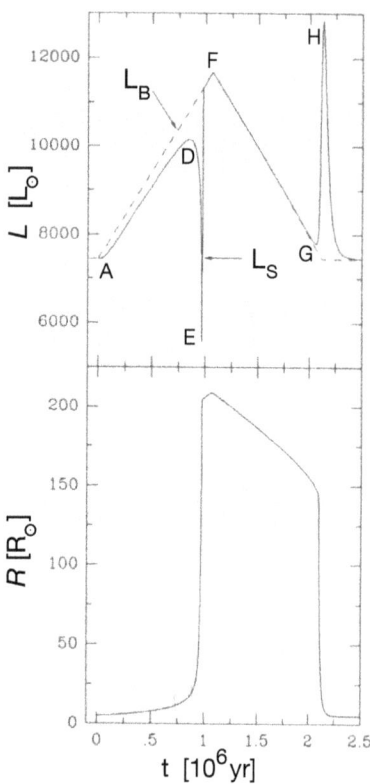

Figure 9.2: Results of a *gedankenexperiment* to show why a star becomes red giant. The top panel shows the evolution in time of the output surface luminosity L_S (solid line) in response to the imposed variation of the input luminosity L_B at the base of the envelope (dotted line). The corresponding evolution of the radius is shown in the lower panel. The figure is adapted from Renzini et al. (1992).

Upon increasing L_B, the envelope adjusts its structure by expanding gently. The energy for the work done is taken from L_B, so that L_S is somewhat smaller than L_B. When the expansion exceeds a limit with a critical opacity, the entire envelope must expand considerably, turning the envelope into that of a red giant star. The energy needed for the work is (again) taken from L_B leading to a drastic decrease in L_S. Once the expansion is complete, a new stable structure is established. For the events after reducing L_B, see the main text of Sect. 9.3.1.

The capital letters refer to phases indicated in the companion Fig. 9.3, and also to those in Fig. 9.4 related with real stars. The letters are here different from the ones given by Renzini et al. because now they match the letters of the phases in further diagrams. The time in this *gedankenexperiment* is close to the reality of a 9 $M_\odot$ star inflating to red giant.

9.3 Why and how does a star become red giant?

Stars from the lower MS as well as from the higher MS become red giant stars. They set out on that evolution in a different way, yet the basic mechanisms are rather similar. One can think of a star at the end of core H-burning as a star with a structural problem (Sect. 9.4; see also Celnikier 1989, his chapter "Model of a star in trouble"). In fact, the star leaves the state of overall stellar thermal equilibrium.

In order to understand why a star becomes a red giant, i.e., why it expands to the gigantic size a red giant star has, Renzini et al. (1992) performed a numerical experiment. Using a software code which calculates stellar models, they considered how the envelope of a star reacts upon an increase in luminosity from below.

9.3.1 A "gedankenexperiment": the gravothermal hysteresis cycle

In Renzini et al.'s *gedankenexperiment* the following luminosities are defined: L_B, the luminosity at the **b**ase of the envelope (the luminosity offered to the envelope from below) and L_S, the luminosity emerging at the **s**urface of the star. When the star is in overall stellar thermal equilibrium (as a normal MS star is), $L_B = L_S$.

In the experiment only the calculations for the structure of the envelope of the star are carried out. Luminosity is offered to the base of the envelope independent of what a real stellar core might have produced. As it turns out, the envelope shows a hysteresis-like behaviour.

Consider what happens when one increases the luminosity L_B at the base of the envelope. Due to the increased luminosity, the temperature of the gas at the base of the envelope will increase a bit leading to some expansion of these gases. The rise in temperature as well as the expansion will lead to some expansion in the next layers up. Also the surface expands a bit and L_S increases as well, the combination leading to a new thermal balance following $L = 4\pi R^2 \cdot \sigma T_{\rm eff}^4$. All these changes are small. For the expansion energy is required, which is taken from L_B, so that (as long

9.3. WHY AND HOW DOES A STAR BECOME RED GIANT?

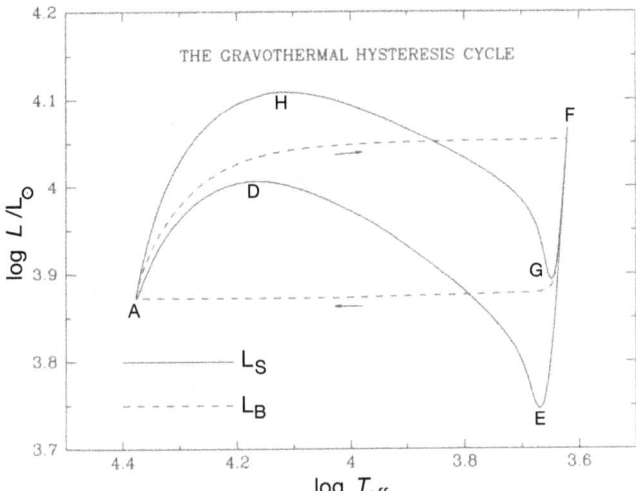

Figure 9.3: Gravothermal hysteresis cycle of the model envelope in an HR diagram showing the variation of L_S (———) versus T_{eff} due to the change of the luminosity at the base of the envelope, L_B (- - - - -).
Note that in Fig. 9.2 the behaviour of L and R was shown in time, here the behaviour of L and T_{eff} are shown. The letters refer to the same phases as in Fig. 9.2. Figure adapted from Renzini et al. (1992).

as work is performed) L_S is somewhat smaller than L_B. If the change in L_B were stopped, the envelope would (after the time needed) get to a new stable situation and L_S will equal L_B again.

The main purpose of the *gedankenexperiment* (carried out in the computer) is to find out how the envelope behaves upon further increasing L_B. This is shown in Fig. 9.2.

An important part of the *gedankenexperiment* is to see the behaviour of the envelope in relation with the opacity. In the atmosphere and outer parts of the envelope the opacity of the gas is rather large, in layers with temperature $10^4 < T < 5 \cdot 10^6$ K at which the metals (Fe, Si, and other elements) are no longer bare nuclei but have at least two recombined electrons. Fig. 4.4 shows the Rosseland opacity, $\overline{\kappa}$, for solar composition gas. For $T > 10^5$ K one has the simplified relation $\overline{\kappa} = \kappa_0 \, \rho \, T^{-3.5}$ (Eq. 4.61) showing clearly that the *opacity of gas is larger at smaller* T.

During the gentle expansion of the envelope the gas is pushed outward so that the surface of a given temperature will move inward through the gas. The hitherto bare Fe, Si, Ar atoms and later the C, N and O atoms will start collecting one or two electrons, thus bringing an increase of the local opacity (again, for the dependence of κ on T see Fig. 4.4). The total mass in the shells having larger opacity increases in this way. With increased opacity, the energy flow is further hampered, the local energy density increases (backwarming; Ch. 2.10.2), inducing local further expansion of the gas. The temperature is increased and the temperature gradient in the outward regions becomes steeper. Thus convection sets in (opacity driven convection; see Ch. 4.3.4). In gas of high opacity, convection will bring the energy faster outward than radiation would do. In this manner a run-away process is triggered, leading to further and further expansion, until the outer envelope layers have sufficiently large R (and have become sufficiently optically thin also *due to dilution*, see Fig. 4.4) to accommodate the transfer of L_B. The behaviour of L_S and R is shown in Fig. 9.2: the gentle expansion between points A and D, the rapid expansion between D and E (requiring energy so $L_S \ll L_B$), to finally reach the new stable state (point F).

In the return process, letting L_B decrease, everything proceeds in the other direction. However, the very extended convective envelope will stay extended even when L_B has decreased well below the level at which in the increasing loop the atmosphere expansion ran away. The atmosphere will shrink slowly along with the shrinking of the base of the envelope (where enhanced conduction plays a role), but the reionization will set in only when convection is no longer maintained (the temperature gradient becomes smaller). The convective envelope collapses, leading to reionization and a drastic reduction in opacity. The excess energy of this contraction will be released in a short time, producing the excess in L_S (point H in Fig. 9.2). Ultimately, the full radiative case is restored. With this *gedankenexperiment* it was found that the envelope shows hysteresis and the entire loop was called the *gravothermal hysteresis cycle*.

The same behaviour is illustrated in Fig. 9.3, showing the parameters familiar for the stellar surface, L and T_{eff}. The letters refer to the same phases as indicated in Fig. 9.2.

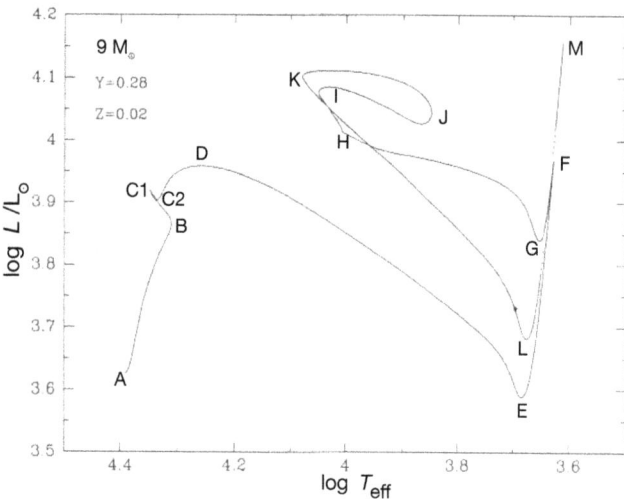

Figure 9.4: Evolution of a 9 $M_\odot$ star in the HR diagram, L versus T_{eff}. The letters refer to similar stages of evolution as shown in the gravothermal hysteresis cycle of Figs. 9.2 and 9.3. Here the true evolution is followed, thus beyond the simple gedankenexperiment. The letters correspond to the ones in the figures mentioned, a resume is given with Fig. 9.5. Figure adapted from Renzini et al. (1992).

9.3.2 The hysteresis cycle and real stars

The behaviour in the cycle and in real stars is very similar. Yet there are differences, which have to do with the evolution of the stellar core. The core behaves of course the way a stellar core has to behave, and it does not show the simple run in L_B as the gedankenexperiment cycle.

Nevertheless, the diagram with L and T_{eff} showing the modelled evolution of a true star (Fig. 9.4) is up to point G immediately comparable with that diagram of the gedankenexperiment (Fig. 9.3). After situation G the evolution is different because a real core evolves in luminosity different from the L evolution in the gedankenexperiment.

In a MS star <u>with</u> an inner convection zone the core will have fusion, with a gradual increase in luminosity (Sect. 9.1.1). The star adjusts itself through modest expansion, slowly proceeding to the TAMS state. After central fusion stops (point B) the core contracts and the H-shell burning is initiated (C). The luminosity L_B increases further while the energy source is farther away from the core and closer to the surface of the star. Due to rapid core contraction in lower mass stars (see Sect. 9.5) additional energy is released and the envelope expansion starts running away (phase D) requiring energy until full expansion is reached (phase E). In higher mass stars L_B simply surpasses the limit the envelope can accommodate (at point D) leading to the runaway expansion (inflation) of the envelope. A new stability will be reached (phase F) and the luminosity L_S is again equal to L_B. At this point the star stays in that phase for a while. In the process, the core gradually becomes isothermal.

Due to the H-shell burning the amount of He in the core has grown and the central T and P have increased, ultimately so much (in stars of sufficient mass) that T_c becomes large enough for He to ignite[3]. The central portion of the star adjusts it structure, L_H and L_He decrease somewhat, the envelope collapses (phase G). This completes the hysteresis cycle (but L_B does in reality not return to the intial gedankenexperiment value).

In a MS star <u>without</u> inner convection the core will have fusion, with a gradual increase in luminosity, as in the more massive stars. Since these stars proceed to shell burning slowly with that shell really close to the centre (see for a 1 $M_\odot$ star Fig. 10.4), while in these stars evolution proceeds at a lower pace anyway, the physics of the gravothermal hysteresis makes itself visible only very slowly. Expansion occurs in the same manner, but only very gradually. There is no run-away situation. These stars will become red giants all the same, growing larger and larger with time. Since the H-burning shell eats its way outward, so that the total amount of matter participating in the fusion (the total mass in the H-burning shell) slowly increases, the luminosity continues to increase. Also, μ in the core increases such that L must increase (Eq. 6.49). For these stars one says that the star 'climbs up the red giant branch' (Sect. 9.6).

---

[3]The ignition of He is gradual or explosive (see Ch. 10.2.3).

9.3.3 A second red giant phase

Stars of sufficient mass ($M_{\text{init}} > 2.5$ M$_\odot$) will evolve to have a second red giant phase as follows.

The H-fusion in the shell adds He to the core. When the He in the stellar core has become sufficiently dense and hot, He can come to fusion in the core. The star adjusts its overall structure and becomes compact again.

Once the He in the core is exhausted, the He-fusion stops and the core will shrink (as with H-burning, Sect. 9.1.1.2). This will lead to the ignition of He in a shell around the core. The process is similar to that after the hydrogen in the core was exhausted.

Because of the similarity, this second inflation is mentioned here but a full discussion is actually part of the further evolution of stars in Chs. 10.3 to 10.5 and in Ch. 13.4.1.2.

At a general level these aspects are related with overall stellar thermal equilibrium (Sect. 9.4).

9.4 The overall stellar thermal equilibrium (STE)

The changes in the structure of the star from the MS-phase to the RG-state have been presented in some detail. The gedankenexperiment helped to see why the envelope of a star must expand (or will reduce its size) depending on the luminosity it has to handle.

This kind of behaviour can also be described at a more general level (see Renzini et al. 1992) which is of relevance for all of stellar evolution. It deals with the question of the overall "stellar thermal equilibrium" of the star (or of parts thereof). Or, what happens to "a star in trouble"?

The run of luminosity in a star in the case of radiative transfer of energy can at each location be given by the two equations

$$\frac{d\,L_r}{d\,M_r} = \epsilon_{\text{nuc}} + \epsilon_{\text{grav}} \tag{9.2}$$

$$L_r = 4\pi r^2\,F_{\text{rad}} = -4\pi r^2\,\frac{4ac\,T^3}{3\,\overline{\kappa}\,\rho}\frac{dT}{dr} \tag{9.3}$$

where Eq. 9.3 is the relation of L and the radiative temperature gradient, Eq. 4.18.

In Eq. 9.2 the term ϵ_{grav} defines the energy production due to the possible workings of gravity (Ch. 4.4.3). When gas expands $\epsilon_{\text{grav}} < 0$ (energy is used for work), and when it contracts $\epsilon_{\text{grav}} > 0$ (energy is liberated).

Every layer of a star tries to accommodate the local luminosity, essentially depending on the parameters $T^3/\overline{\kappa}$ (radiative capacity and opacity). If more energy is offered than can be transferred radiatively (the opacity is, relatively speaking, too large), the energy piling up is used to do work, i.e., the gas expands[4]. If energy can be radiatively transferred easily (low opacity), the gas will cool and thus contract (releasing energy).

If the layers collectively manage to have the star in an almost stationary state, a state where little mechanical energy is converted, the star is said to be in overall *stellar thermal equilibrium* (STE).

Consider a star in the MS-phase. Due to the fusion of H to He the mean molecular weight in the core increases. This leads to a shrinking of the core so that $\epsilon_{\text{grav}}^{\text{core}} > 0$. This extra luminosity has to be taken care of. The envelope will adjust itself and, with $L = 4\pi r^2 \cdot \sigma T_{\text{eff}}^4$, there are two possibilities: T_{eff} may increase, or the entire envelope expands a bit with $\epsilon_{\text{grav}}^{\text{env}} < 0$ (so that $\epsilon_{\text{grav}}^{\text{env}} + \epsilon_{\text{grav}}^{\text{core}} = 0$). Both options, the change in T_{eff} or the expansion, lead to changes in the structure of the gas, in total such that stabilty is maintained. (At the same time, due to the contraction, T_c rises a bit, too, leading to a little faster fusion with its concomittant small increase in L; this bit has to accommodated as well.)

The gedankenexperiment showed that when a little extra luminosity is offered to the envelope the envelope opacity increases due to a change in ionization structure and thus L_r of Eq. 9.3 decreases (so $dL_r < 0$). Since in an envelope $\epsilon_{\text{nuc}} = 0$, the case $dL_r < 0$ allows for STE only when $\epsilon_{\text{grav}} < 0$, in other words, the envelope must expand. Or, the star is momentarily not in STE.

---

[4]The gas may also enter a higher state of ionization, as is the case with pulsations in the atmospheres of stars (see Ch. 11.1.3).

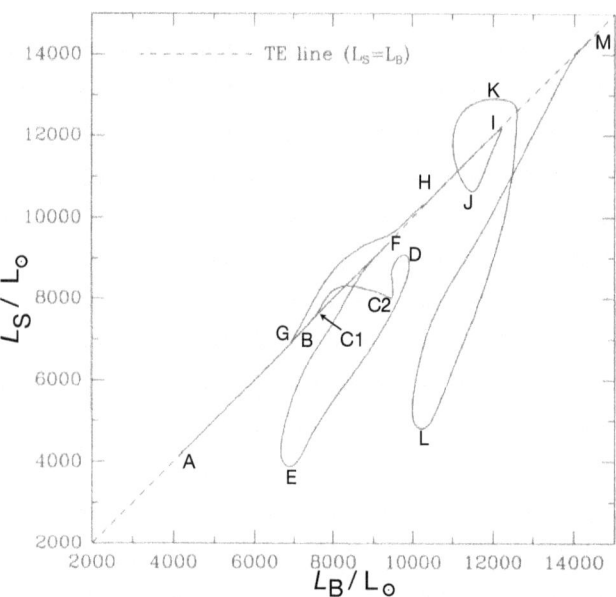

Figure 9.5: The behaviour of the 9 $M_\odot$ star being part of the gedankenexperiment is shown in relation with stellar thermal equilibrium (STE). Figure adapted from Renzini et al. (1992).

As long as $L_S = L_B$, there is equilibrium. In various phases of evolution large deviations occur (notably during envelope expansion) and the star (in this case the envelope) tries to return to the overall stellar thermal equilibrium state. The lettering is the same as that of Fig. 9.4. The letters mean: A: the ZAMS state. A to B: slow MS core shrinking. B: end of core H-burning, onset of strong core shrinking. C1: hydrogen burning reignites in shell around the core. C2: the star has readjusted its structure somewhat after the ignition of shell burning. D: due to Schönberg-Chandrasekhar rapid core contraction or other increases of the core luminosity (like CNO burning; Sect. 9.6) the first critical luminosity is reached and the envelope starts expanding in a run-away process. E: maximum expansion rate and lowest emerging flux at the surface; the star is well out of TE. F: expansion is complete and the star is back in TE. G: the star ignited core He burning (after F) and adjusts its structure by pushing the H-burning shell out and reducing its luminosity; eventually the second critical luminosity is reached and the enevelope deflates. H: after a slight adjustement the star is back in a stable core-He-burning state. I: end of core-He-burning, shrinking of core. J: start of He-burning in a shell around the core. K: critical luminosity is reached and envelope starts to expand. L: maximum expansion rate so largest deviation from TE. M: final phase of He-shell burning.

If, due to changes in the luminosity, Eq. 9.3 does no longer apply because dT/dr has changed so much that convection would set in, the star will be in upheaval. The energy is now transported more efficiently and of the two equations above Eq. 9.3 has to be replaced by the equation for convective energy transport. Now the larger energy throughput has to be taken care of by the gas layers higher up. The stars is at this point far removed from STE which leads to a rapid expansion until a new stability can be achieved.

For a 9 $M_\odot$ star the deviations from STE are shown in Fig. 9.5. The various phases match those of Fig. 9.4 and approximately the phases of the gedankenexperiment (Figs. 9.2 and 9.3).

It has to be mentioned that there is still debate about why really a star becomes so inflated. Renzini et al. note that many modellers have made a sequence of stable models in STE, but often ignore that a star really can be outside STE. Such models in STE can be made both for compact stars being essentialy radiative models, or for inflated ones being models with extended zones of convection. The gedankenexperiment led to the insight that, although models for compact as well as inflated stars exist in STE, during stellar evolution large excursions away from STE occur (see Fig. 9.5). When out of STE, stars will try to restructure to ultimately reach STE again.

Effects of stellar winds have not been included in this discussion. If a star inflated so much that radiation driven winds occur or when a star rotates fast enough for mechanically driven mass loss, the change in overall mass will influence the equilibrium structure as well. These aspects will touched upon in Chs. 10, 13 and 14.

When a star has become red giant, then enters the phase of core He burning, it deflates and becomes bluer. If the second red giant phase occurs, the star then completed a blueing loop in the HRD from the RG called a 'blue loop'. Criteria for its occurrence are given in Ch. 13.4.1.2.

9.5 Isothermal He core and Schönberg-Chandrasekhar limit

MS stars with inner convection build in the course of MS evolution an inert He core, which is nearly isothermal (there is no source of energy in the core). During the H shell burning, He is added to the core and so the core grows. This core behaves in ways which depend on M_{init}.

In stars with $M < 2$ $M_\odot$ the He gas will become degenerate. The core of such RGs behaves as a polytrope with $n = 3/2$ (K is constant since μ is constant, there is only He plus a few heavier atoms; $X = 0$). Thus the radius behaviour is as $R \sim M^p$ with $p < 0$ (see Ch. 6.2.2.3).

In stars with $2 < M <\simeq 6$ $M_\odot$ the He core gas is not degenerate. Adding He means the core size (radius) grows, with slowly increasing pressure. But there is a limit. If the isothermal core mass reaches $\simeq 10\%$ of the total stellar mass, this core gas can no longer support the gravitational pressure of the envelope. This limit is named after Schönberg & Chandrasekhar (1942). Depending on M_{init}, it may be reached even soon after shell H ignition. The core then contracts rapidly, is no longer isothermal, releases energy rapidly, further stimulating the transition to the red giant state. After further H shell burning T_c is ultimately high enough for He to ignite ($T > 10^8$ K).

In stars with $M >\simeq 6$ $M_\odot$ T_c will rise during core growth to high enough values for He to ignite before the Schönberg-Chandrasekhar limit is reached. For further aspects in high mass stars see Ch. 13.4.1.2.

9.6 Luminosity evolution of red giants

Do stars change their luminosity when they structurally become red giant? In general yes, but one has to go into some detail to see how much. These details involve, as in many cases of stellar evolution, the contrast between the high mass and the low mass range.

9.6.1 Red giant luminosity depends on M_{init}

For **high mass stars** the interior temperatures are already so high, that the end of core nuclear burning hardly leads to a change in luminosity (for more see Ch. 13). In general, the higher the mass the smaller the change in luminosity during evolution from the MS to the RG state (Ch. 13).

For very **low mass stars**, the transition to the red giant phase is gradual (as described in Sect. 9.2.1). Since the core becomes ever more compact (degenerate gas; see Ch. 6.2.2.3) the zone with temperatures high enough for fusion grows continuously, and so grows the amount of mass (mass shell) having nuclear fusion. This explains why **low mass stars become more luminous and ascend "up" the RG branch** (evolve upward on the RGB).

Furthermore, with rising temperatures the shell burning may enter the T regime of burning with the CNO-cycle (Ch. 5.1.5), thus enhancing the energy output even more pronouncedly (Fig. 5.6). Thus depends, clearly, on the initial abundance of CNO.

In the **intermediate mass range** these effects act in an intermediate manner (see the evolutionary tracks in Fig. 10.1).

Summarizing one can say that
- higher mass stars evolve <u>from</u> the MS <u>toward</u> the RGB (see Ch. 13),
- lower mass stars evolve <u>from</u> the MS <u>up</u> the RGB (see Ch. 10).

9.6.2 Effects of metallicity

Since the expansion to the red giant state is driven by opacity, it is to be expected that the chemical composition has an influence. Modelling shows that in metal poor stars, with less opacity (being more 'leaky' (Ch. 6.8.2.2), the expansion is less and such stars will not be as red as 'normal' stars (metallicity-dependent colour at given luminosity). The location of the RG branch depends on the opacity of the envelope gas (so is related with the initial chemical composition). Both the He abundance and the abundance of C,N,O and other metals have their influence.

In very metal poor populations the red giant stars are clearly not very red (old star clusters show bluer RG branches).

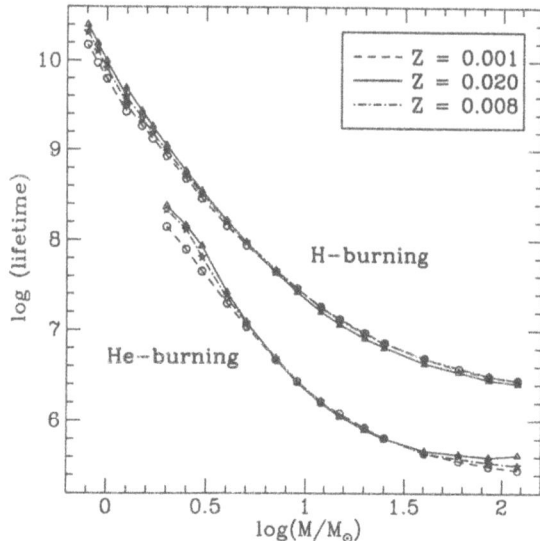

Figure 9.6: The life time in years for the core H-burning phase is shown as function of the stellar mass (both parameters logarithmic). Figure from Schaerer et al. (1993) based on their stellar evolution models.

The effects of chemical composition (Z) are indicated. They are mostly small, but reach 30% for the lower masses.

The later core He-burning phase (see Ch. 10.3) is included (but not for stars with $M < 2$ M$_\odot$ becoming HB-stars).

Note that for stars with $M_{\rm init} < 30$ M$_\odot$ the duration of H burning is $t_{\rm H-burning} = 10^{10} M/L = 10^{10}/M^2$ yr (variables in solar units). In general, $L_{\rm MS} \simeq M^3$ (Eq. 6.49).

For life times see also Table 13.4.

9.7 The core drives the evolution, the envelope follows

The evolution of a star from its MS state to that of a RG is clearly governed by the way the core changes with time. It is the envelope that follows the evolution of the core, adjusting its structure to accommodate both the change in luminosity as well as the different depth at which the luminosity is (predominantly) produced.

In later chapters the dominance of the core in the evolution of a star will (implicitly) be substantiated. It will be elaborated upon in the chapters on evolution of stars in the lower mass range (Ch. 10) and in the higher mass range (Ch. 13). However, if evolution leads to strong loss of mass through stellar winds (Chs. 10.2.2 & 13.3) then, of course, having a less massive envelope must have consequences for the behaviour of the core. With the overall mass reduced the gravitational contraction force is weaker so the core is less condensed and thus cannot be as hot as without mass loss. Still, the structure and evolution of the core dominates the evolution of the star.

9.8 Duration of the main-sequence phase

The burning of H in the core of main-sequence stars will last a time span according the relation $t_{\rm MS} \simeq 10^{10} M/L$ yr, with M and L in solar units.

There is the classical observational relation, theoretically supported in Ch. 6.8.2.1, of $L_{\rm MS} \simeq M^3$ which then leads to $t_{\rm MS} \simeq 10^{10}/M^2$ yr. That relation is also visible in the life times calculated with full stellar evolution models. One so obtains (see Fig. 9.6) the duration of H-burning on the MS as well as the duration of He burning. Effects of metal content can be discerned, too.

References

Celnikier, L.M. 1989, "Basics of Cosmic Structures", Editions Frontières, Gif-sur-Yvette
Renzini, A., Greggio, L., Ritossa, C., & Ferrario, L. 1992, ApJ 400, 280
Schaller, G., Schaerer, D., Meynet, G., & Maeder, A. 1992, A&AS 96, 269
Schaerer, D., Meynet, G., Maeder, A., & Schaller, G. 1993, A&AS 98, 523
Schönberg, M., & Chandrasekhar, S. 1942, ApJ 96, 161

Chapter 10

Stellar evolution: Stars in the lower mass range

10.1 Defining the low mass range

Details of evolution of a star depend strongly on the initial mass, M_{init}. A clear division in the way stars evolve is given by the minimum mass a star must have to become a supernova (SN). This limit is $\simeq 8\,M_\odot$. A star starting below that limit will normally end as a white dwarf (WD). The 'lower mass range' is therefore defined by stars having $M_{\text{init}} < 8\,M_\odot$ (spectral type B3 and 'later'). Of these stars T_{eff} is not very high so they cannot create an H II region around them.

The evolution of a low mass star can be divided in four distinct phases: 1) the star-formation and pre main-sequence phase; 2) the main-sequence phase; 3) the subsequent relatively fast stages of evolution in which the star still has an energy source from nuclear processes; and 4) the final or cooling phase. The pre-MS and the MS phases have been discussed in Chs. 7, 8 and 9. Here the further evolution is the main topic.

In spite of the simple division given above, the evolution of low mass stars is determined by numerous mass limits, both at the start of MS evolution (M_{init}), as well as in the further phases. A summary of mass limits is given at the end of this Chapter (see Fig. 10.24 and Table 10.3).

The contents of this Chapter is organised such, that it follows all lower mass range stars with their successive changes, in sequence of these changes. Clearly, the higher mass stars (Ch. 13) go through (most of) these changes but quicker than the lower mass range stars.

Examples of evolutionary tracks for the surface parameters T_{eff} and L are given in Fig. 10.1, in Fig. 10.2 for the central parameters T_c and ρ_c. These figures are based on models extending the original results of Maeder & Meynet (1989).

10.1.1 The MS-mass limit of $\simeq 1.15\,M_\odot$

Stars with *more than* $\simeq 1.15\,M_\odot$ on the MS possess an inner convection zone (Fig. 6.6). Since the convection zone becomes less extended with evolution and at some moment ceases to replenish the interior with H (see, e.g., Fig. 10.12), the core H-burning stops more or less abruptly because of lack of fuel. In such stars the transition to H-shell burning will be abrupt (as discussed in Ch. 9.2.2).

Stars with *less than* $\simeq 1.15\,M_\odot$ on the MS have no inner convection (Fig. 6.6). Such stars tend to slowly become 'shell burning' stars (see, e.g., Fig. 10.4) and they gradually evolve to become RG (as discussed in Ch. 9.2.1) and evolve up on the RGB (Ch. 9.6).

The mass limit of $\simeq 1.15\,M_\odot$ marks, as also discussed in Ch. 9.2 (see also Fig. 9.1), the distinction in the respective evolutionary tracks: the more massive stars establish a "Terminal Age Main Sequence" (TAMS), the lower mass stars not. This behaviour can again be seen in the evolutionary tracks given in Fig. 10.1.

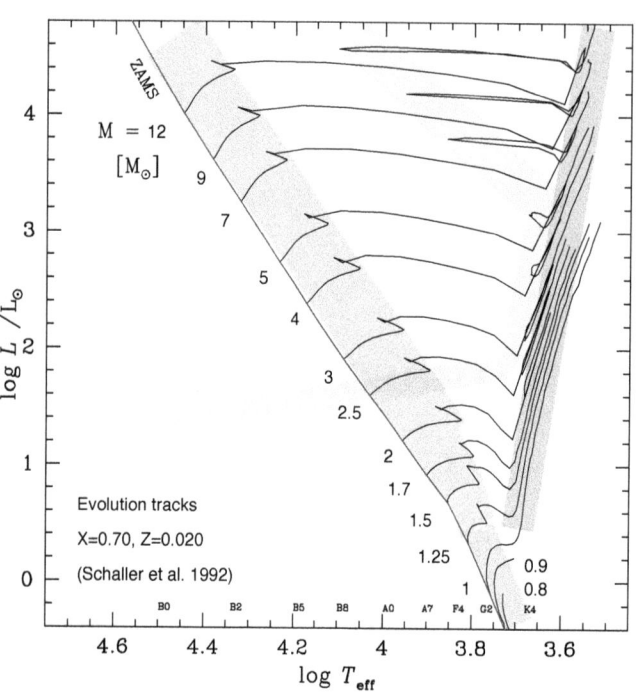

Figure 10.1: HR-diagram showing the evolution of L and T_{eff} of lower mass range stars starting at the ZAMS (Fig. 9.1). Tracks of evolution are from Schaller et al. (1992).

The shading marks three crowded regions of normal HRDs. At left the core H-burning MS region, bounded by the ZAMS and the terminal age MS (TAMS; Fig. 9.1); note the difference in the end of MS evolution for stars with or without inner convection (see Ch. 9.2). At right the shell H-burning red giant branch phase with the considerable rise in luminosity along the RG for the low mass stars (Sect. 10.2). Bending away from the RG to the upper left is the strip of core He-burning "blue loop" stars (Sect. 10.3). The core He burning metal-poor horizontal-branch stars lie in the strip across the MS (for post-RG evolution to the horizontal branch see Fig. 10.6). The change in radius due to evolution can be inferred from Fig. 1.3; for stars of 1, 7 and 3 $M_\odot$ it is given in Figs. 10.4, 10.11 and 10.12 respectively, for a 5 $M_\odot$ star the radius change is shown in Fig. 19.2. Note the *absence of time tic-marks* in the diagram; time information can be read from Figs. 10.4, 10.11 and 10.12, as well as from Figs. 9.6, 10.24, 23.1 and 23.2.

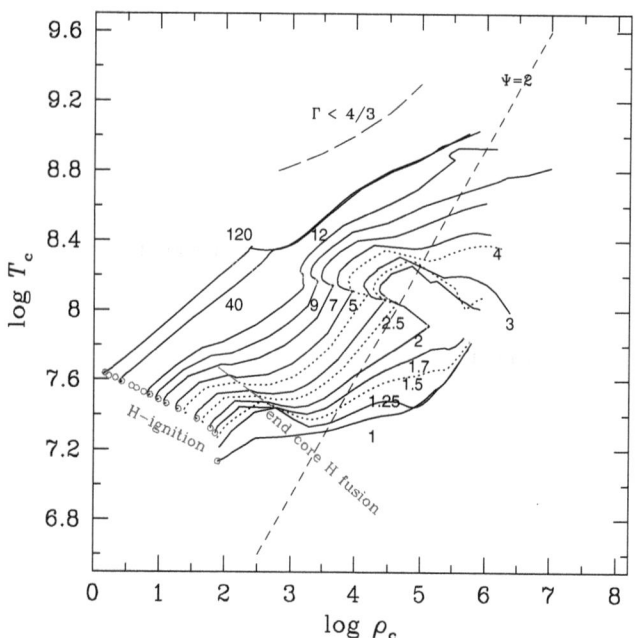

Figure 10.2: Diagram showing the evolution of T_c [K] and ρ_c [g cm^{-3}] for stars of masses ranging from 1 to 12 $M_\odot$ (for more on higher masses see the related Fig. 13.10). Some tracks with overlap are given as dotted lines. Note that there are *no time tic-marks* on the evolutionary tracks. The location on the tracks of H- and of He-ignition are indicated. Evolution after the He flash (stars with $M < 2$ $M_\odot$) is not shown. For stars with $1.15 < M_{\text{MS}} < 7$ $M_\odot$ the end of the MS phase can be recognized by the end of core H fusion, marked by the decrease in T_c. The gas in the centre of lower mass stars becomes at some point degenerate (see the line for the degeneracy parameter Ψ, as in Fig. 6.5; see also Fig. 4.5). Figure using data from Schaller et al. (1992) as in Fig. 10.1.

10.2. H SHELL BURNING: THE RED GIANT PHASE

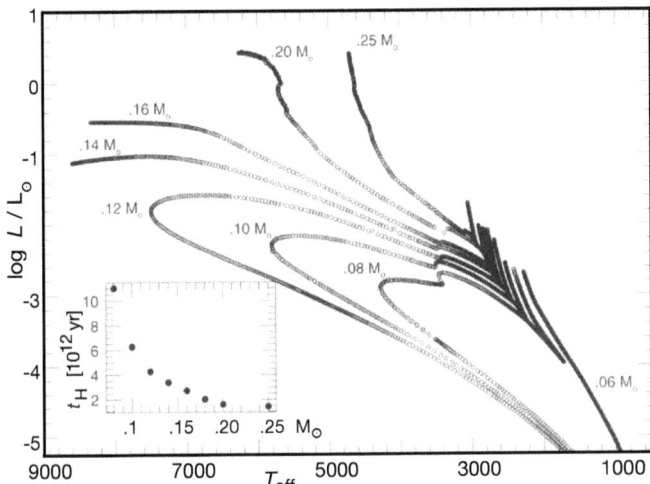

Figure 10.3: HRD with evolution of sub-solar mass stars. The tracks start with the pre-MS phase (see Fig. 7.10). The MS evolution is very slow (see inset with H burning times t_H in units of 10^{12} years!). The stars never evolve to the RG state: when H fusions slows down they contract heating up the surface. After that they become He WDs (the tracks going left and down). extending the data of Fig. 10.20. The $M = 0.06$ $M_\odot$ object becomes BD and then cools at ever lower luminosity. Figure adapted from Laughlin et al. (1997).

However, for stars with $M_{MS} < 1.15$ $M_\odot$, depending on the value of T_c, there will be sufficient diffusion of atoms and ions that also these stars will only change more pronouncedly at the end of the MS phase. To see the details of the evolution of a 1 $M_\odot$ star inspect Fig. 10.4.

10.1.2 The MS-mass limit of $\simeq 0.5$ $M_\odot$

Stars with mass much smaller than that of the Sun have long been regarded as of little interest. Of such stars, t_{MS} is very long and the first such stars of the Milky Way still exist now as MS star even if they were among the first stars having formed in the universe! However, interest increased dramatically with the discovery of the need for extra mass in the Milky Way (dark matter) and of the high mass planets and brown dwarfs (Ch. 8). For both aspects knowledge of the initial mass function (IMF; see Ch. 20) seemed to be of relevance. Furthermore, the IMF indicated that the number of low mass stars must be much larger than the number of high mass stars. Low mass objects would be numerous indeed and the existing ones should be a mix of stars of all ages. Thus the evolution of low mass objects has been studied in more detail.

The very low mass MS stars are of spectral type mid K to late M. The atmospheres may contain considerable amounts of molecules (see Ch. 3.4.3). For more on atmospheres of low mass stars see the review by Allard et al. (1997).

These stars will slowly burn H, are fully convective, and the He produced is distributed throughout the envelope. T_c will not become high enough for He to ignite (as it would in stars with more mass, see Sect. 10.2). When the envelope mass becomes too small, the gravothermal energy release by contraction is too small to maintain fusion temperatures. For details see Laughlin et al. (1997).

Without fusion, the stellar atmosphere will shrink, thereby making the stellar surface hotter and bluer. The star has only its thermal and gravitational energy left and will ultimately slowly cool (see Fig. 10.3) to become a He white dwarf (see Sect. 10.6).

For objects below the stellar mass limit see brown dwarfs (see Ch. 8 and Figs. 8.1 and 8.5).

10.2 H shell burning: the red giant phase

The MS phase of evolution ends with the onset of shell hydrogen burning (see Ch. 9). Without central fusion the core contracts thereby inducing higher temperatures in the H burning shell. This leads to a steepening of the temperature gradient compared to that of the core-burning phase which induces convection. And this, in turn, leads to a restructuring of the entire star in which the envelope expands (Ch. 9.3) and the star becomes a red giant (RG). Ultimately, RG stars are almost fully convective (see Figs. 10.4, 10.11, 10.12, 13.11; for the polytropic description see Ch. 6.2.2.2).

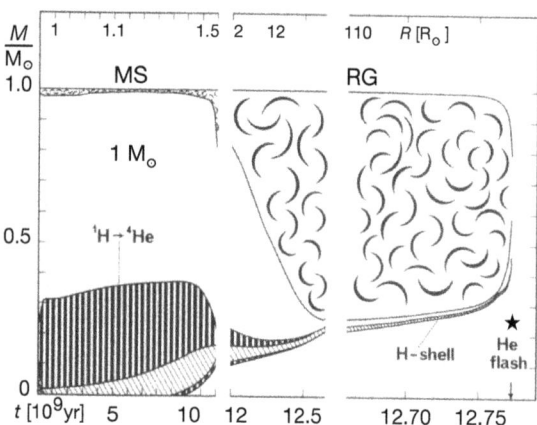

Figure 10.4: The structural evolution of a 1 $M_\odot$ star is shown (up to the He flash). The structure is plotted as fractional mass $M(r)/M_\odot$. At the top radius information is given. Cloudy regions mark convective zones, the hatched regions (||||) zones with nuclear fusion. The zone hatched slanted (\\\\) has the highest energy production ($>10\,\mathrm{erg\,g^{-1}\,s^{-1}}$). Note the gradual transition from H core to H shell burning, the mass loss at the end of the RG phase, and finally the approximate location (in $t, M(r)$) of the He flash (see Sect. 10.2.3). Figure adapted from Maeder & Meynet (1989).

10.2.1 Evolution of the RG core and of the H-burning shell

A red giant star has no fusion in its core and the He of the core is nearly isothermal (no energy source present). Because of the H shell burning, He will be added to the core and the central temperature, T_c, will slowly increase.

Since the temperature in the core will rise slowly, also the temperature in the H-burning shell will be elevated, speeding up H fusion and eventually, in the more mass-rich stars, leading to temperatures fit for CNO cycle H shell fusion. These processes explain why RG stars become more luminous in the course of evolution (see Ch. 9.6). The location of the RG branch (in T_eff) depends on details of the opacity in the envelope (He, C,N,O, metals): larger opacity, redder stars.

Depending in M_init, the evolution of the core differs, e.g., in relation with the Schönberg-Chandrasekhar limit (see Ch. 9.5). And depending on M_init the He ignition (marking the end of the RG phase) is different (see Sect. 10.2.3).

The evolution of the gas parameters in the stellar core (the tracks) are shown in Fig. 10.2.

10.2.2 The RG surface: spectral lines, mass loss and dust

RG stars have extended envelopes due to the efficient convective transport of energy from the interior to the surface. The more luminous a RG star becomes, the more extended the envelope will be. This has several consequences. *First*, the gas at the surface becomes very dilute, so that in the process of spectral line formation the effects of collisional damping (see Ch. 3.1.1.2) becomes very small. The spectral lines become sharp (see Fig. 3.1), except for effects of rotational broadening. *Second*, the large luminosity implies there is radiation pressure acting on the gas of the envelope. The effects are not very large but lead to a slow vertical drift of the outermost layers of the star (see Ch. 4.5). *Third*, the collisions between the atoms in the outer envelope lead to excitation of hydrogen and a *H Lyα emission line* is normally observed in RG stars. The large intensity of the Lyα radiation leads to strong absorption of those photons in the very outer parts of the atmosphere. These atoms pick up some of the momentum and get accelerated. In this way a *stellar wind* is generated, leading to loss of mass (see also Ch. 13.3).

Observations show that the mass loss is more pronounced in luminous than in less luminous stars, while a relation with R and T_eff is likely. One therefore has

$$\dot{M} = f(L, M, T_\mathrm{eff}, Z, ...) \quad . \tag{10.1}$$

The strength of the wind depends, obviously, on the possibilities of transfer of momentum (Ch. 4.5). This in turn depends on the radiation field and on the chemical composition of the atmosphere. *In **hotter** stars*, the atmosphere is ionized, and e.g. Si IV (C IV) is strongly collisionally excited leading to intense emission in ground state resonance lines (see Ch. 13.3) at 1400 Å (1550 Å). These photons transfer momentum and drive the wind. *In **cooler** stars*, dust may form in the atmosphere and here winds are dust driven. For more see the texbook by Lamers & Cassinelli (1999).

10.2. H SHELL BURNING: THE RED GIANT PHASE

Since a RG increases its luminosity in the course of time, the mass loss in the RG phase will also increase with time. The total mass lost during the RG phase depends on both the strength of the wind (which in turn depends on the chemical composition, XYZ) and the duration of the wind phase. The widely used empirical relation (Eq. 4.86) from Reimers (1977) is

$$\dot{M}(L(t), XYZ)\, dt = \dot{M} = -4 \cdot 10^{-13} \frac{L}{g\, R} \quad M_\odot\, \text{yr}^{-1} \qquad (10.2)$$

with L, R as well as g in solar units. A steeper relation may be needed for AGB stars. For stars with $2 < M_{MS} < 6\ M_\odot$ this means $\dot{M} \simeq 10^{-8}$ to $10^{-5}\ M_\odot\,\text{yr}^{-1}$ (see Fig. 13.6). Wind velocities range for K to M giants from 80 to 25 km s^{-1}. For more on the cool star wind properties see the reviews by Dupree (1986) and Willson (2000).

The gases drifting away in the wind of red giants (and later also in those of asymptotic GB stars, e.g., Sect. 10.5.1) will cool and create conditions for the condensation of dust. "Dust" is the general term for molecular conglomerates containing heavy elements. In particular highly fractionate elements condense out, forming $CaTiO_3$, $MgAl_2O_4$, $CaMgSi_2O_6$, etc. (see, e.g., Ferguson et al. 2001), and amorphous C (soot, smoke), being the seed molecules for interstellar dust.

10.2.3 The end of the RG phase: He ignition, He flash

If $M_{init} > 2\ M_\odot$, the growth of the He core will lead to a slow rise in T_c (Sect. 10.2.1). For stars with $2 < M < 6\ M_\odot$, adding He leads to surpassing the Schönberg-Chandrasekhar limit (Ch. 9.5)[1] and the He core collapses, raising T_c drastically. Once T_c reaches $\simeq 10^8$ K He fusion can take place. When this point in the evolution is reached, the helium core contains at least 0.47 $M_\odot$ of He.

Once the *He fusion* (Ch. 5.1.7) sets in, it creates energy in the core and the star has to find a new thermal equilibrium (Ch. 9.4). For stars with $M_{MS} > 2\ M_\odot$ this leads to an expansion of the core, a reduction in L with a collapsing of the atmosphere (see Ch. 9.3.1), and then the star stabilizes as a blue loop core He burning star. For higher mass blue-loop stars see Ch. 13.4.1.2.

Stars with $M_{init} < 2\ M_\odot$ react different. The gas in the RG core is in a degenerate state (see Fig. 10.2). The cooling of the core by neutrinos is most efficient near the centre so that T_c is smaller than T at some distance from the centre (see Fig. 10.5, bottom panel). At some point in the evolution, a shell around the centre with the highest T will reach the ignition temperature ($T \simeq 10^8$ K) for the onset of He fusion. The energy liberated in that shell has several simultaneous effects: the increase in temperature leads eventually to a lifting of the degeneracy, it leads to a steeper outward temperature gradient and so to sudden convection, and it brings expansion once degeneracy is lifted. The latter means also an expansion of the outer layers including expansion of the H burning shell so that its luminosity decreases, while the expansion of the He burning gas brings a reduction of its temperature so that the He fusion stops. This short period of He fusion produces a flash of energy: the first flash (see Fig. 10.5, top panel) reaches $L = 10^{10}\ L_\odot$! This energy is predominantly used to lift the degeneracy in layers above the flash site, so little shows up at the surface. And since the H burning luminosity decreases the star becomes overall less luminous (Figs. 10.5 and 10.6).

After the first flash, the radius of the shell with the highest temperature has decreased. A next flash therefore emerges somewhat closer to the centre, again leading to a burst of convection, etc., as described for the first flash. These flashes continue to occur until the degeneracy is lifted everywhere and the He fusion has established itself as core burning. The star is now a stable zero-age HB (ZAHB) star. The evolution of the surface parameters is given in the HRD of Fig. 10.6.

After the last flash, a total of about 5% of the core's He fuel has been burned. The time elapsed between the first off-core flash and stable core burning is of the order of $1.5 \cdot 10^6$ yr.

The further evolution of the star, now being a *core He-burning star*, depends crucially on M_{init} in combination with the amount of mass lost during the RG phase.

When the RG has very high mass loss the star may not even reach the HB. The He core flash is then quite delayed, perhaps at the top of the WD cooling sequence (Brown et al. 2001).

---

[1] The mass limit of 2 $M_\odot$ results from stellar models; note that different models give somewhat different limits, ranging from 1.8 to 2.5 $M_\odot$ or from 6 to 7 $M_\odot$, respectively.

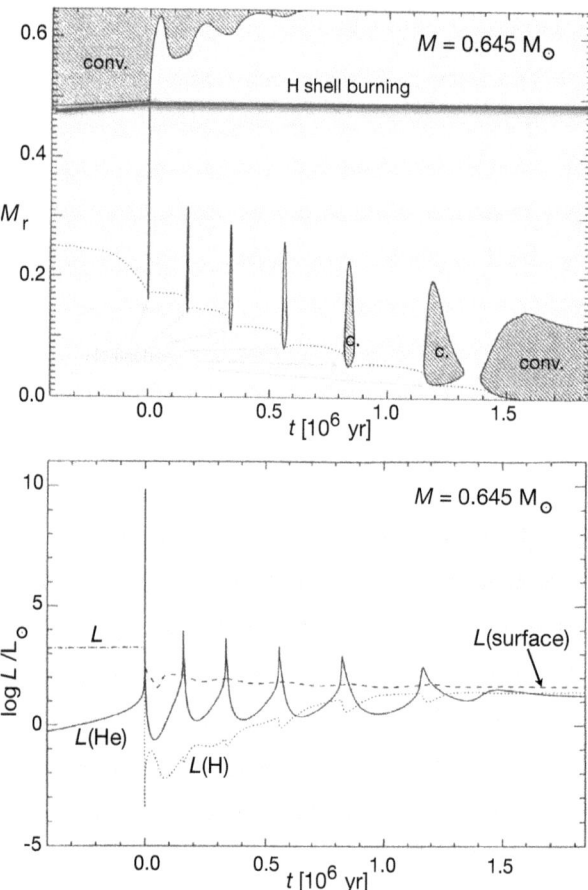

Figure 10.5: Evolution from the RG phase to the HB phase is not smooth. He fusion sets in rapidly off centre, leading both to strong convection as well as to rapid local expansion and thus cooling so that this fusion dies out quickly. Core He fusion thus starts with a set of "flashes". The model shown is for an RG star of 0.645 $M_\odot$.
Top: He flash locations with their immediate upward convection zone (gray). The dotted line gives the location of highest temperature in the core. Convection in the envelope is still present, decreasing with time. The H-shell burning line marks the layer in which the abundance of H has, due to fusion, been reduced to just 50% of its original abundance. This line marks (almost) the shell in which the largest part of $L(H)$ is produced.
Bottom: Pre-flash-luminosity L and changes in $L(He)$, $L(H)$ and $L(surface)$.
The flashes lead to a few evolution loops in the HRD from the RGB to the horizontal branch (see Fig. 10.6). After about $1.5 \cdot 10^6$ yr the star has become a zero-age HB (ZAHB) star.
Figures adapted from Sweigart (1994).

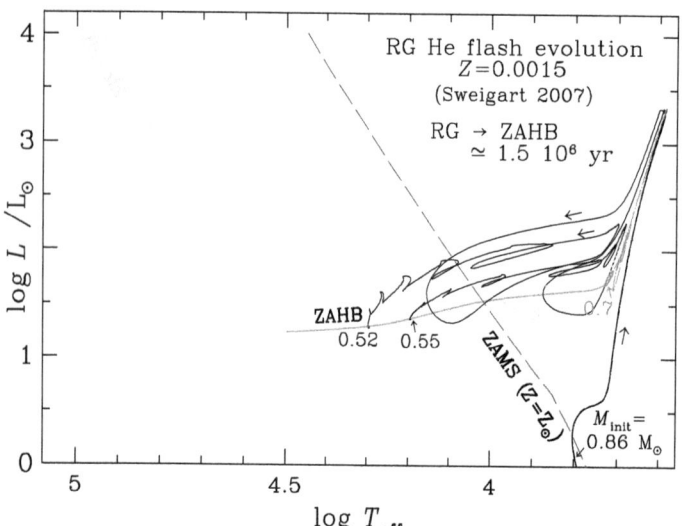

Figure 10.6: The evolution of RG stars due to the onset of core He burning with out-of-centre He flashes is shown using models calculated by Sweigart (2007).
The MS star has $M_{init} = 0.86\ M_\odot$ with $Z = 0.0015$. The different imposed RG branch mass-loss rates lead to zero-age HB (ZAHB) stars of the indicated mass.
He flashes as described in Fig. 10.5 bring repeated changes in L which lead for each flash to a loop in the HRD and ultimately to an overall reduced luminosity. The evolution from the first flash (with large L-T_{eff} loop) through further flashes (with smaller loops) leads to stable ZAHB core He burning. The time from the first to the last flash is $\simeq 1.5 \cdot 10^6$ yr. The further evolution of the ZAHB stars is shown in Fig. 10.9. The ZAMS for solar metallicity stars and the theoretical He-MS (Fig. 6.8) are included for orientation.

10.3 Core He-burning stars

Once He is ignited in the core, the star enters a stable phase of core burning. Its inner structure is that of the He star (see Ch. 6.9.2). Since these stars have a hydrogen envelope with opacity they are redder than the theoretical He MS (Fig. 6.8). They also may have a H-burning shell.

As mentioned above, the amount of He formed in the core during H-burning is 0.5 $M_\odot$ or more.

$$\text{Stars with } M_{\text{MS}} > 2 \text{ M}_\odot \text{ will have } M_{\text{He core}} > 0.5 \text{ M}_\odot,$$
$$\text{stars with } M_{\text{MS}} < 2 \text{ M}_\odot \text{ have only } M_{\text{He core}} = 0.47 \text{ M}_\odot.$$

Since the He main sequence runs more or less parallel to the theoretical H main sequence (see Fig. 6.8 in Ch. 6.9.2), core He-burning stars with $M_{\text{MS}} > 2$ M$_\odot$ lie in the HRD on a line parallel to the MS, but shifted toward lower T_{eff} because their hydrogen envelope with its opacity makes the surface 'red'. The region occupied by the more mass rich core He-burning stars can be found in Fig. 10.1 as the shaded region starting at the RG branch (RGB) and pointing to the upper left. These stars are called *blue loop stars* (see Sect. 10.4.4).

Stars having started with small M_{MS} and which have suffered mass loss during the RG phase have only a thin envelope. Such stars must be bluer than thick envelope core He-burning stars. The stars with small M_{MS} all reach just $M_{\text{He core}} \simeq 0.47$ M$_\odot$, look like pure He stars (Ch. 6.9.2) and their blueness is directly related to the thinness of the envelope. They are found on a line starting near the RGB directed at the theoretical He main sequence at the point of 0.5 M$_\odot$ (Fig. 6.8). This is the so called *horizontal branch* (HB), a name based on the location of these stars in the first observed globular cluster CMDs (see Fig. 10.7). Its locus in L, T space is given in Fig. 10.1 as well as in Figs. 10.6 and 10.9. Thus:

core He-burning stars are found in the HRD along two almost orthogonal stretches.

For the duration of the core-He-burning phase of stars with $M_{\text{init}} > 2$ M$_\odot$ see Fig. 9.6.

10.4 The end of core He burning and on to the AGB

10.4.1 General aspects

10.4.1.1 The end of core He burning

When He is exhausted in the core, the star enters the He shell burning state. Due to the previous central convection as well as the high temperatures required for He burning, this shell lies below the edge of the He core. The star adjusts its structure (much like how a MS star changed into RG; see Ch. 9.3.3), establishes a new thermal equilibrium (see Ch. 9.4) and becomes again a red giant.

In this transition the surface parameters are different from those in the change of MS stars to RG. The road to the RG branch lies in the CMD a bit more to the blue so that the evolution up in M_V joins the RGB asymptotically. That is why these stars are called *asymptotic giant branch stars*, or AGB stars. The AGB phase has a large range of phenomena.

Theory about the evolution of HB and AGB stars can be found in Iben (1974) and Iben & Renzini (1983). The effects of mass loss on the AGB phase were explored notably by Schönberner (1979) and Blöcker (1995). Nucleosynthesis in AGB stars has been reviewed by Busso et al. (1999) and Lattanzio & Wood (2004). See further Kippenhahn & Weigert (1990, Ch. 32.7), Herwig (2005) and the volume dedicated to AGB stars edited by Habing & Olofsson (2004).

10.4.1.2 Envelope thickness, pulses, dredge-up, hot bottom burning, s-process fusion

Once the He core burning has turned into He shell burning, the star expands (as with the H shell burning expansion) to become an AGB star. When the He shell approaches the outside edge of the He core its fusion becomes less, while the temperature at the boundary between He and H rises so H burning becomes stronger. This is the "early-AGB" (E-AGB) phase in which more He is added to the core. In the late phases of an AGB star "thermal pulses" occur and there is strong mass loss so these stars will not reach C-burning.

Very important concepts for the evolution of AGB stars are:
◦ *Envelope thickness*: The envelope of AGB stars is very extended. Characterizing the envelope as thin or thick may be confusing. Thick means a fair amount of mass, thin just a little mass. However, in all cases the envelope is optically thick.
◦ *Thermal pulses*: In AGB stars the He burning shell is close to the H burning shell. Their proximity leads to mutual influence. In the late AGB phase the He burning brightens and fades, a process called thermal pulsing (Sect. 10.4.3).
◦ *Dredge-up*: When a star turns into a giant, the surface convection zone will deepen and extend inward. Thus material from earlier fusion processes may be brought to the surface, a process called "dredge-up". It leads to mixing and to changes in the surface chemical composition. The first dredge-up occurs when a MS star turns into an RG, the second dredge-up when the star turns from a core He burning star into an AGB star. The third dredge-up is described in Sect. 10.4.3.3.
◦ *Hot-bottom burning*: When the shell with H fusion becomes part of the outer convection zone (because convection extends deeper into the envelope), fresh H is transported inward which leads to a larger $L(H)$. This is called hot-bottom burning (see Sect. 10.4.4.3).
◦ *s-process nucleosynthesis*: Due to convective mixing in the layer of nuclear fusion (see Fig. 10.10), heavier nuclei may be exposed again to a H-rich or He-rich environment. Due to this mixing various fusion chains are possible (see Ch. 5.2) including those making free neutrons. Thus the s-process (see Ch. 5.2.4) may become active producing heavy nuclei. *The interior of AGB stars is the primary place in the universe for the build up of heavy elements through the s-process.*

10.4.2 Low mass core He burners ($M_{init} < 2\,M_\odot$): Horizontal-Branch stars

Stars starting with $0.6 < M_{MS} < 2\,M_\odot$ accumulate $\simeq 0.5\,M_\odot$ He in the core. They lose mass in the RG phase and are left with about 0.5 to 1.2 $M_\odot$. These stars lie on the so called *horizontal branch* (HB). The HB has its blue limit at the theoretical He main sequence (see Fig. 10.6) at $L_{HB} \simeq 10^{1.2}\,L_\odot$, the red limit lies in the RG branch region at $L_{HB} \simeq 10^{1.8}\,L_\odot$. HB stars burn He (Ch. 5.1.7) in their core, thus building up C and O.

10.4.2.1 HB stars and the various types

The HB stars are, like the blue loop stars, core He-burning stars. Their luminosity is essentially determined by the He-burning near the core, although in thicker envelope HB stars H-shell burning is present. In the core C+O is accumulated. The core has degenerate gas.

The thickness of the envelope determines the colour of the HB stars.
◦ A *thick* envelope (large amount of material) thus *with large opacity* means the star looks *red*.
◦ A *thin* envelope (small amout of material) so less opacity makes the star look *blue*.
The HB state is reached by stars with $M_{init} < 1.2\,M_\odot$ (the old and thus metal poor stars of globular clusters). Given their different colours and their appearance in globular cluster CMDs, HB stars have a range of type names (Table 10.1).

Table 10.1: Parameters and (spectral) type of HB core He burning stars

T_{eff} [10^3 K]	40	30	20	10	8	6
type	sdOB	sdB	HBB	HBA	RR Lyr	RHB
envelope [$M_\odot$]	< 0.01		0.03	0.12		$\simeq 1.0$
$L(H)$	−	+	+	+	++	++

$L(H)$ = luminosity from H-shell burning (− = none, + = some, ++ = much)

The HB stretches from the He-MS (see Fig. 6.8) to the RGB. Hot HB stars lie *below* the MS and thus are "subdwarf" stars (Sec. 1.7).

The HB passes through the zone where stellar atmospheres have pulsational instability (see Ch. 11). Pulsating HB stars are called RR Lyr stars (Sect. 10.8.1.1). In CMDs they are often not plotted because in (normal) *sequential measuring* different phases are measured at different times leading to 'wrong' colours. This results in globular cluster CMDs to an 'RR Lyrae gap'.

10.4.2.2 Metal content and age of HB stars, morphology of HBs

HB stars are the result of evolution of stars in the lower mass range, $0.5 < M_{\rm MS} < 2$ M$_\odot$. This means that objects now being HB stars but having started with $\simeq 0.8$ M$_\odot$ must be old, the ones with $M_{\rm init}= 2$ M$_\odot$ are therefore much younger. Old stars come from the early phases of star formation in the Milky Way and are therefore metal poor.

The morphology of the HB in the CMD depends apparently also on the metallicity of the star group. The metal poorest globular clusters ([M/H] $\simeq -2.0$), which on average are the oldest, have a limited HB population, only between the sdB stars and the RR Lyr stars. Globular clusters of medium metallicity ([M/H] $\simeq -1.5$) have very blue HBs and no stars redder than HBA type. And the more metal-rich globular clusters ([M/H] $\simeq -0.7$) have rather red HBs, so much so, that they may contain only RR Lyr and RHB stars (or even only RHB stars, as 47 Tuc, see Fig. 10.8). Oosterhoff (1939) recognized that RR Lyrae of the metal poorest clusters have a different pulsation behaviour from the metal-rich ones (later known as the Oosterhoff effect). We know now that they are also slightly different in luminosity, the metal richer ones with a thicker envelope are somewhat more luminous (see also Fig. 11.7).

However, age and metallicity are in their effects intermixed and the details of the cause of the redness or blueness of the HB is still unclear.

There is no established explanation for the *distribution* of the stars of a globular cluster along the HB. An example of a fully populated HB is the globular cluster M 3 (see the CMD in Fig. 10.7). Other examples are given in Fig. 10.8. Since MS stars evolve to the RG phase in line with their initial mass, and given that the RGB phase is passed in line with the initial conditions, how then can HB stars be produced with a range in envelope mass, however small that range for any given globular cluster may be? This problem is indicated by the term *second parameter effect*. Aspects possibly influencing evolution and thus the mass of the HB stars are: 1) metallicity influences the strength of the RG wind; – but how could stars having formed together have different metallicities? 2) stellar rotation (fast or slow) may influence the mass loss during the RG phase thus leading to slightly different masses at which He ignites to make the star into HB star; – but HB stars are all known to be very slow rotators. 3) stronger core rotation may delay the He flash so the star has a more massive core.

10.4.2.3 Evolution of stars on the HB and toward the AGB

HB star evolution is shown in Fig. 10.9. As core He burners they initially behave in a manner having similarities to that of core H burning MS stars. During the He burning (which essentially takes place within the central 3% of mass) in the convective core (extending over the interior $\simeq 0.1$ M$_\odot$) also HB stars slowly become brighter with some expansion leading to a smaller $T_{\rm eff}$ (see Ch. 9.1.1). If the envelope of such stars has enough mass, a H-burning shell will contribute to the stellar luminosity, too.

In the course of the HB core He burning phase central convection will decrease. A "semi-convective" zone will develop in which there is an intricate interplay between the radiative and the convective temperature gradients leading to extended convective motions (see, e.g., Sweigart 1994). Slowly the star will turn to burning He in a shell around the C+O core.

The behaviour of stars after the HB depends on $M_{\rm HB}$.
- $M_{\rm HB} < 0.52$ M$_\odot$: The envelope is very thin and the He-burning shell gets closer and closer (in mass shell) to the (then quite extended) surface, until the fusion can no longer be sustained because of the lower temperature. The star then has only thermal and gravitational energy left and turns into a WD. These stars are also called 'AGB-manqué' stars.
- $0.52 < M_{\rm HB} < 0.55$ M$_\odot$: The envelope is thin. In the early AGB phase the He-burning shell gets closer and closer to the surface, becomes thinner and thus less luminous. Simultaneously, the H shell becomes hotter and quite more luminous. The star ascends the AGB.
- $M_{\rm HB} > 0.55$ M$_\odot$: The He-burning continues and adds C and O to the core. The envelope is sufficiently thick to sustain a well developed He-burning shell. The star evolves further to become an AGB star (Sect. 10.4.3).

150   CHAPTER 10. STELLAR EVOLUTION: STARS IN THE LOWER MASS RANGE

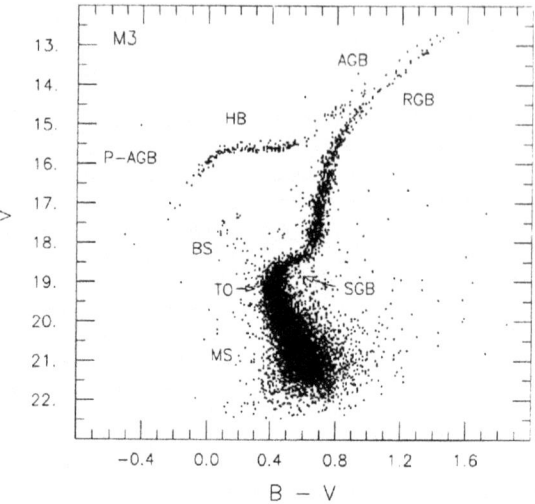

Figure 10.7: The CMD of the stars in the globular cluster M 3 shows all phases of evolution relevant for stars of the lower mass range. The letters mean (in sequence of stellar evolution): MS = main sequence, TO = turn-off, SGB = sub giant branch, RGB = red giant branch, HB = horizontal branch, AGB = asymptotic giant branch, P-AGB = post AGB. The HB has no gap because RR Lyr stars are included. Stars labelled BS are so-called blue stragglers (see Sect. 10.8.4). Figure from Renzini & Fusi Pecci (1988).

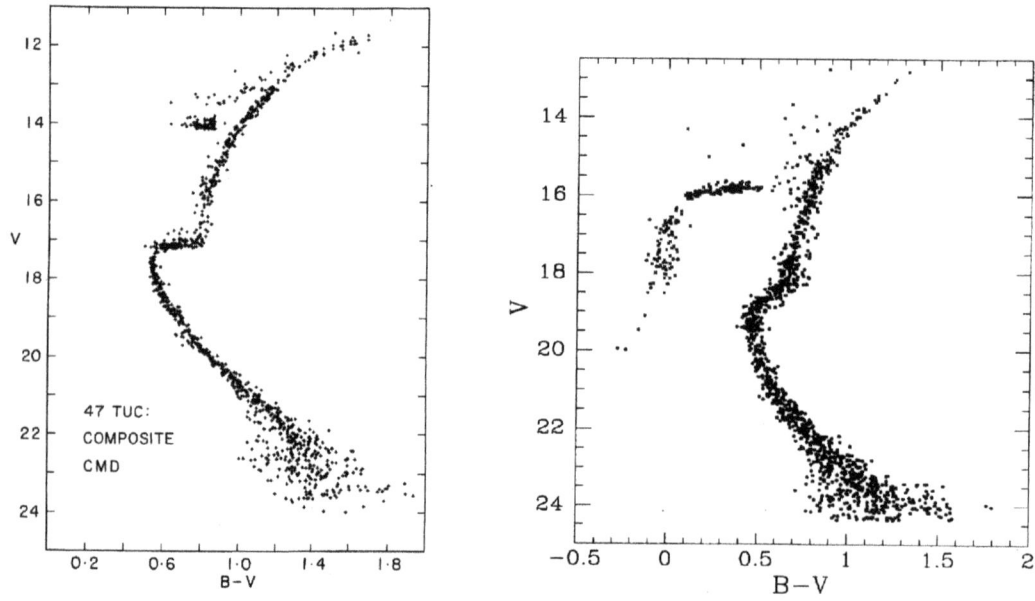

Figure 10.8: The CMDs of two globular clusters show how different the morphology on the horizontal branch (and RGB) can be. At left is 47 Tuc, a metal rich GC (Figure from Hesser et al. 1987). Note the 'stubby' RHB and the AGB stars. At right is M 15, a metal poor GC (Figure from Durrell & Harris 1993). Note the blue HB, including the 'drooping' part. Note also that the shape of the TO, the SGB and the RGB differ from those in 47 Tuc. Compare this figure also with Fig. 10.7 for M 3, a cluster in the middle metallicity range.

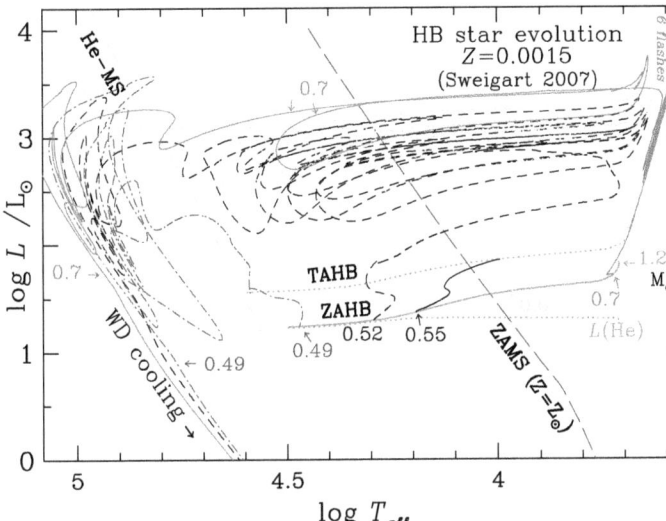

Figure 10.9: Evolution of metal-poor HB stars extending the calculations of Fig. 10.6.
The tracks are for the masses indicated at the Zero-Age HB (ZAHB). The ZAMS and theoretical He-MS (Fig. 6.8) are included for orientation. The ZAHB luminosity from only core He fusion, $L(\text{He})$, is also indicated ($\cdots$). With higher envelope mass, the higher pressure in the H-burning shell (see Fig. 10.5) leads to a larger $L(\text{H}) = L - L(\text{He})$. Evolution toward a second red giant phase (the AGB phase) is based on the transition to He shell burning in which H shell burning is somewhat decreased. Then the He shell at $\simeq 0.2$ $M_\odot$ burns outward through the core and decreases in L while the H shell is heated from below and increases in luminosity. The loops from the tip of the AGB are due to flash-like brightening of He shell burning and their subsequent dimming (thermal pulsing). The number of pulses (here 6 for 0.7 $M_\odot$) depends on the total mass structure thus on how much mass is lost in each phase. The lowest mass HB stars will not reach the AGB (AGB-manqué stars). After a few final H shell flashes at large T_{eff} the stars shrink and evolve to the WD cooling tracks. Evolution of the star with $M_{\text{ZAHB}} = 0.55$ $M_\odot$ is not shown beyond the TAHB to not confuse the top part of the diagram with yet more loops. Data from Sweigart (2007).

10.4.3 AGB stars: structure and evolution

10.4.3.1 AGB star evolution and the CMD

At the end of the core He burning phase, after $t_{\text{HB}} \simeq 10^8$ yr (at the terminal-age HB, TAHB), the He burning will shift into a shell. The degenerate C+O core will shrink further raising the temperature of the hydrogen shell so that $L(\text{H})$ increases (considerably in HB stars with a thicker envelope). This leads to envelope expansion with opacity-driven convection (Ch. 4.3.4) and the star becomes again a red giant (see Ch. 9.3). In the CMD the evolutionary tracks of the middle mass range HB stars approach the (red) giant branch aymptotically, and the star is called an AGB star (for the general structure see Fig. 10.10). Stars with too little envelope to reach the AGB are called AGB-manqué stars, evolving up from the HB and to the left, then to the WD-cooling region.

10.4.3.2 He-shell flashes (thermal pulses) and convection

Near the end of the AGB phase the He burning shell has become nearly extinct. The now luminous H burning shell adds He to the core. This leads to rising temperatures in the core and the thin He shell reignites in a flash like manner (steep ϵ, Eq. 5.14) at almost constant pressure. (Note: the flash at the onset of core He burning takes place because of the degeneracy of the gas; Sect. 10.2.3). The energy production heats the layer, radiation-driven convection (Ch. 4.3.4) occurs, energy escapes and the shell cools to below fusion temperature so the He fusion decays quickly. This rather rapid process, called thermal pulsing, has its origin in the proximity of the shells with He and with H fusion (see Fig. 10.10, a separation of only 0.03 $M_\odot$). A small change in the properties of one of the shells will immediately influence the properties in the other shell. The convection extends not very far upward but transports fusion products into the H shell. During pulsing the star makes a looped evolution in the HRD (Fig. 10.9). Each further flash takes place at larger $M(r)$, of course. The number of flashes and concomittant loops in the HRD is related with $M_{\text{He core}}$ and M_{ZAHB}.

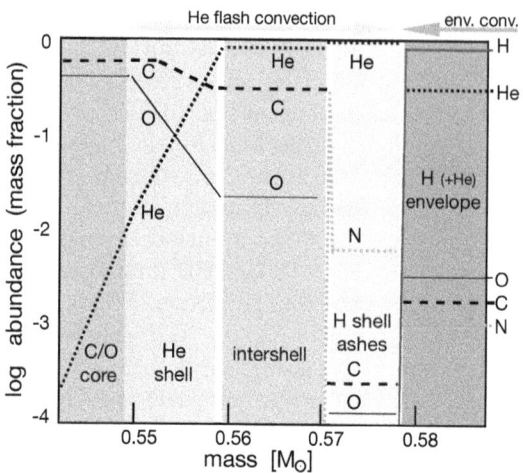

Figure 10.10: Sketch of the structure of the essential internal zone of an AGB star. Depending on M_init and M_AGB some features are different.

AGB stars have very close He and H burning shells (see the mass values of this cross section). Due to this proximity these shells (and changes in them) influence each other strongly.

Nuclear fusion in AGB stars takes place in three zones (white). Most of the luminosity is produced in the H burning and He burning zones.

The chemical profiles of H, ^{4}He, ^{12}C, ^{14}N and ^{16}O are shown. The shading and the shell names are indicative of chemical composition. The "intershell" (containing only $M = 0.01\,M_\odot$) is the thin inert layer between the outer H burning shell and the inner He burning shell. The gas density drops over the thin mass shell between $M = 0.55$ and $0.58\,M_\odot$ from $\rho > 10^4$ to $\rho < 10^{-5}$ g cm^{-3} (see also $\log\rho$ in Figs. 10.11 and 10.12). Such stars have a very extended and tenuous envelope. Note the peak in the abundance of ^{14}N in the zone at the bottom of the H shell ashes. This peak emerges because in the H fusion CNO cycle the reaction from ^{14}N on to other nuclei is a slow one (Eq. 5.10) thus leading to a build up of ^{14}N. This nitrogen is, however, in the intershell region soon consumed in nitrogen burning (Eq. 5.18). At the top the approximate locations of the convective zones are indicated; they vary in depth and may reach the same mass shell but not simultaneously. The He flash convection zone mixes chemically very different layers. Hot bottom burning (HBB) may be present in AGB stars from $M_\text{init} > 4\,M_\odot$ (Sect. 10.4.4.3) in the rightmost H-burning zone. At each further He flash the C/O core grows and the He convection zone moves up in mass. For the larger scale stucture of AGB stars see Figs. 10.12 and 10.11, right panels. This sketch is of a star of $M_\text{AGB} \simeq 1.7\,M_\odot$ (for which $M_\text{init} \simeq 2\,M_\odot$) and was designed after Herwig (2005).

10.4.3.3 Third dredge-up: nuclear fusion and s-process

The He flash leads to convection. This convection extends over a limited extent but far enough to bring products of the He fusion, i.e., C and O, up into layers with abundant He or even H (third dredge-up). Mixing C and O with He and H at the high temperatures available allows low level C-burning (C+He, O+He, Eq. 5.15) building up nuclei like Ne, Mg and Si. Also p-process reactions are possible (see Fig. 5.7). Furthermore, dredged-up C now can come in contact with H with the possibility of slow C burning (C+H, Eq. 5.19) leading to O and free neutrons. With free neutrons s-process fusion (Ch. 5.2.4.1) becomes possible with the building up of numerous heavy elements.

Envelope convection may extend into the shell with H ashes. That shell has an excess of N (from the CNO cycle H burning, Eq. 5.10), so N-burning becomes possible (N+He, Eq. 5.20). Also here free neutrons are produced allowing heavy element formation. The envelope convection then carries fusion products to the surface, the "third dredge-up", thereby changing the original atmospheric chemical composition.

More on the fusion processes can be found in Herwig (2005) and Lattanzio & Wood (2004). The amount of fusion products seen at the surface depends on the efficiency of the dredge-up as well as on the early efficiency of the wind ridding the star of its initially less enriched outer layers.

10.4.3.4 Flashes and mass loss of fusion enriched material

The He flashes occur after more or less constant time intervals and one speaks of the He-flash frequency. During the flashes with the ensuing envelope expansion the mass loss is enhanced. For flash luminosities and the mass loss from a model see Fig. 10.13. Since the core evolves as it has to, while the mass is lost based on the envelope conditions (Eq. 10.2), it is almost undetermined what the actual envelope structure is when the last He flash occurs. The mass lost is chemically

10.4. THE END OF CORE HE BURNING AND ON TO THE AGB

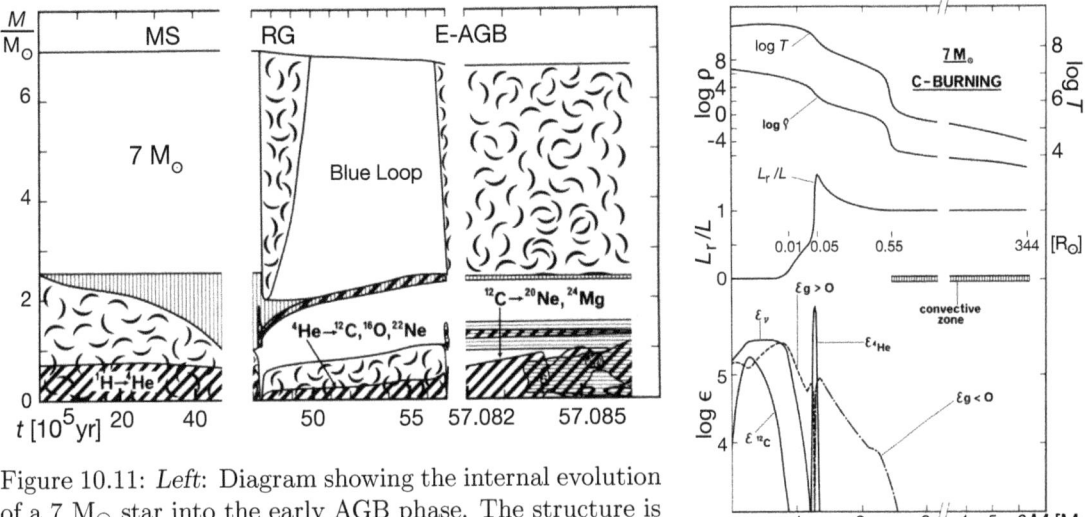

Figure 10.11: *Left*: Diagram showing the internal evolution of a 7 M$_\odot$ star into the early AGB phase. The structure is plotted (vertically) with radial mass $M(r)$ [M$_\odot$]. Cloudy regions represent convective zones. Note the mass loss during the RGB phase. Figure from Maeder & Meynet (1989). *Right*: Diagram showing the run of the important stellar parameters T, L, ρ and ϵ of a 7 M$_\odot$ star in the C-burning phase, at the highest temperature T_c reached (see Fig. 10.2). The structure is plotted against radial mass $M(r)$ [M$_\odot$]. Note the gap in the scale at $M = 3.3$ M$_\odot$ (and the gaps in the curves at that mass). The status shown is that just after the onset of some C-burning (the E-AGB phase, left panel). The burning of H in the shell has stopped. At the outer edge of the He burning shell the luminosity reaches $\simeq 1.5$ of that of the surface (the surface has $L \equiv 1$ L$_*$). About 0.5 L_r/L is taken up by the expansion taking place of the shell between 1.3 and 2.5 M$_\odot$ (there $\epsilon_g < 0$). The core (at $M < 1.2$ M$_\odot$) is contracting (there $\epsilon_g > 0$). Note also the energy loss through outflow of neutrinos (ϵ_ν). The log ρ profile and the tics marking the radius r [R$_\odot$] in the star underline the compactness of the stellar core and the large extent of the envelope. Figure adapted from Maeder & Meynet (1989).

enriched which leads to ample possibilities of enriching the local interstellar medium and the abundant formation of dust (see Sect. 10.5.1).

10.4.3.5 All happenings in a very thin layer

All the action of AGB stars take place in the thin mass shell. For lower mass AGB stars it is between $M \simeq 0.55$ and 0.58 M$_\odot$. In that shell the gas density drops from $\rho > 10^4$ to $\rho < 10^{-5}$ g cm^{-3} (Fig. 10.10). In AGB stars of higher mass the shell is still thinner. AGB stars have a very dense slowly growing core, then the transition layer with flashes, mixing, and build up of elements, overlain by a very tenuous and extended envelope evaporating at the top. There is a certain similarity with the structure of the Earth with a thin layer where all the action is....

10.4.4 Higher mass core He burners: blue loop stars and the AGB

Stellar evolution proceeds from a MS star (with CNO-cycle H burning) to a RG star (all stars of $0.6 < M_{\text{init}} < 8$ M$_\odot$ lose then only a little mass), then into a blueish core He-burning star and back again to an RG (as an AGB) star. This was as such recognized rather late in the development of models of stellar evolution. The blueing was regarded as a surprise and this evolution got named 'blue loop'. During the blue loop, the stars cross the instability strip, becoming a Cepheid variable (see Sect. 10.8.1.2). For more on why a blue loop occurs or not see Ch. 13.4.1.2.

Blue loop stars are those having $\simeq 4 < M_{\text{init}} < 8$ M$_\odot$ (in fact up to 15 M$_\odot$; see Ch. 13.4.1). They have a He core in which He burns to C, O, Ne. H burns to He in a shell so that slowly but continuously He is added to the core. This is a stable phase lasting $10^{8.4}$ to $10^{6.5}$ yr (see Fig. 9.6),

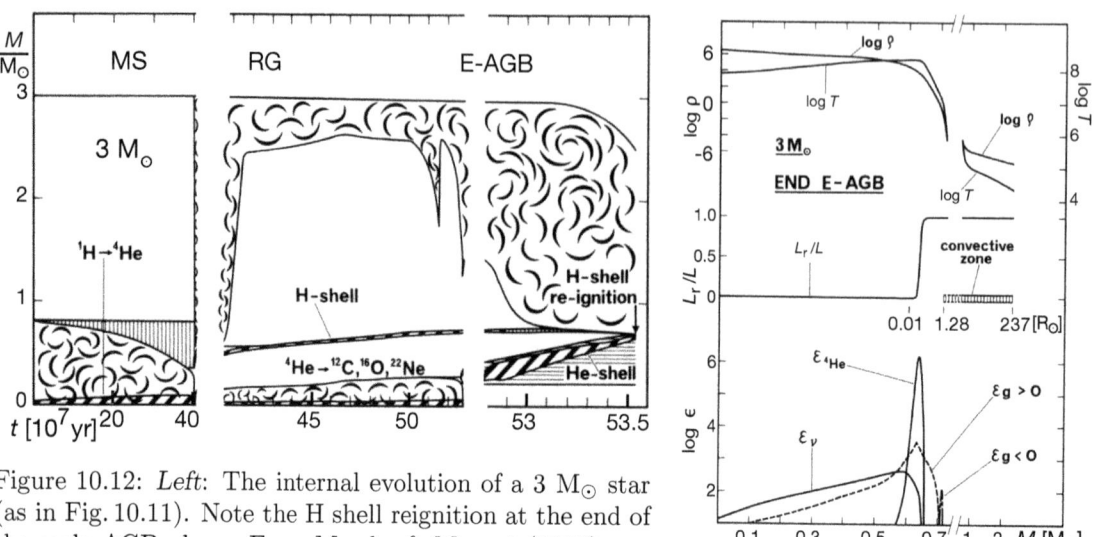

Figure 10.12: *Left*: The internal evolution of a 3 M$_\odot$ star (as in Fig. 10.11). Note the H shell reignition at the end of the early AGB phase. From Maeder & Meynet (1989).
Right: The run of T, L, ρ and ϵ inside a 3 M$_\odot$ star at the end of the He shell burning phase (just before C-ignition). Note the gap in the scale near radial mass $M = 0.75$ M$_\odot$. Gravity contributes ($\epsilon_g > 0$) to the luminosity but the He-burning shell near 0.7 M$_\odot$ expands, so in that layer $\epsilon_g < 0$. Note also the $\log \rho$ and the r [R$_\odot$] information on the compactness of the core and the extent of the envelope. See also the caption of Fig. 10.11. Figure adapted from Maeder & Meynet (1989).

depending on M_{init}. The structure of a 7 M$_\odot$ star is shown in Fig. 10.11. It has to be noted that the mass loss rate of such stars is, in fact, not known from observations.

AGB stars have a very compact core (see the $\log \rho$ profiles and the radius tics in Figs. 10.11 and 10.12, right hand panels). The ultimate WD is, as it were, already present inside the outer layers.

Note again that in AGB stars the shells burning H and He lie close to each other (Fig. 10.10). So changes in one shell may strongly influence the behaviour of the other shell.

10.4.4.1 Stars with M_{init} larger than 7 to 8 M$_\odot$

In the upper mass range (with approximate limit $M_{\text{MS}} \geq 7$ to 8 M$_\odot$) a non-degenerate core of $M \simeq 1.05$ M$_\odot$ of C+O will be built. In a shell the temperature will be high enough for slow carbon burning (Eq. 5.15). The burning processes create high temperature gradients and the envelope is highly convective. These stars are called "super AGB stars".

The more massive AGB stars continue to add C to the core through He fusion. Eventually the CO core might approach the mass limit of 1.4 M$_\odot$, and C+C burning (Eq. 5.16) would start flash like in the core which would result in a run-away fusion process and in the complete disruption of the star (supernova). However, AGB stars loose enough mass that this does not happen.

The limit between quiet fading of the AGB star or its explosive demise as a SN Type II (Ch. 18) lies between 8 and 10 M$_\odot$ (Fig. 10.24).

10.4.4.2 Stars with $M_{\text{init}} = 2$ to 7 M$_\odot$

In the mass range with approximate limits $2 \leq M_{\text{MS}} \leq 7$ M$_\odot$ He core burning is established smoothly. When the central He is exhausted He will continue to burn in a shell. Then the H-burning shell will be pushed outward and its luminosity is reduced.

With He burning in a shell, a core containing C is further built up. The C-core is degenerate and almost isothermal (neutrino cooling). The evolution of the structural parameters for a star with $M_{\text{MS}} = 3$ M$_\odot$ are given in Fig. 10.12. Then He flashes will occur (thermal pulsing).

An example of the He flash pulses in an AGB star of 1 M$_\odot$ (from $M_{\text{init}} = 2$ M$_\odot$) is given in Fig. 10.13. Note how drastic the mass loss is especially in relation with the last He flash.

10.5. THE END OF THE AGB PHASE

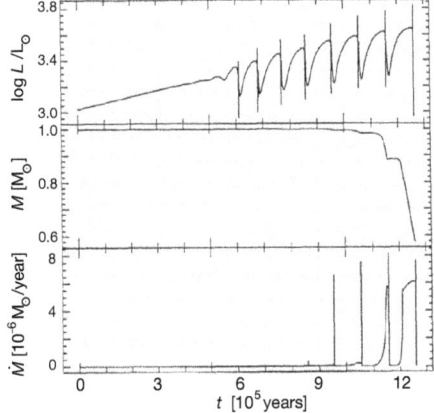

Figure 10.13: At the end of the AGB phase the He shell reignites periodically (Fig. 10.9) leading to He shell flashes and luminosity pulses (thermal pulses).
The model shows the behaviour of an AGB star of $M_{\rm AGB}$ = 1 M$_\odot$ (for which $M_{\rm init} \simeq 2$ M$_\odot$). The luminosity, the mass, and the mass loss rate are shown over the relevant time interval. Mass loss takes place mostly in the end phase with before the last flash an extreme mass loss of 0.4 M$_\odot$ in $\simeq 5 \cdot 10^4$ year. The lost mass disperses around the star, cools, and forms more dust and molecules, leading to the OH/IR phenomenon (see Fig. 10.14). Figure adapted from Wood & Vassiliadis (1992).

10.4.4.3 AGB stars and hot bottom burning

In the more mass rich AGB stars ($M_{\rm init} > 4$ M$_\odot$) the outer convection zone can reach that deep, that it includes the layer with H fusion. This convection efficiently replenishes the H fuel. The phenomenon is called "hot bottom burning" (see Fig. 10.10). In it, build-up of Ne, Na, Mg, and Al is possible in proton-capture reactions (Ch. 5.2.3, Fig. 5.7).

The C dredged up during the He flash convection is (in CNO cycle H burning) transformed into N and the star will show little C at the surface. These AGB stars are more luminous than the lower mass AGB stars. Toward the end of the AGB phase the hot bottom burning becomes less efficient and thus more C can be brought to the surface convectively. The star can now become a "carbon star".

For stars with $M_{\rm init} < 4$ M$_\odot$ deeper convection is not possible since their weaker convection cannot overcome the chemical barrier (mean molecular weight) of the He and C containing layers (Ledoux criterion; Ch. 4.3.2). So in solar metallicity stars hot bottom burning occurs for $M_{\rm init} >$ 4 M$_\odot$, while in zero metal stars even down to $M_{\rm init} = 2$ M$_\odot$. The shell mass involved in this burning is normally just 10^{-4} M$_\odot$. Due to the special conditions in these AGB star shells also Li can be produced and circulated to the stellar surface.

10.4.5 Timescales

A summary of processes in AGB stars and the relevant approximate mass ranges is given Fig. 10.24, the time scales for the various phenomena and the mass lost are summarized in Table 10.2.

Table 10.2: Lower mass stars: summary of time scales in years and of mass lost

phase	MS	RG	blue loop, HB	E-AGB	TP-AGB
duration	$\simeq 1/M^2$ (Fig. 9.6)	$\simeq 0.1 \cdot t_{\rm MS}$	$\simeq 10^{6.5}$ to $\simeq 10^{8.4}$	$\simeq 3 \cdot 10^6$	$\simeq 5 \cdot 10^5$ (Fig. 10.13)
mass lost	-.-	10 to 40%	-.-	few %	up to 70%

10.5 The end of the AGB phase

In the course of the AGB evolution thermal pulsing takes place (Fig. 10.13). Due to the pulsing and the excess luminosity large amounts of gas with molecules (Fig. 3.10) and freshly formed dust are shed (see Ch. 4.5 for mechanisms of mass loss, Ch. 4.5.5 for pulsation-driven winds). That gas is enriched in metals (third dredge-up, etc.; Sect. 10.4.3). These gases leave the cool AGB star atmosphere in surges in line with the thermal pulses, cool further and build further (bigger) dust grains shielding the star from sight. The dust is warm and emits in the infrared (see Fig. 10.14).

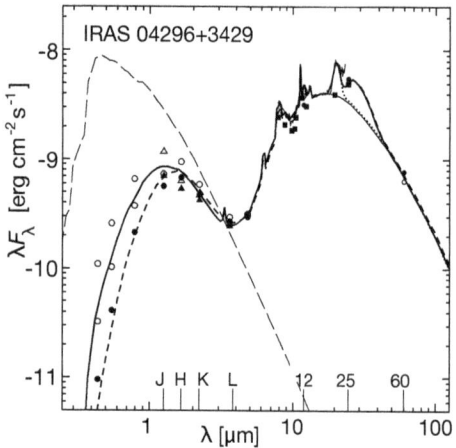

Figure 10.14: The observed spectral energy distribution of an OH/IR star shows how the original stellar spectral energy distribution (long-dash curve at left, star with $T_{\text{eff}} = 6500$ K, $\log g = 0.5$) is heavily extinguished by the dust shell. The IR flux is basically emission by dust. Some model curves have been added. Molecules and grains in the dusty shell produce emission lines. The object was discovered at radiowavelengths and confirmed to be a dusty object in the IRAS all sky IR survey. Figure from Klochkova et al. (1999).

10.5.1 Massive AGB stars: OH/IR stars and pAGB stars

Massive AGB stars come from main-sequence stars in the mass range $M_{\text{MS}} > 4$ M$_\odot$ (up to $\simeq 10$ M$_\odot$). AGB stars present now are therefore relatively young objects and the original MS stars had a nearly solar chemical composition. Note that, during the RG-phase, of course some chemical mixing may have occured (first dredge-up).

Massive AGB stars have massive winds with lots of dust. This material shrouds the star and may become so thick and cool that molecules are formed in the gas. These may become radiatively excited by the AGB star radiation ($L \simeq 10^{3.5}$ to 10^4 L$_\odot$). Particular molecules may produce the so-called microwave amplification of stimulated emission radiation (MASER). One of such molecules known is OH radiating at 18 cm wavelength (1612 MHz). The dust emits most of its radiation in the IR. This explains the name of these dusty objects: OH/IR stars. For reviews on OH/IR stars see Habing (1996), Olofsson (2004) and Habing & Whitelock (2004). The OH-MASER phenomenon does not last long, only 500 - 2000 yr. OH MASERS are known to have "died" (Lewis 2002).

The dust shroud expands away from the star, as has been derived from the double nature (Doppler splitting) of CO and OH emission lines (see, e.g., Olofsson 2004). The material will slowly disperse into interstellar space.

The star remaining has a degenerate He core and H-burning in a thin shell. The star will end as a pAGB star with mass of about 1 M$_\odot$. It will then contract, thereby heating its surface. Ultimately it will become a WD.

10.5.2 Low mass AGB stars: pAGB stars and planetary nebulae

AGB stars with little mass come from stars originally in the lower mass range ($0.6 < M_{\text{MS}} < 2$ M$_\odot$). Some of such AGB stars present now are either somewhat younger than the Sun and thus of near solar composition, or are older than the Sun and thus metal poor (as in globular clusters).

AGB stars have deep convection zones, so that fusion products like C and O and s-process elements get dredged-up to the surface.

At the end of the AGB phase the star, bereft of most of its atmosphere, ceases to have fusion. The star contracts (as in Ch. 4.2.2) and now is a post-AGB (pAGB) star. Its maximum mass, based on models, is 0.65 M$_\odot$. The contraction leads to faster rotation and, at $L \simeq$ constant, to a very hot surface and T_{eff} may reach even 200 000 K. Thereafter the star cools and becomes a WD.

The large flux of high energy photons coming from the hotter pAGB stars can photoionize the surrounding medium. That space contains also material blown away from the AGB star during the thermal pulses, in particular the last pulses. These gases get ionized and all the processes taking place in H II regions can be observed. The radiant nebulae were, in the days of early astronomy, mistaken as 'planetary nebulae' (PNe); that name was kept nevertheless.

The pAGB stars from low mass stars exist in globular clusters. Since their evolution (cooling)

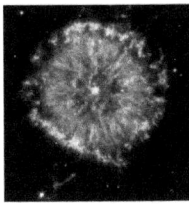

Figure 10.15: The planetary nebula NGC 6751 in Aquila measures 0.3 pc in diameter and has an expansion velocity of $\simeq 40$ km s^{-1}. The CSPN has $T_{\mathrm{eff}} \simeq$ 140000 K. This nebula (HST image) is rather spherical suggesting a spherically symmetric wind (compare with the PN in Fig. 14.7). Some PNe have multiple shells of gas perhaps associated with the surges of mass loss related with the thermal pulsing of the AGB star.

is quick, clusters have at most 2 - 3 such stars (often none) as confirmed from globular cluster CMDs. These stars contribute considerably to the UV output of globular clusters (de Boer 1985).

Since deep convection zones during the last part of the AGB phase may have transported fusion products to the surface, such products also land in interstellar space. AGB winds can be C-rich and carbon dust (soot) will form ('smoking stars'). Spectra of several pAGB stars show strong C IV absorption and C III] emission lines, demonstrating the presence of the fusion products at the surface (see figure in de Boer 1985). The star K 648 in the globular cluster M 15 has a PN.

The Central Star of a PN (CSPN) is always hot. Sometimes the maximum of the spectral energy distribution lies so far into the UV, that the visual magnitude is extremely faint and no CSPN can be seen. However, the nature of the ionization of the nebula and the strength of its emission lines can be used to derive T_{eff} of the CSPN. This method is called the Zanstra method. For details of PNe see Kwok (2000). The shape of the nebula depends on the parameters and history of the CSPN (such as rotation and the nature of the pre-pAGB mass loss) as well as on the local interstellar environment (see, e.g., Balick & Frank 2002). An example is given in Fig. 10.15.

10.6 The end phase: white dwarfs

All stars having started with $0.6 < M_{\mathrm{MS}} < 8$ M$_\odot$ become WDs. The mass of WDs lies in the range of $0.5 < M_{\mathrm{WD}} < 1.4$ M$_\odot$ (see Ch. 17). WDs of much lower mass coming from single star evolution do not exist yet, as they will be the descendants of MS stars with mass < 0.5 M$_\odot$ (see Fig. 10.3) taking more time than the current age of the universe (likely $\simeq 15 \cdot 10^9$ yr). However, in binary evolution also a WD may be stripped of part of the envelope and they may have a mass < 0.5 M$_\odot$. The surface gravity of WDs is $\log g = 7$ to 9.

10.6.1 Classification of WDs

White dwarfs are named this way because the first such stars were seen as 'white' and were thought to be dwarfs (see Ch. 17.1.1). We now know they are much more compact than dwarf (MS) stars. Fig. 10.16 shows a CMD of the solar neigbourhood, demonstrating the presence of very local WDs.

WDs exist over a large range in temperatures and a large range in spectral features (see the summary of the classification of faint blue stars; de Boer et al. 1997). Spectral type DA shows very strong and broad Balmer lines but no He I and metal lines, type DB has He I lines without H and

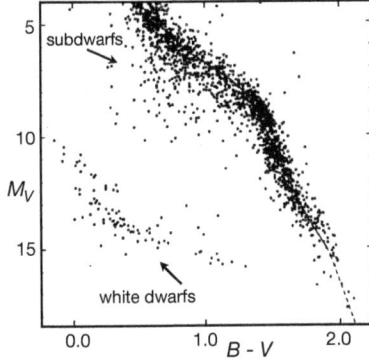

Figure 10.16: CMD of stars in the solar vicinity (within 25 pc). These are stars for which parallaxes could be determined with considerable accuracy. The data show that WDs are abundant. The dotted line is the location of the main sequence. Objects appearing in this diagram to the left of the MS (subjectively below the MS) are cool subdwarf stars (see Sect. 10.8.3). Figure adapted from Jahreiss (1987).

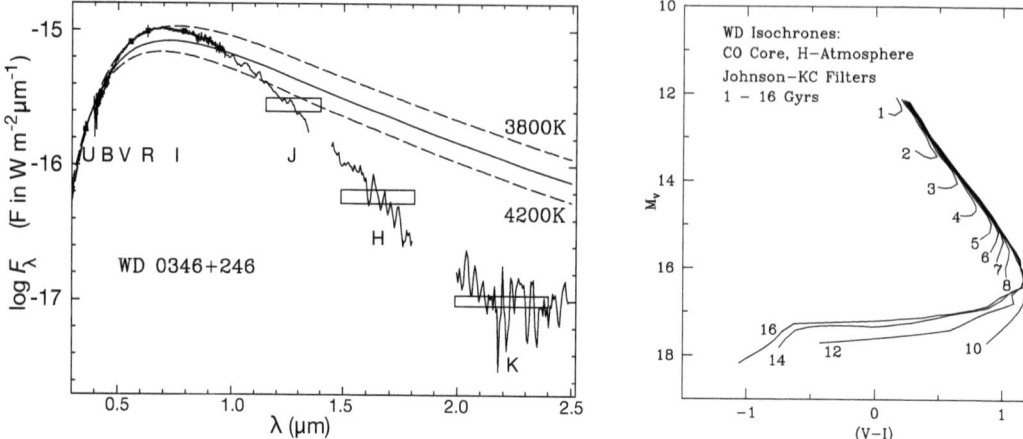

Figure 10.17: *Left:* The cooling of WDs leads to temperatures at which H_2 can form in the atmosphere. The spectral energy distribution is then depressed in the near infrared. The normal photometric bands U, B, V, R, I, and J, H, K are indicated. Such stars have *blue* colour indices between the photometric V and H bands. Figure from Oppenheimer et al. (2001).
Right: The cooling of WDs leads to changes of their colours with time (in units of Gyr) due to the lower temperature (Planck-function) and then at certain wavelengths due to extra opacity from molecules. Note at the late stages the blue $V - I$ index, which can be seen also in the left panel when comparing the measurements with the model. WDs discovered in $V - I$ and thought to be very hot may turn out to be cool after all. Figure from Richer et al. (2000).

metal lines, type DO for WDs with strong He II. Later, further types were introduced: e.g., DC spectra almost without any spectral lines (none deeper that 5%), the type DZ for WDs without H or He but only metal lines and type DQ for spectra with carbon (either atomic or molecular).

Due to the compact nature of the WDs their original magnetic field (if present) has become much condensed. That introduces Zeeman splitting of the Balmer lines (see Ch. 3.5), spectral structures which were originially unexplained. Absorption lines of other elements are mostly weak. This is either due to low intrinsic abundance or due to gravitational settling (see Ch. 3.6) of the heavier atoms. Also metal lines may show magnetic Zeeman-splitting (see Fig. 3.12).

WDs are a mixed bag. Some are the descendants of HB stars, other come from more massive originals (see Fig. 10.24), perhaps having had the OH/IR phase. Their composition differs accordingly.

WDs having been HB or pAGB star have a degenerate C core with a He shell (He WDs), and possibly a non-degenerate thin outer H atmosphere. Stars having started with low MS mass have as WD only a He core and a H atmosphere. The entire interior is degenerate and behaves as a polytrope with index $n = 3/2$. For more about the internal structure of WDs see Ch. 17.1.

WDs radiate without fusion as energy source, so they cool. Cooling to $T \simeq 4000$ K takes $\simeq 10^{10}$ yr, somewhat less for lower, more for higher mass WDs. Several Figs. show cooling tracks.

When the atmosphere of a WD has become cool, molecules may form. Such molecules lead to additional opacity. The spectral energy distribution changes accordingly. In particular H_2 produces a strong depression near 1 - 4 μm (see Fig. 10.17, left), so that the colour index $V - I$ becomes suprisingly blue. In searches in the red for very cool WDs such objects were classified as blue and thus warm WDs or not recognized as WDs at all. Fig. 10.17 (right panel) shows how the colour $V - I$ of WDs changes with cooling.

More on the observational aspects of WDs can be found in Weidemann (1990). For more on cool WDs see Hansen & Liebert (2003). Some WDs pulsate (see Sect. 10.8.1.4).

White dwarfs have become of great interest in relation with questions about the structure and

10.6. THE END PHASE: WHITE DWARFS

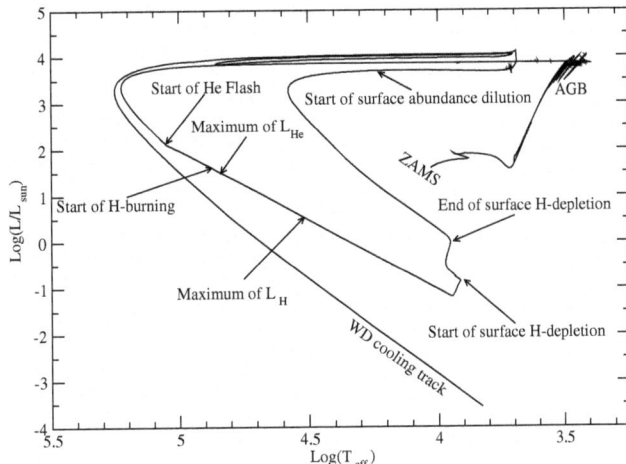

Figure 10.18: A cooling WD may have a last thermal pulse. This produces suddenly an energy source near the stellar surface so that large convective motions establish, leading to expansion of the thin envelope and to a rapid reduction in T_{eff} and L. At the same time, convection transports surface layers with H as fresh fuel into the interior. The star becomes for a very brief period again a supergiant (born again AGB star) and soon turns back to the WD configuration. Along the evolutionary track the onset of He burning and the further evolutionary steps are marked. Figure from Miller Bertolami et al. (2006).

evolution of the Milky Way. There are implications for stellar evolution in general and for the nature of the initial mass function in particular. The present day number of WDs does, after all, reflect the past history so that models for the Milky Way must also in that respect be consistent.

10.6.2 Ultimate fate of WDs

The ultimate fate of WDs is to become very cool and dull objects.

However, if a star was part of a binary system, the more massive one (the primary), evolving a bit faster than the mass poorer one (the secondary), may shed mass onto its companion. The primary will, after further evolution, become WD first. The secondary will in turn evolve and become RG too, shedding mass back. The primary WD may thus accumulate so much mass that it surpasses its physical mass limit (see Ch. 17) of $\simeq 1.4\ M_\odot$ so that a disastrous event sets in: it will become supernova (Type Ia). The transfer aspect will be treated in Ch. 19.

10.6.3 Born-again stars

If WDs are formed as described, then almost all WDs should have a remainder of hydrogen at their surface. However, many a WD shows no trace of H at all in the spectrum.

When a star is well into the pAGB phase, a final thermal pulse may take place (Schönberner 1979, Iben et al. 1983). This spurt of energy production causes so much upheaval and convection that the outer layers (with whatever little remaining H) are transported inward providing these objects with yet more fuel. The star is "born-again" and turns into a viable red giant to red supergiant (however of only 0.6 $M_\odot$). It then again sheds its outer layers, and reverts via the pAGB phase to the WD state, now absolutely without H in the atmosphere (see Fig. 10.18).

Support for this scenario came in 1996 when Sakurai, hunting for comets, discovered a new bright object in Sgr which did not fade like a Nova but continued to brighten. The object, soon called Sakurai's object, turned out to also have a faint PN, and must have been a post CSPN. Its total evolution as born-again star is fast: discoverd as yellow giant in 1996, being red SG in 1998, blow out of material and fading in 2002. Probably a second PN phase will follow.

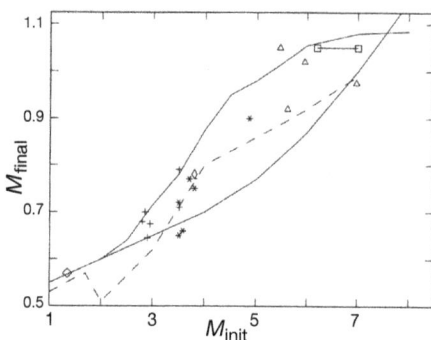

Figure 10.19: The relation between the initial mass of stars, M_{init}, and their final mass as WD, M_{final} (vertical axis), is shown. The symbols refer to data of particular clusters and stars. The lines are empirical relations derived by different groups. All stars which have started with $0.6 < M_{\text{MS}} < 1\, M_\odot$ end with $\simeq 0.5\, M_\odot$. Figure from Weidemann (2000).

These born-again stars may also explain the not seldom seen "double PN", a wide and diffuse faint outer PN with a brighter inner planetary nebula.

10.7 Initial to final mass relation for lower MS stars

Since WDs come from stars having started with $M_{\text{MS}} < 8\, M_\odot$, and since the evolutionary routes are known, one can determine a relation between initial and final mass of such stars. Using results from cluster photometry, such a relation is given in Fig. 10.19. It can be used in the modelling of galactic evolution accounting for delayed return of stellar material then available for star formation.

10.8 Some special stars

A few special kinds of stars need to be discussed. Their importance lies mostly in their significance for the calibration of CMDs and HRDs.

10.8.1 Pulsational variables: RR Lyrae, δ Cepheids, PG 1159 and ZZ Ceti stars

A general discussion of pulsation is given in Ch. 11, which also includes further information on the low-mass pulsators mentioned here.

10.8.1.1 RR Lyrae stars

RR Lyr stars are pulsating HB stars with $M = 0.6$ to $0.7\, M_\odot$. The instability occurs in the upper envelope due to the κ-mechanism (Ch. 11.2.3). The period of pulsation is 0.3 to 1.5 days. The light curve has a characteristic shape thus allowing to identify such stars from just photometry.

Since all RR Lyr stars have nearly the same mass and structure (see Sect. 10.4.2) they have nearly the same luminosity and so nearly the same M_V. Their (nearly) identical luminosity makes them useful as distance determinators. Note that M_V for a variable star is defined by the mean brightness over the light curve.

There is an extensive body of research on RR Lyrae, because of their use as tracers of certain populations, and because of their use of determining their distance (both for the field stars and the globular clusters). To make sure one really understands their nature, their luminosities have to be determined without the help of the very hard to come by accurate parallactic distances.

Using models for the structure of RR Lyr stars and adding models for their atmospheres, one can compare the data of stars with the models to achieve consistency. Note that RR Lyr stars may have a range in age and thus chemical composition. Using Strömgren photometry (see Fig. 11.8a) calibrated through atmosphere models, the variation of stellar parameters during the pulsational cycle can be retrieved (see Fig. 11.8b). With modern techniques the radial velocity variations at the surface can be measured directly (see Fig. 11.10).

10.8. SOME SPECIAL STARS

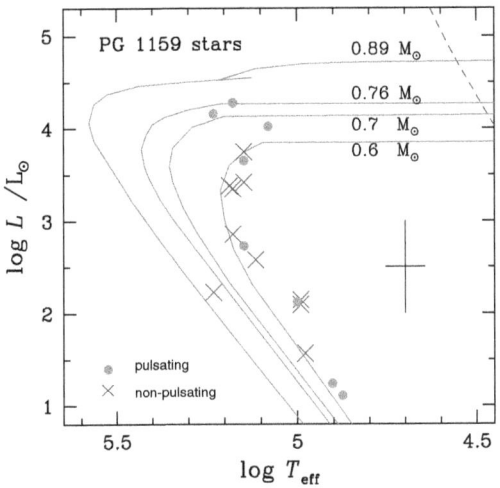

Figure 10.20: Stars on the WD cooling sequence may show pulsation. The data are for stars of the type PG 1159 (see Sect. 10.8) lying in a strip in the hot part of the HRD. Some of the stars pulsate, others are not pulsating. The error cross is characteristic for the data plotted. Figure using data from Werner et al. (1996).
Tracks are given for pAGBs of 0.6 to 0.89 $M_\odot$ from Wood & Faulkner (1986). The cooling tracks extend in a diagram of $\log T_{\rm eff}$ and $\log L$ (double logarithmic) linearly to ever lower values. Note that the higher mass WDs lie to the left of the lower mass ones (degenerate core; see Ch. 17.1.1). The MS is visible in the upper right corner of the diagram (see also Figs. 10.1 and 13.9).

10.8.1.2 δ Cepheid stars

A very important kind of pulsation variable is the group called after δ Cephei. The instability occurs, as with RR Lyr type stars, due to the κ-effect. The δ Cep light curve has a characteristic shape, too, so such a variable is classified easily. δ Cep stars are more massive than RR Lyrae (see above), coming from stars with $4 < M_{\rm init} < 15$ $M_\odot$ (see also Ch. 13), but are otherwise very similar. That their mass is larger follows from the fact that δ Cep stars are members of young clusters and generally found only in the Milky Way disk.

The larger mass and the younger age implies a location higher in the HRD, well above the HB (see Fig. 11.4). Furthermore, since stars with various (large enough) $M_{\rm init}$ evolve across the HRD and cross the pulsational instability strip, the δ Cep stars differ from each other in mass and luminosity. δ Cep stars exhibit the well known "period-luminosity" relation. This aspect, together with their brighter M_V, makes them exceedingly useful as distance indicators, even out into other galaxies.

10.8.1.3 PG 1159 stars

A special case of variability is present in the so called PG 1159 stars (the prototype star is star 1159-035 from the PG catalogue). The stars are pAGB, pre-WDs and can be found on the upper part of the WD cooling sequence, at $T_{\rm eff} \simeq 120\,000$ K (see Fig. 10.20).

The spectra show no hydrogen but all show He, C, O and other metals. In some of the stars, the atmosphere has a κ-effect (ionization zone of C^{+4} to C^{+5}), causing multi-mode non-radial g-mode pulsations (see Ch. 11.6.4) with amplitude of $\Delta M_V \simeq 0.2$ mag. They most likely have only a rudimentary He envelope after a late He flash (see the loops of the late pAGB stage in Fig. 10.9).

10.8.1.4 ZZ Ceti stars (pulsating WDs)

Also WDs may show pulsation. There are three kinds.

ZZ Ceti stars or DAVs (DA-variables) have pure H atmospheres of $T_{\rm eff} \simeq 12000$ K. These stars pulsate in a rhythm of 100 to 15000 seconds and have brightness variations of $\Delta V \simeq 0.003$ mag. Furthermore, there are the DBVs with pure He atmospheres of $T_{\rm eff} \simeq 25000$ K an the DOVs (showing absorption lines of N, C, O and He) which are hotter than 70000 K. The pulsations are likely based on the κ-effect. For more see Ch. 11.3.1.4.

10.8.2 λ Bootes stars

A special class of stars, called after its prototype λ Boo, shows peculiar metal content. The abundances of the heavy elements are reduced *in the same way as in the ISM*.

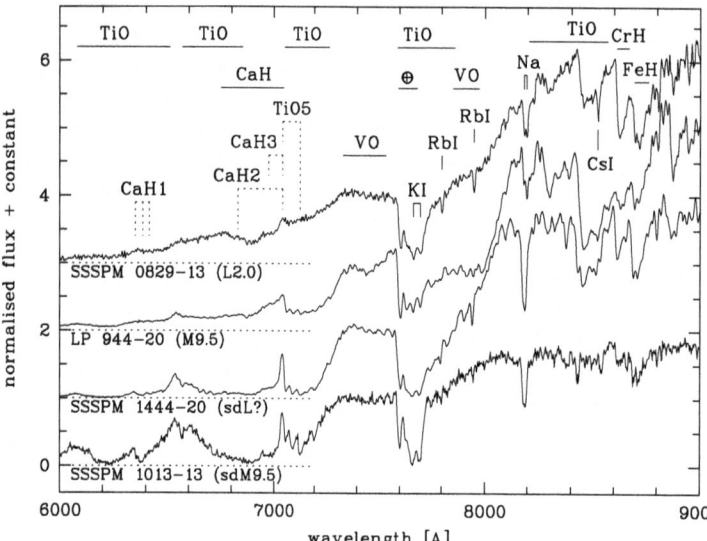

Figure 10.21: Cool subdwarfs are old stars with low metal content and they therefore lie away from the solar-metallicity MS (see Fig. 6.7). The spectra of cool subdwarf stars show strong absorption by molecular bands. TiO, CaH, VO, and FeH indicate the location of absorption bands, some molecular absorption lines are marked. CaH1 etc and TiO5 indicate photometric bands used in some studies. Figure from Scholz et al. (2004).

λ Boo stars are *young MS-stars* and have masses around 2 $M_\odot$. They might, in the phase just before arriving on the ZAMS, have accreted from the ISM only gas but not the dust (radiation pressure kept the dust out). The separation of gas and dust would have taken place in the circumstellar shell. If the stellar atmosphere is stable (no convection, not too hot), the chemical composition of the *gas* of the ISM would remain visible in the atmosphere (see, e.g., Kamp & Paunzen 2002).

One effect of the low abundance of heavy elements in the atmosphere is a considerably reduced opacity. In fact, this is so much so, that the quasi-molecular hydrogen absorption features (Ch. 3.4.2, Fig. 3.9) are seen rather clearly in λ Boo stars.

10.8.3 Cool subdwarf stars

As shown in Ch. 6.9.2, stars having started with a small content in metals will have a structure different from stars with solar composition. In particular old and, therefore, more metal poor and low mass stars should lie on a MS different from that of 'normal' stars (see Fig. 6.7), in fact *to the left the MS* and *at larger L* than solar metallicity stars. It is, however, commonly said they lie 'below' the MS.

Stars from the early phases of formation of the Milky Way are still present, also in the solar neighbourhood. Such stars have been found in field studies based on their low metal content (see Fig. 10.21), their parallax-based absolute magnitude (see Fig. 10.16), or their high proper motion.

These objects are very important to calibrate our knowledge of the evolution of low mass metal poor stars and they thus are important stepping stones in calibrating CMDs of older star clusters.

10.8.4 Blue stragglers

In the CMD of M 3 (Fig. 10.7) stars are seen above the turn-off (TO) of the MS. These stars are called *blue stragglers* (BSs), since they seem to have stayed behind in evolution compared to normal stars. In particular in globular clusters (homogeneous, coeval and closed groups of stars) no MS stars should be present with more mass than the stars at the TO. Yet such stars exist.

BS stars probably are the result of mass transfer in a binary system. In several globular clusters eclipsing binaries have been found, supporting the presence of binaries in GCs. Apparently the transfer is complete (see Ch. 19.5.2.7), because BS stars seem to have a mass ~ 2 times that of stars near the TO. During the merger the material of the star will get thoroughly mixed, so that the star can continue normal H-burning as a (now more massive) MS star.

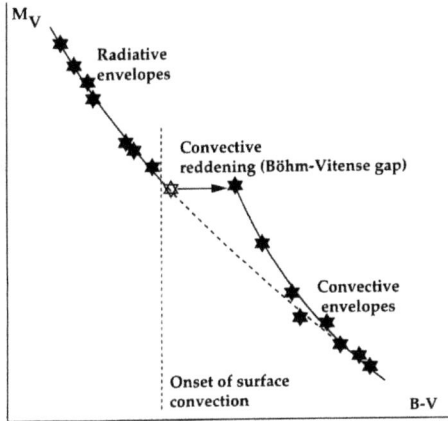

Figure 10.22: The boundary on the MS between stars with $M_{\rm MS} > 1.15$ M$_\odot$ with radiative atmospheres and stars with $M_{\rm MS} < 1.15$ M$_\odot$ with a convective atmosphere was predicted by Böhm-Vitense to lead to a transition in the basic atmospheric structure. This gap has been confirmed using HIPPARCOS data. Figure from de Bruijne et al. (2000).

10.9 Gaps and bumps in the MS, HB, AGB

Along several 'branches' stars (data points) appear not to be distributed homogeneously. Apparently phases of otherwise quiet and stable stellar evolution have slower and quicker parts, too. Also, phase transitions may occur in atmospheric structure.

10.9.1 Gap on the main sequence

Böhm-Vitense (1970) predicted that the boundary on the MS between stars with surface convection and stars without, known to lie near 1.15 M$_\odot$ (see Sect. 10.1.1), should imply a basic difference in atmospheric structure and thus to a gap in a CMD. The principal effect is shown in Fig. 10.22.

Only since very accurate absolute magnitudes from the HIPPARCOS-mission are available could this prediction be verified. The CMD of the Hyades stars (de Bruijne et al. 2000) shows this gap near $B - V = 0.4$ mag (see their paper).

10.9.2 Gaps on the HB

After HB stars had been discovered in larger numbers as field stars, Newell et al. (1969) noted that the HB in his CMDs seemed to have regions devoid of stars. At least two gaps were identified. Newer studies have shown that these gaps are indeed real. For an example of a HB-gap in a globular cluster CMD see Fig. 10.8b (that gap is much bluer than an "RR Lyr gap").

Accurate Strömgren photometry and high dispersion spectroscopy have been used to study this phenomenon in more detail. Apparently, the metallicity in the atmospheres of these stars is reduced due to gravitational settling (see Ch. 3.6). This results in a change in atmospheric opacity and thus to a jump in atmospheric structure when going along the HB to higher temperatures, leading to a 'gap' visible in the photometry. The explanation involves complex physics of very stable atmospheres (see, e.g., Moehler et al. 1999).

Then there is the observational "RR Lyr gap", which is not a true gap (see Sect. 10.4.2).

10.9.3 The RGB and AGB bumps

On the RGB as well as on the AGB the H- and He-burning shell, respectively, burn outward and pass through layers which have different chemical compositions due to effects of mixing by convection or due to special fusion processes (see Sect. 10.4.3). This may cause a temporary slowdown of the evolution. The result is that stars evolving up the RGB and AGB, respectively, move locally slower or faster upwards in the CMD. This results in a 'bump' in the statistics along these branches. For the AGB bump see Fig. 10.23.

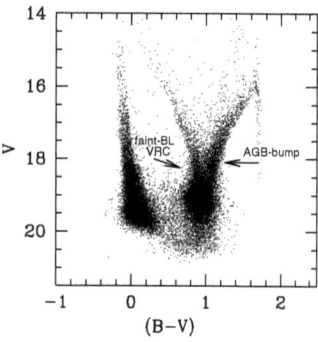

Figure 10.23: Theoretical CMD of the LMC total stellar population with a large mix of stellar ages shows a clump of data points in the red near $V = 19$ mag. This is the so-called **red clump**. At left the MS is visible. At right one has the RGB (pointing to the upper right), the sequence of massive core He-burning stars of the blue loop (pointing to the upper left), and the reddish HB (for the LMC distance at $V \simeq 19$ mag). These three branches intersect and form the red clump (the faint blue loop, BL, stars establishing a vertical-left extension of the red clump, VRC). Also the AGB bump is indicated. Figure from Gallart (1998).

10.10 The Red clump

Due to the various ways of stellar evolution, there is a region in the red part of the HB were several kinds of stars seem to be present. The region is called the **red clump** and it lies at the intersection of the RG and the HB (see Fig. 10.23).

Four kinds of stars may be found here. These are:
1: Stars evolving up the RGB passing this region during the RG evolution;
2: Stars after the RGB phase becoming thick envelope HB stars, the RHB stars;
3: Stars of the lower part of the blue loop of just about 5 $M_\odot$,
4: Stars evolving up the AGB.

The RGB stars are the 'normal' stars in this part of the HRD. The red HB stars form a characteristic clump in older and not too metal poor populations. If such stars are recognized they can help set a limit to the age of such groups. In the younger groups the stars at the lower end of the blue loop are the normal stars there. Recognizing the stars as blue loop stars again allows to set age limits for the group observed.

Often stars covering a range of distances are observed together. In a CMD from such data the red clump stars spread in V vertically due to the distance differences. This vertical spreading is sometimes called the "red plume".

10.11 Summary

The MS is divided in several zones, and the limits have a significance based on limits in stellar structure or limits for the stellar evolution.

All stars will evolve along the same pattern, but depending on the total mass the evolution is either *truncated* or deviates from a main trend. The main characteristic of the lower mass range of the MS is, that all stars end as WDs and that all stars starting with more than 0.6 $M_\odot$ shed mass into the surroundings. If enough material has been blown away and stays near the star it later gets illuminated to be a PN or an H II region.

The fate of WDs may be dull or may be spectacular, depending on the star having been single or been part of a binary system.

The nomenclature of and classification scheme for the spectra of the stars in late evolutionary phases of this chapter are summarized in de Boer et al. (1997).

References

Allard, F., Hauschildt, P.H., Alexander, D.R., Starrfield, S. 1997, ARAA 35, 137; *Model Atmospheres of Very Low Mass Stars and Brown Dwarfs*
Balick, B., & Frank, A. 2002, ARAA 40, 439; *Shapes and Shaping of Planetary Nebulae*
Blöcker, T. 1995, A&A 299, 755
Böhm-Vitense, E. 1970, A&A 8, 283 and 8, 299
Brown, T.M., Sweigart, A.V., Lanz, T., Landsman, W.B., & Hubeny, Y. 2001, ApJ 562, 368

10.11. SUMMARY

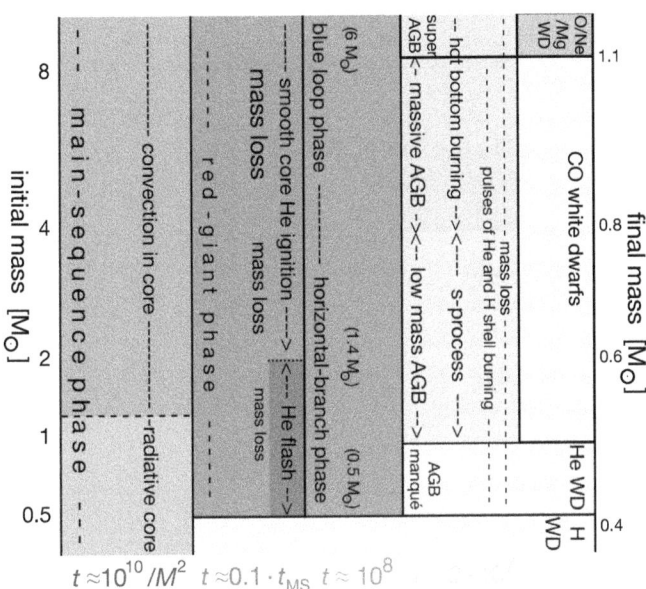

Figure 10.24: The evolution of stars in the low mass range is summarized in a diagram with time (not to scale) running to the right.

All limits of the mass ranges are *approximate*. Several special aspects occur within such ranges. The various phases of evolution are named and their approximate duration is given at the bottom.

All stars with $M_{init} < 8$ M$_\odot$ end as white dwarfs. H WDs have a He core and a H envelope, He WDs have a CO core and a thin He envelope while CO WDs have a thick He envelope, O/Ne/Mg WDs have a core of those elements and further layers of He and H depending on the mass lost.

Busso, M., Gallino, R., & Wasserburg, G.J. 1999, ARAA 37, 239; *Nucleosynthesis in Asymptotic Giant Branch Stars*
de Boer, K.S. 1985, A&A 142, 321
de Boer, K.S., Drilling, J., Jeffery, C.S., & Sion, E.M. 1997, in "Third Conf. on Faint Blue Stars", A.G.D. Philipp et al. (eds.); Davis Press, Schenectady, p. 515
de Bruijne, J.H.J., Hoogerwerf, R., & de Zeeuw, P.T. 2000, ApJ 544, L65
Dupree, A.K. 1986, ARAA 24, 377; *Mass Loss from Cool Stars*
Durrell, P.R., & Harris, W.E. 1993, AJ 105, 1420
Ferguson, J.W., Alexander, D.R., Allard, F., & Hauschildt, P.H. 2001, ApJ 557, 798
Gallart, G. 1998, ApJ 495, L43
Habing, H.J. 1996 A&A Rev 7, 97; *Circumstellar Envelopes and Asymptitic Giant Branch Stars*
Habing, H.J., & Olofsson, H. (eds.). 2004, *Asymptotic Giant Branch Stars*; Springer, Heidelberg
Habing, H.J. & Whitelock, P.A. 2004, in "Asymptotic Giant Branch Stars", H.J. Habing & H. Olofsson (eds.); Springer, Heidelberg, p.411
Hansen, B.M.S., & Liebert, J. 2003, ARAA 41, 465; *Cool White Dwarfs*
Herwig, F. 2005, ARAA 43, 435; *Evolution of Asymptotic Giant Branch Stars*
Hesser, J.E., Harris, W.E., Vandenberg, D.A., Allright, J.W.B., Shott, P., & Stetson, P.B. 1987, PASP 99, 739
Iben, I. 1974, ARAA 12, 215; *Post Main-Sequence Evolution of Single Stars*
Iben, I., Kaler, J.B., Truran, J.W., & Renzini, A. 1983, ApJ 264, 605
Iben, I., & Renzini, A. 1983, ARAA 21, 271; *Asymptotic Giant Branch Evolution and Beyond*
Jahreiss, H. 1987, Mem. Soc. Astron. It. 58, 53
Kamp, I., & Paunzen, E. 2002, MNRAS 335, L45
Kippenhahn, R., & Weigert, A. 1990, "Stellar Structure and Evolution", Springer, Heidelberg
Klochkova, V.G., Szczerba, R., Panchuk, V.E., & Volk, K. 1999, A&A 345, 905
Kwok, S. 2000, "The origins and Evolution of Planetary Nebulae"; Cambridge Univ. Press
Lamers, H.J.G.L.M., & Cassinelli, J.A. 1999, "Introduction to Stellar Winds"; Cambridge U. P.
Lattanzio, J.C., & Wood, P.R. 2004, in "Asymptotic Giant Branch Stars", H.J. Habing & H. Olofsson (eds.); Springer, Heidelberg, p.23
Laughlin, G., Bodenheimer, P., & Adams, F.C. 1997, ApJ 482, 420
Lewis, B.M. 2002, ApJ 576, 445
Maeder, A., & Meynet, G. 1989, A&A 210, 155
Moehler, S., Sweigart, A.V., Landsman, W.B., Heber, U., & Catelan, M. 1999, A&A 346, L1

Table 10.3: Summary of mass limits relevant for lower mass star evolution

mass limit	reason for limit	ref.	Fig. 10.24
M_{init}			
$\simeq 0.5\ M_\odot$	stars with smaller M_{init} become WD after MS phase	Ch. 10.1.2	×
$\simeq 0.6\ M_\odot$	stars with M_{init} below this limit are still on MS	Ch. 10.1.2	
$\simeq 1.15\ M_\odot$	MS limit of inner convection	Ch. 10.1.1	×
$\simeq 2\ M_\odot$	blue loop star or HB star	Ch. 10.3	×
$\simeq 4\ M_\odot$	heavier stars become massive AGB and OH/IR star	Ch. 10.5.1	×
$\simeq 4\ M_\odot$	lower limit for slow C-burning	Ch. 10.4.4.3	×
$\simeq 7\ M_\odot$	T_c cooling at end of MS	Fig. 10.2	
8 to 10 $M_\odot$	upper mass limit for heavy AGB stars	Fig. 10.2	×
$\simeq 8\ M_\odot$	limit between lower and higher mass range	Ch. 10.1	×
M_{HB}			
$\simeq 0.47\ M_\odot$	minimum mass in He core	Ch. 10.2.3	
$\simeq 0.49\ M_\odot$	lower limit of HB stars	Ch. 10.4.2	
$\simeq 0.53\ M_\odot$	lower limit for HB star becoming AGB star	Ch. 10.4.2	×
$\simeq 0.65\ M_\odot$	maximum mass of pAGB for old stars	Ch. 10.5.2	
$\simeq 1.4\ M_\odot$	maximum mass for WD (Chandrasekhar limit)	Ch. 10.6	

Miller Bertolami, M.M., Althaus, L.G., Serenelli, A.M., & Panei, J.A. 2006, A&A 449, 313
Newell, E.B., Rodgers, A.W., & Searle, L. 1969, ApJ 156, 597
Olofsson, H. 2004, in "Asymptotic Giant Branch Stars", H.J. Habing & H. Olofsson (eds.); Springer, Heidelberg, p.325
Oosterhoff, P.T. 1939, Observatory 62, 104
Oppenheimer, B.R., Saumon, D., Hodgkin, S.T., Jameson, R.F., Hambly, N.C., Chabrier, G., Filippenko, A.V., Coil, A.L., & Brown, M.E. 2001, ApJ 550, 448
Reimers, D. 1977, A&A 61, 217
Renzini, A., & Fusi Pecci, F. 1988, ARAA 26, 199; *Tests of Evolutionary Sequences using Color-Magnitude Diagrams of Globular Clusters*
Richer, H.B., Hansen, B., Limongi, M., Chieffi, A., Straniero, O., & Fahlman, G. 2000, ApJ 529, 318
Schaller, G., Schaerer, D., Meynet, G., & Maeder, A. 1992, A&AS 96, 269
Schönberner, D. 1979, A&A 79, 108
Scholz, R.-D., Lodieu, N., & McCaugrean, M.J. 2004, A&A 428, L25
Sweigart, A.V. 1994, in "Hot Stars in the Halo", S.J. Adelman, A. Upgren, & C.J. Adelman, (eds.), Cambridge Univ. Press, p.17
Sweigart, A.V. 2007, priv. comm.
Weidemann, V. 1990, ARAA 28, 103; *Masses and Evolutionary Status of White Dwarfs and their Progenitors*
Weidemann, V. 2000, A&A 363, 647
Werner, K., Dreizler, S., Heber, U., & Rauch, T. 1996, ASP Conf. Ser. 96, 267
Willson, L.A. 2000, ARAA 38, 573; *Mass Loss from cool stars*
Wood, P.R., & Faulkner, D.J. 1986, ApJ 307, 659
Wood, P.R., & Vassiliadis, E. 1992, in "Highlights of Astron." 9, IAU Publ., J. Bergeron (ed.); p. 617

Chapter 11

Stellar pulsation and vibration

Many kinds of stars show some form of pulsation. The use of the word pulsation derives from the notion that stars (their outer layers) apparently can oscillate, leading to variations in the output of light. Such oscillations could be driven by processes of variation in nuclear fusion efficiency in the interior or by "instabilities" in the stellar atmosphere. Strong oscillations may lead to cyclic expansion and contraction of the stellar atmosphere (radial oscillations), easily recognizable in variations in the brightness of the star. Such oscillations are normally referred to as pulsations. Non-radial oscillations of the gaseous sphere normally are referred to as vibrations (Sect. 11.4). These may be triggered by small disturbances in the behaviour of the gas, e.g., by effects of convective motion (present in all stars), by effects of fluctuating nuclear fusion in certain evolutionary stages or by gravitational effects of the companion star of a binary system. When vibrating, one can say that a star "rings like a bell". Or, comparing with a pendulum, that a star swings as a 3-dimensional body in three dimensions (side-ways, foreward, and up and down, or in more complicated patterns). For a thorough treatment of the topic see, e.g., Unno et al. (1979), Cox (1980) or Christensen-Dalsgaard (2003) and for a review including observational aspects Gautschy & Saio (1995, 1996). For details of solar oscillations see, e.g., Stix (2002, 2004).

11.1 Describing a star with oscillations

11.1.1 The formalism

Treating a star as a sphere performing oscillations, the description of its structure is in terms of a self-gravitating fluid. One normally assumes conservation of thermal energy in which the flux is defined by the energy transported through radiation and through convection. Clearly, the full opacity structure and the full equation of state of the gas must be known, as well as the velocity structure of and the flux by convection. The convection related parameters are the most complicated aspect of the theory, and they are in practice neglected, also because the dynamical perturbation time is much shorter than the time of heat exchange. This limits the success of any pulsation theory (see Gautchy & Saio 1995).

One can simplify the description by treating it as a small perturbation about the equilibrium state of a star. The temporal (t) and azimuthal (ϕ; spherical coordinates) dependence of any perturbed quantity can then be represented by $\exp[i(\sigma t + m\phi)]$. The eigenvalue $\sigma = \sigma_R + i\sigma_I$, where σ_R represents the oscillation frequency and σ_I measures the behaviour with time: growth for $\sigma_I < 0$ or damping for $\sigma_I > 0$.

The oscillation eigenmodes can be given (see the reviews by Brown & Gilliland 1994, Gautchy & Saio 1995) as a product of a function of the radius, $\xi(r)$, and a spherical harmonic, Y. The spatial and temporal variation of a perturbation to the star's mean state are then

$$\xi_{nlm}(r,\theta,\phi,t) = \xi_{nl}(r)\, Y_l^m(\theta,\phi)\, e^{-i\omega_{nlm}t} \qquad (11.1)$$

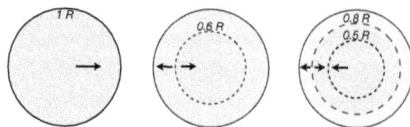

Figure 11.1: Sketches of stars with standing radial waves and nodes (radial pulsations, $l = 0$, with overtones). Left: fundamental mode; middle: first overtone (node at $\simeq 0.6\ R$); right: second overtone (nodes at $\simeq 0.5\ R$ and $\simeq 0.8\ R$). The arrows give the gas motions in the first half of the pulsational phase, in the second half of the cycle the motions are reversed.

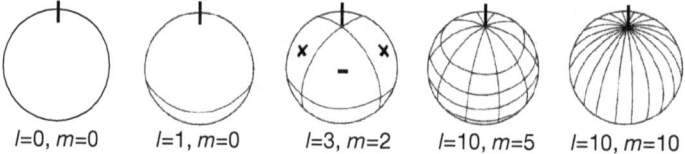

Figure 11.2: Seismic node circles on spheres are shown in a schematic way. The number of these circles equals l, the number of circles crossing the rotation pole equals m. Excursions of the surface of the sphere happen between the node lines (see the middle example: + is outward, − is inward motion, reverse in the other half of the oscillation period). Figure after Stix (2002).

in which ξ is any scalar perturbation associated with the mode (such as, e.g., the radial displacement), and r, θ, ϕ, and t are the radial coordinate, the co-latitude, the longitude, and the time, respectively. ω_{nlm} is the frequency of the oscillation.

The mode's *radial order* n is usually identified with the number of nodes in the eigenfunction that exist between the centre of the star and its surface (see e.g. Fig. 11.1). Since the nodes lie inside a star, n normally is not accessible for observation.

11.1.2 Oscillations and limiting frequencies

In the case of a non-rotating spherically symmetric star and when treating the problem in the adiabatic approximation, the spherical harmonics Y are eigenfunctions of the operator $\nabla_\perp^2$

$$\nabla_\perp^2 Y_l^m(\theta, \phi) = -l(l+1) Y_l^m(\theta, \phi) \tag{11.2}$$

and the angular dependencies of perturbed quantities can be expressed by a single $Y_l^m(\theta, \phi)$. The equations describing the dynamical behaviour of non-radial motions are thus reduced to an ordinary differential equation with radial dependence only. Then $l(l+1)$ is related to the horizontal wavenumber of the oscillation, $[l(l+1)]^{1/2}/r$. The different radial overtones, i.e., the different number of nodes n of the eigenfunctions in radial direction, are called *radial orders*.

The following parameters are used for the description of the behaviour of the star:
• The quantity l is referred to as the **angular** or **spherical degree** of a mode (the number of node circles). $l = 0$ signifies spherical symmetric oscillations (radial pulsations). The angular degree l is the product of the stellar radius $R_\star$ and the total horizontal wavenumber of the mode. The **azimuthal order** m is the projection of l onto the star's equator, so that $m \leq |l|$; m gives the number of node lines passing through the rotation axis. For examples see Fig. 11.2. Note that $l = 0$ is a radial, $l = 1$ is a dipole, $l = 2$ is a quadrupole, $l = 3$ is an octapole oscillation, etc.
• In non-radial pulsations the high frequency **p-modes** can be called sound or accoustic waves and the restoring force comes from the compressability of the star gas. These waves occur predominantly in the outer layers of a star.
• In non-radial pulsations the low frequency **g-modes** can also be called sound waves for which the restoring force is the buoyancy (see Ch. 4.3) force. These are mostly found in the interior.

11.1. DESCRIBING A STAR WITH OSCILLATIONS

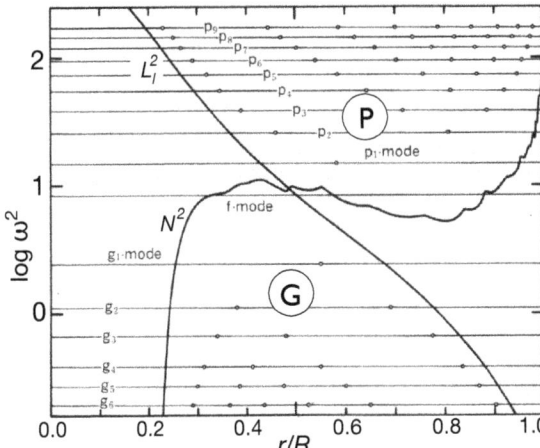

Figure 11.3: Values for the Lamb (L_l) and the Brunt-Väisälä (N) frequencies, defining the limits for oscillations to be possible (see Sect. 11.1.2, Eq. 11.5), calculated for a 10 $M_\odot$ MS star. For each possible g and p value (leading to a frequency ω) the horizontal lines with the nodes (•) indicate the range in r within which the oscillation can take place.
Both g- and p-mode oscillations are possible within the criteria of Eq. 11.5, the regions fulfilling the criteria are labelled with P and G. In this model also an f-mode is possible, intermediary between g- and p-modes.
Figure adapted from Unno et al. (1979).

One can define two frequencies for the particular gaseous medium, which govern which oscillations can take place.

One is the **Lamb frequency**, L_l, given by the ratio of one horizontal wavelength and the sound speed c_S (Eq. 11.8)

$$L_l^2 = \frac{l(l+1)\, c_S^2(r)}{r^2} \quad . \tag{11.3}$$

For pressure or gravity to work as restoring force, the frequency of the oscillation should be higher (in the case of pressure) or lower (in the case of gravity) than L_l.

The other is the **Brunt-Väisälä frequency** (N) with which a bubble of gas may oscillate adiabatically (see also Eq. 4.56), defined by

$$N^2 = g\left(\frac{1}{\Gamma_1}\frac{d\ln p_0}{dr} - \frac{d\ln \rho_0}{dr}\right) \quad , \tag{11.4}$$

with the adiabatic exponent $\Gamma_1 = (d\ln p/d\ln \rho)_{\rm ad}$ (see also Eq. 4.21) and where the subscripts 0 refer to the unperturbed state of the gas. Allowed frequencies must be larger than N for acoustic and lower than N for gravity waves.

Thus a disturbance with frequency σ will set up an oscillation

$$\begin{aligned} &\text{in p-mode when} \quad \sigma > L_l \text{ and } \sigma > N \\ &\text{in g-mode when} \quad \sigma < L_l \text{ and } \sigma < N \end{aligned} \tag{11.5}$$

Fig. 11.3 shows the run of L_l and N in a star and the ensuing possible p- and g-mode frequencies.

11.1.3 The driving forces of oscillations

Oscillations may be driven by effects of nuclear burning (the ϵ-effect). The time scale of the oscillation and the effect it has on ϵ (see Ch. 5) are critical parameters for the establishment of oscillations. One example is the He-reignition flickering at the end of the AGB phase (see Figs. 10.9 and 10.13). Also post explosion relaxation in novae is a (damped) pulsation (Fig. 19.14).

Oscillations are mostly driven by effects of opacity (the κ-effect). E.g., the boundary of ionization regimes may be influenced by minute pressure fluctuations (possibly due to convective flow noise), which may lead to an oscillation in the level of ionization (P_e in Eq. 3.27), triggering κ fluctuations. In fact, the star is then locally not in thermal equilibrium (see Ch. 9.4). Since the opacity in star gas is dominated by the contributions from H and He (if present), the opacity of H, He and He$^+$ plays a crucial role. In particular the He$^+$/He^{2+} ionization zone (see Fig. 4.2 for the ρ, T combination of that ionization boundary in Pop. I star gas) may have such oscillations.

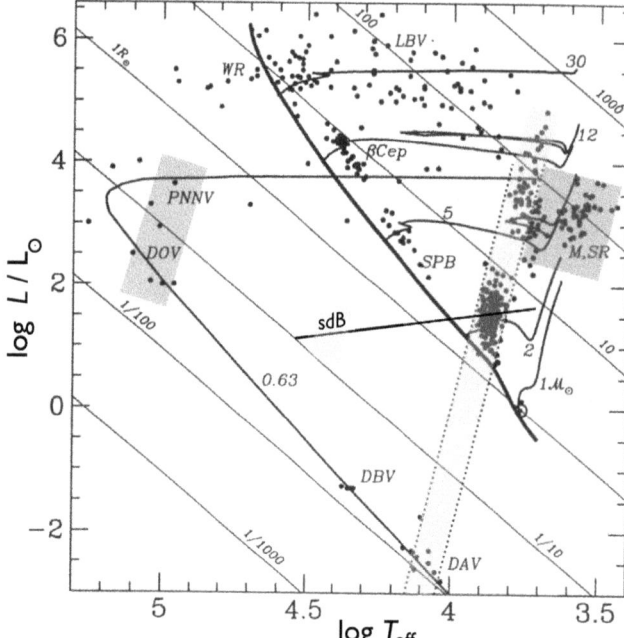

Figure 11.4: HRD showing locations of stellar pulsation. The heavy line is the MS. The thinner lines are tracks of evolution of stars of 1, 2, 5, 12, and 30 $M_\odot$, the evolution of pAGB stars bending down into the WD cooling line of 0.63 $M_\odot$, and the horizontal branch (HB).
SPB: slowly pulsating B stars,
β Cep: upper MS star variables,
M: Mira variables,
SR: semiregular variables,
LBV: luminous blue variables,
WR and sdB: Wolf-Rayet and sdB variables,
PNNV: PN nuclei variables,
DOV, DBV, DAV: variable WDs.
Not labelled are the variables of type δ Cep, W Vir, RR Lyr, δ Sct in the instability strip (the dotted line strip).
Adapted from Gautschy & Saio (1995).

These lead to standing waves (Eq. 11.4) if this ionization boundary is at the appropriate depth in the stellar envelope (see Christensen-Dalsgaard 2000).

Furthermore, convective motions lead to 'noise' that may suffice to trigger one of the possible oscillation frequencies (which then may be maintained by the κ-effect). Here it is as with a trumpet: a broad-band input spectrum excites oscillations through stochastic resonances.

11.2 Spherically symmetric radial pulsations

11.2.1 Formalism for radial pulsation

The "cleanest" example of oscillations are the radial pulsations (so $l = 0$; Fig. 11.1). The equation for the displacement vector field can be much simplified and leads to oscillations with eigenvalue σ^2. The eigenvalue associated with the n nodes is represented by σ_n^2. These eigenvalues form a sequence $\sigma_0^2 < \sigma_1^2 < \sigma_2^2 < \ldots$. Thus the period of the oscillation is smaller when the number of nodes is larger because the period is approximately the sound travel time between two adjacent nodes (Hansen 1972). One can derive that

$$(3\,\overline{\Gamma} - 4)\,(-E_{\text{grav}}/I) > \sigma_0^2 > (3\,\overline{\Gamma} - 4)\,4\pi G\,\overline{\rho}/3 \tag{11.6}$$

where $\overline{\rho}$ and $\overline{\Gamma}$ are the mean density and adiabatic exponent respectively, E_{grav} the gravitational potential energy and I the moment of inertia of the star. Note that $-E_{\text{grav}}/I$ is proportional to the mean density of the star so that the period of the fundamental mode ($2\pi/\sigma_0$) is inversely proportional to the square of the mean density. This relation states the *Period - Mean density* relation of stellar pulsations.

The logic to obtain that relation is as follows. In the fundamental mode the period, Π, of the wave is given by the travel time of a pressure wave to the centre of the star and back, i.e.

$$\Pi \sim \frac{2\,R}{c_{\text{S}}} \tag{11.7}$$

where c_{S} is the velocity of sound

$$c_{\text{S}} = \sqrt{\Gamma \frac{P_g}{\rho}} \tag{11.8}$$

11.2. SPHERICALLY SYMMETRIC RADIAL PULSATIONS

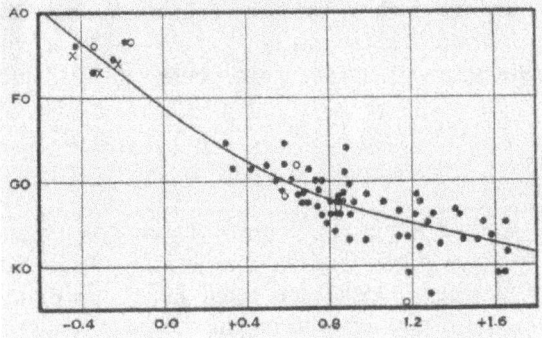

Figure 11.5: A presentation by Shapley (1927) of the instability strip shows the spectral type of these stars plotted against period as $\log \Pi$. Note that at that time all stars from the red-giant variables through the δ Cepheids and the RR Lyr stars were collectively called "Cepheids". The stars at left of type A are, of course, the RR Lyrae. The run of stars from left to right is one of increasing mass and thus also of increasing period.

with $\Gamma \simeq 5/3$ for monoatomic gas. At any place in the star the pressure must balance the weight of the overlying material, so that for the centre

$$P_g \simeq \overline{\rho}\, R\, \overline{g} \simeq \overline{\rho}\, R\, \frac{GM}{R^2} = \overline{\rho}\, \frac{GM}{R} \qquad (11.9)$$

with G the gravitational constant and M the mass of the star. Thus $\overline{P_g}/\overline{\rho} \sim \frac{GM}{R}$ and one then finds for the period

$$\Pi \sim \frac{2}{\sqrt{\Gamma\, G}} \sqrt{\left(\frac{R^3}{M}\right)} \qquad (11.10)$$

and thus $\Pi = \text{const}\, \rho^{-1/2}$.

Since the density is related with the total mass and the mass with the luminosity (for stars of similar structure) this leads to a *Period - Luminosity relation*. In the instability strip a large luminosity points at a large radius (and at a smaller mean density). To cover the depth of a large radius star, a pressure wave needs more time: the period is longer.

Actual stellar oscillations are inevitably non-linear. Analytic solutions are possible only for idealized systems so that attempts at hydrodynamic simulations have been made. These are, however, cumbersome and require excessive computing time. The complexities of stellar atmospheres severely limit the making of good oscillation models. For more mathematical and physical detail see Unno et al. (1979) and the review by Gautschy & Saio (1995).

Pulsational instabilities occur in various types of star (see Fig. 11.4)[1].

11.2.2 Atmospheric radial pulsations

As discussed in Ch. 9.3 "How does a star become a red giant?", the driving force behind the expansion of the stellar envelope is the increase in luminosity from the interior together with the instability of the stellar envelope to accommodate this increase in the state it has at that moment. The resulting big change there is that, with expansion, the outer layers change their opacity leading to a run-away process of ever further expansion, until a new stable state is reached.

The pulsation of a stellar atmosphere is the result of recurring phase transitions. In fact, the transition has a time behaviour like hysteresis: a change occurs in one direction, the new condition persists while slowly changing, until the limiting condition is reached and the reverse phase transition sets in, again followed by a slow adjustment of conditions. It is indeed much like the hysteresis occurring in the "Gedankenexperiment" described in Ch. 9.3.1. The hysteresis is also visible in the colour of pulsating stars, (see the colour loop of an RR Lyr variable, Fig. 11.9).

For pulsation to occur, the physical conditions in the relevant atmospheric layer must be near a "critical point", performing repeated phase transitions[2]. Such a critical point in gas may be the condition at the boundary of two neighbouring ionization states of some element. If, e.g., the

---
[1] Note that in the regions in the HRD where stars can pulsate, also non-pulsating stars can be present. Their different behaviour may be due, e.g., to differences in age (as with SPB variables; see Briquet et al. 2007) or [M/H].
[2] For the general importance of such phase transitions see Buchanan (2000), p. 74-81

element has in each of these states sufficiently different opacity, then the transition must have consequences for the outward flow of energy. Furthermore, if the gas layer above the layer with the phase transition is not thick enough (both mass wise and opacity wise), there will be little in terms of restoring forces.

11.2.3 Details of the κ mechanism

The κ-mechanism can be explained in the same way as Eddington's thermodynamic heat engine. Suppose an atmospheric layer moves inward. It becomes denser and its opacity increases so that heat is withheld. Thus pressure is built up strongly leading to expansion, reduction of opacity and to release of heat (radiation). Now the layer can move inward again to repeat the cycle.

The opacity of the layer acts much like the "valve" in a steam engine. The essential step is the assumption that, upon compression, the opacity increases. The opacity of star gas has a complicated dependence on T, in that κ decreases on both sides of $T \simeq 10^{4.5}$ K (see Fig. 4.4). So gas compression will in the deeper interior generally not lead to an increase in opacity ($\kappa \sim \rho\, T^{-3.5}$, Eq. 4.61) while in the outer layers it will increase the opacity.

In stellar envelopes the behaviour of ionization is essential. During compression most of the excess energy is used to further ionize the gas so that T hardly changes, while ρ increases so that κ increases (Fig. 4.4). Likewise, during expansion, recombination brings energy into the gas so that its temperature is almost unchanged. Due to ionization, the density determines the opacity.

The details of the structure of the stellar envelope will thus determine whether an oscillation in density will maintain itself or whether it will dampen out. Examples of ionization driven oscillations are those due to the partially ionized layers of: $H \leftrightarrow H^+$ at $T \simeq 1.5 \cdot 10^4$ K, $He \leftrightarrow He^+$ at $T \simeq 3 \cdot 10^4$ K, $He^+ \leftrightarrow He^{2+}$ at $T \simeq 4 \cdot 10^4$ K, $C^{+3} \leftrightarrow C^{+4}$ at $T \simeq 2 \cdot 10^5$ K, etc.

The oscillation can only operate when the relevant layer has enough mass (is deep enough in the atmosphere) for a sizeable opacity. It means that the He^+ layer lies deep enough to be active in the oscillatory way for low T_{eff}. For high T_{eff} this He^+ layer is too close to the surface.

11.3 Types of pulsational variables

The overview of pulsational variables given here provides rudimentary information only. For more and for access to the literature see Gautchy & Saio (1996). The names of the variable stars types are in most cases based on the name of the prototype variable.

11.3.1 The instability strip: δ Cep, W Vir, RR Lyr, δ Sct, DA variables

A well-known location of pulsational variability in the HRD is the so-called "instability strip" from $T_{\text{eff}} \simeq 5000$ K for high luminosity low gravity stars to $T_{\text{eff}} \simeq 14000$ K for low luminosity high surface gravity stars (see Fig. 11.4). It contains (from high to low luminosity) red giant pulsators, the δ Cep, the W Vir and the RR Lyr stars, the near MS δ Scuti variables, as well as the DA WDs or ZZ Ceti stars. Several of these were mentioned in Ch. 10. This region of the HRD was first identified by Shpaley (1927) and at that time all these stars were called "Cepheids". A diagram from Shapley (1927) is reproduced in Fig. 11.5 showing spectral type versus pulsation period.

In the instability strip the pulsation is due to opacity effects at the ionization boundary of He to He^+ or He^+ to He^{2+}. The pulsations are radial ones ($l = 0$).

The width and location of the instability strip is not uniquely defined. For Population II stars (older, lower metallicity) the red edge may lie further to the red and the strip may thus be wider than for Population I stars. The physical cause may be different metallicity (opacity) and/or different structure of convection zones.

11.3.1.1 Cepheids

Cepheid variability occurs when a more massive star crosses the instability strip during its evolution. There are two kinds of Cepheids, those of Population I (being young and rather mass rich)

11.3. TYPES OF PULSATIONAL VARIABLES

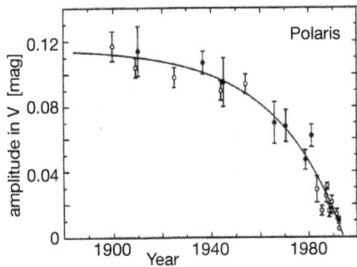

Figure 11.6: The behaviour of the Cepheid Polaris is shown against time. The amplitude decreased to very small levels over 100 yr, originally thought to indicate evolution of the star out of the pulsational instability strip (figure adapted from Fernie et al. 1993). However, the amplitude seems to have become steady (Kamper & Fernie 1998).

and those of Population II evolving up and away from the HB.

Pop I Cepheids (δ Cep stars): Since "Cepheid" pulsation occurs in stars coming from a range of M_{init} (see Ch. 10.8.1.2) and thus $\bar{\rho}$, their pulsations show the *Period - Mean density* relation. Stars with $M_{\text{init}} > 2.5\,M_\odot$ evolve with blue loops (see Ch. 10.4.4) and can enter the instability strip. δ Cep stars are found in the HRD (see Fig. 11.4) at $10^{2.5} < L < 10^4\,L_\odot$. Clearly 'blue loop' stars will become Cepheid variable: two, possibly even three times! The highest Cepheid luminosity seen (e.g., in a star group) shows the highest mass for which blue loops occur (see Ch. 13.4.1.2).

Masses estimated from the pulsation period were in the past smaller than evolution theory predicted. However, the new "OPAL" opacity values (see Fig. 4.4) brought agreement between the two kinds of determination because metals contribute more opacity than previously known (one speaks of the *Z-bump* in the opacities of gas near 10^5 K).

There are single (fundamental) and double (fundamental + overtone) mode Cepheids. The light curve shows a secondary maximum in the faint part, whose location in phase depends on the mass of the star. This secondary maximum is attributed to the accidental 2:1 ratio of the period of overtone and fundamental mode.

Polaris (α UMi) is a special Cepheid (see Fig. 11.6). Its light amplitude diminished over the past 100 years. It probably evolves toward the edge of the instability strip (but there are also doubts about that explanation). It is likely one of the few stars where evolution can be "seen".

Pop II Cepheids (W Vir stars): Stars *evolving up and away* from the HB toward the AGB (see Fig. 10.9) may cross the instability strip. These stars have $0.52 < M < 1.0\,M_\odot$ but a luminosity larger than at the HB. They oscillate with periods of 0.8 to 30 days like Pop. I Cepheids.

11.3.1.2 RR Lyr

RR Lyrae stars (see Ch. 10.8.1.1) have a mass in the range of $0.6 < M < 0.7\,M_\odot$ and pulsation periods of $\simeq 0.3$ to $\simeq 1.5$ d. There are three kinds of RR Lyr (see Fig. 11.7).

→ The RRab pulsators oscillate in the fundamental mode. Their light curves are highly asymmetric (see, e.g., Fig. 11.8, left).
→ The RRc oscillate in the first overtone and have sinusoidal light variations.
→ The RRd are double mode RR Lyrae. The first overtone has a higher amplitude than the fundamental mode.

These RR types differ (see Fig. 11.7) in that the RRa amplitudes are larger than those of the RRb stars while the RRc stars have the smallest amplitude and the shortest periods. This sequence

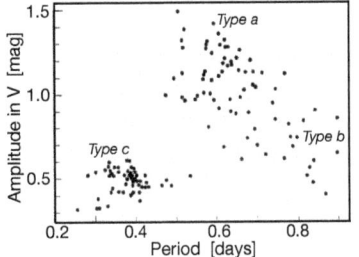

Figure 11.7: There are three types of RR Lyr stars.
Type RRab (in the past seen as seperate types RRa and RRb) are fundamental mode pulsators, with RRa having the largest brightness amplitude, RRb a smaller amplitude with longer period. RRc have overtone pulsations with small amplitude and short period. The RRd are double mode pulsators, exhibiting both RRab and RRc characteristics.
Figure adapted from Ledoux & Walraven (1958).

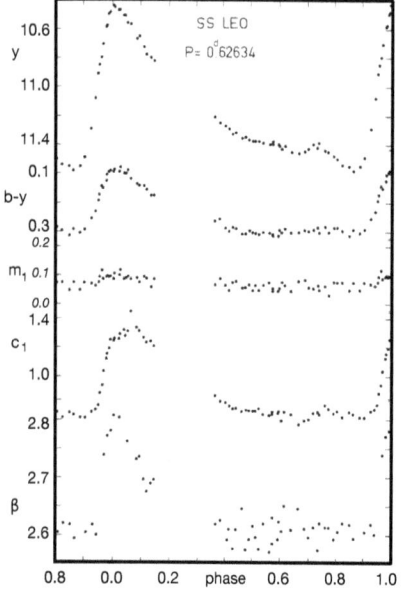
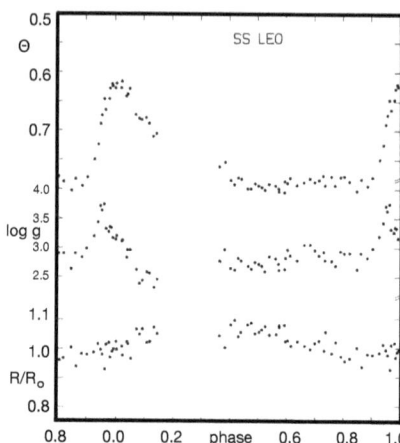

Figure 11.8: **Left:** The change in colours of RR Lyr stars due to the pulsation of the atmospheres can be followed with photometry. The example given is of the star SS Leo using Strömgren photometry (y, $b-y$, m_1, c_1, and β). The data are plotted against pulsation phase.
Right: Combining such data with atmosphere models one can derive the atmosphere parameters over the entire cycle. Plotted are $\Theta = 5040/T_{\rm eff}$, $\log g$, and R/R_0. Note that when the star is brightest, the atmosphere is hottest and the radius smallest. The pulsation is due to the atmospheric κ-effect. Figure from van Albada & de Boer (1975). See also Fig. 11.9.

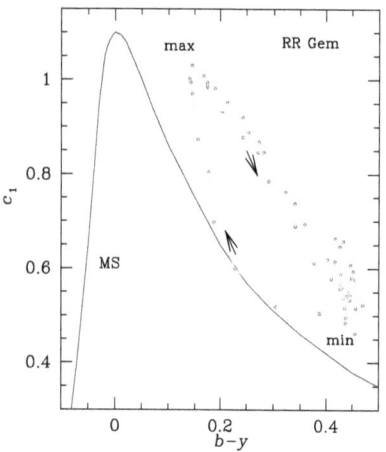
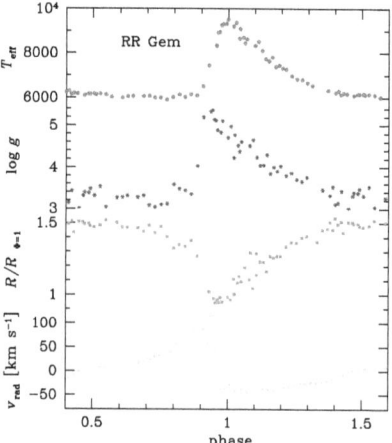

Figure 11.9: **Left:** The pulsation of an RR Lyr star leads, in the colours c_1 vs. $b-y$ to a "hysteresis", which follows the hysteresis effects of phase transitions inside the relevant atmospheric layers (see also Ch. 9.3.1). Data are for RR Gem (Maintz & de Boer 2008) based on simultaneous multi-band photometry. The location of the MS is indicated (compare with Fig. 3.5).
Right: Run of $T_{\rm eff}$, $\log g$, $R/R_{\Phi=1}$ and $v_{\rm rad}$ for RR Gem using the data of the left panel. The star is during about 60% of the period cool ($6000 < T_{\rm eff} < 6500$ K). Note that $\log g$ has its highest value just before $T_{\rm eff}$ starts to increase. The variation in R is found from $L = 4\pi R^2 \sigma T^4$ where L is based on y and the distance. The variation in R can be used to calculate $v_{\rm rad}$ of the atmosphere as seen by an observer.

11.3. TYPES OF PULSATIONAL VARIABLES

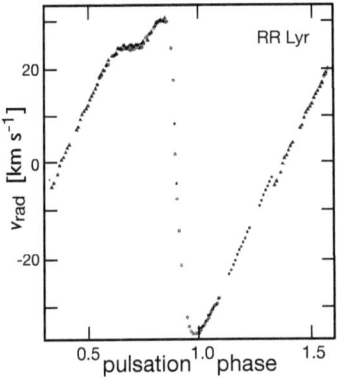

Figure 11.10: The pulsation of the atmospheres of RR Lyrae stars can be followed with present spectroscopic techniques in great detail. Shown is the variation of the radial velocity of the atmosphere of RR Lyr itself, derived from the shifts of one Fe absorption line, as seen by the observer. The symbols refer to data from different observing sessions.
At $\Phi \simeq 0.7$ the contracting atmosphere meets an outward going shock, halting temporarily the inward motion. At $\Phi = 0.9$ the atmosphere is most compact ($v_{\rm rad} = 0$ km s^{-1}) and the expansion takes over ($v_{\rm rad} < 0$). Note that the phase values are based on the variation of brightness. Figure adapted from Chadid (2000).

may be related with age and metallicity (see Ch. 10.4.2.2, location of the HB).

Furthermore, there is the so-called Blazhko effect (Blazhko 1907) which is that over longer time spans variations occur in the light curve. These may be due to interference between the normal pulsations, but other explanations have also been proposed (see, e.g., Breger & Kolenberg 2006).

RR Lyr stars have pulsation driven by opacity. As mentioned in Sect. 11.2.2, this causes hysteresis effects, as shown for RR Gem in Fig. 11.9. In the quiet phase the atmosphere is still shrinking slowly with increasing opacity until the temperature inside the atmospheric layers has risen enough to shift the ionization balance of He toward higher ionization levels. This results in a decrease of opacity and thus extra flux passing to the surface (the star suddenly gets brighter), which heats up the atmosphere ($b - y$ gets smaller; see also Fig. 11.8 left), exciting H more to the Balmer level, thus to more opacity in the Balmer continnuum (so an incraese in c_1). The new situation then slowly decays to the former compact structure. The radius changes in such a cycle by > 30%, the surface temperature (Θ) by > 30% with $T_{\rm eff}$ large when R small. The expansion goes along with changes in $v_{\rm rad}$ (for RR Lyr itself see the curve in Fig. 11.10). Consider, using Figs.11.8 to 11.10, at which point in the pulsation cycle the atmosphere is most compact and most extended as well as when the largest radial or $T_{\rm eff}$ changes take place.

For more on RR Lyr stars see, e.g., Smith (1995).

11.3.1.3 δ Sct

In the region where the instability strip intersects the MS, δ Scuti stars with $1.5 < M < 2.5$ M$_\odot$ are found. They are stars on the MS ascending the lower giant branch, have periods of 0.02 to 0.25 days, and oscillate with low-order radial and nonradial p-modes of low spherical degree but also of high degree (spectroscopically). The driving force is the κ-mechanism in the He II zone at $T \simeq 50000$ K.

δ Sct stars form the transition between high amplitude pulsators (like δ Cephei) and non-radial pulsators. They show both type of pulsation, p- and g-modes (see Fig. 11.3). For a review on δ Sct stars see, e.g., Breger (2000) and in general Breger & Montgomery (2000).

11.3.1.4   DA variables or ZZ Cet stars

The DA type WDs (see Ch. 10.8.1.4) are at the faint end of the instability strip. Variability is $\Delta V < 0.2$ mag and periods are from 100 to >1000 s. The available pulsation models lead, assuming $l = 1$, to a left-over hydrogen surface layer of just 10^{-4} M$_\odot$ around the central He (and perhaps C+O) zone. The oscillation may be driven by the ionization zone of H or perhaps may have to do with the onset of the surface convection zone (see Ch. 17.1.2) but models are still inconclusive. These objects are also subject of investigation with asteroseismology (see Ch. 11.6).

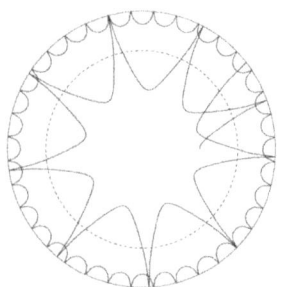
Figure 11.11: Acoustic ray paths of two solar p modes are shown. Depending on the frequency of the wave, they are bent more or less strongly toward gas at smaller density (sound speed gradients). The increase in density toward the interior results in more severe bending, leading to an oscillation free central cavity. Sound waves thus may provide information about gas at different depths in the solar interior. (Such ray paths occurr also in the Earth, tracing the propagation of earthquake waves.) The dashed circle marks the base of the convection zone in the Sun. Figure from Stix (2004).

11.3.2 Main-sequence variables

The δ Sct variables of the lower MS have been discussed with the instability strip (Sect. 11.3.1).

F main sequence stars. F-type stars have been found showing cyclic *radial velocity variations* and low amplitude light variability (e.g., γ Dor stars). The cause for the variations is based in opacity effects (perhaps partial He II ionization), but possibly also in convection.

SPB stars. Slowly pulsating B stars (B3 - B8) are multiperiodic variables with periods between 1 and 3 days. The first of this kind was discovered as a line-profile variable object. Pulsation is probably in the g-mode for the cooler stars and in the p-mode for the hotter. Brightness variations are small (see, e.g., Baade 1998).

β Cep stars. These stars are short period variables (< 0.3 days) of the upper (B-type) MS. Oscillations are detectable in radial velocity as well as in brightness. The driving force most likely is the κ-mechanism based on a locally enhanced opacity by metals (Z-bump).

11.3.3 Red variables: Miras

Semi-regular red variable stars cover a fair range of pulsation parameters. Since this part of the HRD can be occupied by stars from very different origins such as age and evolutionary state (see Fig. 10.23), a unique assignment is difficult.

Mira stars are semi-regular long period ($P > 80$ d) variables of relatively low mass at low temperature and low luminosity. They are most likely AGB stars. The driving mechanism is probably the combined action of partial H and He ionization in the atmosphere. For Mira itself both a first overtone as well as the fundamental mode have been proposed.

Mira stars lose mass with 10^{-7} to 10^{-4} $M_\odot$ yr^{-1}, in part stimulated by the oscillations. The mass lost may enshroud the remaining object and then the entire structure becomes an OH/IR star (see Ch. 10.5.1).

At the end of the evolution up the AGB the H shell may reignite. This leads to surges in luminosity and thus to oscillations (see Fig. 10.13). This is a case of the ϵ-effect pulsations.

11.3.4 Massive variables (LBVs)

LBVs. In the middle top part of the HRD one finds the Luminous Blue Variables (see Ch. 13.2.4), stars in relatively rapid evolution. These apparently oscillate and show heavy mass loss.

11.4 Vibrations

A gaseous sphere cannot be perfectly stable in its shape. In particular the more compact, cooler stars turn out to be *vibrating*. The science of vibrating stars is called *asteroseismology*, in borrowing from the science about oscillations of and waves running through the Earth. For a graphic example of ray paths see Fig. 11.11, for a review of the topic see Brown & Gilliland (1994).

If waves running through a star establish an integer number of nodes a standing pattern emerges and vibration is established. Thus the vibration of a sphere is quantized (as indicated in Sect. 11.1).

11.5. HELIOSEISMOLOGY

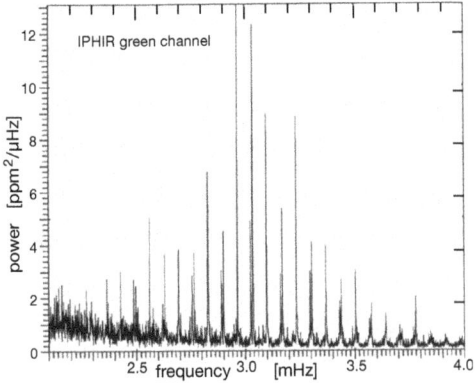

Figure 11.12: The power spectrum of solar brightness variation shows in particular oscillations in the 2500 to 4000 μHz domain (data from IPHIR on the *Phobos 1* satellite). These are the so-called "5 minute oscillations" of the Sun in the p_{15} mode (the fine splitting is in part due to the rotation of the solar surface). For mode sketches see Fig. 11.2. A spatial representation is given in Fig. 11.13. Data obtained with a photometer integrating the solar disk. Figure from Toutain & Fröhlich (1992).

Figure 11.13: The Sun oscillates in the p_{15} mode, with the "5 minute oscillations" of $l=20$ and $m=16$. They can be represented in a 3-dimensional sketch. The vibrations are seen at the surface and note the oscillation cells extending into the interior. The stippled region is the zone with convection. The solar core is bright (the fusion zone). The cells will have slightly different radial velocities due to the rotation of the Sun while there is also a latitude effect. This leads to modulation of the signal as seen in the fine structure in the frequencies of Fig. 11.12. Figure from the Nat. Opt. Astron. Obs.

Vibration has modes: the fundamental mode and the so-called "overtones". Most of what is known about vibration of stars has been derived from the Sun (*Helioseismology*).

The vibrations (or pulsations) observed in the Sun are essentially *sound waves*, termed p-modes[3]. Here pressure gradient forces provide the largest part of the restoring force, driving the redress of the spatial deviation.

11.5 Helioseismology

Although the Sun is normally not thought of as a variable star, very small variations in brightness are present. These were recorded in particular with the IPHIR full-disk photometer on board of the *Phobos* spacecraft on the way to Mars in 1989, since 1995 with the *SoHO* satellite and with observations from Antarctica in its 'summer'. A frequency analysis of the IPHIR solar brightness resulted in a power spectrum with great detail (see Fig. 11.12). From these the "5 minute oscillations" were derived, displayed as the "p_{15} mode" in Fig. 11.13. It shows the p-modes, with frequencies between 2500 and 4000 μHz. The largest peaks represent intensity variations of only $\delta I/I \simeq 3 \cdot 10^{-6}$.

The vibration data can be used to obtain information about the interior structure of the Sun (see e.g. Gough & Toomre 1991). It involves in particular information about the sound speed c_S (see Basu 1997), the internal rotation (see, e.g., Sekii 1997), and the depth of the atmospheric convection zone. (Many of the effects of stellar rotation are addressed in Ch. 14.) In this manner the relation of the gas density with depth could be derived (even latitude dependent information), showing the downward diffusion of He in the solar atmosphere (see diffusion, Ch. 3.6). For more on these aspects see, e.g., the proceedings of the conference on "Helio- and Asteroseismology" (Christensen-Dalsgaard & Frandsen 1988) and on "Sounding solar and stellar interiors" (Provost

---

[3]The giant planets are also accessible for seismology investigations. For Jupiter a few oscillation frequencies are known (see Mosser 1997).

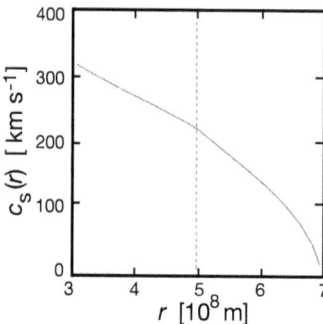

Figure 11.14: Using data from Helioseismology, the depth of the solar convection zone follows from the change in the slope of the depth related values of the sound speed $c_S(r)$. Compare with the sketch of Fig. 11.11. Figure adapted from Stix (2004). Surface convection zones are present in MS stars of $M < 1.1\,\mathrm{M}_\odot$ (see Fig. 6.6).

& Schmider 1997); for the history of this research field see Leibacher et al. (1985). Summaries are given by Stix (2004) and Thompson et al. (2003).

The source of energy for the p-modes most likely is *acoustic noise* generated by high-speed convective motions in the solar surface convective layer. If so, then all stars of the lower MS should exhibit such p-mode vibrations.

The depth of the solar convection zone followed from the depth of penetration of the p-mode oscillations. The run of possible p-mode frequencies depends on the *run of the temperature gradient*, which depends on the mechanism of energy transport (radiation or convection)

$$\left(\frac{dT}{dr}\right)_{\mathrm{rad}} = -\frac{3\overline{\kappa}\rho}{16\sigma T^3}F \quad,\quad \left(\frac{dT}{dr}\right)_{\mathrm{conv}} \simeq -\frac{\gamma-1}{\gamma}\frac{T}{H_\mathrm{P}} \quad, \tag{11.11}$$

where $\overline{\kappa}$ is the Rosseland mean opcity, ρ the density, σ the Stefan-Boltzmann constant, F the energy flux density, and H_P, the pressure scale height (see for all these parameters Ch. 4.4). Using the p-mode frequencies, one can derive the run of the value of the sound speed $c_S(r)$ with depth, showing a clear change in the slope (see Fig. 11.14), at the radius of the convection boundary layer at $r = 0.713 \pm 0.001\,\mathrm{R}_\odot$. The outer 29% in radius of the Sun is convective but note that the concomittant mass fraction of convective gas is much smaller (see Fig. 6.6).

The modelling of the p-mode frequencies requires also knowledge of the density structure. Detailed modelling showed that a fit can only be obtained if there is a radial gradient in the atmospheric He abundance. This gradient is induced by gravitational settling (see Ch. 3.6).

Very important are these investigations for the solar neutrino problem (see also Chs. 5.3.2 and 5.3.4). Since the low-l p-modes penetrate rather close to the solar core (but not quite; the centre of the Sun is an "acoustic cavity"; see Figs. 11.11 and 11.13), their frequencies may be used to test for the physical effects that might account for the paucity of ^{8}B neutrinos from the Sun. The latest solar oscillation models provide good agreement with the earlier structure models. Thus also the observed neutrino flux (see Shibahashi & Takata 1997) agrees with the early solar models (Ch. 5.3.6.3), in particular once the "neutrino oscillations" (see Ch. 5.3.5) were taken into account.

11.6 Asteroseismology

Asteroseismology of stars has, since very accurate measurements are necessary, been limited to that of bright stars. Moreover, long photometric sequences are required. These can be accomplished by, e.g., a consortium of observatories, like the one called "Whole Earth Telescope" (WET).

Vibrations go along with surface displacements and thus *radial velocity variations*. Non-radial pulsations produce localized small v_rad variations which are of the order of only a few cm s^{-1} so that successful Doppler-shift measurements require very high spectral resolution (consider that 15 cm s$^{-1} \to \delta\lambda/\lambda = 5\cdot 10^{-10}$) and low noise in the data.

Many stars have also outer convection zones so that the convective cells produce convective radial velocity noise. Given the need for high spectral resolution and very good signal to noise, only the visually brightest lower MS stars have been investigated.

11.6. ASTEROSEISMOLOGY

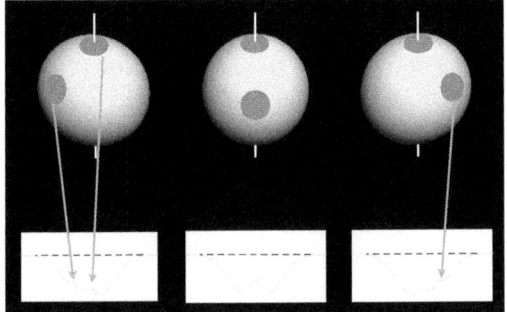

Figure 11.15: Model showing how starspots cause features in the spectra. A star with two (cool) spots rotates. Absorption in a selected spectral line may be less from the cool spot gas than from the normal warm surface gas, seemingly leading to some emission in the profile (see arrows). With the rotation, the lack of absorption gets Doppler-shifted. Using time series of spectra one may reconstruct the nature of the spotty stellar surface.

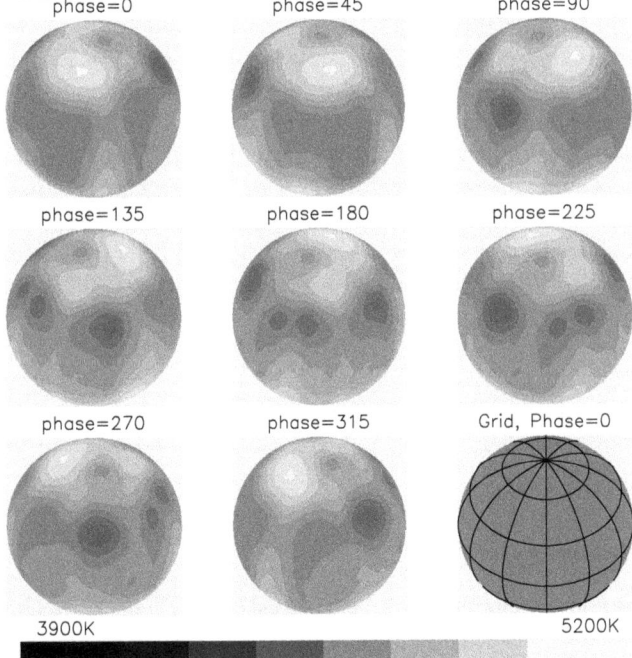

Figure 11.16: Surface structure of the spotted star PW And as reconstructed from Doppler imaging. PW And is a cool star (spectral type K 2) in its evolution just approaching the ZAMS. The grey scale (see the bar at the bottom: 3900 - 5200 K) indicates the range of temperature.

Observational time series allow the reconstruction as shown at the indicated orbital phases. For the reconstruction, one starts with a reasonable guess of the surface structure which is subsequently iterated to match the features in the observations. Figure from Strassmeier & Rice (2006).

To appreciate the method of Doppler-shift asteroseismology consider first the simpler case of "Doppler imaging" of spotted stars.

11.6.1 Doppler imaging and spotted stars

The method to obtain information about the surface structure of a spatially unresolvable star is called Doppler imaging.

If a star has surface structure due to dark spots (temperature effects, see Fig. 11.15), cool regions may cause in well selected spectral lines less strong absorption than warmer regions. Thus, depending on the location of the cool area in relation with the surface rotation of the star, these regions approach us or receede from us. This may lead to an orbital phase-dependent and thus time-dependent variation in the shape of a spectral line. Thus the time-dependent strength variations are localized in λ inside the profile (see Fig. 11.15). A time series of very accurate spectra can, after detailed modelling (see, e.g., Rice 2002), lead to a reconstruction of the features on a stellar surface. An example of such a reconstruction is given in Fig. 11.16.

11.6.2 Doppler-shift asteroseismology

A more complex case of Doppler imaging is that of Doppler-shift asteroseismology. The example shown in Fig. 11.17 is that of a star in non-radial vibration with modes $l = 6$, $m = 6$. It will

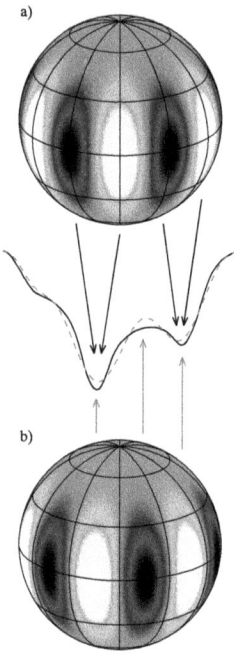

Figure 11.17: Model for the pulsation and its phase dependent spectral line changes in a rapidly rotating star. Figure from Kochukhov (2004).
Top map: The vertical component of the surface velocity field due to the $l = 6$, $m = 6$ non-radial pulsation mode; dark areas receede, the brighter areas approach the observer.
Bottom map: The surface temperature structure is shown (same phase) with hotter regions bright, cooler regions dark.
Middle: The changes in the shape of a particular spectral line due to effects of pulsation and temperature (the arrows connect surface features with resulting spectral structure). These are radial velocity shifts (from the upper map), temperature structure (from the lower map) and radial velocity effects (due to fast rotation).
Using time sequences of spectroscopic data of a pulsating star it is possible to reconstruct the nature of the pulsation of such stars. Depending on the pulsation (phase and mode) and the rotation of the star (for the various parts of the surface the observed radial velocity differs), larger portions of the stellar surface move toward the observer or are receeding, causing surface location related doppler-shifted absorptions. Note that also effects of limb darkening have to be included in the modelling. The concomittant changes in the spectral line shape allow to retrieve the overall behaviour of the stellar atmosphere.

exhibit vertical strips having different (alternating) temperatures as well as surface velocities. With rotation, these structures create time-dependent changes in the shape of the spectral line profiles (as indicated in Fig. 11.17).

11.6.3 Photometric asteroseismology

In particular WDs (g-mode) and δ Sct (p-mode) variables have been investigated. The WDs show substantial brightness variations (up to 0.3 mag) in fairly short periods (100 s to 2000 s), the δ Sct similar amplitudes but with longer periods. Moreover, they show many excited modes.

Various δ Sct stars show widely different behaviour, such as in the amplitudes, the number of periods seen and the number of modes (Breger & Audard 1997).

The WDs include all those mentioned in Sect. 11.2.2. They oscillate in g-modes, usually of high radial order n (in the WD literature the order is mostly given as k instead of n). The driving force is the κ-effect, although in DOVs there may be an atmospheric H-burning instability (ϵ-effect).

The structure of WDs is as follows (see also Ch. 17.1). Around an isothermal and degenerate C+O core one finds a thin ($M < 10^{-2}$ M$_\odot$) non or partially degenerate envelope consisting mostly of He. On top there is perhaps an even thinner ($M < 10^{-4}$ M$_\odot$) layer of H. Due to the high gravity, the atmosphere is strongly stratified, with the heavier elements separated down by gravitational settling (see Ch. 3.6). Only such a structure can explain the various modes observed, which, in reverse, confirms the structure.

Details of the fusion processes which took place during the evolution of a star determine also in which way seismologic effects become visible. E.g., the settling of ^{22}Ne (from N-burning, see Ch. 5.2.2) may have considerable effects on the structure and appearance of WDs (Deloye & Bildsten 2002), such as their location in the HRD. For more on asteroseismology of WDs see Vauclair (1997).

Some very massive DAVs may start to crystallize (see Ch. 17.1.3) in the instability strip, having effects on the excitable acoustic modes.

Also sdB stars (at the left end of the HB) have recently been found to show vibrations. Periods range from 10 to 30 min with amplitudes of 0.01 mag in the visual. To investigate such stars photometrically a simultaneous multiband photometer is very useful (Cordes 2004).

11.6.4 PG 1159, sdB, and DB variables

PG 1159 stars. These stars are found on the hot part of the pAGB/pre-WD track (Ch. 10.8.1.3) at $T_{\text{eff}} > 70 \cdot 10^4$ K having $M \simeq 0.6$ M$_\odot$. Oscillations are short (7 - 30 min) and probably with $l = 1$ in high order g-mode. The driving mechanism may be the partial ionization of C^{+4} to C^{+5} and O^{+6} to O^{+7} (K-shell ionization). More can be found in the review by Rauch & Werner (1997).

sdB variables. Fast and small multimode oscillations have been discovered in sdB-type stars (at the blue end of the HB). Periods run from 80 to 600 s at amplitudes of up to 0.025 mag. The driver for the pulsation is an opacity bump due to Fe and other metals at $T_{\text{eff}} \simeq 2 \cdot 10^5$ K in the envelope (Charpinet et al. 1997). One example, PG 1605+072, may be found in Falter et al. (2003).

Variable WDs (type DB). Variables of DB type are known with $21500 < T_{\text{eff}} < 24000$ K. Periods are between 140 and 1000 s. The driving force probably is the zone with ionization from He$^+$ to He^{2+}. DA variables have been discussed above with the instability strip (Sect. 11.3.1).

11.7 The Solar cycle of 11 years; effects on climate

It is long knwon that the Sun has a sunspot cycle: the number of sunspots waxes and wanes in a rhythm of approximately 11 years. It is, in fact, a 22 year rhythm, since the magnetic field polarity of the sunspots switches after 11 years. Sunspot counts are available since the 17th century. The reason for the cycle is unclear. Successive cycles have varying activity: there may even be a "beat" phenomenon in the cycle amplitude.

In the early 17th century sunspots were almost absent and no periodicity can be seen. This epoch was later related with the cold wheather phase in Europe and elsewhere, the "little ice age".

An anti-correlation exists between sunspot activity and the intensity of cosmic rays. The active Sun has stronger magnetic activity than a quiet Sun and then appears to shield the inner heliosphere from cosmic rays coming from interstellar space. Since charged particles in the Earth atmosphere may provide condensation nuclei for water vapour, a connection with the weather was suspected (Svensmark & Friis-Christensen 1997).

The period of the solar cycle varies. In periods with short solar cycles of $\simeq 10$ y the heating of the Earth climate is more pronounced than in periods with long cycles of $\simeq 12$ y (see Friis-Christensen & Lassen 1991). There is also a correlation of the activity cycle with the "Solar constant", the total energy radiated by the Sun. Friis-Christensen and others in Denmark have been active in uncovering the influence of the Sun on the Earth climate. Access to the topic can be obtained through the proceedings of the conference "The Sun as a variable star" (Pap et al. 1994) and through the book by Calder (1997).

References

IAU181 below means: Provost, J., & Schmider, F.-X. 1997 (see below)

Baade, D. 1998, in IAU Symp. 185, "New Eyes to see Inside the Sun and Stars", F.-L. Deubner, J. Christensen-Dalsgaard, & D. Kurtz (eds.); Kluwer; p. 347

Basu, S. 1997, in IAU181, p. 137

Blazhko, S.N. 1907, Astron. Nachr. 173, 325

Breger, M. 2000, in "Delta Scuti and Related Stars - Reference Handbook"; M. Breger & M.H. Montgomery (eds.), ASP Conf. Ser, 210, p.3

Breger, M., & Audard, N. 1997, in IAU181, p. 387

Breger, M., & Kolenberg, K. 2006, A&A 460, 167

Breger, M., & Montgomery, M.H. (eds.). 2000, "Delta Scuti and Related Stars - Reference handbook"; ASP Conf. Ser. 210

Briquet, M., Hubrig, S., De Cat, P., Aerts, C., North, P., & Schöller, M. 2007, A&A 466, 269

Brown, T.M., & Gilliland, R.L. 1994, ARAA 32, 37; *Asteroseismology*

Buchanan, M. 2000, "Ubiquity", Phoenix, London

Calder, N. 1997, "The Manic Sun", Pilkington Press, London

Chadid, M. 2000, A&A 359, 991

Charpinet, S., Fontaine, G., Brassard, P., Chayer, P., Rogers, F.J., Iglesias, C.A., & Dorman, B. 1997, ApJ 483, L123

Christensen-Dalsgaard, J. 2003, http://www.phys.au.dk/~jcd/oscilnotes

Christensen-Dalsgaard, J. 2000, in "Delta Scuti and related Stars - Reference Handbook"; M. Breger & M.H. Montgomery (eds.), ASP Conf. Ser, 210, p.187

Christensen-Dalsgaard, J., & Frandsen, S. (eds.) 1988, IAU Symp 123, "Advances in Helio- and Asteroseismology"; Reidel

Cordes, O.-M. 2004, PhD Thesis Univ. of Bonn

Cox, J.P. 1980, "Theory of Stellar Pulsation", Princeton Univ. Press

Deloye, C.J., & Bildsten, L. 2002, ApJ 580, 1077

Falter, S., Heber, U., Dreizler, S., Schuh, S.L., Cordes, O., & Edelmann, H. 2003, A&A 401, 289

Fernie, J.D., Kamper, K.W., & Seager, S. 1993, ApJ 416, 820

Friis-Christensen, E., & Lassen, K. 1991, Science 254, 698

Gautschy, A., & Saio, H. 1995, ARAA 33, 75; *Stellar Pulsations Across the HRD: Part 1*

Gautschy, A., & Saio, H. 1996, ARAA 34, 551; *Stellar Pulsations Across the HRD: Part 2*

Gouch, D., & Toomre, J. 1991, ARAA 29, 627; *Seismic Observations of the Solar Interior*

Hansen, C.J. 1972, A&A 19, 71

Kamper, K.W., & Fernie, J.D. 1998, AJ 116, 936

Kochukhov, O. 2004, A&A 423, 613

Ledoux, P., & Walraven, Th. 1958, in Handbook of Physics, Vol. LI, p. 353

Leibacher, J.W., Noyes, R.W., Toomre, J. & Ulrich, R.K. 1985, Sci. Am. 253 (Sept.) p.34

Maintz, G., & de Boer, K.S. 2008, in prep.

Mosser, B. 1997, in IAU181, p. 251

Pap, J.M., Fröhlich, C., Hudson, H.S., Solanki, S.K. (eds.) 1994, "The Sun as a Variable Star", Cambridge Univ. Press

Provost, J., & Schmider, F.-X., eds. 1997, IAU Symp 181, "Sounding Solar and Stellar Interriors"; Kluwer, Dordrecht

Rauch, T., & Werner, K. 1997, in "Third Conf. on Faint Blue Stars", A.G.D. Philipp et al. (eds.), Davis Press, Schenectady, p. 217

Rice, J.B. 2002, AN 323, 220

Sekii, T. 1997, in IAU181, p. 189

Shapley, H. 1927, Circ. Harv. Obs. No 313

Shibahashi, H., & Takata, M. 1997, in IAU181, p. 167

Smith, H.A. 1995, "RR Lyrae stars", Cambrigde U. Press

Stix, M. 2002, "The Sun", Springer, Heidelberg

Stix, M. 2004, "Helioseismology", Rev. Mod. Astronomy 17, 51

Strassmeier, K.G., & Rice, J.B. 2006, A&A 460, 751

Svensmark, H., & Friis-Christensen, E. 1997, J. Atmospheric Terrestrial Physics 59, 1225

Thompson, M.J., Christensen-Dalsgaard, J., Miesch, M.S., & Toomre, J. 2003, ARAA 41, 599; *The Internal Rotation of the Sun*

Toutain, T., & Fröhlich, C. 1992, A&A 257, 287

Unno, W., Osaki, Y., Ando, H., & Shibahashi, H. 1979, "Nonradial Oscillations of Stars", Univ. of Tokyo Press

van Albada, T.S., & de Boer, K.S. 1975, A&A 39, 83

Vauclair, G. 1997, in IAU181, p. 367

Chapter 12

Stellar coronae, magnetic fields and sunspots

Special phenomena have been discovered in the investigation of the best studied star, the Sun.

During solar eclipses diffuse radiation was found from gas layers well above the solar atmosphere. The lower part of this region was called "chromosphere" (at 5000 to 15000 km above the solar atmosphere, $T_{\rm chr} \simeq 10^4$ K) which has line emission, detectable during a solar eclipse. The higher part the "corona" and its spectrum showed emission lines, which at the time of discovery in the late 19th century could not be assigned to a known atom. Thus for the coronal emission an element "Coronium" was proposed (the periodic table still had many gaps!). Later it was understood that the emission is due to forbidden lines of highly ionised Fe (and other elements) at $T_{\rm cor} \simeq 10^6$ K.

Furthermore, the Sun shows dark spots which turned out to be locations of polarised radiation where magnetic fields are strong at the solar surface. Both these features can be studied essentially only for the Sun (stars are too distant to be resolved), so treatment of these will be kept rather brief. However, for some stars spottyness has also been detected, predominantly by analysing long times series of photometry, methods related with asteroseismology (see Ch. 11.6 and Fig. 11.17).

12.1 Stellar coronae

Many low mass stars may have a hot and tenuous outer envelope. The corona of the Sun can, because of light scatted in the earth atmosphere, in the visual be seen only during a solar eclipse when the moon covers the bright solar surface. Since the satellite "Solar and Heliospheric Observatory" (SoHO) is available, the corona can be seen continuously (see Fig. 12.1).

With a spectrograph the multitude of emission lines can then be made visible. Using the

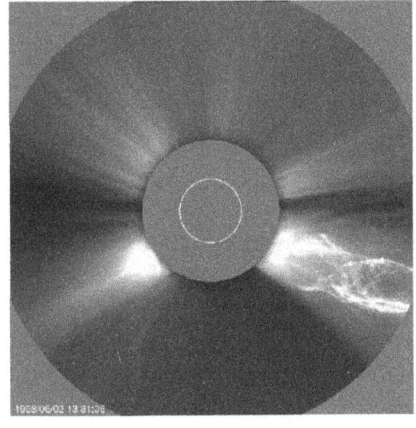

Figure 12.1: The corona of the Sun as seen with SoHO. The disc of the Sun is blocked out so that the tenous luminous outer gas structures can be seen in visible light. The white circle represents the size and position of the Sun. The LASCO C2 coronograph image shows a twisting, helical-shaped magnetic flux tubes having a coronal mass ejection spinning off from the Sun.
Image from http://sohowww.nascom.nasa.gov

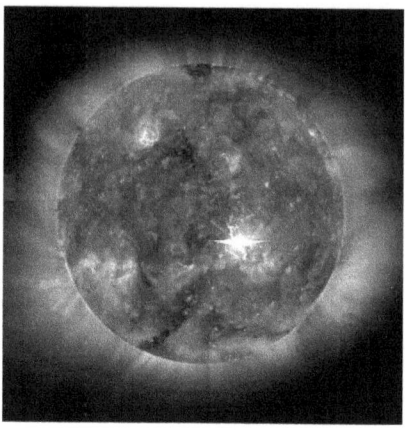

Figure 12.2: The Sun at 195 Å as seen with SoHO. A bright solar flare (just a little below and to the right of the centre) was captured by the EIT 195 instrument on 2 May 1998 (for prominences and flares see also Fig. 12.6).
Note that the cool solar surface is dark, hot gas is luminous while the gas in the flare is hot and dense, radiating strongly at 195 Å, the wavelength of a Fe XII emission line (characteristic for gas of $T \simeq 1.6 \cdot 10^6$ K). The variations in the intensities are due to the particular radiation transport conditions (see Sect. 12.2).
Image from http://sohowww.nascom.nasa.gov

emission-line strengths one can, with the help of a calculation of the excitation state (as in Ch. 3.2.1) of high ions (e.g., Fe IX - Fe XIV), derive the temperature of the gas (for the Sun $\simeq 2 \cdot 10^6$ K) and so predict which emission lines may be present in other wavelength ranges (de Boer et al. 1972).

At UV wavelengths the surface of the Sun is faint so that the emission lines from the very hot coronal gas stand out in any UV observation, even outside of an eclipse. UV (and X-ray) measurements have shown the existence of large and bright regions in particular near active areas of the solar surface. Other cool stars have coronal and chromospheric emission lines as well, as studied by, e.g., Ayres & Linsky (1980) with the IUE satellite (sse also Linsky 1980).

12.2 Effects of radiation transport

Which structure is seen at what wavelength is determined by the actual optical depth τ_ν (see Chs. 2 and 3) in the gas. And τ depends also on the amount of the particular material available. These τ-effects are the cause for all that is seen in Figs. 12.2, 12.4 and 12.6.
1. At some wavelength less absorption capacity may be available in the cooler gas than in hotter gas, e.g., because the ion producing an absorption line in the hotter gas is predominatly neutral in the cooler gas. This applies mostly to wavelengths where the spectral continuum is bright.
2. In faint parts of the spectral continuum, like in the (far-)UV, the actual excitation conditions may lead to recognizable emission lines even from cool gas. Emission lines can also be strong due to magnetic forces.

12.3 Magnetic fields

The Milky Way has a magnetic field which will be included in a star during the star formation (see Ch. 7.3.3). Thus each star has a magnetic field. Depending on the strength of the magnetic field

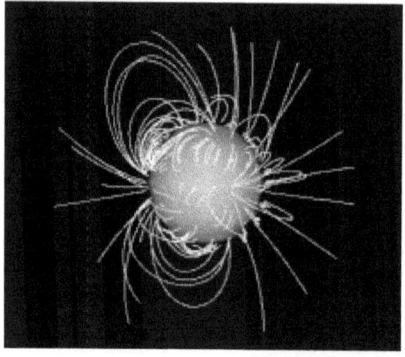

Figure 12.3: Model of actual magnetic field lines of the Sun from a calculation based on measurements and some adequate three-dimensional data as the initial value. The three-dimensional magnetohydrodynamic (MHD) simulation code is based on the concept of MUSCL (of 3rd-order, van Leer-type) and TVD (with linearized Riemann solver) schemes developed for these purposes (see Hayashi 2005; Hayashi et al. 2006). Image from http://sun.stanford.edu/~keiji/gallery.html

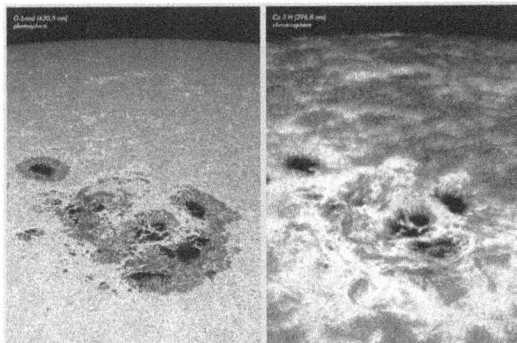

Figure 12.4: The active region AR10486 photographed Nov. 2003 with the Dutch Open Telescope (DOT). *Left:* image taken in the G-band (see Ch. 3.4.1) sampling the photosphere. *Right:* image in the Ca II H line sampling the chromosphere. The effect of solar magnetic fields becomes more important in the chromosphere due to magnetic heating. This is evident in the areas around the sunpots and the "hedges" of brightness towards the solar limb. Image from http://dot.astro.uu.nl/promotion/images

Figure 12.5: Sunspot of the Sun as seen with the DOT in the light of the Ca II K line. On 4 November 2003, just before the active region AR10486 rotated out of sight over the limb to the backside of the Sun it displayed during a few hours this bright stalk-like feature, called a surge. It then sent off the largest flare on record, which likely resulted from magnetic reconnection. The sunspots appear as flat dark pancakes amidst bushes of chromospheric fibrils. The foreground shows reversed granulation in the mid photosphere. Image from http://dot.astro.uu.nl/promotion/images

there may be shifts of spectral lines (see Ch. 3.5). Large effects are normally only seen in compact old objects like white dwarfs. But of course also the areas on the solar surface with large field strength like the sunspots show effects of the magnetic field on spectral lines.

Since the Sun is near, its surface can be spatially resolved and so one can measure the light from individual structures. Using data from SoHO it is possible to make a model for the magnetic field as present at the time of the observations. The result of such a calculation is shown in Fig. 12.3.

12.4 Sunspots

Spots are essentially cool stellar surface regions. The spectral energy distribution is, simplistically, similar to that of a lower surface temperature star. This lower temperature also means that the gas excitation conditions in the spot are different. Observations of the Sun and its sunspots (see, e.g., Fig. 12.4) in the narrow wavelength range of well chosen spectral lines thus often show different aspects of the spatial structure of the Solar surface (as explaned in Sect. 12.2).

Sunspots are understood as due to effects of the magnetic field penetrating the stellar surface. That there is spatial structure can be seen in Fig. 12.5, showing a sunspot with vertically extended tubes of gas.

The gas in a sunspot is cooler than the normal surface. For the Sun mean parameters are: $T = 4040$ K at $\tau = 0$ with similar values above the spot. As of $\tau < -5$ the temperatures are higher being $T \simeq 10^4$ K at $z \simeq 2000$ km. The magnetic field in a sunspot has of strength of 2-4 kG.

12.5 Prominences and flares

Structures protruding from the solar surface are called prominences. They are visible due to optimum optical depth and radiation transport effects. The structures are most likely due to

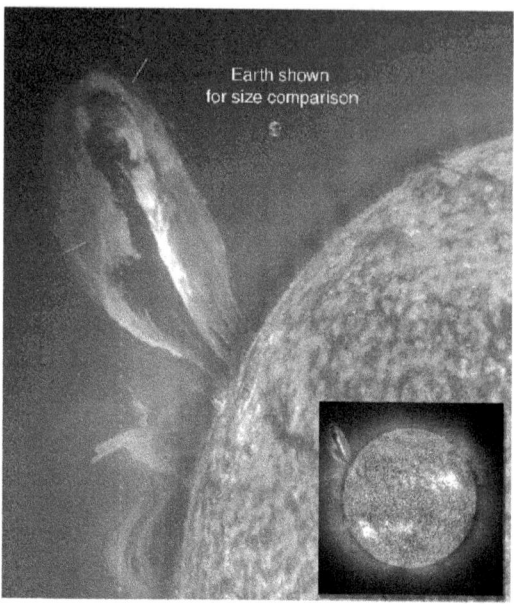

Figure 12.6: A prominence and a flare of the Sun as seen with SoHO. The large, eruptive prominence is seen in emission of He II at 304 Å. An image of the Earth is added for size comparison. This prominence from 24 July 1999 is particularly large and looping, extending over 35 Earths out from the Sun. Erupting prominences can (when directed earthward) affect communication and navigation systems, even power grids, while also producing auroras visible in the night skies. The spottyness of the solar surface is the so-called "granualtion". It shows the effects of convection in the solar envelope (see Fig. 6.6). The upwelling cells bring warm gas to the surface, gas thus being brighter than the surrounding gas (true at almost all wavelengths).
Image from http://sohowww.nascom.nasa.gov

magnetic fields extending away from the surface. Field strengths are of the order of 20 G. Huge prominences may open up and become a flare (see Fig. 12.6).

A solar flare is a sudden, rapid, and intense variation in brightness, lasting from a few minutes to a few hours. It occurs when magnetic energy that has built up in the solar atmosphere is suddenly released, launching material outward at millions of km per hour. The structured solar magnetic field tends to restrain itself and forces the buildup of tremendous energy (as in twisted rubber bands). At some point, the magnetic lines of force merge and cancel in a process known as magnetic reconnection, causing plasma to forcefully escape from the Sun. The charged particles may influence the Earth atmosphere (leading to Auroras).

12.6 Relevance of the structures for stellar evolution

The magnetic field and its effects on the stellar surface (spots, prominences, etc.) are not well understood yet. The magnetic field will have some effect on convective motions, like the twisting of field lines and breaking of motions (see Ch. 14.5) and on rotation (see Ch. 14.3.3). Effects on the overall evolution are in the process of being uncovered.

Much of the text related with the figures has been taken from the explanations on the indicated webpages. For more on the Sun see, e.g., Stix (2002), on the solar corona, its magnetic fields, sunspots, as well as prominences and flares Aschwanden et al. (2001) and for tabular material Cox (1999). For chromospheres and coronae of other stars see, e.g., Dupree (1986) and Güdel (2004).

References

Aschwanden, M.J., Poland, A.I., & Rabin, D.M. 2001, ARAA 39, 175; *The New Solar Corona*
Ayres, T.R., & Linsky, J.L. 1980, ApJ 235, 76
Cox, A.N. 1999, "Allen's Astrophysical Quantities", Springer, Heidelberg
de Boer, K.S., Olthof, H., & Pottasch, S.R. 1972, A&A 16, 417
Dupree, A.K. 1986, ARAA 24, 377; *Mass Loss from Cool Stars*
Güdel, M. 2004, A&A Rev 12, 71; *X-Ray Astronomy of Stellar Coronae*
Hayashi, K. 2005, ApJS 161, 480
Hayashi, K., Benevolenskaya, E., Hoeksema, T., Liu, Y., & Zhao, X.P. 2006, ApJ 636, L15
Linksy, J.L. 1980, ARAA 18, 439; *Stellar Chromospheres*
Stix, M. 2002, "The Sun", Springer, Heidelberg

Chapter 13

Stellar evolution: Stars in the higher mass range

13.1 Defining the high mass range

Details of the evolution of a star depend strongly on initial mass. A clear division in the way stars evolve is given by the minimum mass a star must have to become a Supernova (SN) at the end of single star evolution. This limit is $M_{\text{init}} \simeq 8$ $M_\odot$. The 'high mass range' is thus defined as stars having $M_{\text{init}} > 8$ $M_\odot$ on the main sequence (spectral type B2 and 'earlier'). Such stars also are able to create H II regions. The first evolution steps of high mass stars proceed as described in Chs. 9 & 10 for the low mass range, up to and including the stage of core-He burning (Ch. 10.3).

Observational information about types of high mass stars and various effects relevant for the nature of high mass stars will be presented before proceeding to how such stars evolve.

13.2 Types of high mass stars

High mass stars have a high central temperature leading to a high luminosity, and they thus have a short life. Therefore they are always said to be young. In the Milky Way (MW) they are present only in the disk, mostly in the vicinity of regions with dense gas (see Ch. 7) and are easy to spot (if there is not too heavy extinction).

Essentially all high mass stars show strong emission lines in their visual spectra indicating the existence of stellar winds which expel mass from the stellar photosphere at a high rate. The winds are caused by the high luminosity and thus relatively high radiation pressure (Ch. 4.5).

The observational parameters including stellar distances lead to placement of these stars in the upper part of the HRD or CMD. When the stars are in the MS phase, they have O or early B spectral type. Later they evolve into other kinds of luminous stars, such as B, A, or red supergiants, or into Wolf-Rayet stars. Several luminous stars exhibit variability in brightness mostly related with mass loss. Common types of high-mass stars are:

OB Stars of sepctral type **O** or **B**. With O stars the spectral type sometimes has a extra letter "f" signifying the presence of special emission lines in the visual of helium, nitrogen, and silicon.
RSG Very red high mass stars, the **red supergiants**, have in their spectra relatively narrow absorption lines. They show often chromospheric emission in the centre of strong atmospheric absorption lines. To see those structures, high dispersion spectra are needed.
LBV A further type of luminous stars is that with very strong wind lines and with spectral type BIe...FIe. They are often variable: the **luminous blue variables**. Their visual spectra exhibit extremely strong "P-Cyg" emission profiles, named after the lines of the prototype star **P Cygni**.
WR A special category is formed by the hot **Wolf-Rayet stars** (WR stars), named after Wolf & Rayet (1867), exhibiting emission lines mainly of helium without signs of hydrogen.

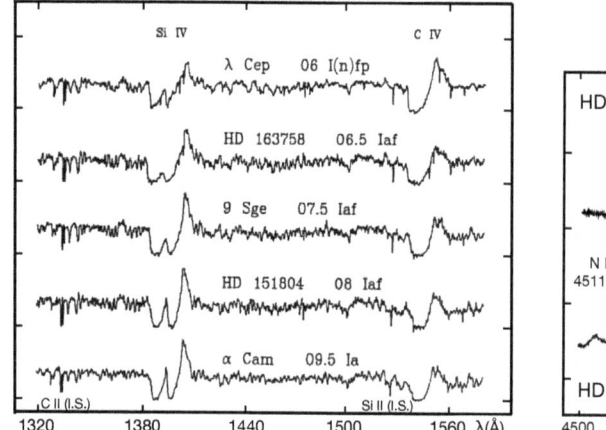 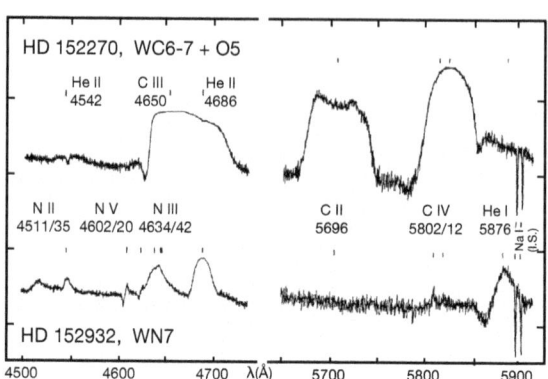

Figure 13.1: Spectra of 5 Of stars in the UV (left) and of two WR stars in the visual (right).
Left: The UV spectra (recorded with the IUE satellite) of Of-type stars show P-Cyg-like lines (emission plus blue-displaced absorption; see Fig. 13.7) in the resonance doublets of C IV 1548/50 Å and Si IV 1393/1402 Å (resolved in the absorption components!). Note the sharp interstellar absorption of C II and Si II. Figure from Walborn & Nichols-Bohlin (1987).
Right: Top: The spectrum of the WC star shows extremly broad emission lines of He II (e.g. at 4686 Å) and of C III/IV. Bottom: In the WN star spectrum lines of He I,II and N III,V are prominent. Note the sharp interstellar absorption in the Na I doublet. (Spectra from Seggewiss; unpublished).

13.2.1 The O and Of-type stars

13.2.1.1 Determining the temperature of O stars

The classical MK sequence (see Ch. 3.8) of stellar spectra starts with spectral type O4 V: He II absorption lines are strong, but He I lines are weak. Running the temperature sequence up from O9 to O4 the He II/He I line ratio increases because at higher temperature He is more and more ionized (spectral type is always based on temperature through spectral lines; Ch. 3.2.2!). At type O4 V weak emission of the N III triplet at 4640 Å appears, but N IV, which has the high ionization potential of 77 eV, is definitely absent in emission as well as in absorption

In 1971 Nolan Walborn discovered 4 stars in the Carina Nebula where he saw even weak N IV 4058 in emission (ionization stage above N III). For those stars he introduced the spectral type O3. Later, Walborn et al. (2002) found a couple of stars in compact OB associations of the Milky Way and the Magellanic Clouds where in their spectra the N IV line is much stronger than the N III triplet thus indicating an even higher temperature than previously assumed for O-type stars. Accordingly, the MK sequence had to be extended to spectral type O2. In addition, very sophisticated spectral criteria allow to classify luminosity classes from O2 V via O2 III to O2 I.

It is interesting to note that among the 45 stars known to be of spectral class O2 and O3 only 10 stars belong to the MW (in the northern MW one has only a pair in Cyg OB2), one star is found in the SMC, but 34 stars belong to the LMC, of which 22 are associated with the gigantic 30 Dor nebula region! Since O2 stars must have more mass than O4 stars, they must be very young.

Parallel to the discovery of 'earlier' and 'earlier' O-type stars the letter "f" (and various modifications) was introduced to mark special emission features in the stellar spectra (Table 13.1).

In the meantime it became clear that the "f" character is an excellent indicator of the luminosity of the O-type stars which is independent from temperature, and thus from the spectral subtype. In practice both the luminosity class and the f character are given. Compare the following sequence of stars with decreasing luminosity in the Carina nebula (note the f type!): HD 93129A ⇒ O2 If*, HD 93128 ⇒ O3.5 Vf$^+$, HD 93204 ⇒ O5 Vf, HD 93222 ⇒ O7 III(f), HDE 305612 ⇒ O9 V. The near UV spectrum of Of-type stars is dominated by the P-Cyg lines of Si IV and C IV; see Fig. 13.1.

The way the physical parameters of stars in the upper left part of the HRD/CMD are determined

13.2. TYPES OF HIGH MASS STARS

Table 13.1: Definition of " f " in O star spectral classification

((f))	N III 4640 in emission, He II 4686 in absorption
(f)	N III 4640 in emission, He II 4686 filled in
f	N III 4640 and/or He II 4686 in emission
f+	Si IV 4089/4116 in emission
f*	N IV 4058 in emission

elucidates how on the one hand the stellar properties are intimately intermixed in their effects and on the other hand how observations and theory have to support each other. A more detailed analysis may illustrate this "process" as follows.

Absolute magnitudes M_V of O-type stars can be fairly easily determined because most stars are members of young clusters and associations whose distances and reddenings can be derived by isochrone fitting. (But a warning is necessary: isochrones are based an theoretical stellar models; and the theories change with better insight into the physical processes; see Ch. 21.) The problem is to obtain the luminosity (or the bolometric correction BC, which normally is taken from model atmospheres). The effective temperature $T_{\rm eff}$ (and thus for a coarse approximation also the spectral type) is needed to select the appropriate model. The brightest star so far known in the Milky Way is HD 93129A, O2 If*: $M_{\rm bol} = -10.6$ and $L = 3.1 \cdot 10^6$ L$_\odot$.

13.2.1.2 Determining the mass of O stars

Determination of the mass of O-type stars is still problematic because there is a 'mass discrepancy'. Masses have been derived in three different ways.

(1) Direct mass determination is only possible using the few well-observed eclipsing binary systems. But no mass was found larger than 60 M$_\odot$. Highest masses are known for the components of the O6:+O7If binary system HD 47129 (Plaskett's star) in the Milky Way (51+43 M$_\odot$) and of the O3 Vf*+O6 V binary R 136-38 in the LMC (57+23 M$_\odot$).

(2) If L and $T_{\rm eff}$ are known one can plot the star in the HRD (e.g., in Fig. 13.9) and then read the *evolutionary mass* of the star from the corresponding evolutionary mass track. But especially in the high-mass range of the HRD, tracks of different masses intersect each other and the determination of evolutionary masses is therefore not unique. The highest evolutionary mass is claimed for the "Pistol Star" (P Cyg type) in the Quintuplet cluster near the Galactic centre. Figer et al. (1998) found a mass in the order of 200 to 250 M$_\odot$.

(3) Spectroscopists like to derive masses from their atmospheric models if the stellar luminosity L is known. By fitting the observed line profiles to model profiles, it is possible to determine the effective temperature $T_{\rm eff}$ and the surface gravity g. Then using the equations given in Ch. 1.6 ($L = 4\pi R^2 \cdot \sigma T^4$ and $g = G\,M/R^2$), the *spectroscopic mass* is calculated.

However, stellar masses derived from the spectroscopic method are systematically smaller (for Of-type stars up to a factor of 2) than the evolutionary masses. E.g., Herrero et al. (2000) derived

Table 13.2: Properties of main-sequence OB-type stars (Massey & Meyer 2002). $t_{\rm MS}$ is the hydrogen burning time on and near the main sequence.

Spectral type	mass	$t_{\rm MS}$	$T_{\rm eff}$	log L	R
O3 V	120 M$_\odot$	2.56 Myr	53 300 K	6.25 L$_\odot$	16 R$_\odot$
O4 V	60 M$_\odot$	3.45 Myr	48 200 K	5.73 L$_\odot$	10 R$_\odot$
O8 V	25 M$_\odot$	6.51 Myr	37 900 K	5.29 L$_\odot$	6.5 R$_\odot$
B0 V	12 M$_\odot$	16.0 Myr	28 000 K	4.01 L$_\odot$	4.3 R$_\odot$
B2 V	8 M$_\odot$	38.0 Myr	22 000 K	3.76 L$_\odot$	3.7 R$_\odot$

a spectroscopic mass of $51 \, M_\odot$ for a very young O4If$^+$ star in Cassiopeia (HD 15570) but for its evolutionary mass they found $139 \, M_\odot$, with then $M_{\text{init}} = 142 \, M_\odot$.

This problem seems to have been solved by several groups (e.g., Martins et al. 2002). One added line blocking or blanketing (see Ch. 2.10.2) of the continuum flux by the absorption lines into the atmosphere models. This blanketing occurs mainly in the blue and near UV spectral region and leads to backscattering effects in the IR. The inclusion of blanketing leads to lower T_{eff}, typically by ~ 4000 to 1500 K from O3 to O9.5 dwarf stars, and via the Eqs. of Ch. 1.6 to higher masses. Now, spectroscopic and evolutionary masses seem to be in fair agreement. Both are in good accord to directly (binary stars!) determined masses, too. But here we have the problem that $60 \, M_\odot$ is the upper limit of all available observations whereas the *indirect* methods lead to values between 60 and $200 \, M_\odot$ for O3 – O2 stars (see Walborn et al. 2002, their Table 1).

13.2.1.3 Oe/Be stars

Rotational velocities of stars are derived from the rotational broadening of spectral lines (Ch. 3.7). Looking along the main sequence, $v \sin i$ values start with $\simeq 220 \, \text{km s}^{-1}$ for O and B-type stars and go down to $\leq 10 \, \text{km s}^{-1}$ for stars of spectral types G to M.

About 10 % of all late O and early B-type stars reach the break-up velocity at their equator, i.e., the centrifugal forces overcome the gravitational forces and material is expelled which settles into ring-like structures in the star's outer equatorial plane (see Ch. 14). The rings emit radiation which appears as emission features superimposed mainly on the Balmer and He I absorption lines. The emission is often accompanied by central absorption dips due to self absorption in the rings. These rapidly rotating early emission-line stars are designated as Oe- and Be-type stars. Cooler Be stars fall in the low mass range (see Ch. 7.7.4).

13.2.1.4 Summary O type stars

Table 13.2 gives a summary of the currently known properties of OB-type MS stars.

All luminous Of stars have strong stellar winds (see the review by Kudriztki & Puls 2000) with high mass loss rates (see Sect. 13.3).

13.2.2 B type stars

MS stars of spectral type B are less massive than the O stars, have cooler surfaces, but are otherwise normal stars. Note that the *higher mass range of stars* in terms of stellar evolution ends (downward) at $8 \, M_\odot$ (Sp.T. B2). Stars of type B2 and 'earlier' produce sufficient amounts of Lyman continuum radiation to create H II regions around them.

The properties of early B type stars are listed in Table 13.2, along with those of the O types.

13.2.3 Wolf-Rayet (WR) stars

WR stars have strong emission lines, mainly those of helium, *but not those of hydrogen*. Their spectral types are subdivided according to the relative strength of emission lines of N, C and/or O, into a WR nitrogen (WN), a WR carbon (WC) and a WR oxygen (WO) sequence, respectively (see Table 13.3). WN2, WC4 & WO1 are hotter than WN11, WC9 & WO4, in accordance with the level of ionization of N, C & O ions seen. For a typical WC and WN spectrum see Fig. 13.1.

Table 13.3: Definition of WR star types (WN, WC, WO); see van der Hucht (2001)

WR type	species seen in spectrum in emission		
WN2 - WN11	N V - N II	He II, He I	
WC4 - WC9	C IV - C II	He II, He I	C III, O V weak
WO1 - WO4	O VIII - O V	C IV - C II	no C III

13.2. TYPES OF HIGH MASS STARS

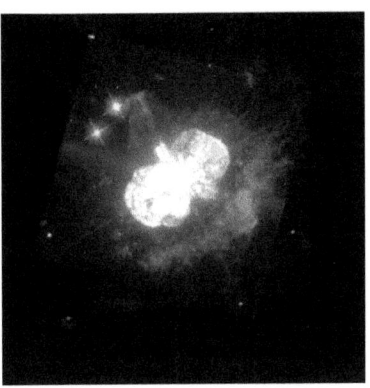

Figure 13.2: The star η Carinae is the most luminous star known in the Milky Way. Its eruptions have led to a huge and luminous nebula around the star. Figure from Weis (2001). For an attempt to describe the effect the evolution of massive stars with an LBV phase have on the stellar environment, see García-Segura et al. (1996).

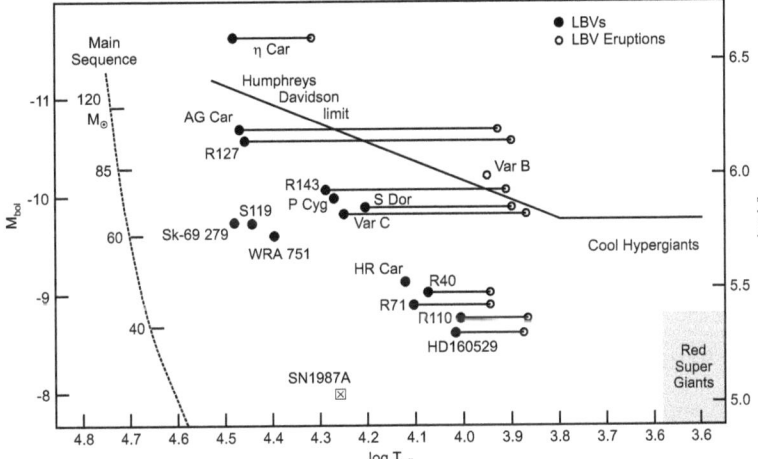

Figure 13.3: Location of known luminous blue variables (LBVs) in the HRD. Stars are labelled with their names. Lines connect data points of the eruptive and non-eruptive states. Note the location of the most luminous galactic star known: η Car (see Fig. 13.2). The progenitor star of SN 1987A is added for reference. Figure from Weis & Duschl (2002).

Generally, $T_{\text{eff}} > 50000$ K. Masses of WR stars are in the order of 10 M$_\odot$ as deduced from WR stars in eclipsing binary systems. Maximum wind velocities v_∞ (see Sect. 13.3) run from $\simeq 2000$ km s^{-1} for WN8 and WC9 types up to $\simeq 3000$ km s^{-1} for WN3 and $\simeq 4000$ km s^{-1} for WC4 types. Mass loss rates lie in the range of 1 to $8 \cdot 10^{-5}$ M$_\odot$ yr^{-1} (see the review by Abbott & Conti 1987).

The number of WR stars known in the Milky Way is over 200, including 127 WN stars, 87 WC stars, and 13 stars of mixed type. The Magellanic Clouds have been surveyed completely. In the LMC there are 135 WR stars (110 WN, 25 WC), and in the SMC 9 WR stars (8 WN, 1 WC). For details see the reviews by van der Hucht (1993, 2001).

13.2.4 Luminous blue variables: LBVs; P-Cygni stars

A small category of luminous stars exists in the intermediate temperature range, $8000 < T_{\text{eff}} < 30000$ K, at very high luminosities, $L \simeq 2$ to $5 \cdot 10^6$ L$_\odot$. They are also known as Hubble-Sandage variables, as luminous blue variables (LBVs), or as S Doradus variables (after the prototype S Dor in the LMC). The best known stars of this category are P Cyg and η Car in the MW, S Dor, R 71 and R 127 in the LMC, and AE And and AF And in M 31.

P Cyg is the spectroscopic prototype of stars showing stellar wind phenomena in the spectrum. For examples of profiles see Fig. 13.1. Such profiles are also seen in lower mass stars. The shape of these special line profiles is explained in Sect. 13.3 and Fig. 13.7.

The stars show erratic pronounced brightness variations of perhaps 1 to 7 mag, such as the star η Car had in the 19th century. For an image of η Car with its nebula see Fig. 13.2. (For a review on the many questions about η Car see Davidson & Humphreys, 1997.) These brightness variations are explained as sudden increases in opacity due to expanding, cooling and dust forming

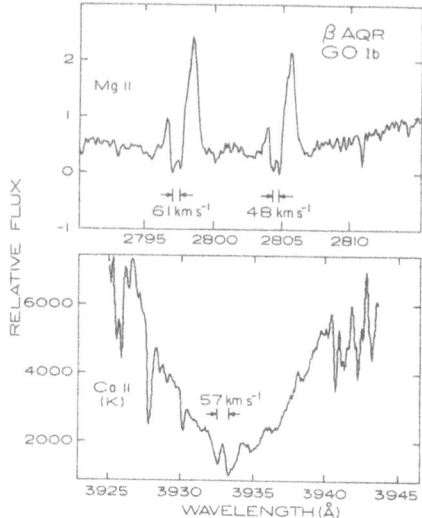

Figure 13.4: Spectral lines of red supergiants with signs of stellar wind. The red G0 Ib supergiant β Aqr shows in Ca II K and in the Mg II doublet a complicated spectral structure. The photosphere produces strong absorption, the chromosphere shows emission in those lines. Part of the emission is absorbed (see the dip in the emission peak) at slightly blue-shifted wavelengths. From the spectra a wind velocity of $v_{\text{wind}} \simeq 55$ km s^{-1} can be derived. Note that interstellar absorption may be superimposed in these lines. Figure from Dupree (1986).

shells. These states are also called eruptive and non-eruptive states.

The location of the LBVs is given in Fig. 13.3. Note that their luminosity is so high that, when these stars are in an expanded state, the outer atmosphere is no longer gravitationally bound.

13.2.5 Red supergiant stars

In high dispersion spectra of red supergiants (G I ... M I) emission structures can be detected in the centre of strong absorption lines. This was originally seen in the Na I D and Ca II K lines, and became with satellite UV spectroscopy also evident in the Mg II doublet near 2800 Å. Examples of such spectra are shown in Fig. 13.4.

Cool stars have above the photosphere a *chromosphere*, a layer of higher temperature and lower density, in which the excitation level of almost all atoms or ions is different from that in the photosphere. A chromosphere normally shows several lines in emission, lines which are wide and strong in absorption in the photosphere. Further out this emission may be absorbed again. For a model of the photosphere-chromosphere region of a red supergiant, see Fig. 13.5.

A full understanding of the complicated spectral line structure requires a complete treatment of radiation transfer. Clearly, the absorption in the outer layers leads to transfer of momentum (see below) and thus to outward acceleration of those layers. Observed wind velocities in G to M supergiants range from 60 to 10 km s^{-1}. Mass loss in α Ori (M1-2 Ia-Iab) may be as high as $\dot{M} \simeq 3 \cdot 10^{-6}$ M$_\odot$ yr^{-1}. For more on the mass loss from such stars see the reviews by Dupree (1986) and by Willson (2000). The material lost will cool further and will form dust (see Ch. 10.2.2) around these stars. The dust possibly leads to shrouding and formation of OH/IR objects and MASERS (see Ch. 10.5.1).

When the atmospheres are sufficiently cool, molecules will form, which can contribute considerably to the opacity in the stellar atmosphere (see e.g. Ch. 3.4.3 and Fig. 3.10).

A cooling stellar wind, in particular of metal rich material such as in AGB stars, is the source of dust nuclei in the universe. In dense interstellar clouds these dust particles may grow further, accumulating icy mantles (frozen-out molecules). This in turn will darken molecular interstellar clouds and create conditions favourable for star formation (Ch. 7).

13.3 Expanding envelopes, luminous winds

The spectra of Of, WR and P Cyg stars with their strong wind lines, and the line-core emission of red supergiant stars point to heavy mass loss. Indeed, in most of the luminous stars the maximum

13.3. EXPANDING ENVELOPES, LUMINOUS WINDS

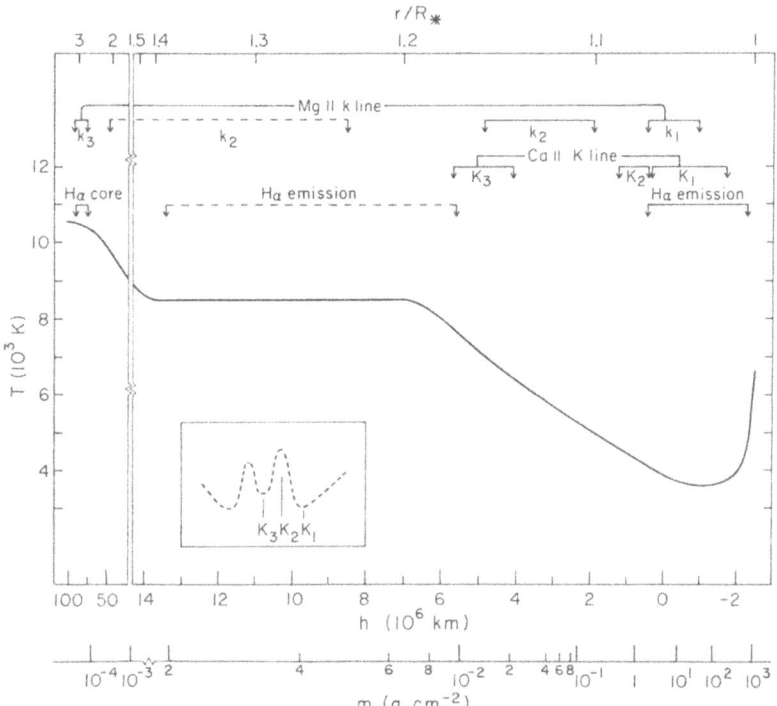

Figure 13.5: Model structure of a red supergiant outer atmosphere showing the nature of the stratification of the temperature (T) versus radius (r) in units of stellar radius R_*, hight h [km] above the photosphere, and column of material m [g cm^{-2}]. Note the break in the height scale. The range of T in the photosphere ($T \simeq 4000$ K) through the chromosphere ($T \simeq 8500$ K) to layers further out has implications for where which spectral features of Ca, Mg, and H are produced. These features have complicated profiles (as in Fig. 13.4) consisting of three components, numbered 1,2,3. Note the example for the Ca II K line in the inset. Figure from Dupree (1986).

velocity in the wind lines, v_∞, which is measured from the "bluest" wing of the emission lines or from the blue-displaced absorption of the P-Cyg profiles, respectively, exceeds the escape velocity $v_{\rm esc}$ from the stars (see Ch. 4.5). Using Eq. 4.80 one has

$$v_{\rm esc} = \sqrt{2G\frac{M}{R}} \quad \text{equal to} \quad v_{\rm esc} = 618\sqrt{\frac{M}{R}} \quad \text{km s}^{-1} \tag{13.1}$$

with mass M and radius R in solar units in the second relation. The well-known O7f star 29 CMa, e.g., would have $v_{\rm esc} = 650$ km s^{-1}, while the observed $v_\infty = 1480$ km s^{-1}. These stars have an optically thick wind and the mass loss is so large that they almost have an expanding envelope of material. In the case of WR stars the expanding envelope may even obscure our view of the stellar atmosphere. Fig. 13.6 gives an overview of mass loss rates in relation with location in the HRD.

13.3.1 Processes of radiation acceleration

Stellar winds are foremost radiation driven. Of luminous stars the radiative intensity is that large, that the acceleration enforced on atmospheric material leads to acceleration (for radiation pressure see Ch. 2.2.1.3; for general aspects of stellar winds see Ch. 4.5).

Photons with energy $h\nu$ may transfer momentum $h\nu/c$ onto an atom or ion followed by re-emission of the photon. Normally, the photon field in stars is essentially isotropic and no net

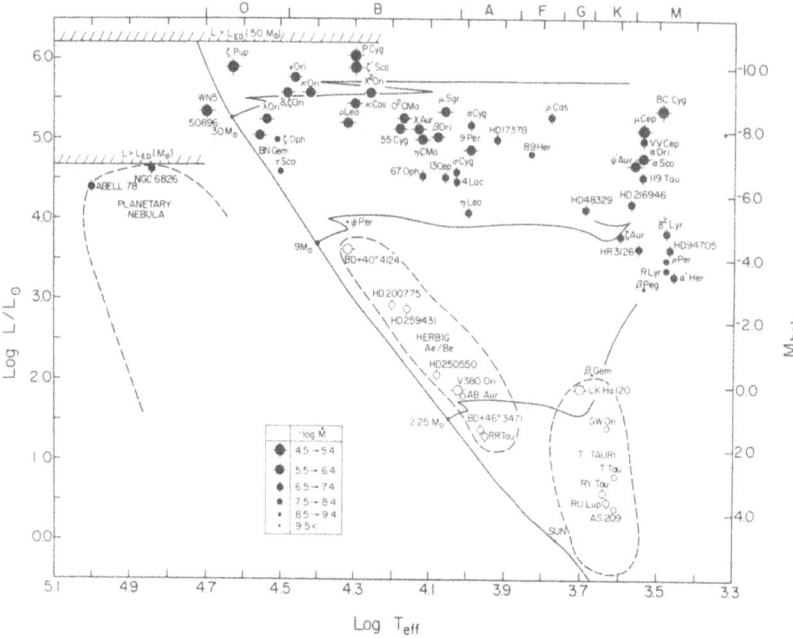

Figure 13.6: Diagram showing which kind of stars are known to lose mass. The symbols refer to particular mass loss rates (note that the actual values given are values of $-\log \dot{M}$). Clearly, mass loss is strongest at high luminosity (in the upper part of the HRD) and for stars with large radii (to the upper right in the HRD). Figure from Cassinelli (1979).

acceleration will arise. Since near the stellar surface the radiation flux is highly asymmetric (low τ outward, thus net outward flux) material can be accelerated indeed.

13.3.1.1 Radiative acceleration by the continuum

Photon momentum may be transferred to free particles. The acceleration can be given by

$$a_{\rm r.c.} = \frac{\sigma_e}{m_f} \frac{F}{c} = \frac{\sigma_e}{m_f} \frac{L}{4\pi R^2 \, c} \qquad (13.2)$$

with $a_{\rm r.c.}$ the radiative acceleration by the continuum, $\sigma_e = 6.65 \cdot 10^{-25}$ cm^2 the Thomson (electron) scattering cross section, $F = \sigma T_{\rm eff} = L/(4\pi R^2)$ the surface flux, and m_f the mass per free electron. In completely ionized hydrogen m_f is practically the mass of a proton, m_p; in a real hot atmosphere it is in the order of 2 to 2.5 m_p. Then the ratio of $a_{\rm r.c.}$ to the surface gravity due to mass, $g_{\rm grav} = (GM)/R^2$, is

$$\frac{a_{\rm r.c.}}{g_{\rm grav}} = \frac{\sigma}{4\pi \, 2.5 \, m_p \, c \, G} \frac{L}{M} \simeq 2 \cdot 10^{-5} \frac{L}{M} \qquad (13.3)$$

with L and M in solar units in the second term. The atmosphere is stable if $a_{\rm r.c.} \leq g_{\rm grav}$. The limiting case is the so-called Eddington limit

$$L_{\rm Edd} = 5 \cdot 10^4 \, M \, . \qquad (13.4)$$

If the stellar luminosity is above this limit (L and M in solar units) the radiation pressure in the continuum leads to very heavy mass loss and thus to expanding envelopes.

13.3.1.2 Radiative acceleration through spectral lines

Momentum transfer through line absorption by excitable atoms brings an acceleration of

$$a_{\rm r.l.} = \frac{\pi e^2}{m_e c} f \, \frac{F_\nu}{c} \frac{1}{m} \qquad (13.5)$$

with f the oscillator strength (see Ch. 3.1.2) of the transition, m_e the electron mass, F_ν the radiative flux at the spectral line frequency and m the mass of the ion doing the absorption. To find the

13.3. EXPANDING ENVELOPES, LUMINOUS WINDS

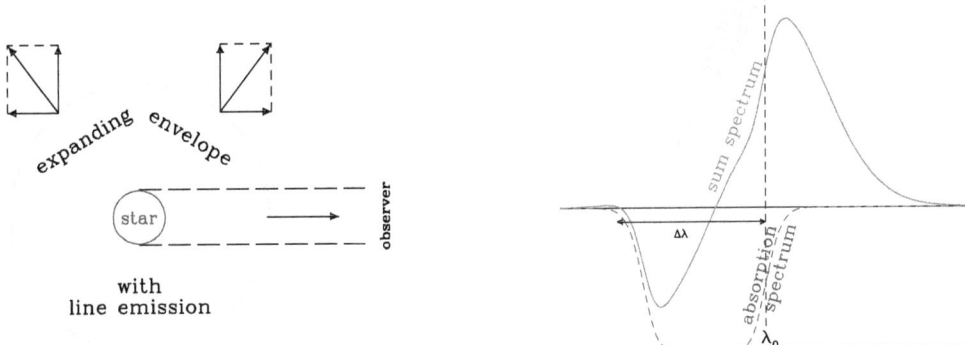

Figure 13.7: Geometry of an expanding shell leading to the formation of a P-Cyg spectral line profile. **Left**: Gas in the expanding shell around a luminous star will show line emission with a profile shape set by the velocity profile (see the decomposition of the outer velocity vectors). **Right**: The emission by the entire expanding shell (see envelope emission) and the light of the star with superimposed absorption by shell gas (see absorption spectrum) is recorded. The sum of these spectra is the P-Cygni profile, the full drawn line (compare with Fig. 13.1). With weaker envelope emission the absorption part of the P-Cyg profile will be better visible. The maximum wind speed, the 'terminal velocity' (v_∞, see Eq. 13.11 and Fig. 13.8) is derived from $\Delta\lambda$.

effective momentum transfer, one has to calculate the population of all excitable levels (see Ch. 3.2) as well as the frequency dependent optical depth of the transition (Ch. 3.1.2).

13.3.2 Making a P-Cyg profile

Stars with expanding envelopes develop the so called (see Sect. 13.2.4) P-Cygni profiles.

Imagine a luminous star with an expanding envelope (Fig. 13.7 at left). The expanding envelope has a certain velocity profile, in general larger velocities at larger stellar distance (see Fig. 13.8). Line emission observed from the expanding shell thus has a certain profile shape.

The star produces continuum light on which line absorption by the shell gas is superimposed, with a shape according to the velocity profile (Fig. 13.7 at right). Emission and absorption combine, leading to the P-Cyg profile. Note that the ratio of emission and absorption strength is set by the proportionality of stellar brightness and emission line brightness, the latter being a function of the mass and size of the entire envelope as well as the abundance and excitation level of the absorbing species. Since the excitation is based in the stellar flux, there is an intricate interplay between all these factors in the process of P-Cyg profile formation.

13.3.3 Mass loss

Suppose we have a symmetrically expanding envelope (see Fig. 13.7) with constant mass loss. At distance r from the stellar centre ($r > R$) the wind may have density $\rho(r)$ and velocity $v(r)$. At the surface $4\pi r^2$ a mass of $\rho(r)v(r)$ streams per unit of time through a unit of area. Summing up over the whole surface we obtain the mass loss rate ("continuity equation")

$$\dot{M} = 4\pi r^2 \rho(r) v(r) \quad . \tag{13.6}$$

The assumption of symmetric and constant mass loss is apparently not always fulfilled. In binaries the envelopes are clearly asymmetric. Narrow emission and absorption structures of variable strength can be observed on small time scales along with the broad emission lines; these point to the existence of blobs of material in the wind. Also stellar rotation affects the strength and structure of the wind (see Ch. 14.6). This complicates considerably the derivation of the velocity profile from observations as well as of the density profile of the expanding envelopes.

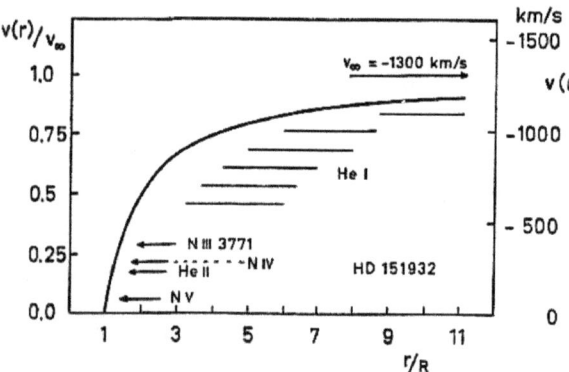

Figure 13.8: Velocity-radius relation as derived from several P-Cyg profiles in the spectrum of the WN7 WR star HD 151932. Observed spectral lines are named, each providing information out to a certain optical depth in the WR wind. The derived v_∞ (see Eq. 13.11) is indicated. Figure from Seggewiss (1977).

13.3.3.1 Velocity profile

If the total outward acceleration $\vec{a}_{r.c.} + \vec{a}_{r.l.} = \vec{g}_{rad}$ in such stars is larger than the gravitational pull g_{grav}, i.e.,

$$\Gamma = \frac{|\vec{a}_{r.c.} + \vec{a}_{r.l.}|}{|\vec{g}_{grav}|} > 1 \quad (13.7)$$

(implying proper use of the directional signs), radiation will drive gas away from the star. Assuming Γ is constant (it most likely is since F and g_{grav} both behave as $1/r^2$; but it is not constant near the surface of rotating stars, see Ch. 14.2.2) then the equation of motion for distances r $(r > R)$ is

$$v\frac{dv}{dr} = (\Gamma - 1)\, g_{grav}\, \frac{R^2}{r^2}\ . \quad (13.8)$$

After integration and with $v(R) = 0\ \mathrm{km\,s^{-1}}$ one finds

$$v(r) = \left[2(\Gamma - 1)\, g_{grav}\, R\left(1 - \frac{R}{r}\right)\right]^{1/2} \quad (13.9)$$

so that for very large r

$$v_\infty = 2(\Gamma - 1)\, g_{grav}\, R = \sqrt{2R\,|(\vec{a}_{r.c.} + \vec{a}_{r.l.} - \vec{g}_{grav})|} \quad (13.10)$$

with v_∞ the maximum or 'terminal velocity' (Ch. 4.5). Thus the distance-velocity relation found is

$$v(r) = v_\infty\, \sqrt{1 - \frac{R}{r}}\ . \quad (13.11)$$

Comparing this result with Eq. 4.82 it means that there the exponent $\beta = 1/2$.

An example of such a velocity relation, the result of the analysis of observations of a Wolf-Rayet star, is given in Fig. 13.8.

13.3.3.2 Density profile

The run of density with radius $\rho(r)$ is best obtained from measurements in the radio continuum. That radiation is free-free radiation, the scattering of radiation on free electrons or ions. One thus must consider radiation transport (following, e.g., Wright & Barlow 1975)

$$I(\nu, T) = \int_0^{\tau_{max}(q)} B(\nu, T)\, e^{-\tau}\, d\tau \quad (13.12)$$

with $\tau_{max}(q)$ the total optical depth for a shell with projected radius q. Assuming the simplified case of optically thick emission, spherical geometry, complete ionization of the envelope, constant electron temperature, T_e, constant velocity flow, $v_{\exp}$, and an electron density distribution varying as r^{-2} above the stellar atmosphere, the radio flux is

$$S_\nu = 5.12 \, \nu_{10}^{0.6} \, T_e^{0.1} \left(\frac{\dot{M}}{\mu/1.2 \, v_{\exp}} \right)^{4/3} Z^{2/3} \, d^{-2} \quad \text{mJy} \qquad (13.13)$$

with ν_{10} the frequency in units of 10 GHz, T_e in 10^4 K, μ the mean atomic weight per electron, $v_{\exp}$ in 10^3 km s^{-1}, Z the average ionic charge, d the distance of the source in kpc, and finally $\dot{M}$ in units of 10^{-5} M$_\odot$ yr^{-1}. Note that $\tau_\nu \simeq 1$ occurs for different ν at different radii. But, as it turns out, the description seems to be a realistic approximation.

13.4 Evolution and the HRD

The evolution of high mass stars has been modelled numerous times. Initally simple models were calculated. The discovery of 'new' processes, such as convective overshoot (Chs. 4.3.5 and 16), mass loss (Sect. 13.3), and effects of stellar rotation (Ch. 14.8), as well as improved data (new opacity tables, better fusion reaction networks), made the models ever better. In particular, 'free' parameters in models were adjusted to match existing data like CMDs of star clusters. Thus theoretical work has its own fast 'evolution' and more recent diagrams may not be quite consistent with older diagrams. Note therefore that *all mass limits given below are approximate only*.

Although massive stars evolve relatively fast, evolution changes have been observable only in very few stars, one being P Cyg (Sect. 13.4.4).

13.4.1 General nature of evolution of high mass stars

The evolution of the central temperature and density is given in Fig. 13.10. Close inspection shows that stars with $M_{\text{init}} > 8$ M$_\odot$ will reach central C burning (lower mass stars do not) and that stars with $M_{\text{init}} > 15$ M$_\odot$ have a very smooth development of the central parameters T_c and ρ_c.

Fusion proceeds smoothly from H to He burning. At ever higher temperatures and higher densities, fusion processes beyond He burning become possible like C, Ne, O, and Si burning (Chs. 5.2.1 and 5.2.3). The fusion enriches the interior material with heavy elements. The mean molecular weight in the core increases and the core contracts. These processes provide the star continuously with energy balancing the gravity (Chs. 1.2 and 6.8.2.1). For larger and larger M_{init} the successive fusion stages in the core follow each other with ever shorter time intervals. This shortness is due to the ever more limited supply of fusion material (it has to be produced in previous fusion stages anyway) and the fact that each next fusion step generates less energy per reaction (smaller mass deficit, Fig. 5.1) while requiring larger T. Moreover, neutrinos produced provide efficient core cooling and winds have driven most of the H+He envelope away meaning a smaller envelope opacity, so both of these aspects make the star more leaky (Ch. 6.8.2.2).

The H-, He- and C-burning times are given in Table 13.4 and Fig. 9.6. Note that due to very high T_c and inner convection, mixing may reach out to considerable radii or mass shells (Fig. 6.6). At all times the star tries, of course, to maintain (or return to) thermal equilibrium (Ch. 9.4).

The evolution of the surface parameters of high mass stars is shown as evolutionary tracks in the HRD of Fig. 13.9. Evolution in the MS phase and out to the red giant phase proceeds as described in Ch. 9. Evolution is at first fairly similar to that of lower mass stars, but after the MS phase we see mass-dependent differences in the tracks (compare Fig. 13.9 with Fig. 10.1).

• Stars with $8 < M < 14$ M$_\odot$ perform blue loops (Chs. 10.3 & 10.4.4, Fig. 10.1) after the RG stage. Depending on various model parameters other mass limits may apply (see Sect. 13.4.1.2). Blue loops cross the Cepheid instability strip (Ch. 11.3.1); there the star will pulsate.

In the later evolutionary phases the core proceeds from fusion step to next fusion step. The spectral evolution of these stars runs like

$$\text{O} \rightarrow \text{B} \rightarrow \text{F I} \rightarrow \text{Cepheid} \rightarrow \text{G I} \rightarrow \text{Cepheid} \rightarrow \text{A I} \rightarrow \text{Cepheid} \rightarrow \text{G I} \rightarrow \text{SN}.$$

• Stars with $15 < M < 40$ M$_\odot$ become red supergiants (Fig. 13.9) and likely stay red supergiants (for the possible blue loops see Sect. 13.4.1.2). They loose mass through a wind but perhaps not

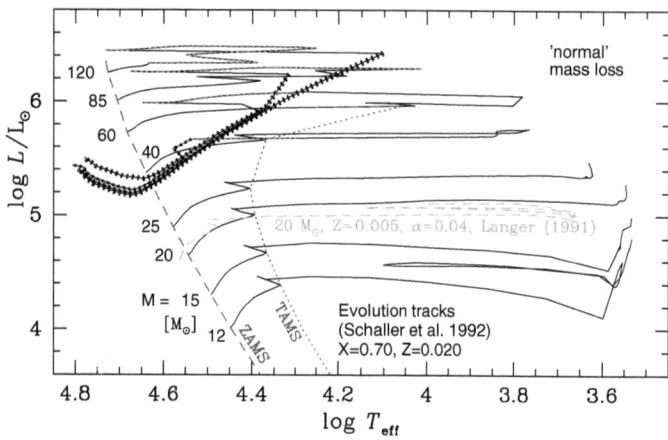

Figure 13.9: Evolutionary tracks for $12 < M_{\text{init}} < 120$ M$_\odot$ stars with $Z = 0.020$ up to the end of C burning (models from Schaller et al. 1992) which include mass loss and convective overshoot. The mass loss is 'normal' and taken from Nieuwenhuizen & de Jager (1990), for the WR stages $\dot{M}$ is taken from Langer (1989).
In these models, stars starting with $M > 14$ M$_\odot$ do *not* evolve with a blue loop, stars below that mass do (but see Sect. 13.4.1.2). However, the dashed line gives the model for a 20 M$_\odot$ star with different composition and convection parameters, which does evolve to have a blue loop (from Langer 1991).

In very high mass stars H- and He-fusion takes place simultaneously in the core; the TAMS lies far to the right of the ZAMS and is ill defined. The most massive stars soon have heavy mass loss (the heavy and the dotted portions of tracks). Such mass loss phases lead to the spectroscopic Wolf-Rayet nature with P-Cyg profiles (Fig. 13.7). Note that the mass loss is such, that the tracks for stars of $M > 40$ M$_\odot$ nearly merge into one (dotted tracks toward lower L and larger T_{eff}).

The first WR stage is the "WNL stage" with N enhancement and still strong Lyman α emission (the red/heavy part of the tracks) where the stars again become blue (less hydrogen envelope thus less opacity). Due to the outer convection zones heavy elements reach the surface, increasing the opacity leading to stronger and stronger wind phenomena. The further WR stage toward the WC phase is called the "WNE stage" and the stars have an almost pure He atmosphere. In this second WR stage, the "WC stage", the stars lose lots of matter (with fusion products). This part of the evolution is marked by the dotted portions of the tracks, with big dots for the 120 M$_\odot$ track and ever smaller dots down to the 40 M$_\odot$ track.

With core C-burning, the stars strive toward the theoretical C-MS (of Fig. 6.8). Note the *absence of time tic-marks* in the diagram; time information can be read from various later figures in this chapter, from Table 13.4 as well as from Figs. 23.1 and 23.2.

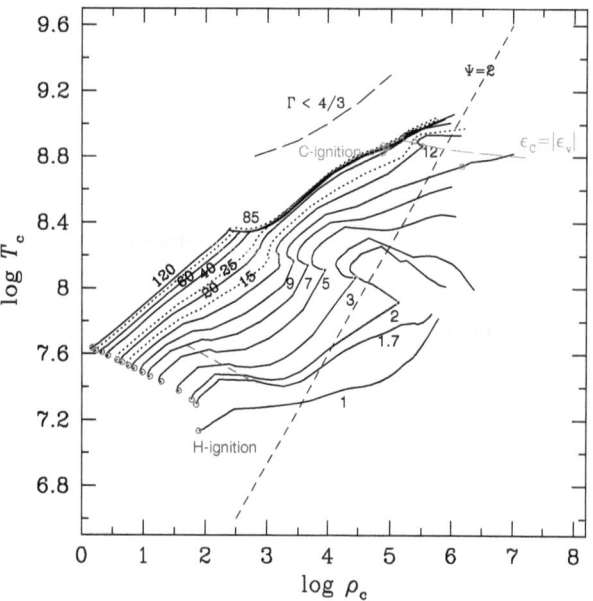

Figure 13.10: Evolution of T_c and ρ_c for stars with $1 < M < 120$ M$_\odot$. Some tracks are given as dotted lines. There are **no time-tick-marks** on the evolutionary tracks. $\Psi = 2$ indicates the level of degeneracy of the gas. The region with $\Gamma < 4/3$ is the region of pair instability (see Ch. 15). The line labelled ϵ_C signifies the limit above which energy loss by neutrinos (ϵ_ν) becomes more important than the energy produced by C burning.
Stars with $M_{\text{init}} > 8$ M$_\odot$ have no temperature drop at the end of central H exhaustion. The tracks of evolution of stars with $M > 40$ M$_\odot$ merge due to very heavy mass loss (WR stage) into the one of a 40 M$_\odot$ star (see also Fig. 13.9). Data from Schaller et al. (1992). For more on lower mass stars see Ch. 10 and Fig. 10.2.

13.4. EVOLUTION AND THE HRD

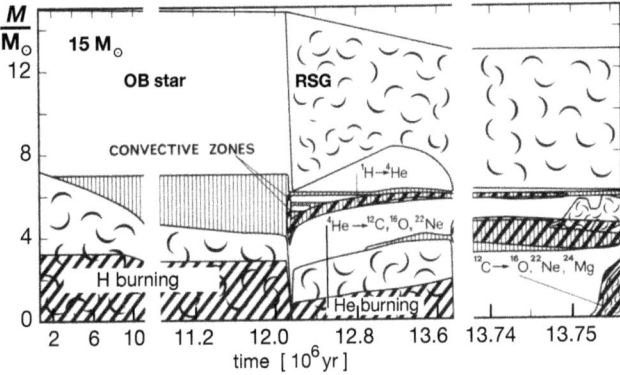
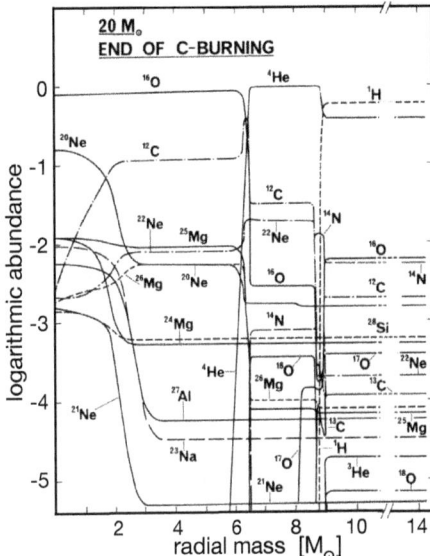

Figure 13.11: *Left*: Evolution of a $M_{\rm init} = 15$ M$_\odot$ star into the C-burning phase. Shown with time (note the scale gaps) are: the extent of the fusion zones (///, $\epsilon > 10^3$ erg g^{-1} s^{-1}) including the shell burning, the convective regions (curls), the He-enriched region (due to central convection) and the loss of mass. Figure adapted from Maeder & Meynet (1987).
Right: The internal chemical composition of a $M_{\rm init} = 20$ M$_\odot$ star is shown at the end of C-burning. Note that the mass shells are indicated from the core (at left) to the surface (at right), also showing that at this stage 6 M$_\odot$ has been lost. The jumps in chemical composition near $M \simeq 9$ M$_\odot$, near $M \simeq 6$ M$_\odot$ and near $M \simeq 2.5$ M$_\odot$ are due to the zones with H-, He- and C-burning, respectively. Figure from Maeder & Meynet (1989).

excessively (but from observations little is known about the actual mass loss rates of RG stars; see Reimers 2007) while acceleration by dust becomes important, too (e.g., van Loon 2000).

After core He ignition the inner convection zone persists leading to smooth chemical composition gradients (see Fig. 13.11 right panel). Note that the central convection during core H burning decreases in extent, the one during core He burning increases (Fig. 13.11 left panel). These stars ultimately reach the SN stage (SN Type II; Ch. 18).

- Stars with $M_{\rm init} > 40$ M$_\odot$ evolve into the red supergiant area of the HRD. High mass loss (due to a large luminosity) peels off the H rich layers more and more and uncovers the nuclear processed stellar interior (He, N, C, O, etc.). The stars gradually change into Wolf-Rayet stars and have no H absorption lines in their spectra (Sect. 13.2.3; all H is lost). The WR phases last about $0.3 \cdot 10^6$ yr (Schaerer et al. 1993). An ever hotter surface is exposed and the stars enter the hottest region of the HRD. For further details see the explanation of the 60 M$_\odot$ star evolution below.

- The most mass rich stars, with $M > 80$ M$_\odot$, never reach the red supergiant phase due to heavy mass loss (expanding envelopes, Sect. 13.3). They exist a while in the LBV regime (see also Sect. 13.2.4) of the HRD. The mass loss is so large, that they end up as stars with 40 M$_\odot$, irrespective of their $M_{\rm init}$, and soon become WR stars.

13.4.1.1 Evolution of stars of 15 – 25 M$_\odot$

The evolution of the central parameters and the surface parameters of stars in the 15 to 25 M$_\odot$ range can be read from Figs. 13.10 and 13.9 (but see Sect. 13.4.1.2).

Fig. 13.11 (left) shows the evolution of internal structure of a $M_{\rm init} = 15$ M$_\odot$ star, Fig. 13.11 (right) the internal composition of a $M_{\rm init} = 20$ M$_\odot$ star at the end of core C-burning.

Note the following. Winds driving out envelopes have strengths depending on metallicity (see Ch. 9.6.2). Also, the sequence of evolutionary steps in the interior is rapid, depends on many variables and may not be identical (a chaotic system) for stars having started with identical initial conditions. Thus, when the core reaches the conditions leading to a supernova, the envelope of the star may be of any type between blue (e.g., Sp.T. A I) and red supergiant!

13.4.1.2 When does a star evolve with a blue loop?

In Ch. 9.3 the question was addressed why a star becomes a red giant and why it perhaps contracts (first and second half of the hysteresis cycle, Ch. 9.3.1). These changes are determined by the interplay of the luminosity from all nuclear fusion offered to the base of the envelope, L_B, with the luminosity the envelope can accommodate before a run-away expansion (inflation) occurs or when the envelope deflates. The latter luminosity limit, following Renzini et al. (1992) called L_{loop}, is set by the envelope and its energy transport conditions. Deflation leads to a blueing of the star, which may occur when with further RG evolution L_B becomes smaller than L_{loop}.

Consider a red giant. After core He burning sets in, the overall luminosity of a star decreases somewhat due to core expansion and the ensuing outward push on the H burning shell thus the lowering of its T and L_H. If due to these changes L_B ($= L_H + L_{\text{He}}$) drops below L_{loop}, the envelope will deflate (point G in Figs. 9.2 to 9.5). In the course of further evolution L_H and L_{He} increase so that eventually L_B becomes so large that the envelope must expand (point K in Figs. 9.4 and 9.5) and the star becomes red giant a second time. In short, this evolution produced a blue loop.

Several factors influence L_{loop} and the core's L_H and L_{He} (see Renzini et al. 1992).
Core convective overshoot: a large overshoot (large $\alpha = l/H_P$) leads to a larger He core and so to larger L_{He}. Models with large overshoot (thus large L_{He}) may not reach the condition $L_B < L_{\text{loop}}$.
Envelope convective overshoot: the depth of the RG surface convection zone (first dredge-up) determines at which $m(r)$ a (modest) chemical discontinuty is established. When the H shell burning approaches this discontinuity, L_H becomes smaller, perhaps leading to $L_B < L_{\text{loop}}$ (if L_{He} is not too large). Thus depending on the location of this discontinuity the deflation of the envelope comes early or late in the shell H burning phase.
Chemical composition: the opacity of the envelope depends on Z so L_{loop} depends on Z. However, the level of the opacity may be strongly influenced by the recombination of He. Generally, L_{loop} depends on the leakyness of the envelope (Ch. 6.8.2.2). Also rotation plays a role (see Ch. 16).
Mass loss: if mass loss is large, the mass M_B from which the fusion energy comes is, compared to M, proportionally larger. For stars with larger M_B/M, generally L_B/L_{loop} is larger and so blue loops are suppressed. The effect is similar to that of core convective overshoot.

For solar metallicity stars the blue loop is present for $M \simeq 2$ to $\simeq 10$ $M_\odot$ (Chs. 10.2.3 & 10.4.4).

Stars with $M > 10$ $M_\odot$ *need not evolve with blue loop*. In models it depends on the level of core convective overshoot. For more metal poor stars (all depending on the other 'free' parameters) blue loops show up (see, e.g., Langer 1991), as given for a 20 $M_\odot$ star in Fig. 13.9. Using the observed luminosity of Cepheids (Ch. 11.3.1.1) one can find for which mass the blue loop occurs.

For high mass stars (> 40 $M_\odot$) the blue loop occurs when their surface is at high T_{eff} (Fig. 13.9).

13.4.1.3 Evolution of a 60 $M_\odot$ star

A typical example for high mass evolution is that of a 60 $M_\odot$ star, given in the Figs. 13.12 & 13.13.

From Fig. 13.12 (left) it is apparent that the central H burning lasts 3.7 Myr, but the core-He burning (^{4}He $\to$ ^{12}C, ^{16}O, ^{20}Ne) only 0.6 Myr (see also Fig. 9.6). The core-C burning phase (^{12}C $\to$ ^{16}O, ^{20}Ne, ^{24}Mg; which extends a bit beyond the right boundary of the figure) finishes after just $\simeq 5000$ yr (see Table 13.4). Strongs winds cause a steady mass loss pushing away $\simeq 40$ $M_\odot$.

After 3.7 Myr, at the *end of core H burning*, when the star becomes a red supergiant, He can be more abundant than H at the stellar surface (see Fig. 13.12) due to the mass loss and the He transported up by the deep inner convection zone already present in the early MS state. Just 0.2 Myr later the H-burning zone disappears because H is completely wind-expelled from the surface. The star appears as a WR star of type WN and N (enhanced during H burning by the CNO process; Eq. 5.10) as well as He dominate the surface composition. The star evolves into the hottest part of the HRD (see Fig. 13.9) thereby changing from 'late' WN type (WNL; strong N III emission – besides He I,II) to 'early' WN (WNE; strong N V emission). It has a structure akin to that of a He MS star (Fig. 6.7). Convection flushes C and other products of He burning to the surface (again Figs. 13.12 and 13.12), whereas N is easily destroyed by α capture (Ch. 5.2.2). Therefore, the WN type WR star changes into a WC type.

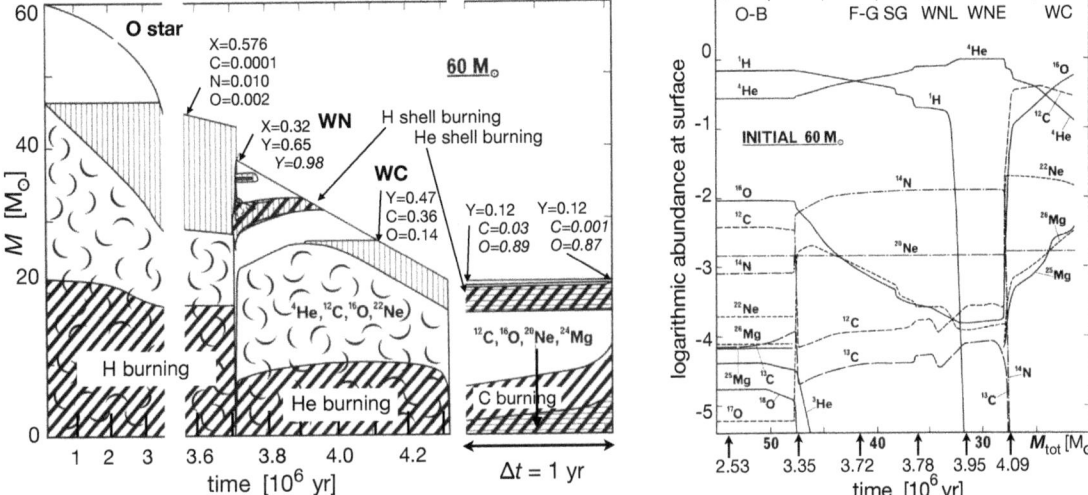

Figure 13.12: *Left*: Evolution of a 60 $M_\odot$ star up to central C exhaustion. The changes in its appearance (spectral type) are given. Figures adapted from Maeder & Meynet (1987).
Left: Internal structure. Fusion zones (core and shell) of H, He, and C (heavy diagonals) the convective regions (curly hatching) and zones with convection induced gradients in He and C content (vertical hatching) are indicated. There is dramatic mass loss (see upper envelope of total mass). Actual mass fractions of the most abundant elements (X = H, Y = He, C, N, O) are listed: upright values are for the surface (see also right panel), *slanted* for the centre.
Right: Evolution of the *chemical composition at the surface* of a M_{init} = 60 $M_\odot$ star (plotted against remaining mass M_{tot}) related to the evolutionary age given in the bottom labelling.

During C burning (Eq. 5.15), in Fig. 13.12 left panel at the far right, the star has a core like the carbon MS star of Fig. 6.8. The core is radiative because neutrinos contribute considerably to cooling making the temperature gradient more shallow. O becomes for a short period the most abundant element throughout the star (Fig. 13.12, right panel at the far right), before further rapid fusion processes lead to an unstable iron core ready for a SN explosion.

Due to the sequential nuclear fusion stages the interior of high mass stars will evolve to become more compact. At the same time, the stellar wind will carry away a large portion of the envelope. The result is that a 60 $M_\odot$ star will become an object of 20 $M_\odot$ (see Fig. 13.12) and ultimately is more compact than a MS star of 20 $M_\odot$ while being more luminous (see Fig. 13.9). Again, the basis for that is gravity and the increase in mean molecular weight μ (see Chs. 1.2 and 6.8.2.1).

The profile of the internal chemical composition at the end of C burning is shown in Fig. 13.13 (left). The drastic abundance changes in the shells near $\simeq 3\,M_\odot$ and $\simeq 18\,M_\odot$ are due to the inner C-burning and the outer He-burning zones, respectively. Note again that the H-burning zone has disappeared due to wind mass loss. In fact, this model results in a WC star (see Fig. 13.12).

It should be mentioned that nuclear reaction rates (notably that of C) are still uncertain mainly due to uncertainties in the fusion cross sections. Thus, the abundance profiles in Figs. 13.12 and 13.13 may change with future better understanding.

Finally, attention is drawn to the distribution of the physical parameters temperature, density, and energy production (Fig. 13.13 right). The energy of the star is produced by the He-burning shell and by C burning in the core. Contributions by gravitational energy, ε_g, and by neutrino loss, ε_ν, are indicated. The energy unhampered carried away by neutrinos, ε_ν, prevents the formation of a convective core during C burning.

The phases of the evolution of a 60 $M_\odot$ star can be summarized by the (spectral) sequence:
$$O \to B \to F\,I\,,\,G\,I \to WN \to WC \to SN\,.$$

After all the efforts of a 60 $M_\odot$ star, little remains. With the strong stellar winds about 40 $M_\odot$

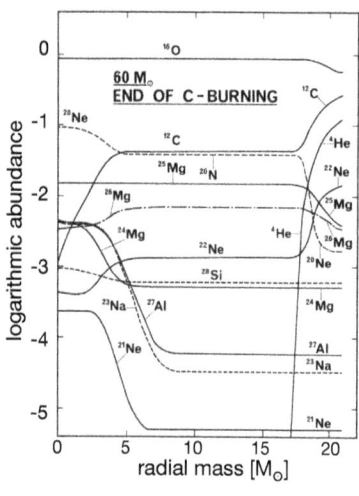
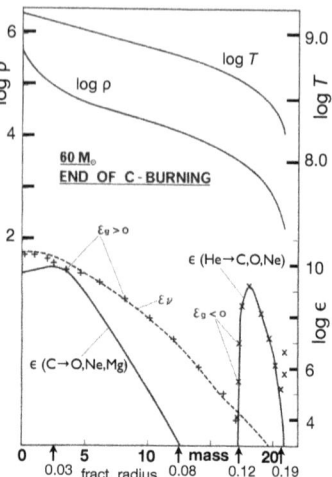

Figure 13.13: The *internal structure and composition* of a $M_{\text{init}} = 60$ M$_\odot$ star *at the end* of C burning in units of radial mass. The mass is indicated from the core (left edges) in the direction of the surface. At this stage $M_{\text{tot}} \simeq 22$ M$_\odot$) so the star has lost $\simeq 38$ M$_\odot$. The figures are adapted from Maeder & Meynet (1987).
Left: Chemical composition. Note the isotopes present.
Right: Physical parameters. Temperature T and density ρ are given in the upper part of the diagram, the energy productions ε of the He-burning shell (full line at right) and the C-burning core (full line at left) are given in the lower part. The actual fractional radius is indicated in the right panel along with $m(r)$. Contributions by gravitational energy, ε_g, and by neutrino loss, ε_ν, are indicated. The core region is contracting ($\varepsilon_g > 0$) while the He-burning shell is expanding ($\varepsilon_g < 0$). Note that in the central zone the energy lost due to neutrino cooling is larger than the enegry generated by C burning.

is carried away. The ultimate fate of such a star is a supernova, throwing $\simeq 19$ M$_\odot$ back into the ISM, and leaving a neutron star. A **60 M$_\odot$** star gives thus in total $\simeq 59$ M$_\odot$ back to the ISM. This material is in part considerably enriched in heavy elements due to the deep convection zones in the various evolutionary stages while the SN ejecta contain further fusion products.

The duration of the H-, He- and C-burning phases are given in Table 13.4 (H and He are shown in Fig. 9.6). The duration of the phases does not vary much with mass (for $M_{\text{init}} > 40$ M$_\odot$ rather $L \simeq M^2$), while these stars lose so much mass that they become like a 40 M$_\odot$ star.

13.4.2 Evolution and effects of metallicity

The details of stellar evolution in the high-mass range depend also on metal content.

The location of the MS is further to the blue in metal-poor stars (for examples see Ch. 16.1.2), but not as much as with stars in the lower mass range (as the metal-poor subdwarf stars; Ch. 10.8.3).

The initial chemical composition influences the build-up of the He core. E.g., with low abundance of CNO there will be only a slow CNO cycle. A smaller amount of metals means a somewhat lower total opacity, governing the colour of the star at any point in time. Furthermore, opacity drives the expansion of the envelope when the star changes from MS star into red giant (see Ch. 9.6.2). In stars with lower metallicity, the mass loss is less strong and the evolution of the highest mass stars stays therefore more to the red (with $T_{\text{eff}} < 2 \cdot 10^4$ K), thus without the decrease in luminosity to high T_{eff} for the C-burning stars (as the dotted track in Fig. 13.9). On the other hand, less metals makes a star more leaky and thus its T_c higher.

Looking at various evolutionary tracks one sees that low metallicity stars have
1) more pronounced blue loops,
2) not so red giant branches, but no very blue loops evolving to the left of the MS,
3) the possibility of a supernova exploding from a *blue* supergiant.

Tracks for various metallicities (Solar, LMC, SMC and even less) can be found in Arnett (1991). These metallicity possibilities were probed by Arnett to explain the blueness of the progenitor star of SN 1987A. Note that metallicity also interacts with convection (Sect. 13.4.1.2).

Other examples of effects of metallicity are given in Fig. 16.3.

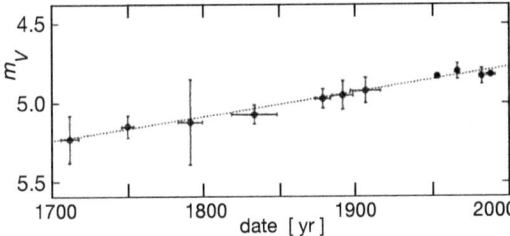

Figure 13.14: The evolution in brightness of P Cyg is shown over a period of 300 years. The dotted line gives a linear fit. P Cyg evolves apparently toward larger size R and lower surface temperature T_{eff}. Figure from Lamers & de Groot (1992).

13.4.3 Evolution and effects of rotation

Rotation may have pronounced effects on the structure of stars. Most prominent are the effects of mixing and of magnetic fields. Effects of rotation are addressed in Ch. 14.

13.4.4 See a star evolve: P Cygni

The star P Cyg is the prototype star for wind-mass-loss spectral profiles and a typical S Dor star. The Dutch cartographer Willem Janszoon Blaeu discovered in 1600 it is a variable star. The star has been extensively studied. It has an apparent magnitude of $V \simeq 4.8$ mag and is member of an open cluster at the distance of 1.8 ± 0.1 kpc. Using that information, the star's characteristics are $T_{\text{eff}} = 19300 \pm 700$ K, $\log(L/L_\odot) = 5.86 \pm 0.10$, $R = 76 \pm 8$ $R_\odot$.

With these values one can place P Cyg in the HRD (e.g., Figs. 13.3 and 13.6). The location of P Cyg has evolutionary tracks crossing the Hertzsprung gap in both directions (compare with Fig. 13.9), to the blue and to the red. Therefore the question is, which track does apply, in which direction does P Cyg evolve now? And, what is the mass and current evolutionary state of P Cyg?

Over the years it appears P Cyg has become slowly brighter *in the visual*, from $V = 5.23$ mag in 1712, through $V = 4.97$ in 1878, to $V = 4.83$ mag in 1953 and staying approximately constant into 1988 (in part mean values). After excluding effects of circumstellar extinction Lamers & de Groot (1992) conclude that:
→ Because stars of this kind evolve with *constant* luminosity L, thus on a horizontal track in the HRD, the brightening in the visual can only be due to a slow decrease of the surface temperature; at constant L the flux maximum shifts (horizontally in Fig. 1.1) to lower temperatures! The change in T_{eff} amounts to -6 ± 1 % per century.
→ P Cyg evolves in the HRD towards the red. This defines the evolutionary track: P Cyg has started with about $48\,M_\odot$. It lost about 12 - 15 $M_\odot$ due to strong stellar winds. Now its core contracts and the outer layers expand. P Cyg is evolving to become a red giant.

P Cyg is (with 'born again stars', Fig. 10.18, and Polaris, see Fig. 11.6) one of the very few stars in which evolutionary changes have been observed.

13.5 Nuclear fusion times and endphases of high mass stars

High mass stars live only short lives. After the core H-burning phase the remaining phases last only about 1/10 of the H-burning phase (see Table 13.4). Stars which had $M_{\text{init}} > 30\,M_\odot$ exhibit the wind-WR phenomenon most of that time. All the high mass stars end as supernova (see Ch. 18).

Depending in M_{init} as well as on the mass lost the last stages of evolution may differ considerably. What matters is which values of T_c and ρ_c are reached which determine how the stellar core will behave. Evolution in the core may be that quick that the stellar surface cannot follow all these changes. So massive stars with a large range in surface characteristics may become supernovae.

References

Abbott, D.C., & Conti P.S. 1987, ARAA 25, 113; *Wolf-Rayet Stars*
Arnett, D. 1991, ApJ 383, 295

Table 13.4: Lifetimes of H-, He-, and C-burning in stars with $Z = 0.008$ (Schaerer et al. 1993)

M_{init} $M_\odot$	H-burning	He-burning	C-burning
	time given in units of 10^6 yr		
120	2.78	0.31	0.0053
85	3.12	0.34	0.0055
60	3.70	0.39	0.0053
40	4.81	0.43	0.0051
25	7.17	0.64	0.0095
20	9.11	0.81	0.0141
15	12.94	1.12	0.0270
10	23.87	2.25	0.8304

Cassinelli, J. 1979, ARAA 17, 275; *Stellar Winds*
Davidson, K., & Humphreys, R.M. 1997, ARAA 35, 1; *Eta Carinae and its Environment*
Dupree, A.K. 1986, ARAA 24, 377; *Mass Loss from Cool Stars*
Figer, D.F., Najarro, F., Morris, N., McClean, I.S., Geballe, T.R., Ghez, A.M., & Langer, N. 1998, ApJ 506, 384
García-Segura, G., Langer, N., & Mac Low, M.-M. 1996, A&A 316, 133
Herrero, A., Puls, J., & Villamariz, M.R. 2000, A&A 354, 193
Kudritzki, R.-P., & Puls, J. 2000, ARAA 38, 613; *Winds from Hot Stars*
Lamers, H.J.G.L.M., & de Groot, M.J.H. 1992, A&A 257, 153
Langer, N. 1989, A&A 220, 35
Langer, N. 1991, A&A 252, 669
Maeder, A., & Meynet, G. 1987, A&A 182, 243
Maeder, A., & Meynet, G. 1989, A&A 210, 155
Martins, F. Schaerer, D., & Hillier, D.J. 2002, A&A 382, 999
Massey, P., & Meyer, M.R. 2002, *Stellar Masses*, www.lowell.edu/users/massey/recentpubs.html
Nieuwenhuijzen, H., & de Jager, C. 1990, A&A 231, 134
Reimers, D. 2007, in "The Milky Way Halo", K.S. de Boer & P. Kroupa (eds.);
    http://www.astro.uni-bonn.de/~mwhalo/proceedings
Renzini, A., Greggio, L., Ritossa, C., & Ferrario, L. 1992, ApJ 400, 280
Schaerer, D., Meynet, G., Maeder, A., & Schaller, G. 1993, A&AS 98, 523
Schaller, G., Schaerer, D., Meynet, G., & Maeder, A. 1992, A&AS 96, 269
Seggewiss, W. 1977, in IAU Coll. 42, "The Interaction of Variable Stars with their Environment",
    R. Kippenhahn, J. Rahe, & W. Strohmeier (eds.); Publ. Sternwarte Bamberg; p. 633
van der Hucht, K.A. 1993, A&A Rev 4, 123; *Wolf-Rayet Stars*
van der Hucht, K.A. 2001, New Astron. Rev. 45, 135
van Loon, J.T. 2000, A&A 354, 125
Walborn, N.R., & Nichols-Bohlin, J. 1987, PASP 99, 40
Walborn, N.R., Howart, I.D., Lennon, D.J., Massey, P., Oey, M.S., Moffat, A.F.J., Skalkowski, G., Morrell, N.I., Drissen, L., & Parker, J.W. 2002, AJ 123, 2754
Weis, K. 2001, Rev. Modern Astronomy 14, 261
Weis, K., & Duschl, W.J. 2002, A&A 393, 503
Willson, L.A. 2000, ARAA 38, 573; *Mass Loss from Cool Stars: Impact on the Evolution of Stars and Stellar Populations*
Wolf, C.J.E., & Rayet, G.A.P. 1867, Comptes Rendus 65, 292
Wright, A.E., & Barlow, M.J. 1975, MNRAS 170, 41

Chapter 14

Rotation and stellar evolution

14.1 General aspects of rotation

It has been known for long that stars rotate (see Ch. 3.7). Some rotate slowly (the Sun takes about 27 days for one revolution), others fast (many OB stars have a $v \sin i \simeq 300$ km s^{-1}), none rigid. Rotation causes a deformation of the spherical shape to a certain level of oblateness. This has an effect on the local surface brightness and thus on the local surface temperature of a star so that, depending on the viewing angle i, a star may look bluer or redder (see Fig. 3.14). However, since the mid 1990s it has become clear that the internal structure may be affected by rotation in a more profound way and thus that stellar evolution may be affected by rotation. An up-to-date compilation of the field can be found in Maeder & Eenens (2003).

Several parameters related with rotation come into play such as: rotation-deformation and its effects on the parameters of the gas at the surface (their actual ones and those derived from observations), rotation and mass loss, rotation inducing internal gas flows, rotation and effects of evolution on angular momentum, rotation and effects of magnetic fields. Some of these work all by themselves, most of them interact and thus make for more complicated analyses. In the following each of the principle workings will be presented while the interactive aspects will be addressed only in a general way.

The rotation frequency of a star is given as Ω. This normally is the rotation frequency at the equator of the star. Since stars do not rotate as solid bodies one uses the generalized expression $\Omega(\theta, r)$ with θ the latitude as seen from the star's centre and r the radial distance from the centre with the stellar radius given as $R_\star$.

14.2 Rotation and effects of deformation

14.2.1 Rotation and variation in T_{eff}

Let us first consider only rotation induced effects on the geometry of a stellar surface with the consequences for T_{eff}. The surface will become somewhat oblate and this leads to latitude dependent differences in the flux passing through the surface.

Assume that for oblate stars the latitude dependence of radius results at the poles in

$$R_{\text{pole}} = q \cdot R_{\text{equator}} \quad \text{with } q < 1 \ . \tag{14.1}$$

As seen from the core, each solid angle crossing a surface element has $L = 4\pi R^2 \sigma T_{\text{eff}}^4$ so that with conservation of flux per solid angle an element with a radius different by a factor q must have a T_{eff} different by $1/\sqrt{q}$. Thus the poles of a star have a higher T_{eff} than the equator. For radiative stars this must lead to differences in colour (see Fig. 3.14) as well as in the spectrum (and in limb darkening). In general,

$$T_{\text{eff}}^4(\theta) \simeq T_{\text{eff}}^4(\text{pole}) \cdot (1 - \omega^2 \cos(\theta)) \tag{14.2}$$

with $\omega = v_{\text{rot equator}}/v_{\text{crit}}$ in which $v_{\text{crit}} = \sqrt{GM/R_{\text{equator}}}$, factors accounting for the rotation induced deformation of the stellar surface.

Furthermore, spectral lines are generally rotation broadened (see Fig. 3.13). Lines in the spectrum of a star taken pole-on will *not* be broadened by rotation while a spectrum of that star seen on the equator will show the maximum possible rotational broadening of its spectral lines.

14.2.2 Rotation and effective gravity

If a star (a gaseous body) rotates, the surface with equal gravitational potential inside a star will no longer coincide with the surfaces of equal pressure (isobaric surface) and equal temperature (isothermal surface). This is so, because rotation is a cylindrical while gravity is a radial phenomenon. Rotation leads to a cylindrical gravity force, $g_{\text{rot}} = g_{\text{rot}}(s)$, with s the distance from the rotation axis, while mass leads to a radial gravity force, $g_{\text{grav}} = g_{\text{grav}}(r)$. Radiation adds a factor g_{rad} (the sum of continuum- and line-driven acceleration, $a_{\text{r.c.}}$ and $a_{\text{r.l}}$; see Ch. 13.3), which is perpendicular to surfaces of equal T (equivalent to iso-opacity surfaces). This gravity factor depends on the local spectral energy distribution and thus also depends on chemical composition and ionization. These vectors point in different directions as illustrated in Fig. 14.1.

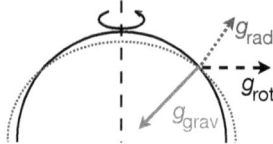

Figure 14.1: Schematics of the vectors making up g_{tot} of a rotating star. The surface of a rotating star becomes spheroidal ($\cdots$). The approximate directions of the vectors based on gravity (mass; ——), radiation (optical depth surface, $\cdots$) and rotation (- - - -) are sketched.

Transforming $g_{\text{rot}}(s)$ into $g_{\text{rot}}(r,\theta)$ and including $g_{\text{rad}} = g_{\text{rad}}(T,\tau)$ one gets

$$\vec{g}_{\text{tot}} = \vec{g}_{\text{grav}}(r) + \vec{g}_{\text{rot}}(r,\theta) + \vec{g}_{\text{rad}}(T,\tau) \quad . \tag{14.3}$$

One can easily visualize that the equipotential surfaces of g_{tot} and P will differ. This leads to different gradients of energy transport. For a full description of effects of rotation and the workings of the various potentials see, e.g., Kippenhahn & Weigert (1991; their Ch. IX).

14.3 Possible effects of rotation on structure

Various effects of rotation may become active inside a star with rotation.

14.3.1 Rotation and meridional circulation

A simple consequence of Eq. 14.3 is that an upward flow must occur in the equatorial zone so that a downward flow is induced in the polar region (a simplfied version compared to the circulation in the earth's atmosphere). The ultimate result is that meridional circulation is established (see Fig. 14.2).

Meridional circulation will induce vertical mixing. However, the strength of the vertical flow depends also on the buoyancy force (see Fig. 4.1). If the radial structure of the rotating star is a very stable one in relation with the temperature gradient, then the meridional circulation may not reach very deep. Furthermore, if there is a density gradient between shells (e.g., because a deeper layer has larger μ due to fusion products) also this gradient will hamper establishing meridional circulation over large radial ranges. Thus, the radial depth of the meridional circulation depends on the material parameters in the respective layers. It may mean that the meridional flow will result in predominantly horizontal flows as in Fig. 14.2. When then $\Omega(r,\theta) \simeq \Omega(r)$ this is called "shellular" (Zahn 1992) circulation.

Summarizing, rotation will, in general, cause an upward flow in the equatorial region and a downward flow near the poles. Along the surface there is a flow from the equator to the poles.

If meridional circulation is established, the gas will be subjected to the Coriolis force and a directional deviation from the original motion will be imposed.

14.3. POSSIBLE EFFECTS OF ROTATION ON STRUCTURE

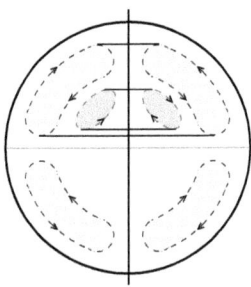

Figure 14.2: Rotating stars develop meridional circulation. With rotation, surfaces of equal gravitational potential do not coincide any more with surfaces of equal pressure. This leads to radially upward flows at the equator and consequently downward flows near the poles, creating a torus with meridional curculation. The radial extent of the meridional circulatory cells depends on the internal thermal and density (chemical composition) structure. When $\Omega(r,\theta) \simeq \Omega(r)$ this circulation has been named "shellular rotation" since it takes place whithin a shell. In high-mass stars counter-rotating cells are established in the interior.

14.3.2 Rotation driven instabilities

Several rotation driven instabilities are of relevance for stellar structure (see Kippenhahn & Weigert 1991, Maeder 1995, Heger et al. 2000, Talon 2003).

14.3.2.1 Brunt-Väisälä oscillations

If a parcel of gas is displaced in a stable shell (also in stars without rotation; see Ch. 11.1.2), and if the displacement is adiabatic, then the volume experiences a restoring force $-g\,(\rho_{\text{parcel}} - \rho_{\text{local}})$ which may lead to oscillatory motions (see Ch. 4.3.2) with frequency

$$N^2 = N_T^2 + N_\mu^2 = \frac{g}{H_P}\left((\nabla_{\text{ad}} - \nabla) + \nabla_\mu\right) \quad (14.4)$$

in which N^2 is the so-called Brunt-Väisälä frequency based on a thermal (N_T^2) and a molecular weight (N_μ^2) contribution (see Eqs. 4.56 and 11.4).

14.3.2.2 Solberg-Høiland instability

In a cylindrical rotating gas (neglecting gravitation) pressure gradients will balance centrifugal acceleration so that

$$-\frac{1}{\rho}\frac{dP}{ds} + s\Omega^2 = 0 \quad . \quad (14.5)$$

If the radius s of a torus gets enlarged to some distance $s + \xi$ then, with conservation of angular momentum, this leads to a restoring force (Solberg-Høiland). This may cause oscillatory motions with the Rayleigh (or epicyclic) frequency

$$N_\Omega^2 = \frac{1}{s^3}\frac{d}{ds}(s^2\Omega)^2 \quad . \quad (14.6)$$

If gravity is taken into account the stability condition becomes

$$N^2 + N_\Omega^2 \geq 0 \quad . \quad (14.7)$$

14.3.2.3 Baroclinic instability

In a rotating gaseous sphere the surfaces of constant pressure, constant density, constant entropy and constant gravity do not coincide. This is called a **baroclinic state**. These conditions have as a consequence, that displaced parcels of gas will be subjected to restoring forces depending on the direction into which they are displaced.

Normally the thermal diffusivity K_T is larger than the viscosity ν (where ν essentially controls angular momentum diffusion), so that the stabilizing effect of the density stratification is reduced. This may lead to the so-called GSF-instability (named after Goldreich, Schubert and Fricke).

Stars do not rotate cylindrically and they are not homogeneous. So gradients in μ contribute to stratification. This leads to a further refinement and the Solberg-Høiland criterion becomes

$$\frac{\nu}{K_T}(N_T^2 + N_\mu^2 + N_\Omega^2) \geq 0 \quad . \quad (14.8)$$

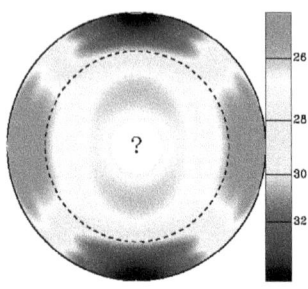

Figure 14.3: Rotation of the solar envelope as derived from SoHO data (helioseismology, see Ch. 11.5). The shading represents a rotational period from 24 to 35 days. The Sun does not rotate as a solid body. At the surface, the slower velocity at the poles is due to a flow from the equator to the poles with a deviation due to the Coriolis force. Note that it is possible to derive the internal rotation periods too, except for the core region. Figure from http://sohowww.nascom.nasa.gov; see Thompson et al. (2003).

14.3.2.4 Shear instability

The minimum energy state of a rotating fluid is solid body rotation. A differentially rotating star, in which layers have some horizontal velocity, u, can liberate energy if the velocities become homogeneous. For that a density stratification must be overcome which, in the adiabatic case, leads to the stability criterion with the "Richardson" number Ri (see Maeder 1995) based on N^2 as in Eq. 14.4,

$$Ri = \frac{N^2}{(du/dz)^2} \quad , \tag{14.9}$$

in which du/dz is the so-called **shear**. If $Ri > 1/4 = Ri_{\text{crit}}$ then the gas is shear unstable.

• **Dynamical shear instability:** There is no restoring force along isobars and thus shear is unstable where there is horizontal differential rotation. Thus horizontal shear leads to a large turbulent viscosity.

• **Secular shear instability:** Both the thermal and the molecular weight stratifications will hinder the growth of instabilities. But thermal diffusion and horizontal shear will diminish those effects leading to a stability criterion

$$\frac{\Gamma}{\Gamma+1} N_T^2 + \frac{\Gamma_\mu}{\Gamma_\mu+1} N_\mu^2 < Ri_{\text{crit}} \left(\frac{du}{dz}\right)^2 \tag{14.10}$$

where $\Gamma = vl/K_T$ and $\Gamma_\mu = vl/K_\mu$ (Talon & Zahn 1997) with v the flow velocity and l a length parameter.

14.3.3 Rotation of the Sun

Measurements of the Sun showed that it does not rotate as a solid body. Instead, depending on latitude, rotation is faster or slower than the average. Rotation of the surface can easily be derived using the doppler effect in spectra of local regions. The equatorial zone has a rotation period of about 24 days, the polar region only of 35 days (see Fig. 14.3).

Using helioseismology (Ch. 11.5) it became possible to derive how internal layers of the Sun rotate as well as information about its internal structure (see Fig. 11.14). For a review see Thompson et al. (2003), in which the frequency splitting of oscillation modes (as visible in Fig. 11.12) as well as many general aspects due to rotation are explained. Further into the interior there is shellular (Zahn 1992) rotation (see Fig. 14.3) down to the bottom of the convection zone.

14.3.4 Convective flows will be turbulent

In a rotating star each element of gas has some angular momentum. If a pocket rises it will therefore have reduced horizontal velocity.

In layers without convection, this does not cause problems. However, when vertical motions are present (due to convection or due to the circulatory patterns induced by meridional circulation) they will deviate from vertical and get a slanted direction of motion. Thus rotation will hamper convection and convective streams will reduce the contrast of shellular rotation.

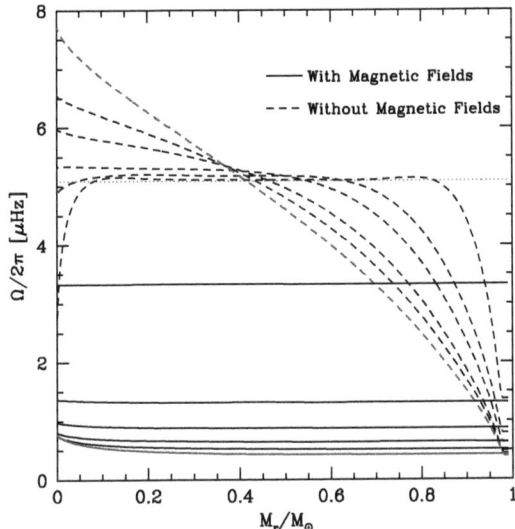

Figure 14.4: The evolution of rotation frequency $\Omega/2\pi$ inside a 1 $M_\odot$ star is modelled. Initial solid body rotation on the ZAMS starts with $v_{\rm rot} = 20$ km s^{-1}, shown by the dotted horizontal line. The further curves show the rotation frequency for 0.1, 0.5, 1, 2, and 3 Gyr, as well as the final one at 4.75 Gyr.
Without magnetic field the angular momentum will, during MS evolution, be redistributed due processes like shrinking of the core, envelope expansion, shear, meridional circulation.
If a magnetic field is present it will force the envelope to follow the core which leads to a decrease of Ω and thus a slowing down of rotation with time, ending at $\Omega/2\pi \simeq 0.45$ μHz = 26 days.
Figure from Eggenberger et al. (2005).

14.4 Braking internal rotation

14.4.1 Stabilizing forces

Rotation will induce turbulent circulatory patterns, unless stabilizing forces counteract sufficiently.

One of these counteracting forces may be the stability against buoyancy (see Fig. 4.1), which is based on the classic criterium for convection (see Ch. 4.3). Another stabilizing force is that due to boundaries of molecular weight zones (see Ch. 4.3.2), also called μ-barriers. Also magnetic fields may stabilize (see section 14.5), since flows will cause the magnetic tension to increase, leading to braking forces.

14.4.2 Redistribution of angular momentum with evolution

During the MS evolution, the core of a star will slowly shrink while the envelope will slowly expand (see Ch. 9). This means that, with conservation of angular momentum, the core must gradually rotate faster while the envelope will gradually rotate slower. This should cause an increased level of shear inside the star. The evolutionary effects are relative easy to model. The result for the Sun is shown in Fig. 14.4 by the dashed lines.

After the red giant phase, with the loss of the outer layers and the earlier redistribution of angular momentum, the star itself will become compact again. This must mean that it will now have faster surface rotation than before the red giant phase. However, magnetic fields counteract this redistribution (see Section 14.5).

14.5 Magnetic field and rotation

14.5.1 Rotation makes a magnetic field stronger

Consider gas in a star with radiative energy transport. Assume further that the star has a weak, radially oriented magnetic field of strength B_r. When the star rotates differentially, any magnetic field present, however small, will be stretched horizontally and the field lines will be wound. This leads already after a few turns to a strong horizontal magnetic field, $B_\phi \gg B_r$ (Spruit 2002). This process has been called the "Spruit dynamo" (see Maeder & Meynet 2003, 2004, 2005).

14.5.2 Rotation braking by magnetic fields

Magnetic fields will tend to drag gas along with rotation due to the coupling of the field to the charged particles. This means that the redistribution of angular momentum due to evolution in the MS phase will be restrained. Model calculations (see Fig. 14.4) show that magnetic fields tend to reduce the overall rotation speed. Models calculated for a 1 $M_\odot$ star indicate that the solar interior must rotate almost as a solid body. This is in agreement with results from helioseismology below the outer convective zone (see Fig. 14.3 and Ch. 11.5).

14.5.3 Loosing angular momentum

During star formation and with conservation of angular momentum with contraction, stars will be rotating fast. Since low-mass stars do not rotate excessively fast, it has long been regarded a problem to find a mechanism to get rid of this angular momentum.

Magnetic fields can reduce the internal rotational shear considerably. Due to the Spruit dynamo, strong and wound-up magnetic fields may reconnect and release (magnetic) energy as heat. Also magnetic braking with low level mass loss contributes.

14.6 Rotation and mass loss

The mass loss rate of a star is given by the empirical relation $\dot{M} \sim L/gR$ (Eq. 4.86), thus by the interplay between L, R, and g. Due to rotation, the surface of a star may show a latitude dependent strength of the surface gravity (Sect. 14.2.2). Near the equator g will be smaller than near the poles which may lead to enhanced mass loss there. On the other hand, due to the oblateness of the star F will be larger near the poles.

Consider for an oblate star the empirical mass loss relation (Eq. 4.86). Inserting $L = 4\pi R^2 \sigma T_{\text{eff}}^4$ and $g = GM/R^2$ (the linear forms of Eqs. 1.16 and 1.19) one finds that Eq. 4.86 transforms to

$$\dot{M} \sim (M\, T_{\text{eff}}^4)/R \quad . \tag{14.11}$$

Thus, inserting this in Eq. 14.1, a surface element at the poles would have

$$\dot{M}_{\text{pole}} \sim (\sqrt{q})^{-4}/q\, \dot{M}_{\text{equator}} \sim \frac{1}{q^3}\, \dot{M}_{\text{equator}} \quad . \tag{14.12}$$

The conclusion must be that, for an oblate star, mass might be easier lost at the poles than at the equator. But in this simple derivation, the centrifugal effect of rotation on the envelope gas (which of course will be very noticeable at the equator) has been ignored.

Actually, the latitude dependent mass loss rate will depend on the interplay of the so-called effective gravity, $\vec{g}_{\text{eff}} = \vec{g}_{\text{grav}} + \vec{g}_{\text{rot}}$, at each latitude and the gravity mimicked by the radiation pressure, $\vec{g}_{\text{rad}}$. The total gravity is, of course, the sum of all: $\vec{g}_{\text{tot}} = \vec{g}_{\text{eff}} + \vec{g}_{\text{rad}}$, as given by Eq. 14.3.

14.6.1 Mass loss disks

Mass loss normally is radiation driven. Its size will be related with the actual opacity of the wind, in fact the opacity of appropriate resonance spectral lines (Ch. 4.5, Fig. 13.7). Gas leaving the rotating surface vertically at some latitude will eventually cross the stellar equatorial plane and will there form a "mass loss disk" (see Fig. 14.5). For details see, e.g., Lamers & Cassinelli (1999).

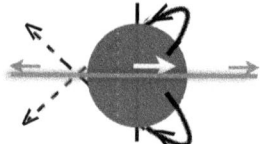

Figure 14.5: With a rotating star (white arrow), mass lost from every latitude sets out (based on $\vec{g}_{\text{tot}}$) into an orbit spiralling outward. In the course of this orbit, the gas tries to cross the equatorial plane but will collide there with gas coming from the opposite latitude. This leads to the formation of a gas disk with a net outward velocity (black arrows).

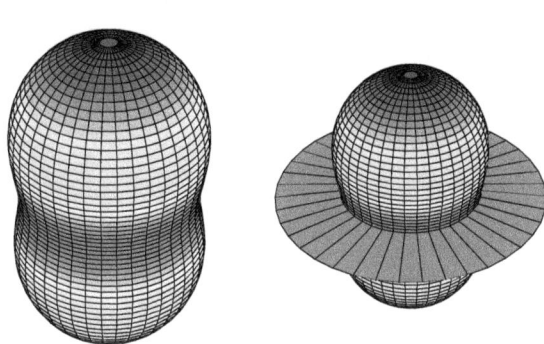

Figure 14.6: Rotation causes latitude dependent mass loss. Shown are iso-mass-loss contours (the 3-D grid) of a 120 M$_\odot$ star with $L = 10^6$ L$_\odot$ rotating at 0.8 of the break-up or critical velocity. The colours represent the range in surface temperature of the two stars. Figure from Maeder & Desjacques (2000).
At left: star with at the poles $T_{\text{eff}} \simeq$ 30000 K having prolate iso-mass-loss contours (largest mass loss at the poles).
At right: the same star with a polar $T_{\text{eff}} \simeq$ 25000 K, so a larger equatorial optical depth and thus an equatorial mass-loss disk.

Since the star will be cooler at the equator than at the poles, also the wind opacity will vary with latitude. For a rotating star with a cooler hydrogen envelope the wind opacity will come from hydrogen with its Lyman and Balmer absorption series. For hot stars it will be mostly opacity in lines of highly ionized elements based on lines of Si IV, C IV and the like (see Fig. 13.1). Such stars may have a fast, hot wind at high latitudes and a slower, cooler wind in the equatorial region.

Hot rotating stars may exhibit a mass loss disk. As an example, Fig. 14.6 shows the modelled effect for a star of $M \simeq 120$ M$_\odot$.

Several observational examples support this notion. One is the star η Car, which has a double mass-loss blob around it (see Fig. 13.2), the star for which the modelling of Fig. 14.6 was made. Another is the now bright gas ring around SN 1987A (see Fig. 14.7), bright due to the collision of ejected gas with gas previously blow away in an equatorial ring around the red giant progenitor. Also the planetary nebula Hubble 5 shows such a structure (Fig. 14.7). Note that the direction of evolution in the late stellar phases will influence the shape of the luminous nebula. In the case of the PN the central star shrinks, leading to enhanced rotation, while T_{eff} increases. SN precursor stars may evolve at high luminosity leading to changes in T_{eff} and/or rotational speed.

14.6.2 Mass loss and loss of angular momentum

When stars lose mass, they also lose angular momentum. This may have a considerable effect on their rotation. Stars with > 10 M$_\odot$ essentially get rid of their angular momentum by loosing just 10% of their mass (Langer 1998).

14.7 Chemical effects of rotation: mixing

Rotation will have effects on the star in many ways. The meridional circulation means there are vertical flows, which lead to chemical mixing.

If circulatory patterns develop, the gas inside a star will be mixed over long path lengths

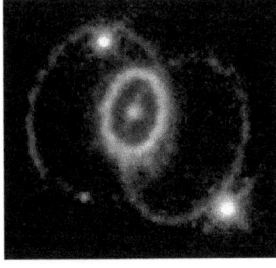
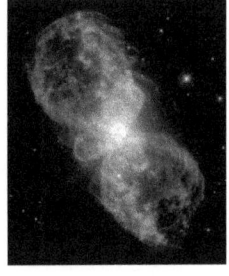

Figure 14.7: Two nebula having shapes going back to the spatial structure of mass loss contours (as in Fig. 14.6 right). At left SN 1978A with an equatorial ring and two high latitude structures. At right the planetary nebula Hubble 5 with an equatorial gaseous disk and two lobes extending to high latitudes. Both images were obtained with the HST (courtesy NASA/ESA).

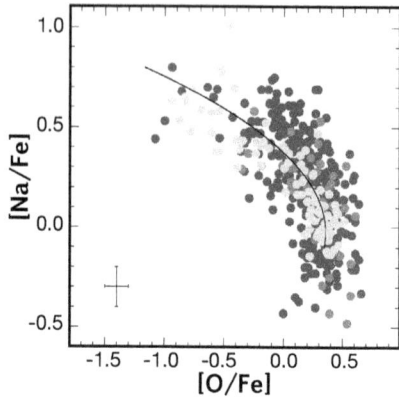

Figure 14.8: Observations of stars in globular clusters show a rather large variation in the O and Na abundance, relative to Fe, while the Fe abundance itself is the same in all stars of one cluster. The dark dots are for stars in various individual clusters (blue for RG stars, red for MS stars) while the gray dots (green) are for RG stars in NGC 2808. The full line shows the mean general anticorrelation in the sense that with more O there is less Na. Figure from Carretta et al. (2006).

Figure 14.9: High mass stars with rotation produce particular chemical patterns in their winds and in the rotation-induced mass-loss disk. Successive stages of evolution disperse elements into the surroundings which will be included in the next generation of stars. This may lead to the abundance scatter observed in globular clusters (as shown in Fig. 14.8).

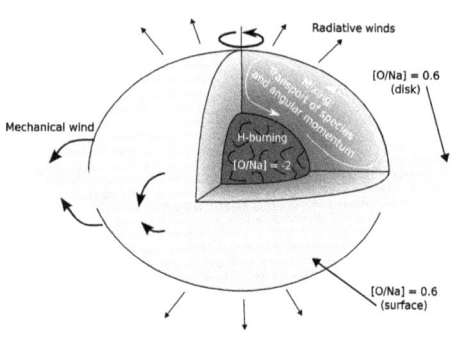

The sketches of the stars with their mass-loss disks are from Decressin et al. (2007b).

Top: Structure of a MS star with slow mass loss into an equatorial disk as well as having radiative mass loss (see Fig. 14.6 for iso-mass loss contours). In the core [O/Na]= −2.

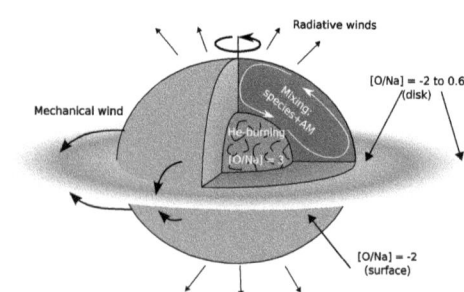

Middle: In the core He burning stage (a good portion of mass has already been lost) the mass loss disk has a range in [O/Na] composition much like the one observed in low mass cluster stars. In the core [O/Na]= +3.

Bottom: After considerable further mass loss the rotation of the star is reduced and no longer in the critical range so the equatorial mass loss stops and the disk will escape.

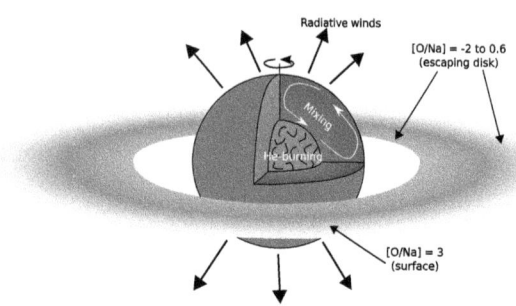

Colours: The colours available in the figure indicate the following.
Green = original chemical composition (the surface and the mass loss disk of the star in the top panel).
Blue = material loaded with products of H burning (the interior of the top panel star as well as the surface of the star in the middle panel and its mass loss disk).
Red = material with products of He burning (the core of the star in the middle panel and the entire star in the bottom one).

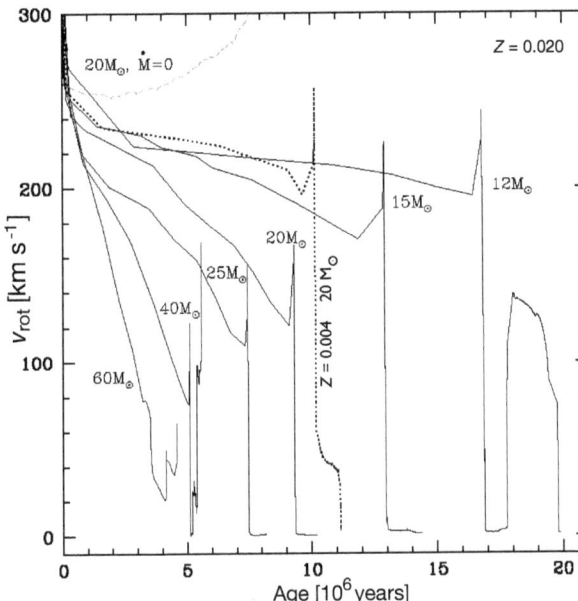

Figure 14.10: Evolution of the rotation of stars of solar composition is modelled, with initially $v_{rot} = 300$ km s^{-1}.

On the MS, mass loss transfers angular momentum away from the star. Expansion to the red giant state reduces the rotation at the surface (compare with evolutionary tracks in Fig. 14.11 paying attention to radius changes). Only for $M < 15$ M$_\odot$ is there the "blue-loop" evolution.

For 20 M$_\odot$ two additional evolutions are given to show differences depending on the choice of parameters. In a 20 M$_\odot$ metal poor star ($Z = 0.004$) fewer metals mean less wind and thus less mass loss and also a less extended red giant state (see also Ch. 9). The dotted evolutionary track (at the top) is for a 20 M$_\odot$ model with zero mass loss. Figure from Meynet & Maeder (2000).

(provided it can overcome the buoyancy force). It has become clear that also rotating MS stars have rotation induced mixing. For more on mixing see, e.g., Pinsonneault (1997).

Another effect of rotation induced mixing is the enhanced surface abundance of N in higher mass stars. High mass stars in the MS-phase burn H also through the CNO cycle. This cycle has a bottle neck in that the reaction ^{14}N + ^{1}H $\longrightarrow$ ^{15}O + γ (see Eq. 5.10) is a slow one, leading to an overabundance of N in the interior. Rotational mixing brings this N to the surface to such an extent, that SMC hot O stars, which intrinsically have [M/H] $\simeq -1$, show an almost solar surface N abundance (see, e.g., Meynet et al. 2006).

Variations in the element abundance seen in globular cluster red giants as well as MS stars (see Fig. 14.8) cannot be explained by any fusion and mixing process in the low-mass stars themselves. These variations are most likely due to local variations in the chemical compositions of the clouds having produced these stars.

A viable explanation was put forward by Decressin et al. (2007a, 2007b). The cloud chemical compositions have been influenced by the material lost by the first generation of high-mass stars in that cluster. Since also globular cluster high-mass stars most likely rotated fast, considerable internal rotation mixing may have taken place together with high mass loss in the manner shown in Fig. 14.6. Elements created in the α process (Fig. 5.7) will be circulated upward and be lost in the stellar wind, thereby polluting the surrounding gas (Fig. 14.9) in which the later generation of stars is formed, stars observed now with special Na/Fe and O/Fe abundance ratios.

14.8 Rotation and mass loss affect high mass star evolution

Rotation affects the evolution of high mass stars in a pronounced way (see Figs. 14.10 and 14.11). The effect is in part intertwined with mass loss because rotation may enhance the stellar winds (through g_{rot} and through higher metallicity in the atmosphere). For details see Meynet & Maeder (2000), Maeder & Meynet (2000), and Meynet et al. (2006). The main aspects are:

1. Inside massive stars, rotation converges from equal v_{rot} at all radii to near equilibrium with two main cells as sketched in Fig. 14.2.
2. Rotation implies strong mixing of material. This leads to changes in the surface metal abundance (more N, less C) and in opacity.

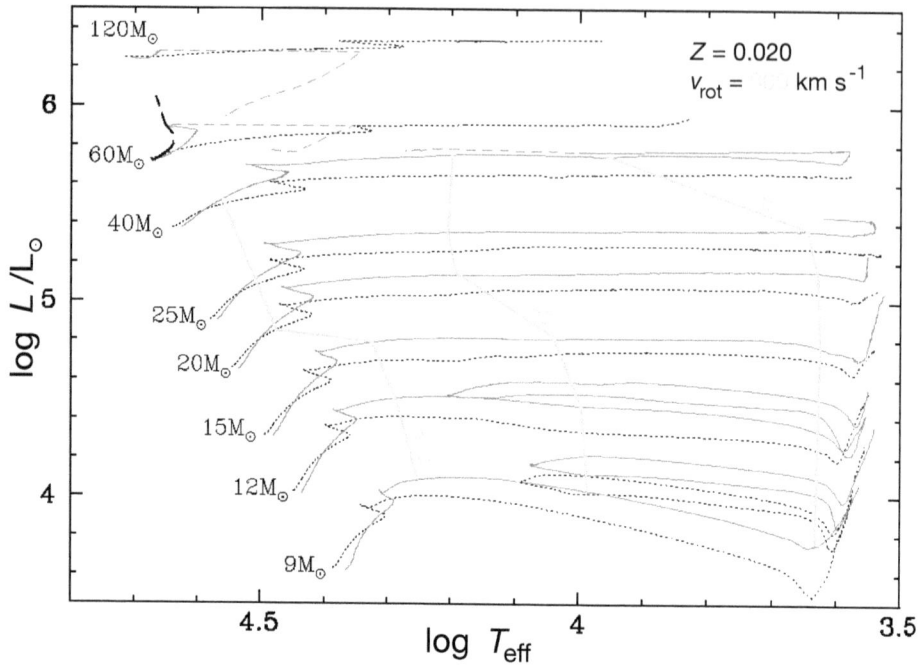

Figure 14.11: Evolution of high mass stars with rotation is shown as evolutionary tracks. The models have initial $v_{\rm rot} = 300$ km s^{-1}. Full lines give evolution with rotation, dotted lines give evolution without rotation (same data as in Fig. 13.9). Note that with rotation, L soon is larger! The full tracks of the highest masses turn into dashed tracks, signifying the WR stage (the tracks bend in part toward lower L due to heavy mass loss). Evolution for a 60 $M_\odot$ star with $v_{\rm rot} = 400$ km s^{-1} (just 100 km s^{-1} more than the other tracks) is shown as the very short track at left. Locations of equal $v_{\rm rot}$ are indicated with value (150, 50, 10 km s^{-1}). The reduction of $v_{\rm rot}$ at this stage is clearly due to the expansion to red giant (conservation of angular momentum). For details of the evolution of $v_{\rm rot}$ see Fig. 14.10. Figure adapted from Meynet & Maeder (2000).

3. Fast rotation enhances the possibility of mass loss (Ch. 14.6) because the rotation enhances mass loss through its centrifugal action. And faster mass loss favours entering the WR phase.
4. Rotation leads to deformation of a star and thus to surface colour effects (Ch. 3.7 and section 14.2).
5. The deformation means that the polar regions are hotter with larger flux.
6. Fast rotation of hot stars leads to polar winds while in stars of more moderate temperature it leads to equatorial winds perhaps with the formation of an equatorial disk (see Fig. 14.6). This may result in dust formation.
7. The transport of angular momentum leads to somewhat lower $v_{\rm rot}$ which in turn means higher ρ_c and thus higher fusion rates and larger L. However, an important effect of rotation is increased mixing leading to enhancement of helium so that the opacity is changed.
8. Mass loss is large in the most heavy stars (high L) and less in stars of lower mass (smaller L). Thus the most massive stars lose proportionally more mass and they slow down their rotation faster (Fig. 14.10).
9. The evolution in the HRD (Fig. 14.11) runs on tracks quite different from those of non-rotation models. This is especially conspicuous in the highest mass range.
10. Mass loss in massive very luminous stars exposes rather quickly interior layers, so that a chemical composition becomes visible which is completely different from the initial one (see e.g. Fig. 13.12).
11. With progressive mass loss and enhanced mixing, the visible atmosphere is richer in heavy

elements. Perhaps this further drives the wind by radiation pressure in spectral lines but the exact mechanisms are still under debate.
12. The stars of $8 < M_{\text{init}} < 15$ M$_\odot$ lose relatively little mass in the MS phase, and may therefore shed considerable mass at the end of the MS phase, leading to thick shells, and perhaps mimmic the Ae/Be star phenomenon (i.e. of intermediate mass stars that are still in the contracting phase onto the MS, see Ch. 7.7.4).
13. When after the red supergiant phase the star contracts, the surface speeds up rotation, perhaps even to critical values (Heger & Langer 1998).
14. Mass loss from evolved stars leads, if the atmosphere became enriched with products from the fusion, to enrichment of the interstellar medium in heavy elements relatively soon after star formation. It so leads perhaps to "pollution" of the IS environment in which further stars are still in the formation process. Those newer generation stars can then be expected to show 'peculiar" abundances.

14.9 Rotation and mass accretion affect WD evolution

During normal stellar evolution, stars lose most of their angular momentum. Stars from $0.6 < M_{\text{init}} < 8$ M$_\odot$ end their evolution as WDs. These WDs rotate very slowly.

However, stars in binary systems may accrete mass from the companion (see Ch. 19). In that case also angular momentum is accumulated and this may lead to a considerable speeding up of rotation. Since mass accumulating WDs may come again to nuclear processes, the speeding up of rotation has special consequences. These are discussed in Ch. 19.4.5.

References

Carretta, E., Bragaglia, A., Gratton, R.G., Leone, F., Recio-Blanco, A., & Lucatello, S. 2006, A&A 450, 523
Decressin, T., Meynet, G., Charbonnel, C., Prantzos, N., & Ekström, S. 2007a, A&A 464, 1029
Decressin, T., Charbonnel, C., & Meynet, G. 2007b, A&A 475, 859
Eggenberger, P., Maeder, A., & Meynet, G. 2005, A&A 440, L9
Heger, A., & Langer, N. 1998, A&A 334, 210
Heger, A., Langer, N., & Woosley, S.E. 2000, ApJ 528, 368
Kippenhahn, R. & Weigert, A. 1991, "Stellar Structure and Evolution"; Springer, Heidelberg
Lamers, H.J.G.L.M., & Cassinelli, J.A. 1999, "Introduction to Stellar Winds"; Cambridge Univ. Press
Langer, N. 1998, A&A 329, 551
Maeder, A. 1995, A&A 299, 84
Maeder, A., & Desjacques, V. 2000, A&A 372, L9
Maeder, A., & Eenens, P. (eds.). 2003, IAU Symp. 215, "Stellar Rotation"; ASP
Maeder, A., & Meynet, G. 2000, ARAA 38, 143; *The Evolution of Rotating Stars*
Maeder, A., & Meynet, G. 2003, A&A 411, 543
Maeder, A., & Meynet, G. 2004, A&A 422, 225
Maeder, A., & Meynet, G. 2005, A&A 440, 1041
Meynet, G., & Maeder, A. 2000, A&A 361, 101
Meynet, G., Eckström, S., & Maeder, A. 2006, A&A 447, 623
Pinsonneault, M. 1997, ARAA 35, 557; *Mixing in Stars*
Spruit, H. 2002, A&A 381, 923
Thompson, M.J., Christensen-Dalsgaard, J., Miesch, M.S., & Toomre, J. 2003, ARAA 41, 599; *The Internal Rotation of the Sun*
Talon, S. 2003, in IAU Symp. 285, "Stellar Rotation"; A. Maeder & P. Eenens (eds.), p. 336
Talon, S. & Zahn, J.-P. 1997, A&A 317, 749
Thompson, M.J., Christensen-Dalsgaard, J., Miesch, M.S., & Toomre, J. 2003, ARAA 41, 599; *The Internal Rotation of the Sun*
Zahn, J.-P. 1992, A&A 265, 115

Chapter 15

The first stars

15.1 First stars have very low metal content

After the big bang the universe contained only H, D and He as well as a sprinkling of Li, B and Be. Without metals, interstellar gas cannot cool very well so that star formation is much more difficult. Still, stars must have formed but the mass function of this first generations of stars may have been different from the one valid today (see, e.g., Ch. 20 and Fig. 20.4). First ideas about these first stars were formulated by Schwarzschild & Spitzer (1953). Without metals, their evolution must have been quite different from the "normal" stars described in earlier chapters.

How can the first stars have formed? Are some still observable? They must be very metal poor, much poorer in metals than the so-called Population II stars (of, e.g., old globular clusters; see Ch. 10.4.2.2). The stars much poorer in metals have therefore become known as Population III stars (see also Ch. 7.1.1). Pop. III stars still present now are by definition of low mass!

Very metal-poor stars have been detected (see Beers & Christlieb 2005). They have [Fe/H] $\simeq -4$ (meaning $\log N(\text{Fe})/N(\text{H}) \simeq -8.5$!) or even [Fe/H] < -5 while they have at the same time [C/Fe] $\simeq +4$ (see, e.g., Norris et al. 2001 and Aoki et al. 2006). Neutron-capture elements are present too with, e.g., [Sr/Fe] $\simeq +1$. These abundance values provide clues about the enrichment processes in the early ISM. The review by Bromm & Larson (2004) gives a good overview of the investigations and of what is known about the formation of the first stars. Beers & Christlieb (2005) review the discovery of and the analysis methods for galactic metal-poor stars.

15.2 Making a star in metal-free gas

The metals in the interstellar medium provide good cooling agents. Collisionally excited low lying states of atoms and ions subsequently decay, emitting a photon from a "forbidden" transition, a photon which effectively cannot be absorbed. Metals are also present in composite silicates as dust particles and the dust surface is the only place where the molecule H_2 can be formed in a relatively easy manner. Active cooling and molecule formation are essential requirements to come to dense interstellar gas, gas in which ultimately stars may form.

However, in the early phase of the universe, its expansion had not progressed very far and the density of matter was therefore larger than it is today. This implies that the so-called Jeans criterium may be reached more easily, except for the fact that the IS gas could not cool very well.

For stars to form in such metal free gas, cooling must take place. Initially, cooling is through the Lyα line which is slow cooling, since these photons will be reabsorbed and re-emitted continuously. Such photons must leak out of the denser gas region into the more dilute exterior space. Once the gas has cooled to $T < 10^4$ K, H_2 may form which (in the high density early universe) is possible in H-H encounters. H_2 may cool the gas through collisional excitation. For an H_2 cooling function see, e.g., Santoro & Thomas (2003).

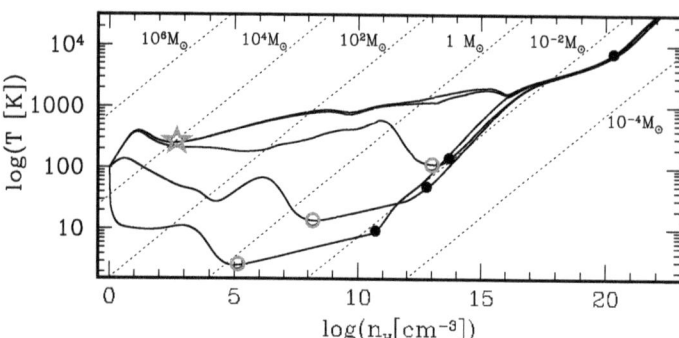

Figure 15.1: The behaviour of the cooling of gas under self-gravity is related with metal content. The graph shows that in metal free gas the first (and only) instability to occur leads to very massive objects and thus to very high mass stars ($\star$). Gas with metals cools faster an will fragment into smaller clumps leading to lower mass stars. The $\star$ and $\circ$ mark the location in n, T of the first stable fragments, showing that in very metal poor gas these contain hughe amounts of mass. The filled dots mark the location of hydrostatic cloud cores. The curves are, from top to bottom, for gas with $Z = 0$, 10^{-6}, 10^{-4}, 10^{-2} and 1 $Z_\odot$. The slanted dotted lines mark constant Jeans mass for spherical clumps. Figure adapted from Omukai (2000) and Schneider et al. (2002).

The curve describing the n, T-behaviour of interstellar gas has a shape which depends on the available amount of species capable of cooling (the coolants). The cooling leads to a "thermal instabilty" thus to local density enhancements which proceed to what is called the "fragmentation" of the gas cloud. Then a collapse must happen and more an more material can be accumulated.

Since the nature of the cooling depends on the content of coolants one has Z-dependent fragmentation (see Fig. 15.1).

$\circ$ In zero metal gas, so without good coolants, cooling is slow, and the instability would occur at relatively high temperature and pressure. That there are hardly fragments leads to large condensing blobs, and so only high mass stars form. This leads to stars with masses in the 10^3 $M_\odot$ range.

$\circ$ In gas containing heavy elements, so with coolants, the cooling will be faster and so the fragmentation leads to smaller clumps. This leads to stars with masses well below the high mass objects of zero metallicity gas, stars with mass in the range of present day stars.

The transition composition between the two regimes is thought to lie near $Z \simeq 10^{-5}$. Note that in metal-free gas, any metals returned by supernovae may enhance the metal content locally from $Z = 0$ to perhaps even $Z \simeq 10^{-3}$ $Z_\odot$. Thus, at the time of formation of first stars their actual metal content may already show considerable variation in Z.

In present-day metal-rich material radiation pressure on dust will at some point halt the inflow of matter (Wolfire & Cassinelli 1987). Perhaps accretion onto first stars is also halted by the formation of an H II region or by the emerging radiation pressure of trapped Lyα photons (see the review by Bromm & Larson 2004). Since star formation itself is not well understood yet, it is clear that modelling the formation of the first stars is still more difficult.

In the contraction phase of first stars the global structure is that of a polytrope (see Ch. 6.2) with $n = 3$ at the stability limit. Given that small deviations from stability may induce a collapse, even the environment of such supermassive stars has to be included in the modelling.

15.3 Evolution of first stars

Once Population III high mass stars have formed in cooled down post big-bang material they will evolve very fast. Without metals, they are very leaky balls of gas (Ch. 6.8.2.2) and gravity thus leads to stronger compaction. In early stars of high total mass, T_c will be so high that simultaneous with H, He will burn, too. Evolutionary tracks from models of such massive first stars and the internal development of the fusion zones are given in Fig. 15.2. The evolution of the central parameters ρ_c and T_c are given in Fig. 15.3. After about 10^5 years the further fusion processes will set in and ultimately one has a very massive star with the usual layered structure

15.3. EVOLUTION OF FIRST STARS

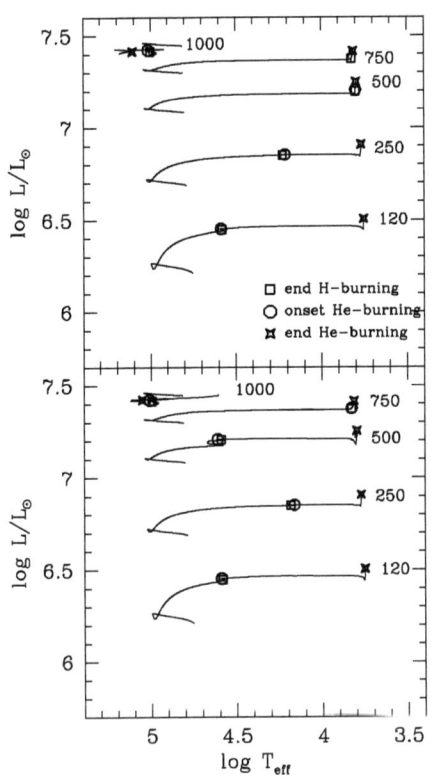

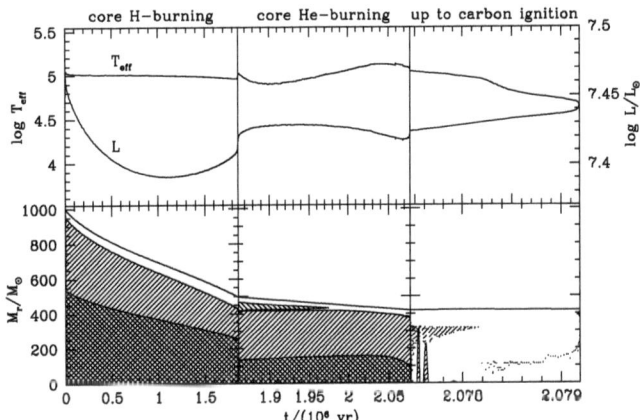

Figure 15.2: Evolution of 120 M$_\odot$ < M < 1000 M$_\odot$ stars. They take $\simeq 2 \cdot 10^6$ a to become supernova.
Left: The top panel shows evolution with winds driven only by radiation, the bottom panel with winds plus effects of stellar rotation. Note that the times of end and onset of H and He burning differ between the two cases.
Below: Evolution of the internal structure of a 1000 M$_\odot$ star with radiation- and rotation-driven mass loss. Heavy shading marks zone with fusion, light shading that of convection. Figures from Marigo et al. (2003).

of fusion zones. For the internal composition of a 500 M$_\odot$ star just before it becomes supernova see Fig. 15.4.

In high mass first stars, the central temperatures get so high ($T_c > 10^9$ K, Fig. 15.3) that various very high energy processes take place. Among these are (see, e.g., Rees 1984)

1. *Comptonization.* If photons scatter on electrons with higher energy ($kT_e > h\nu$), the photons gain energy such that the spectral energy distribution may approach the Wien shape.

2. *Nuclear statistical equilibrium (NSE).* At very high temperatures and densities, there will be a continuous exchange of particles with nuclei, the latter possibly continuously dissolving and building based on "nuclear-statistical equilibrium" as mentioned in Ch. 5.2.4.

3. *Pair production.* Electrons in the velocity distributions with $E > 0.5$ MeV may interact with photons and in that process electron-positron pairs can be created ($\gamma \to e^+, e^-$). Such electron-positron pairs are part of the system and can be annihilated again (creating a γ photon). However, after pair production the number of photons is reduced which is equivalent to cooling and this allows the gas to become denser, leading to a "pair-instability" supernova.

The run of internal T for the models of 500 and 10^3 M$_\odot$ of Ohkubo et al. (2006) are shown in Fig. 15.5. It illustrates that roughly 20% of the interior material is at $T > 4 \cdot 10^9$ K and is in NSE. Investigations about the nature and evolution of very massive (and supermassive, $M > 10^5$ M$_\odot$) stars has not advanced very far yet. Much is unexplored, like the mass loss of such stars and the effects developing jets have on the evolution. For details see, e.g., Ohkubo et al. (2006).

Supernovae of "first stars" come in two kinds.
○ Supernovae of stars with $300 < M_{\text{init}} < 10^5$ M$_\odot$ are core-collapse supernovae. This collapse sets in through Fe-decomposition (see Ch. 18.3.1) followed by electron capture and thus neutron formation. Since these stars have a very large core mass, the remnants turn into black holes immediately, containing about 1/2 of the M_{init}.

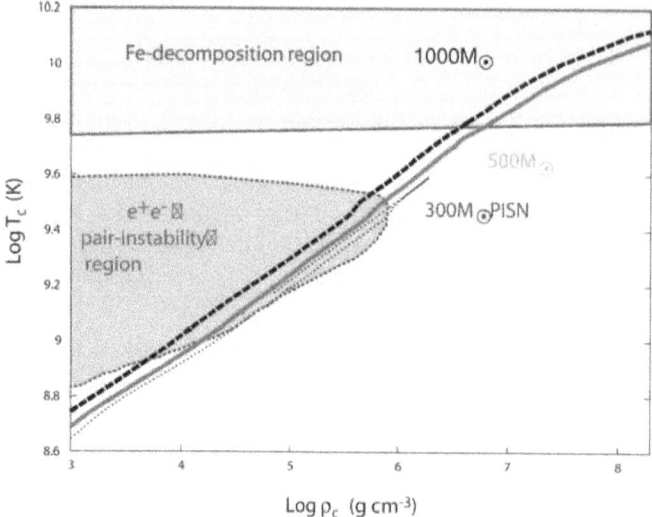

Figure 15.3: Evolution of T_c and ρ_c for stars of 300 M$_\odot$ (thin dotted line), 500 M$_\odot$ (thick solid line) and 1000 M$_\odot$ (thick dashed line). Note the e^+e^- pair instability region, through which all very high mass stars evolve but the 500 and 1000 M$_\odot$ stars continue to evolve to the Fe-decomposition region. The 300 M$_\odot$ star becomes a pair-instability supernova, PISN. Figure from Ohkubo et al. (2006).

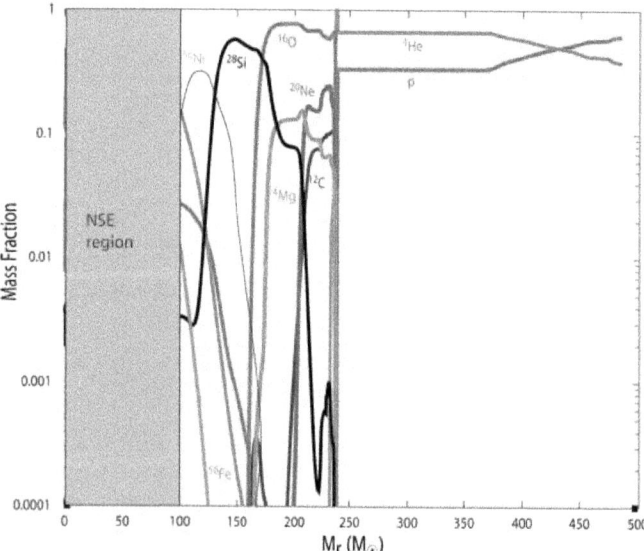

Figure 15.4: The internal composition of a 500 M$_\odot$ star is shown just before it becomes supernova. Note that in the inner 240 M$_\odot$ all H and He has been transformed into heavy elements. The shaded region at left is the nuclear-statistical equilibrium region (see Ch. 5.2.4). Figure from Ohkubo et al. (2006).

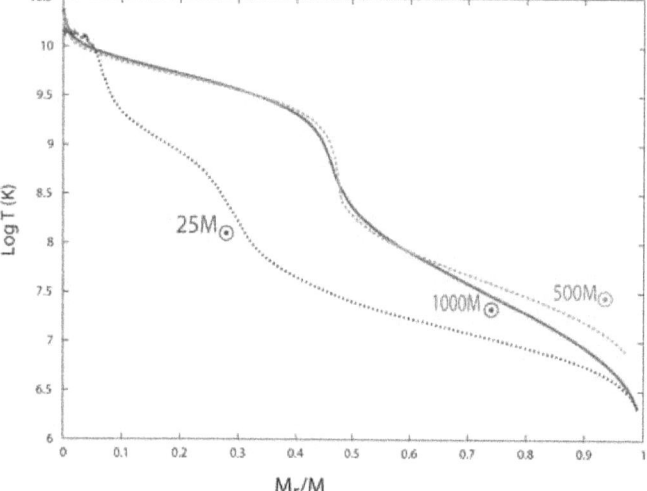

Figure 15.5: The run of internal temperature versus radius in a star of 25 M$_\odot$, 500 M$_\odot$, and 1000 M$_\odot$. Note the large mass fraction inside which $T > 10^9$ K in the very massive stars. At $T > 4 \cdot 10^9$ K the material is in NSE (compare with Fig. 15.4). Figure from Ohkubo et al. (2006).

15.4. NUCLEOSYNTHESIS IN POPULATION III STARS

◦ Stars with $130 < M_{\text{init}} < 300$ M$_\odot$ become pair-instability supernova (PISN) because the reduction of the numnber of photons (in the creation of e^+, e^- pairs) means a reduced radiation pressure which leads to contraction, in fact a run-away contraction. These stars explode completely, dumping all material back into interstellar space.

A summary of the fate of first stars, in particular related with their supernova like demise, is given in Fig. 18.9.

These first supernovae pullute the interstellar medium with metals enabling better cooling so the formation of lower mass second generation stars (Sect. 15.2).

15.4 Nucleosynthesis in Population III stars

Population III stars have to fuse H to He and then He into C before the CNO cycle can operate. Also, building heavier elements through neutron capture requires that elements capable of capturing neutrons first have to be made in the star itself. Since in mass rich Pop. III stars evolution proceeds very fast there is no time for the slow neutron capture to produce notable amounts of elements. The result of the fusion in these stars is therefore that basically just "α"-elements from CO-burning (Ch. 5.2.1) are produced (see Fig. 15.4).

Because convection is strong, the fast rotation of these stars will lead to strong meridional circulation. Thus mixing provides all layers of the star soon with all fusion products.

The final collapse of very high mass first stars leads (see Fig. 18.9) to a range of objects of different nature as well as to the production of lots of heavy elements through rapid neutron capture. These elements are included in the next stars to be formed. The elements heavier than Fe have radioactive decay and that aspect can be used to derive the age of very early stars based on the currently observed abundance ratio of those isotopes. That method is called nucleochronometry (see Ch. 22.3.3).

15.5 Lithium in first stars

In first stars, the content in lithium should be the same as that of the universe just after the big bang (see Ch. 5.2.6). Once some very massive stars had thrown metals into interstellar space, star formation became easier. Stars formed at that time are still very poor in metals and still should have the original Li abundance. The spectroscopic studies of such first stars also aim at finding the Li abundance since its value in stars may corroborate the models for the big bang.

First thorough investigations were those of Spite & Spite (1982a, 1982b). They showed that a sample of stars with low [Fe/H] (by definition old and thus with a mass below 1 M$_\odot$) had a Li abundance at a constant level. An early review of the topic is available from Spite & Spite (1985).

The constant abundance (at the level [Li/H] $\simeq$ 2.1 for $-3.3 <$ [Fe/H] < -2.6, the so-called "Spite plateau") indicates only little consumption of Li during the MS phase of the studied stars. Thus how is it that the Li abundance stays at that level over so long a time while all kinds of stellar processes (e.g., convection, mixing due to rotation, gravitational settling) tend to reduce the abundance of Li at the surface of a star? The study by Bonifacio et al. (2007) presents a thorough discussion of the problem but none of the ideas leads to an acceptable solution.

References

Aoki, W., Frebel, A., Christlieb, N., Norris, J.E., Beers, T.C., & 13 al. 2006, ApJ 639, 897
Beers, T.C., & Christlieb, N. 2005, ARAA 43, 531; *The Discovery and Analysis of Very Metal-Poor Stars in the Galaxy*
Bonifacio, P., Molaro, P., Sivarini, T., Cayrel, R., Spite, M., Spite, F., Plez, B., Andersen, J., Barbuy, B., Beers, T.C., Depagne, E., Hill, V., Francois, P., Nordström, B., & Primas, F. 2007, A&A 462, 851
Bromm, V., & Larson, R.B. 2004, ARAA 42, 79; *The First Stars*
Marigo, P., Chiosi, C., & Kudritzi, R.-P. 2003, A&A 399, 617

Norris, J.E., Ryan, S.G., & Beers, T.C. 2001, ApJ 561, 1059
Ohkubo, T., Umeda, H., Maeda, K., Nomoto, K., Suzuki, T., Tsurita, S., & Rees, M.J. 2006, ApJ 645, 1352
Omukai, K. 2000, ApJ 534, 809
Rees, M.J. 1984, ARAA 22, 471; *Black Hole Models for Active Galactic Nuclei*
Santoro, F., & Thomas, P.A. 2003, MNRAS 340, 1240
Schneider, R., Ferrara, A., Natarajan, P., & Omukai, K. 2002, ApJ 571, 30
Schwarzschild, M., & Spitzer, L. 1953, Observatory 73, 77
Spite, M. & Spite, F. 1982a, Nature 297, 483
Spite, F. & Spite, M. 1982b, A&A 115, 357
Spite, M. & Spite, F. 1985, ARAA 23, 255; *The Composition of Field Halo Stars and the Chemical Evolution of the Halo*
Wolfire, M.G., & Cassinelli, J.P. 1987, ApJ 319, 850

Chapter 16

Models and variation of "free" input parameters

When making stellar models one has to deal with various unknown or unquantified aspects of stars. Or, one likes to make a set of models in which such parameters cover a range of values to see how the models compare with reality.

The aspects involved include the way convective overshoot is treated (see below), the way the initial metal content (or rather, chemical composition) changes the basic structure (see Ch. 6.9.2), the level of mass lost through stellar winds affecting later evolutionary states (see, e.g., Ch. 13), the speed of rotation of the new-born star which leads to weaker or stronger internal circulation (see Ch. 14), and perhaps more.

All this is of relevance for the understanding of stars. In a sense, stars are not understood completely. These aspects also have a bearing for the use of isochrones, to be addressed in Ch. 21.

16.1 Effects on models and evolution

16.1.1 Complications with convection

Convection (see Ch. 4.1.4.2) plays an important role at the surface of many types of star. In massive stars it is an essential aspect influencing stellar evolution.

Convection consists of a radial acceleration of cells of material due to temperature or density gradients in a hydrostatic medium and a resulting counter-movement of other cells. Effectively, convection also leads to mixing of gas, possibly mixing of gas with different composition.

Two important criteria for convection are (see Ch. 4.3)

- the criterion defined by *Schwarzschild*. It compares the radiative with the adiabatic temperature gradient and so mainly evaluates the thermal structure.

- *Ledoux's* criterion for convection, which compares different gas densities and also takes into account chemically inhomogeneous gas mixtures. It also handles possible sinking of material, e.g., at some boundary layer.

When convection occurs, the gas cells do not stop moving where the criterion for convection is no longer fulfiled but they move further, until their motion is damped. This motion beyond the limits of what the formal gradient allows is called *convective overshoot* (see Ch. 4.3.5). Convection with overshoot leads to a MS-phase longer than without overshoot.

For the treatment convection, the *mixing length* parameter $\alpha = d/H_P$ is defined. H_P is the pressure scale height (see Eg. 4.57) and d is the length inside the star over which the convection takes place (see Ch. 4.3.5). In fact, α is used as a free parameter in the modeling of convection and model makers explore in practice a range of values of α (see, e.g., Fig. 16.1).

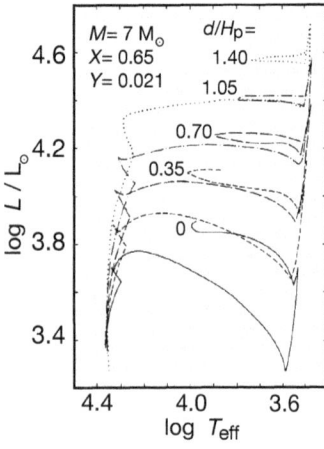

Figure 16.1: HRD showing evolutionary tracks for a star of 7 $M_\odot$ with $X = 0.650$, $Z = 0.021$, and various amounts of convective core overshooting length d over scale height H_P, d/H_P. Models are shown for $\alpha = d/H_P = 0$ (no overshoot) and $d/H_P = 0.35$, 0.70, 1.05 and 1.40. Depending on α the blue loops have different lengths (see Ch. 13.4.1.2). Note that the vertical scale ($\log L$) is in this figure much expanded in comparison with that of Fig. 10.1. The comparison of observed CMDs with theoretical ones for a range of α has been used to empirically fix α. Figure adapted from Chin & Stothers (1991).

Fig. 16.1 illustrates how the evolution of a 7 $M_\odot$ star depends on the parametrization of the overshoot. The more overshoot there is, the larger the region of the star providing fuel for the central fusion. As a consequence, the He core will easily become more massive than the nominal 0.5 $M_\odot$ of the low mass stars. This augmented He mass has quite an impact on the later stages of evolution.

The parametrization of the convection and the overshoot is done quite differently in different models, which leads to rather differing results. It may mean the difference between the occurrence or non-occurence of a blue loop (See Ch. 13.4.1.2).

Fig. 16.2 shows that different models reach quite different values for L and T at the end of the core hydrogen burning phase, at the *terminal age main sequence* (TAMS, see Chapter 9.1.1.2). Note that there is no TAMS for stars with $M_{MS} < 1.15$ $M_\odot$ as convection plays no role in their central parts.

16.1.2 Effects of metal content

Stars starting with a chemical composition different from that of the Sun will establish a main sequence in a different location than the standard MS. This aspect has been discussed in Ch. 6 and examples for the location of a non-solar MS are given in Fig. 6.7. Stars with lower metal content are somehwat bluer and somewhat brighter in the MS phase (they are more 'leaky').

Also the evolution will be somewhat different in that the RG phase is somewhat bluer. This affects the location of the RG branch, which is of relevance for the modelling of isochrones for older stellar populations (see Ch. 21).

Examples of tracks of evolution of stars of 1.5 and 20 $M_\odot$ showing the mentioned effects are given in Fig. 16.3. Further examples can be found in Fig. 16.4.

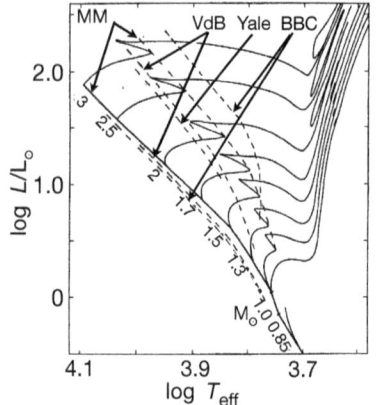

Figure 16.2: Comparison of the location of the ZAMS and of the terminal age main sequence (TAMS; see Fig. 9.1) for stellar structure and evolution models from various groups:
MM = Maeder & Meynet (1989a);
VdB = VandenBerg (1985);
Yale = Green et al. (1987),
BBC = Bertelli et al. (1985).
Models differ in opacity tables, convection criterium and overshoot. Note that the BBC models have the largest separation between ZAMS and the TAMS. Figure adapted from Maeder & Meynet (1989b).

16.2. EFFECTS OF COMBINED PARAMETERS

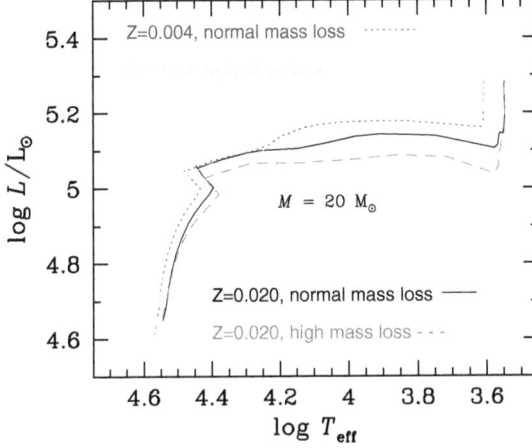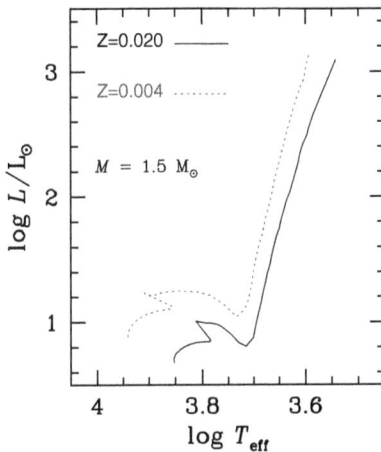

Figure 16.3: Evolutionary tracks for stars of 20 M$_\odot$ (left) and of 1.5 M$_\odot$ (right) for metallicities of $Z = 0.020$ (Pop. I) and $Z = 0.004$. Lower metal content makes the star more luminous and its surface hotter (see Fig. 6.7). For 20 M$_\odot$ different mass-loss rates are compared as well (mass loss is very small for a star of 1.5 M$_\odot$). Results are from models by the Geneva group (see Schaller et al. 1992, Charbonnel et al. 1993, Meynet et al. 1994).

The differences in the model evolution are due to different parameter settings, leading to differences in parameters derived from a comparison with observations. Or, conversely, observations of in particular star clusters may help to set model parameters.

16.1.3 Effects of mass loss

Mass loss affects stellar evolution too and different models treat mass loss in a different way. Depending on the adopted mass-loss rate (the often used "Reimers law", Eq. 4.86; or some other relation), the evolutionary tracks will either slowly or more pronouncedly tend toward tracks of stars of lower mass. And the mass loss rate may depend on the metal content of the stellar envelope.

A strong effect was visible in the evolution of stars in the higher mass range in that stars with $M_{\text{init}} > 40$ M$_\odot$ lose so much mass that they become effectively stars of 40 M$_\odot$ (see Fig. 13.9) but have by that time had more fusion than a star of just 40 M$_\odot$. Effects of mass loss in post MS evolution of a 20 M$_\odot$ star can be seen in Fig. 16.3 (left).

Another example of effects of mass loss can be seen in Fig. 10.6. There a low-mass star turns into an HB star after the He flash and the nature of the HB star is determined by the amount of mass lost. Its further evolution depends on the M_{HB} reached as can be seen in Fig. 10.9.

16.1.4 Effects of rotation

Various effects rotation has on stars, stellar models and stellar evolution have been addressed in Ch. 14. The main aspect is that rotation induces internal mixing and thus may lead to a prolonged MS phase (replenishment of fusion material). Fig. 16.4 shows the effects of rotation (and metallicity) on stellar evolution using stars of 3, 9, 20 and 60 M$_\odot$. Rotation and metallicity may even prevent blue loop evolution in stars of the 2 to 10 M$_\odot$ mass range.

A summary of the effects of rotation was given in Ch. 14.8. Numerous aspects are still under investigation.

16.2 Effects of combined parameters

All parameters may play a role. In particular with the question of the occurence of blue loop evolution (Ch. 13.4.1.2) it became clear that the interplay of the parameters will determine the

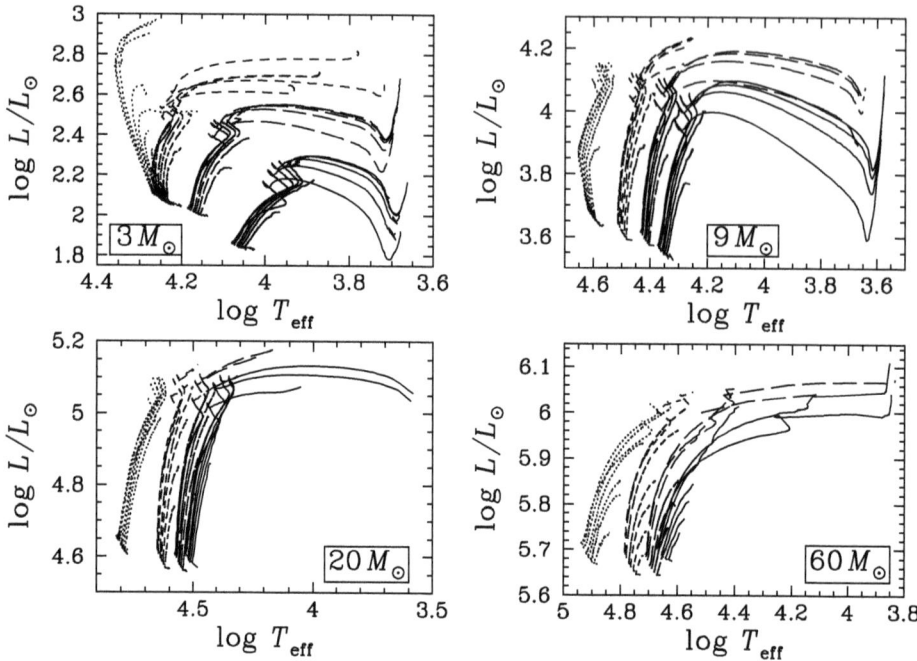

Figure 16.4: Stellar evolution models for four values of M_{init} (3, 9, 20 and 60 $M_\odot$) show the effects of different initial metal content and of initial rotation speed. The different metallicities used are $Z = 0$, 10^{-5}, 0.002 and 0.020 leading to the four different locations of the ZAMS position, the metal poorest being the bluest (as in Fig. 6.7). The evolution shown by each track is for the range in angular momentum adopted: $\Omega/\Omega_{crit} = 0.1, 0.3, 0.5, 0.7, 0.9$ and 0.99, leading to the tracks starting, from right to left, at each of the metallicity ZAMS locations (the largest rotation leading to a surface with lower T_{eff}). The tracks stop when Ω_{crit} is reached or at the onset of central He burning. These models include the effects of convective overshoot (with $\alpha = 0.1 \cdot H_P$). Figure from Ekström et al. (2008).

course of the evolution. Many of the effects have been summarized by Mowlawi et al. (1998). But several critical parameters are often not well known from reality.

Ultimately, all variations are of relevance for the modelling of stellar populations and the comparison with observed stars groups. The examples given demonstrate that it is not straightforward to derive the initial conditions of a star from its current observable parameters.

Star groups can be understood with the help of relatively simple data, such as photometry. Comparison of photometry with model stars is mostly done using isochrones (see Ch. 21).

References

Bertelli, G., Bressan, A.G., & Chisoi, C. 1985, A&A 150, 33
Charbonnel, C., Meynet, G., Maeder, A., Schaller, G., & Schaerer, D. 1993, A&AS 101, 415
Chin, C.-W., & Stothers, R.B. 1991, ApJS 77, 299
Ekström, S., Meynet, G., Maeder, A., & Barblan, F. 2008, A&A 478, 467
Green, E.M., Demarque, P., & King, C.R. 1987, in: "The Revised Yale Isochrones and Luminosity Functions", Yale University, Observatory, New Haven
Maeder, A., & Meynet, G. 1989a, A&A 210, 155
Maeder, A., & Meynet, G. 1989b, A&AS 89, 451
Meynet, G., Maeder, A., Schaller, G., Schaerer, D., & Charbonnel, C. 1994, A&AS 103, 97
Mowlavi, N., Meynet, G., Maeder, A., Schaerer, D., & Charbonnel, C. 1998, A&A 335, 573
Schaller, G., Schaerer, D., Meynet, G., & Maeder, A. 1992, A&AS 96, 269
VandenBerg, D.A. 1985, ApJS 58, 711

Chapter 17

Degenerate stars: WD, NS, BH

At the very end of stellar evolution the matter in stellar remnants is of special nature. A white dwarf (WD) contains ions, which make up the mass, the degenerate electron gas providing the pressure and a magnetic field. A neutron star (NS) clearly consists of neutrons. Each of these kinds of stellar remnants has its own physics. A special case of a NS is a black hole (BH).

Clearly, stars are normally in hydrostatic equilibrium (Ch. 9.4). There is a balance between gravitational contraction and repulsive inner pressure (Ch. 6.3). This internal pressure is determined by the state of the gas as described by the equation of state (Ch. 4.4.2). All terms may contribute but in degenerate stars it is foremost the degenerate pressure which keeps the star in shape. If the conditions change, phase transitions may occur. Such possibilities are addressed here.

WDs are the endproduct of low mass star evolution. BDs (Ch. 8), lower MS stars (Ch. 10.1.2) middle mass MS stars (see Fig. 10.9) and all stars of up to $M_{\text{init}} \simeq 8\,M_\odot$ (Chs. 10.5 and 10.6) end as WDs. WDs may develop into NSs and NSs possibly into BHs. But these steps take place only if the circumstances are right (which normally happens only in binary systems; Ch. 19).

17.1 White dwarfs

17.1.1 Internal structure of WDs

The first object recognized as WD was the faint companion to Sirius, Sirius B. Its $T_{\text{eff}} \simeq 27000\,\text{K}$ together with its brightness could, with the help of its parallax, be converted into a luminosity, $L \simeq 10^{-1.5}\,L_\odot$. This implied that the radius of Sirius B was very small, $R \simeq 10^{-2}\,R_\odot$ ($\simeq 10\,000$ to $30\,000$ km). From the period of the orbit of Sirius B around Sirius A and with the mass of Sirius A derived from its type, the mass of Sirius B was found to be $\simeq 1.05\,M_\odot$, packed within $10^{-2}\,R_\odot$!

Based on these parameters the internal conditions can be easily estimated. With simply $\overline{\rho} = M/V$ and using the Sun as comparison one arrives at

$$\overline{\rho_{\text{WD}}} = \frac{M_{\text{WD}}}{M_\odot} \frac{R_\odot^3}{R_{\text{WD}}^3} \overline{\rho_\odot} \simeq 10^6\,\overline{\rho_\odot} \quad . \tag{17.1}$$

Since also a WD must be in hydrostatic equilibrium we find, using the simplified expression $P/M \simeq -GM/(4\pi R^4)$ and again comparing with the Sun, that

$$P_{\text{WD}} = \frac{M_{\text{WD}}^2}{M_\odot^2} \frac{R_\odot^4}{R_{\text{WD}}^4} P_\odot \simeq 10^8\,P_\odot \quad . \tag{17.2}$$

A WD contains degenerate gas. In the case of non-relativistic degenerate electron gas (see Chapter 6.2.2.3), this leads to

$$R \sim M^{(1-n)/(3-n)} \quad . \tag{17.3}$$

The details of the structure depend on the chemical composition since μ influences the polytrope constant, K. With non-relativistic electron gas ($n = 3/2$) one has $R \sim M^{-1/3}$ (see Fig. 17.5).

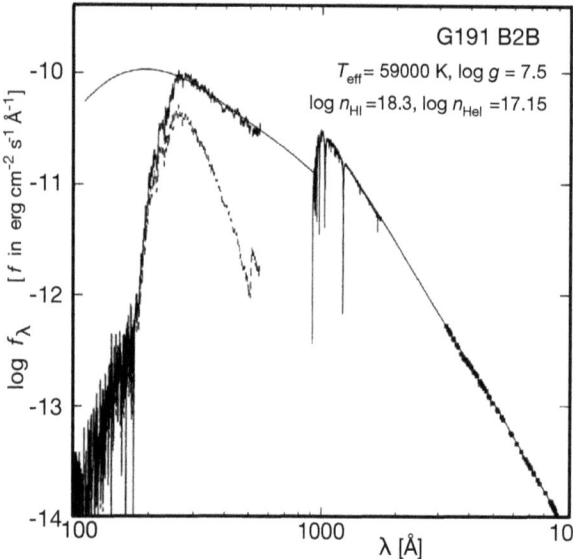

Figure 17.1: The observed spectral energy distribution of the hot WD G191 B2B (the lower spectral data curve). The thin full line shows the appropriate atmosphere model.

The hydrogen ionization edge at 912 Å can be seen in the spectrum for two reasons. One is that, even at $T_{\rm eff} = 59\,000$ K, the atmosphere is dense enough to still contain some neutral H (Saha equation). The other is that there is also absorption by interstellar gas.

The original data shown the depressed spectrum between 250 and 550 Å. The upper spectrum is the one corrected for IS absorption. Figure from Vennes & Lanz (2001).

17.1.2 Atmosphere of a WD

Around the dense, fully degenerated interior with conduction as means of energy transport, a thin shell of 'normal' gas (still high density, but non-degenerate) is present. In it, the heat is transported either by radiation or convection. The thickness of this layer is typically 1% or less of the WD radius. With the high density, some of the hydrogen is still neutral up to high temperatures (use the Saha equation of Ch. 3.2.2 for $T_{\rm eff}= 59000$ K). Fig. 17.1 shows the spectral energy distribution of a hot WD, where the hydrogen ionization edge and some Lyman lines are still visible. Several observational aspects of WDs have been discussed in Ch. 10.6 and Ch. 11.6.

The gravity at the surface of WDs is that large, that spectral lines are relativistically shifted ($\Delta\lambda$) to the red by $(\Delta\lambda)/\lambda \cdot c \simeq 50$ km s^{-1} (in addition there is gravity broadening of the spectral lines). This is of relevance for the determination of the stellar radial velocity.

Due to the high gravity at the surface ($\log g \simeq 9$), the atoms and ions in the atmosphere will settle according to the balance between kinetic energy and gravitational pull (see Ch. 3.6). In many WDs *no metal absorption lines* are seen because the metals have been pulled down into the atmosphere in this manner.

Depending on the origin of the WD, some show still H in the atmosphere, others just He (a He WD), or even only C+O (a CO WD). In the very dense and cooling gas of H-containing WD-atmospheres, molecular hydrogen, H_2, and other molecules may form (see Ch. 3.4.3, for the effect of H_2 see Fig. 10.17).

WDs have an outer convection zone. It starts forming when the atmosphere has cooled to below 15000 K. Associated with it are the ZZ Cet like pulsations (see Ch. 11.3.1). At further cooling the convection zone reaches deeper until it approaches the edge of the degenerate layer.

In the degenerate interior energy transport is dominated by electron conduction (Ch. 4.4.1) which is more efficient than convection. Thus the atmospheric convective layer will not progress further inward. Instead, the cooling allows the degenerate interior to grow at the cost of the extent of the atmosphere.

With the convective zone in direct contact with the conductive region a strong coupling is established between the isothermal core (essentially at T_c) and $T_{\rm eff}$ (through the convection). The atmosphere changes also during these phases (recombination and change in opacity) and modelling is rather complex.

Due to the low luminosity and high surface gravity, WDs may accrete material from interstellar space. That would add heavy elements to the atmosphere, which would have been metal poor due to gravitational settling. Adding metals changes the opacity leading most likely to complex changes. For more on these aspects see Hansen & Liebert (2003).

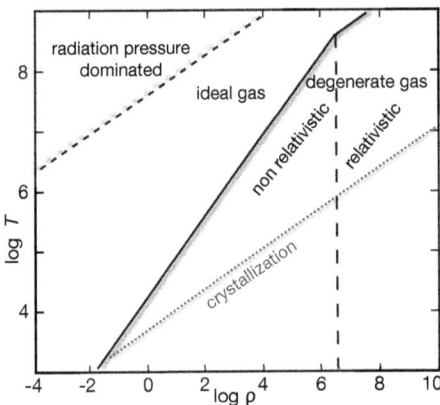

Figure 17.2: The behaviour of very dense and degenerate gas in relation with "crystallization", a process occuring in WDs and NSs.
The figure is similar to Fig. 4.5, but here a dotted line is added, marking the boundary conditions beyond which (toward the lower right) crystallization sets in.

17.1.3 Cooling and crystallization of a WD; cooling time

The core of a WD is basically isothermal and energy is transported mostly through electron heat conduction (see Ch. 2.8). The heat is then transferred to the convective atmosphere.

The luminosity of an object is $L = 4\pi R^2 \cdot \sigma T^4$. Combining this with the simple radius-mass relation of WDs gives

$$L \sim M^{-2/3} \, T_{\text{eff}}^4 \qquad (17.4)$$

so that WDs of a given mass (having no energy source from fusion) lie in an theoretical HRD on a straight line, the **cooling line**. These lines are ordered by mass (see Figs. 10.9 and 10.20).

One defines a "cooling time", the time in which the object cools to half its temperature. The cooling time depends on the material, the mass and the actual luminosity at some point in time. Cooling lines are thus not entirely straight. The relation with M and L in solar units and A the mean atomic weight (carbon=12 is the reference) as derived by Iben & Tutukov (1984) is

$$t_{\text{cool}} \simeq 10^7 \left(\frac{12}{A}\right) \left(\frac{M^5}{\mu^2 \, L^5}\right)^{1/7} \quad [\text{yr}] \, . \qquad (17.5)$$

In order to see how matter inside a WD behaves during cooling one has to consider (at such high densities) effects of *electrostatic interaction* (see Kippenhahn & Weigert 1990, their Ch. 16.4). The ions no longer move freely but tend to form a rigid lattice, minimizing their total energy. This occurs when the thermal energy $3kT/2$ becomes comparable with the Coulomb energy per ion of charge $-Ze$. One can define a parameter $\Gamma_{\rm C}$ to represent the ratio of these energies as

$$\Gamma_{\rm C} = \frac{(Ze)^2}{r_{\text{ion}} kT} \qquad (17.6)$$

in which r_{ion} is the mean separation between the ions[1]. For $\Gamma_{\rm C} < 1$ the electrostatic energy plays a minor role, for $\Gamma_{\rm C} \gg 1$ the kinetic energy no longer is the dominating entity and the gas goes through a "phase transition". The 'gas' settles into a lattice, a process called **crystallization**.

A more detailed consideration shows that the transition lies rather near $\Gamma_{\rm C} \simeq 100$. One can define the transition temperature (also called "melting temperature", $T_{\rm m}$) to be

$$T_{\rm m} \simeq \frac{(Ze)^2}{r_{\text{ion}} k \Gamma_{\rm C}} = 2.3 \cdot 10^3 Z^2 \left(\frac{\rho}{\mu_0}\right)^{1/3} \qquad (17.7)$$

in which $V_{\text{ion}} = 4/3 \, \pi r_{\text{ion}}^3$ and $\rho = m_{\text{ion}} n_{\text{ion}}$ were used, and μ_0 the atomic mass unit. The boundary line for crystallization is shown in Fig. 17.2.

---

[1] This factor $\Gamma_{\rm C}$ is very similar to the exponent in the Planck function, $\Delta E/kT$, which indicates the relative importance of the energy difference in photon processes and the actual temperature of the gas.

17.1.4 Chandrasekhar limit, maximum mass of a WD

For a non-relativistic degenerate gas the mass radius relation (Eq. 17.3) shows that for larger and larger mass the radius will become smaller and smaller. This implies a larger and larger central density, which in turn means that the degeneray will become relativistic. In that case the equation of state in the interior would have to be described by a polytrope with index $n = 3$. But in the case of such a gas, the relation between radius and mass would lead to a singularity, because then $R \simeq M^{-\infty}$.

A full treatment of the equations for stellar structure, which includes a general change from degenerate gas with $n = 3/2$ into relativistically degenerate gas with $n = 3$ leads to the limiting mass for a WD as

$$M = (2/\mu)^2 \, 10^{33.5} \text{ g} \tag{17.8}$$

so that for WD gas (no H but He + metals, so $\mu \simeq 2$; Eq. 4.75) the maximum mass possible is

$$M \simeq 1.4 \text{ M}_\odot \, . \tag{17.9}$$

This limit is called after **Chandrasekhar** (although independently found by Landau in 1932 and by Chandrasekhar in 1935). In this simple analysis the possibly complicating effects of rotation, inhomogeneous composition, magnetic fields, etc., were not included.

If the object were a WD, the above derivation shows what the maxium mass of a WD is. If the object were a higher mass star, the implosion leads to a supernova of Type II (see Ch. 18.3.1).

17.1.5 Transfer of mass onto a WD; Eruptions

Most stars are part of binary (or multiple) systems (see Ch. 19). In a binary system containing stars in the lower mass range the more massive one (the primary), evolves a bit faster than the mass poorer one (the secondary). The primary may in its RG phase shed mass onto its companion (Roche lobe overflow, see Ch. 19.3.1) and will, after further evolution, become WD first. The secondary will in turn evolve and become RG too, shedding mass back.

At this point in the evolution, material passes through the Roche Lobe and accumulates around the WD in an *accretion disk*. This disk gas cools and becomes denser, enters lower orbits and plunges onto the WD. Since the surface gravity of a WD is very high, this material will fall onto the WD with high speed. At impact very high temperatures are reached and **nuclear fusion may take place near the surface** of the WD. This leads to a surge in brightness in this group of objects; the group consists of **cataclysmic variables** and **Novae** (see Ch. 19.4.4).

If the mass transfer brings the total mass of the WD near (or over) the Chandrasekhar limit, the internal material cannot hold up the pressure and the WD will come to explosive C fusion, disrupting the entire star. This explosive event is a **supernova of Type Ia** (see Chs. 18.3.2 & 19.4.4).

17.1.6 Can a WD become NS?

WDs will all by themselves *not* change into NSs. For that their mass is not large enough since lower mass stars evolve into WDs having not more than $\simeq 1$ M$_\odot$ of material (see Figs. 10.19 and 10.24). However, in some cases enough mass can be gained in a double star system (as indicated above) to ultimately have a WD with mass larger than 1 M$_\odot$.

Given that many stars exist as double stars, numerous WDs may accumulate so much mass from the neighbour that they will surpass the physical mass limit, the Chandrasekhar mass limit. If this limit is reached, the material in the nucleus will go to a phase transition.

Perhaps the electrons are able to combine with the protons (p) of the nuclei (A) thereby becoming neutrons

$$e + p \rightarrow n + \nu \quad \text{and} \quad A(Z) + e \rightarrow A(Z-1) + \nu \, . \tag{17.10}$$

The released neutrinos then would cool the central region. This would allow the interior to contract and to neutronize. However, that reaction is not very likely.

17.2. NEUTRON STARS

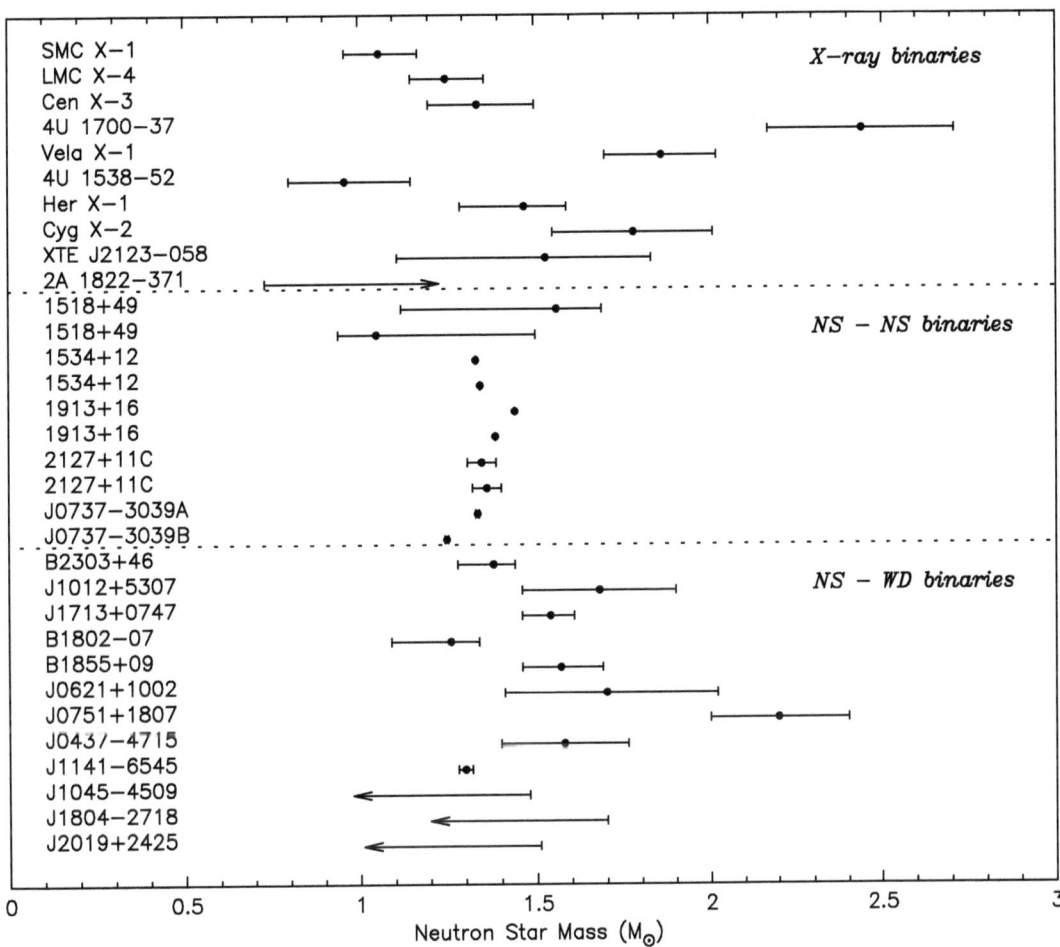

Figure 17.3: For several neutron stars in binary systems the mass has been derived. Given are, from top to bottom, X-ray binaries, NS-NS-binaries and NS-WD-binaries. How these binaries come into existence is given in Chs. 19.5.2.4 & 19.5.2.5. Error bars are 1 σ errors. The fourth object from the top, 4U 1700-37, could be a BH. Figure from van der Meer et al. (2007).

Instead, the nuclei in the WD core (carbon and oxygen, or perhaps helium in the case of stars from very smal M_{init}) may themselves come to fusion (Eq. 5.16), leading to a nuclear runaway process (see Ch. 18.3.2). The energy released is so large and so sudden, that the WD most likely is destroyed (deflagration) and that there is no steller remnant after the explosion.

17.2 Neutron stars

17.2.1 Two ways for stars to become NS

Neutron stars come from two possible evolutionary scenarios:
- Stars starting in the *upper mass range* eventually become SN of Type II or Type Ib. They shed their outer layers, and leave a rapidly spinning NS (a pulsar) as final object.
- Stars of the *lower mass range* become WDs. Those in a binary system may accumulate mass, as described above, and become SN Type Ia and perhaps a NS remains.

The mass of a neutron star is the same as or larger than the WD Chandrasekhar limit. Observational evidence from binary systems supports this (see Fig. 17.3).

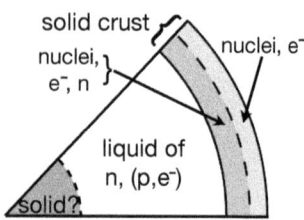

Figure 17.4: Speculative sketch of the internal structure of a neutron star of 1.4 M$_\odot$ (radius $\simeq$10 km). It probably has a solid crust of about 1 km thickness in which the material makes (from the surface downward) the transition from 'normal' (nuclei + e$^-$) to fairly neutronized matter. Below that lies the neutron liquid ($\rho > 2 \cdot 10^{14}$ g cm^{-3}) while the core may again be solid consisting perhaps of hyper-dense quark-like material. However, what really goes on inside NSs is not known.

17.2.2 Structure and mass of neutron stars

The structure of a neutron star is clearly that of a sphere full of neutrons, having a thin outer layer of likely different (not neutronized) material (see Fig. 17.4). The balance between kinetic energy of the particles and the gravitation, as described with the WDs above, holds in the same way.

One can give the relation between radius and mass set by the behaviour of the gas (polytrope with index p)

$$R = 2/\text{G const}/m_X \ M^p \quad \text{with} \quad -1/3 < p < -3 \tag{17.11}$$

where for WDs $m_X = m_e$ and for NSs $m_X = m_n$. Since $m_e/m_n = 5 \cdot 10^{-4}$, one obtains that

$$R_\text{NS} \simeq 5 \cdot 10^{-4} \ R_\text{WD} \ . \tag{17.12}$$

The radius of a NS is somewhere between 10 and 15 km only! The mass-radius relation is similar to that of WDs (see also Fig. 17.5).

Free neutrons should decay with a half life of 636 s into an electron, a proton and an antineutrino. However, due to the very high density in NSs this decay does not take place since the decay electrons would have no place in phase space. There is little further knowledge about the real interior structure of a NS. As all fusion-free objects, also neutron stars cool (see Yakovlev & Pethick 2004).

Considerations of the (not well known) equation of state of NS gas lead to an estimate of a maximum mass $M_\text{NS}(\text{max}) \simeq 3$ to 5 M$_\odot$, much like the Chandrasekhar maximum mass for WDs. If the mass were to increase above this limit a phase transition would take place. The behaviour of matter and the phase transition locations are graphically illustrated in Fig. 17.5.

The total mass of a NS may grow further (e.g., in a binary system) or may be larger to start out with (see Ch. 18.5). The exact nature of a phase transition depends on the equation of state of the particular neutron matter. If it is "stiff", with accretion the matter just stays neutrons and R grows; if it is "soft" R will decrease leading to the phase transition and the material may turn into quarks, or so (Shapiro & Teukolsky 1983). The density at which such a transition occurs is 3 - 5 10^{15} g cm^{-3} (depending on the details of the equation of state). Further aspects of the behaviour of such matter is the domain of particle physics.

17.2.3 The surface layers of a NS

The atmosphere of a NS probably consists of normal material. Diving into a NS one quickly reaches densities $\rho \simeq 4 \cdot 10^{11}$ g cm^{-3}. That density is similar to that in WDs and the material here must be in the same way highly degenerate and crystallized as in WDs. Further down, more and more neutronized nuclei exist until one reaches the real neutron fluid below (see Fig. 17.4).

For more on neutron star surface radiation see Yakovlev & Pethick (2004).

17.2.4 Behaviour of neutron stars: pulsars

Neutron stars do lose energy through radiative processes. Some of that energy surfaces as "beamed radiation", which we see as 'pulses', like we see the rotating light beam from a lighthouse at the seaside. This radiation apparently comes from 'active spots' on the surface, perhaps related with strong magnetic fields. The discovery of such pulsed radiation in the radio domain led to the name

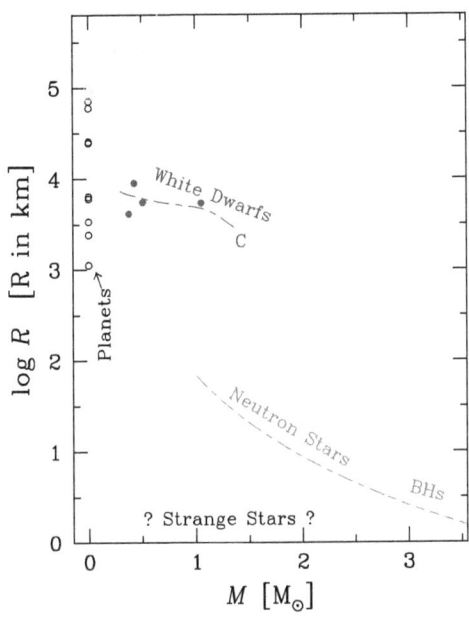

Figure 17.5: Mass-radius relation for spherical objects without an inner energy source. It includes planet like objects (BDs and planets), WDs and NSs. Each of these have a particular mass-radius relation which is defined by the state of the material in the interior (state of degeneracy, see Ch. 4.4.2.3).
The WDs plotted are from Lang (1992). If a WD surpasses the Chandrasekhar limit (marked by the letter C) its matter structure changes and the star deflagrates (Sect. 17.1.6).
Also a NS likely has such a phase transition.
The high mass planets (∘, data from Lang 1992) are very similar to cooled down Brown Dwarfs (⋄, data points from Fig. 8.6) having degenerate electron gas in the interior (Ch. 6.5).
The location of the non-degenerate MS stars is indicated at the top (==).

pulsar. There is, however, no established theory yet about the mechanism producing the radiation from pulsars.

Energy is gained from thermal energy remaining, but also from sudden restructuring of the crystal lattice in the NS to an overall lower energetic state. This causes a sudden shake up, much like an earthquake. The gain in energy is visible in the gain in angular momentum: pulsars tend to jump once in a while to a slightly faster spin and thereafter slow their spinning in the same manner as before the jump. Such jumps are called "pulsar glitches".

The visual brightness of pulsars is very small. For a moving pulsar at 330 pc, Mignani et al. (2002) find $V \simeq 25$ mag or $M_V \simeq 18$ mag. Pulsars are also seen in X-rays. Grindlay et al. (2002) presented data on numerous X-ray pulsars observed by CHANDRA in the globular clusters 47 Tuc and NGC 6397. Their X-ray brightness is not related with the visual brightness. It is most likely due to accretion of material.

Pulsars are used to determine the electron density in the ISM by exploiting the pulsed nature of the radiation. Radiation disperses in the IS plasma: the arrival time of EM waves is a function of the frequency of the radiation and the line of sight integral $\int n_e dl$. Since the radiation from a pulsar is pulsed, one can measure the difference in the arrival time for different frequencies (pulsar dispersion measure).

17.3 Strange (quark) stars

When a neutron star surpasses its material stability limit (see Sect. 17.2.2) the phase transition leads perhaps to matter in the form of some sort of quark soup. Since this is unclear, this exceedingly dense state of matter is called "strange matter" and in the case of stars called *strange stars*. Their mass-radius range is indicated by that name in Fig. 17.5. This kind of material is investigated theoretically. No objects of this kind are known to exist but are surmised. They probably would be black holes anyway.

17.4 Black holes

Black holes have been predicted as possible objects long ago, in fact since the concept of the Schwarzschild radius (see below). Originally it was thought a black hole could be made through the implosion of a massive star. It turns out that this is not very likely or is possible only for the

Table 17.1: Observational evidence for stellar black holes (from Wijers 1995 and Casares 2007)

Name(s)	Sp.T.	P_{orb} [days]	Δv_{rad} [km s^{-1}]	M_{star} [M$_\odot$]	M_{BH} [M$_\odot$]
bright X-ray sources:					
LMC X-3	B3Ve	1.70	235	6	6 - 9
LMC X-1	O7-9III	4.23	68	13	> 4
Cyg X-1 (V1357 Cyg)	O9.7Iab	5.5996	75	33 ± 9	10 ±3
X-ray Novae [XN Constellation year (other names)]:					
XN Cyg 89 (V404 Cyg)	K0IV	6.4714	208	0.5 - 1.0	10 - 15
XN Sco 94 (GRO J1655-40)	F5IV	2.613	227	??	6.3 ± 0.3
XN Mus 91 (GRS 1124-683)	K3V	0.4326	406	0.56 - 0.87	4.2 - 6.5
XN Vul 88 (GS2000+25, QZ Vul)	K3-7V	0.3441	520	??	5.3 - 8.2
XN Mon 75 (A0620-20, V616 Mon)	K3-5V	0.323	443	0.2 - 0.7	5.1 - 17.1

Δv_{rad} = amplitude of v_{rad} curve.
v_{rad} and the masses are not (cannot be) corrected for the systems' orbital inclination, $\sin i$.

largest M_{init} (see Fig. 18.4). Much 'easier' is the formation through accretion by material spiralling in onto a neutron star. This infalling material then emits X-rays, giving evidence for the presence of the BH (see the names of the BH-objects in Table 17.1; several are also known by the name of the visible stars). Making a black hole is, in fact, the domain of binary stars (see Ch. 19.4.1).

17.4.1 Schwarzschild radius

If the total mass of an object is very large and the radius is very small, the gravitational pull at the surface prevents even photons from escaping. Consider for that the relation for the *relativistic* wavelength shift

$$1 + \Delta\lambda/\lambda = 1/\sqrt{1 - \frac{2G}{c^2}\frac{M}{R}} \quad . \tag{17.13}$$

The BH mass limit follows from the simple condition that the redshift is infinitely large once

$$\frac{2G}{c^2}\frac{M}{R} > 1 \quad \text{or} \quad 2G\frac{M}{R} > c^2 \tag{17.14}$$

so that no light can leave the object. It means this is true for objects with radius given by

$$R < \frac{2GM}{c^2} = 3\,M \quad \text{km} \tag{17.15}$$

with M in M$_\odot$. This radius is called the *Schwarzschild radius*.

If no light can leave the object, then for an observer there is a minute and entirely **black** circular area on the sky, since also light from behind a BH is pulled in by the gravitational field. The word **hole** comes from the graphical description of the shape of the gravitational field of the BH which is, in fact, a very deep potential well (see half of Fig. 19.3), which mathematically has a singularity (∞) at the centre.

At the stellar level, *a BH is nothing more than a supermassive NS*. Such a neutron star has an internal density ρ large enough for the NS surface to be within its particular Schwarzschild radius. The lowest mass NS being a BH has roughly $M \simeq 3$ M$_\odot$, for which then $R \simeq 9$ km. The behaviour of neutron star material is not well known because it is not accessible for observation. For a NS being also BH its matter cannot be much different from 'normal' NS matter. But at much higher densities the material may become strange matter (as mentioned above).

Relativistic aspects of material falling into a BH and the questions related with the event horizon are beyond the scope of a text on stars and stellar evolution.

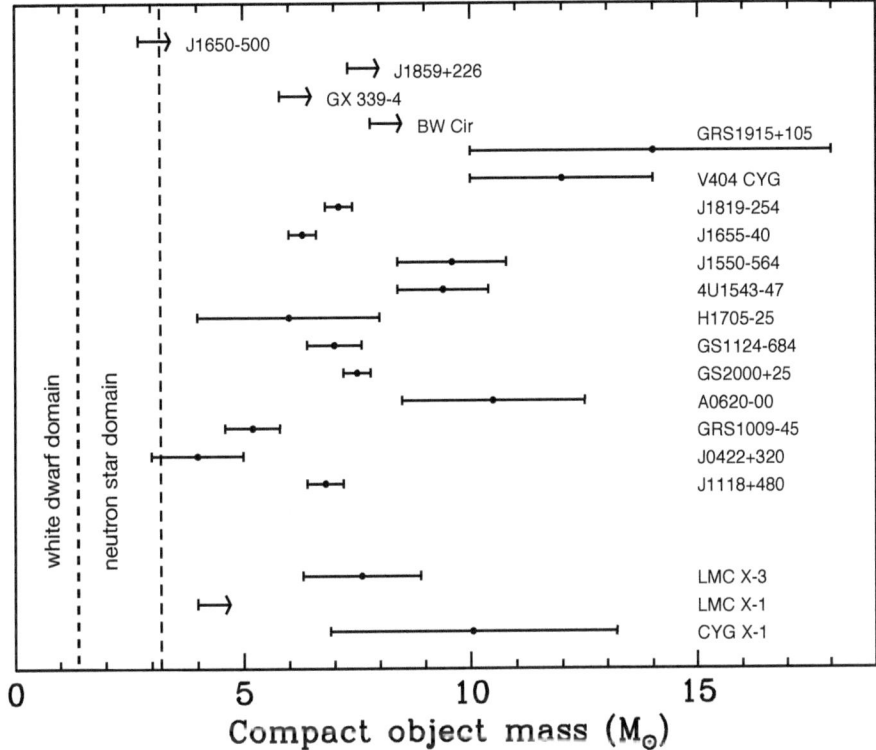

Figure 17.6: Mass derived for the invisible companion of stars with peculiar characteristics, stars which turned out to be binary systems (see Ch. 19) containing a black hole. The object names are given at right. For some of the objects data on the visible star of the binary is given in Table 17.1. The black holes have a mass in the range of 3 to 15 $M_\odot$. The vertical area at left is the mass domain of neutron stars (see Fig. 17.3) and of white dwarfs. Figure adapted from Casares (2007).

17.4.2 Observational evidence for the presence of stellar black holes

Since one cannot *see* a BH, its presence must come from circumstantial evidence.

Several bright X-ray sources in the sky turned out to be in the optical a luminous star, however a star with a wobble. Apparently these stars had companions, which were *not* detectable in the UV, visible, or near-IR. From the period of the orbit one then could estimate the mass of the companion object. These masses turned out to be of the order of $M_{BH} \simeq 4$ to 20 $M_\odot$. Since objects with such a mass being companion to a normal star should be stars themselves, they clearly should be detectable as stars. But since they were not 'seen' in such binaries systems these cases gave proof that BHs do exist indeed (see Tab. 17.1; more data are given in Fig. 17.6). See Ch. 19 for more on binaries and the systems leading to black holes.

The origin and nature of the BH in the centre of our Galaxy, having a mass of $\simeq 15 \cdot 10^6$ $M_\odot$ (Genzel et al. 2000), is completely different.

17.5 Nobel prize 2002 for X-ray astrophysics

The Nobel prize 2002 was shared between Neutrino research and X-Ray astronomy.

Riccardo **Giacconi** was honoured for his initiatives in the beginnings of X-ray astronomy and the further pursuits with X-ray satellites. With rocket experiments and satellites (UHURU, EINSTEIN) he and his team discovered both X-ray point sources as well as the diffuse X-ray background. Some of the point sources were then interpreted as being BHs, the X-rays coming from gas streaming from the companion star (see X-ray binaries; Ch. 19.5.2.4) into the BH's gravitational well.

Davis and Koshiba were honoured for their work on neutrino astronomy (see Ch. 5.4).

References

Casares, J. 2007, in IAU Symp. 238 "Black Holes from Stars to Galaxies – Across the Range of Masses", Karas, V. & Matt, G. (eds.), p. 3

Genzel, R., Pichon, C., Eckart, A., Gerhard, O.E., & Ott, T. 2000, MNRAS 317, 348

Grindlay, J.E., Camilo, F., Heinke, C.O., Edmonds, P.D., Cohn, H., & Lugger, P. 2002, ApJ 581, 470

Hansen, B.M.S., & Liebert, J. 2003, ARAA 41, 465; *Cool White Dwarfs*

Iben, I., & Tutukov, A.V. 1984, ApJ 282, 615

Lang, K.R. 1992, "Astrophysical Data: Planets and Stars"; Springer, Heidelberg

Kippenhahn, R., & Weigert, A. 1990, "Stellar Structure and Evolution"; Springer, Heidelberg

Mignani, R.P., de Luca, A., Caraveo, P.A., & Becker, W. 2002, ApJ 580, L147

Shapiro, S.L., & Teukolsky, S.A. 1983, "Black Holes, White Dwarfs, and Neutron Stars"; Wiley, New York

van der Meer, A., Kaper, L., van Kerkwijk, M.H., Heemskerk, M.H.M., & van den Heuvel, E.P.J. 2007, A&A 473, 523

Vennes, S., & Lanz, T. 2001, ApJ 553, 399

Wijers, R.A.M.J. 1995, in "Evolution Processes in Binary Stars", R.A.M.J. Wijers et al. (eds.); Kluwer, Dordrecht; p. 327

Yakovlev, D.G., & Pethick, C.J. 2004, ARAA 42, 169; *Neutron Star Cooling*

Chapter 18

Supernovae

The majestic fate of stellar evolution is the supernova. The star explodes somehow and the outer layers rush outward. The radiating surface increases drastically so that (with $L = 4\pi R^2 \cdot \sigma T^4$) the luminosity increases tremendously. In comparison with the pre-SN state, previously inconspicuous stars may become visible for the naked eye. In some cultures one spoke of a "guest star".

After the approximate nature of supernovae (SNe) as exploding stars became understood and, in relation with the discovery by radioastronomers that non-thermal radiation from gas entities in the Milky Way represent supernova-remnants as well as the recognition that the Crab Nebula expands at such a rate that all its gas must have been in the point of origin about 9 centuries ago, people started looking for optical information about previous supernovae. This resulted in a list of "historical supernovae" (Table 18.1).

18.1 Historical supernovae, supernova rate

Supernovae are named by the calendar year and a letter to order the discoveries in that year (except the old ones, which have names of historic origin). The scientifically very important and bright SN 1987A (in the LMC) was the first SN reported in 1987. Another well known relatively bright supernova was SN 1993J in M 81. The first SN recognized in another galaxy (the Andromeda galaxy) in 1885 is still called by its discovery name as a variable: S And.

How often are supernovae visible? Or rather, how often does a star explode as supernova in our Milky Way? The answer to these questions can only be imprecise. To estimate the number of visible SNe the past record is too sketchy. Evidently there is no homogeneous time coverage nor probably homogeneous sky coverage in the data up to well into the 19th century. Given that the last SN visible from the northern hemisphere was in 1604 (or perhaps another one in 1680, see Table 18.1) one might expect a next 'optical' SN soon.

The rate of SNe has been estimated from the number of supenova remnants detected in radio continuum measurements. A crude estimate for the entire Milky Way leads to 1 per 25 yr.

18.2 Observed types of supernovae

Supernovae come in different types, the types being defined mostly by *signatures in the spectrum* (Table 18.2). The main types are SN Type I and SN Type II, or shortened to SN I and SN II. There are numerous texts about the classification of supernovae (e.g., Harkness & Wheeler 1990, Filippenko 1991),

The *shape of the light curve* may be a further discriminator, but here the problem is that SNe often are discovered only many days after they light up. The shape of the light curve near maximum and immediately thereafter has informative structure (see Fig. 18.1).

The lightcurves of SNe Ia are all very similar, so that this can be used as a discriminant for this type of SN. For the cosmological importance of this aspect see Sect. 18.7.

Table 18.1: Historical visible supernovae (see Clark & Stephenson 1977)

year	location, name	type	records from/by
185	Centaurus, RCW 86		China
386	Sagittarius		China
393	Scorpius		China
1006	Lupus		Europe, China, Japan, Korea
1054	Taurus, Crab Nebula	II?	China, Japan
1181	Cassiopeia	I?	China, Japan
1572	Cassiopeia	I	Tycho Brahe
1604	Ophiuchus	I	Johannes Kepler
1680	Cassiopeia, 3 Cas ?		John Flamstead [a]
1885	M 31 / S And	?	Ernst Hartwig
	> 200 SNe in other galaxies		
1987	LMC / SN 1987A	II	Ian Shelton
	> 300 SNe in other galaxies		search programmes with automatic telescopes

[a] Flamstead listed stars in each constellation sorted by brightness. The star 3 Cas of his catalogue is no longer visible, but its position is the centre of a supernova remnant.

It is important to know the location of the SN in space (where in the galaxy, in what kind of galaxy). This is of relevance to estimate what the nature of the *progenitor* of the SN likely was, to which stellar population type it belonged. For none of the observed SNe (except SN 1987A and SN 1993J) information about the progenitor star is available.

From observations of other galaxies the following statistics of occurrance emerged:
→ SN Ia: in all kinds of galaxies, but more in spirals than in ellipticals.
→ SN Ib: always in late type galaxies (Sc), normally near H II regions; none in E, S0, Sa or Sb's.
→ SN II: in spiral galaxies, in or near spiral arms, some in Irr galaxies. Never in ellipticals.
It appears that SN II and SN Ib are rather associated with gas-containing galaxies, thus being SN from relatively young stars. SN Ia occur in all galaxies and thus likely are from old stars.

Note that the older classification scheme [SN II, Pop I, hydrodynamic, massive progenitor] versus [SN I, Pop II, thermonuclear, low mass progenitor] is not accepted any more.

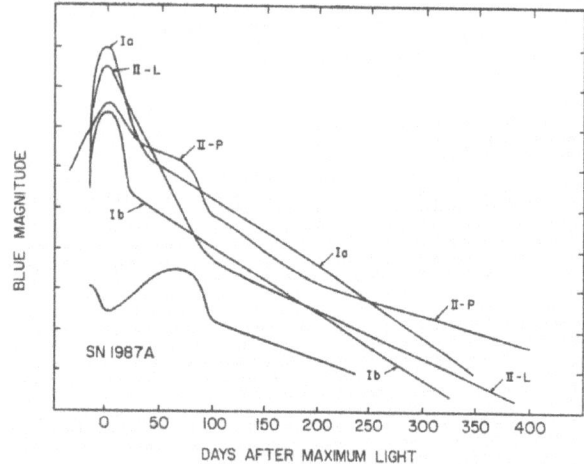

Figure 18.1: Supernovae have different light curves (here blue brightness against time). The light curves can also be used for classification. Types indicated are:
Type Ia and Type Ib,
Type II L (for "linear" behaviour) and
Type II P (for the "plateau" behaviour).
SN 1987A does not seem to fit with standard types, but perhaps with Type II P.
Figure from the review by Filippenko (1997) where also many SN spectra are shown.

Table 18.2: Supernova classification scheme (see Harkness & Wheeler 1991)

main type[a]	main criterium	subtype	further criteria
SN II	hydrogen lines & Ca & other metals	II L	"linear" light curve
		II P	"plateau" light curve
SN I	**no** hydrogen	I a	Si (etc.), and He, CNO
		I b	He, no Si
		I c	no He, no Si

[a] For typical lightcurves of supernovae see Fig. 18.1

18.3 Theories about supernovae

Only after processes of nucleosynthesis were understood could theorists start to explore mechanisms which might lead to the explosion of stars. An overview is given by Woosley & Janka (2005). Two types were proposed, but making a distinction in two types of mechanisms is, in reality, not quite fair. Both mechanisms have at its root a run-away nuclear process.

○ For the *hydrodynamic SN* (Sect. 18.3.1), modelling the evolution of the SN requires considerable hydrodynamics to follow the effects of the explosion.

○ In the *thermonuclear SN* (Sect. 18.3.2) the run-away nuclear process leads only to modest hydrodynamic effects.

However, the distinction in two kinds is also made because the mass of the SN progenitors is very different for the two.

18.3.1 Hydrodynamic (core collapse) supernovae

A hydrodynamic supernova attributes its name to the events *in the shell* around the stellar core. The core resulting from the evolution of a star in the higher mass range collapses because it surpasses the critical Chandrasekhar limit (Ch. 17.1.4). The final core of stars with $M_{\text{init}} > 10$ $M_\odot$ consists mainly of Mg-Fe, of stars with $8 < M_{\text{init}} < 10$ $M_\odot$ of Ne-O. The core mass approaches $M_c \simeq 1.44$ $M_\odot$, the core radius $R_c \simeq 0.01$ $R_\odot$ (as in WDs).

18.3.1.1 Onset of the collapse

During the collapse one has *electron capture* or *photodesintegration* or both. For stars with $M_{\text{init}} < 15$ $M_\odot$ it is mainly electron capture, stars with $M_{\text{init}} > 15$ $M_\odot$ start with photodesintegration followed by electron capture. The transitions are, like all processes at the nuclear level, governed by a shift in the nuclear statistical equilibrium (Ch. 5.2.4), in this case at high densities.

- The typical example for photodesintegration (endothermic) in Fe-core stars is

$$^{56}\text{Fe} + 13\gamma \rightarrow 13\,^4\text{He} + 4\,\text{n} - 124 \text{ MeV} \quad . \tag{18.1}$$

- Electron capture is

$$^4_2\text{He} + 2e^- \longrightarrow 4\,\text{n} + 2\,\nu_e \quad . \tag{18.2}$$

The sum of these reactions is

$$^{56}_{26}\text{Fe} + 13\,\gamma + 26\,e^- \rightarrow 56\,\text{n} + 26\,\nu_e - 124 \text{ MeV} \quad . \tag{18.3}$$

In the process, a huge number of neutrinos is produced. SN 1987A was the first and single supernova thusfar of which neutrinos have been detected (see, e.g., Fig. 18.8). This event surely contributed to the Nobelprize being awarded 2002 in part for neutrino research (Ch.5.4).

Photodesintegration is an endothermic process while neutronization leads to a reduction of the number of charged particles. Thus both lead to a reduction in pressure. This immediately drives contraction, leading to inward motion of outer layers, and to an increase in pressure.

18.3.1.2 The collapse

The interior matter density increases, sped up the by the ongoing electron capture. When a stellar interior **neutronizes**, reducing the number of particles by a factor 2, but more importantly, eliminating the electric charges, the implosion generates a very densely packed core of neutrons of typically $R_N \simeq 10$ km. The collapse time (the free-fall time) is about 1 s.

At the same time, shell material is pulled in and falls onto the core. Gravitational energy is released to the amount of (see Ch. 4.2.2)

$$\Delta E_G = G\, M_c^2 \left(\frac{1}{R_N} - \frac{1}{R_c} \right) \simeq \frac{G\, M_c^2}{R_N} \simeq 3 \cdot 10^{53} \quad \text{erg}. \tag{18.4}$$

Note that the total observed energy of such a SN outburst (kinetic energy, electromagnetic energy plus energy in neutrinos) is roughly 10^{53} erg.

Above $\rho \simeq 5 \cdot 10^{11}$ g cm^{-3} neutrinos are captured. Their free path is reduced to about 200 km (for neutrino energies of 20 MeV; see Ch. 5.3).

18.3.1.3 End of the collapse and rebounce

The collapse ends when the density becomes high enough ($\rho_X \simeq 2.4 \cdot 10^{14}$ g cm^{-3}) for a sudden rise of the adiabatic index. Then the infall "bounces" and condenses to $\rho \gg \rho_X$, followed by a "rebounce": the infalling material rushes back and outward.

Perhaps a shock wave emerges, somewhere halfway in the collapsed core, where the rebounced material meets the infalling outer material at speeds above the sound speed. However, the energy available appears not to be sufficient to blast the stellar shell out of the potential well. The huge amount of neutrinos created provide apparently the further source of energy (see Ch. 5.3.1).

Using full hydrodynamic codes including all the possible nuclear reactions such explosions have been modelled. For graphic examples of the hydrodynamic effects see, e.g., Kitaur et al. (2006).

18.3.1.4 The explosion

The shock wave propagates outward and expels the outer layers. The surface of the progenitor starts to expand very fast, increasing the total surface with relatively slower cooling, so that the luminosity increases rapidly ($L = 4\pi R^2 \cdot \sigma T^4$). This then explains the rise of the light curve (see Fig. 18.1 and Fig. 18.2 left).

18.3.1.5 Decay of luminosity

After the fast rise in luminosity a maximum will be reached, followed by a decline in brightness. Note that this statement refers to a particular colour band (the visual one, or the blue one as in Fig. 18.1), and the dimming is foremost the effect of decreasing temperature and concomitant shift of the Planck function maximum to longer wavelengths (see the example of SN 1987A in Fig. 18.2). The luminosity stays at a high level for a long time. Much later, SNe start to radiate also in the radio domain with a non-thermal (bremsstrahlung) spectrum due to electrons spiralling in the magnetic field of the supernova remnant. For more on the radio emission see Weiler et al. (2002).

18.3.1.6 Endothermic nuclear reactions and light curve bump

At the end of the collapse and at rebounce (high density), endothermic fusion processes are possible, creating elements heavier than Fe, such as Ni, Co, U, etc. through rapid neutron capture (r-process; see Ch. 5.2.4). All this material will ultimately be expelled.

These SN light curves show the plateau phase (SN Type II P). The plateau is the excess luminosity due to the *radioactive decay* of elements having formed in the endothermic part of the SN II explosion process as well as of those being decay products. These elements include ^{56}Ni with a half life of 8.8 d, ^{56}CO with a half life of 77 d, ^{57}Co with a half life of 391 d, and ^{44}Ti with a half life of 78 yr. X-ray spectroscopy showed the presence of photon emission from the decay process of some of these elements.

18.3. THEORIES ABOUT SUPERNOVAE

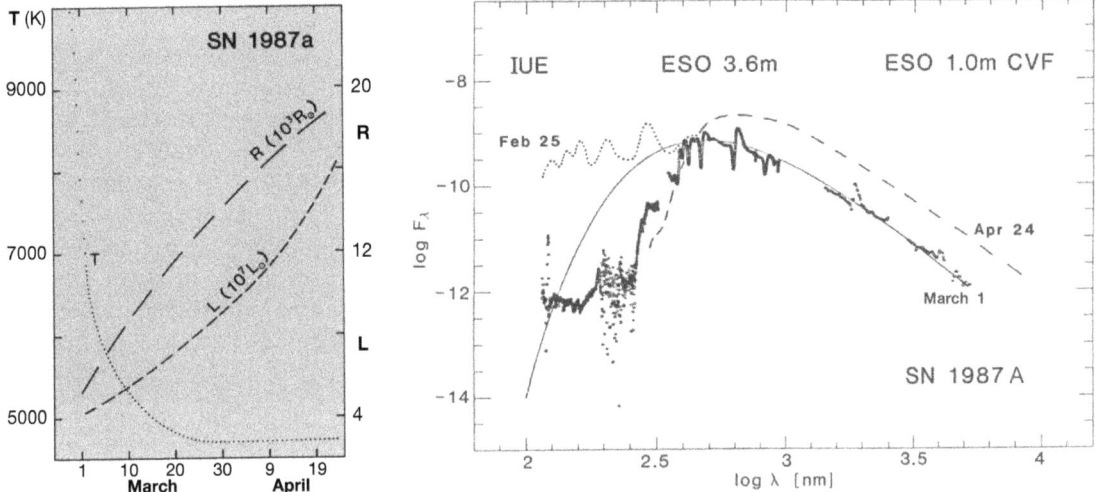

Figure 18.2: The evolution of SN 1987A in the LMC up to 60 days after the implosion of the core. Data from various sources. Figures from de Boer & Richtler (1987).
Left: Plotted is the run of the surface temperature T as well as the radius R of the expanding shell and the luminosity L (these in scaled solar units). The progenotor star to SN 1987A had $T_{\text{eff}} \sim 16000\,\text{K}$, $L \sim 1.1 \cdot 10^5\,L_\odot$, and $R \sim 20\,R_\odot$.
Right: Spectral energy distributions of supernova SN 1987A obtained on Feb 25: $\cdots$; March 1: full spectral data; April 24: ----. The data show the decreasing temperature (for the increase in luminosity see the left panel). The thin line is a Planck spectrum for March 1, allowing to see the opacity effects in the ultraviolet. Note the emission features in the near IR and visual.

18.3.1.7 Deceleration

The expansion velocity of the ejected material is observed to decrease. Two effects play a role.
1. The spectral lines used to measure the velocity are formed (as everywhere) in layers with $\tau \simeq 1$. Because of the expansion, the gas cools and dilutes, and $\tau \simeq 1$ moves *inward* into the expanding material. Thus with time, $\tau \simeq 1$ occurs in material having an ever smaller speed of expulsion.
2. Expelled material will at some level interact with ambient gas and is then decelerated.

18.3.2 Thermonuclear supernovae

A thermonuclear supernova derives its name from the events in the core: a sudden onset of C&O fusion (Eq. 5.16, Ch. 5.2.1) in a run-away process. This happens in WDs which acquire mass from a companion star (as already discussed in Ch. 17.1.5) thereby increasing the central T, ρ. ultimately surpassing the stability limit of the interior material. Then C ignites in the core. Because the temperature dependence of ϵ_C is very steep (see Ch. 5.2.1) the C-burning proceeds in an explosive manner and the star "deflagrates" (Fig. 18.3).

Two possibilities for the mass accretion are considered.
- A CO-WD has a large accretion rate of hydrogen gas, $\dot{M} \simeq 10^{-7}\,M_\odot\,\text{yr}^{-1}$. With a large accretion rate the Nova mechanism (see Ch. 19.4.4) is suppressed. Once the critical mass of $1.4\,M_\odot$ is reached (Chandrasekhar limit) C ignites explosive.
- A CO-WD possibly with He shell accretes He (no Nova events possible) with $\dot{M} \simeq 3 \cdot 10^{-8}$ $M_\odot\,\text{yr}^{-1}$. At some point He ignites in the shell (shell burning), which leads to a shockwave directed into the interior so that C ignites. The C-burning is explosive.

In all cases, the WD core material is very dense (degenerate gas with $\gamma = 4/3$, $n = 3$), the ignition is rapid and fusion "flames" proceed to the surface. Roughly half of the C is burned at so high temperatures and densities that further fusion processes even lead to the formation of Fe (at

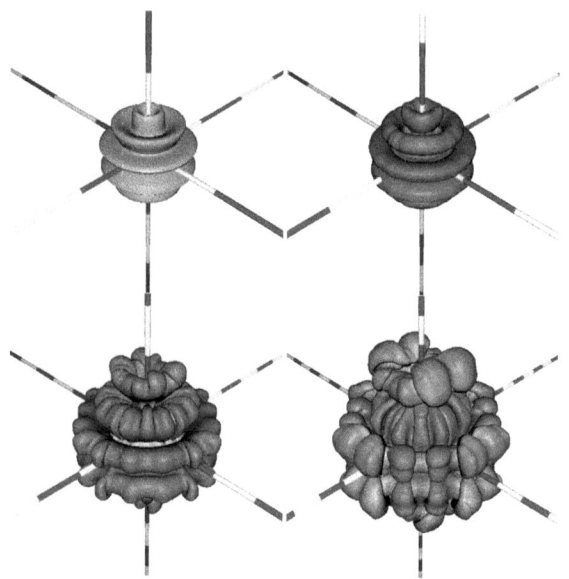

Figure 18.3: A supernova Type Ia is an example of a *thermonuclear explosion*, in which the star (the WD) is completely destroyed (deflagration) because of the explosive onset and speedy runaway of central C-burning. The panels show the evolution of the burning front at four points in time, at 0, 0.2, 0.4, and 0.6 seconds after ignition. Each yellow/blue unit along the axes represents 10^7 cm ($R_{\rm WD} \simeq 0.01$ $R_\odot \simeq 7\ 10^8$ cm). Modelling supernova explosions requires large computers because of all the plasmaphysics involved in such events. Depending on the scenario, the nuclear yields will differ. Figure from Travaglio et al. (2004).

least 0.35 $M_\odot$ of Fe is produced). The remainder of the star burns to intermediate elements. Due to the speed of the fusion, the star is completely disrupted (deflagration), leading to expansion velocities of ~ 5000 km s^{-1}. Note that this is a spatially chaotic process (see Fig. 18.3).

These thermonuclear SNe are the result of mass transfer in *low mass binaries* (see Ch. 19.4.4). Also because of the importance of SN Type Ia for cosmology (see Sect. 18.7 below) there is extensive literature (also in conference proceedings) about thermonuclear SNe; see e.g., Woosley (1990), Ruiz-Lapuente et al. (1997) and Hillebrandt & Niemeyer (2000).

For problems with the above WD - SN Ia scenario related with angular momentum accretion (leading to $M_{\rm WD} > 1.4$ $M_\odot$) see Ch. 19.4.5.

18.3.3 Other mechanisms to make SNe

A few other mechanisms have been proposed which lead to a SN explosion.

One is that very massive stars which evolve very fast become pair-instabilty SNe leaving perhaps a BH (see with Hypenovae, Ch. 18.5). This can especially occur with 'first stars" (Ch. 15).

In another scenario, two WDs (binary system) lose energy through gravitational interaction. This leads to reduction of their separation and ultimately to a merging of the two stars. Its structure may not be stable and a thermonuclear reaction may set in: SN Ia.

18.4 Supernovae and their progenitors

An attempt to summarize which stars become what kind of supernova was made by Filippenko (1991). However, the picture is still not fully clear among experts. For the evolution leading up to the SN event see chapters for the relevant stellar mass range.

- $M > 35$ $M_\odot$. These stars have very high mass loss during the LBV and WR stages (see Ch. 13.2.4). They lose all their outer hydrogen and end with a final mass of $\simeq 5\,M_\odot$: thus they become SN Ib or perhaps Ic.
- $35 > M > 25$ $M_\odot$. The stars have less heavy mass loss and may end up as regular SN Type II, and are perhaps the candidates to become BHs.
- $25 > M > 10$ $M_\odot$. The stars proceed to core collapse and become SN II.
- $10 > M > 8$ $M_\odot$. These stars do not develop degenerate CO cores (see also Fig. 10.24). If they would become SN, then as Type II. If the star is in a binary system, the mutual mass transfer may lead to SN Ib (and perhaps to SN Ia).

18.5. HYPERNOVAE / GAMMA-RAY BURSTS

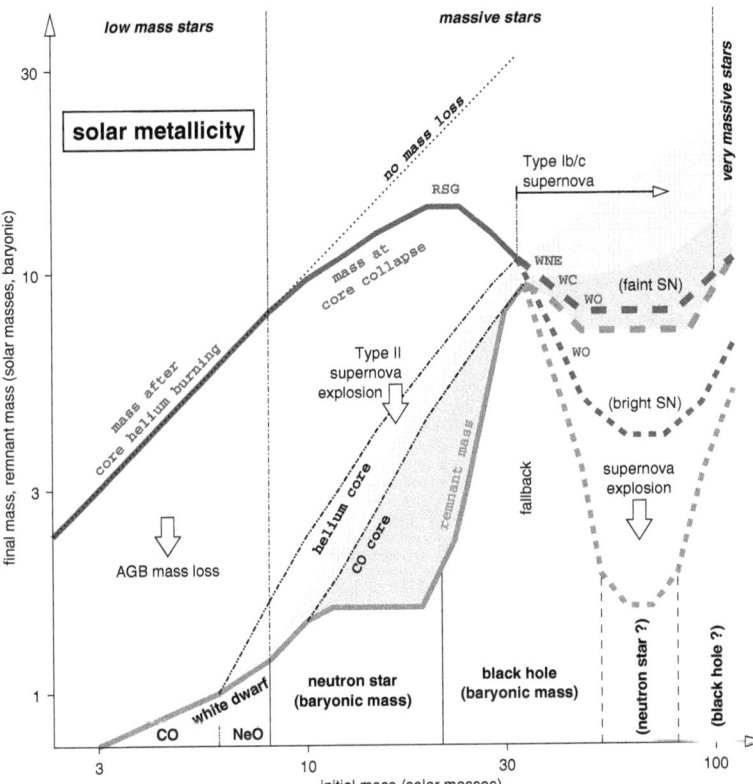

Figure 18.4: Calculation of stellar evolution up to the supernova phase. Shown is $M_{\rm init}$ verus the final mass and the remnant mass. The two thick lines show the mass at explosion and the remnant mass, as well as other information. With shading the mass released in metals is indicated. Note the mass range for SN Type II (after little mass loss) and for Type Ib/c (after heavy mass loss, and without H!). SN from large $M_{\rm init}$ may have quite different spectral apperance (Type Ib and/or Ic). Figure from Heger et al. (2003).

- $8 > M > 1$ M$_\odot$. These stars have large red-giant mass loss. They do *not become SN*. However, when such stars are part of close binary systems, there may be mass transfer and the WD becomes SN Ia.

Note: **Stars in close binary systems** develop, depending on their mass, the mass ratio, and separation in a large variety of ways (for examples see Ch. 19), many of which lead to a SN.

More on the relation of $M_{\rm init}$ to $M_{\rm final}$ is given in Sect. 18.6.

18.5 Hypernovae / Gamma-ray bursts

In the early 1980s flashes or bursts of γ-ray light were detected with space probes from diverse directions in the sky. It then also became known that the earlier US military satellites Vela (1967-1984), designed to detect the γ-ray flash from atomic bomb explosions, had seen such bursts. With the "Burst and Transient Source Experiment" (BATSE), on board of the "Compton Gamma Ray Observatory" (CGRO), launched 1991, many more such flashes were detected. The distribution over the sky of the γ-ray bursts (**GRBs**) appeared to be homogeneous so it was most likely they were extragalactic events. Speculations about the origin of these flashes abounded. The brevity of the flash, the time needed to retrieve and analyse the satellite γ-ray data, and the impossibility to get an accurate position on the sky prevented useful follow-up observations from the ground.

The Italian-Dutch Beppo-SAX satellite (since 1997) was specifically designed to get already on board accurate positions of the γ-ray flashes. Using the now well defined position of the first one so measured (GRB 970228) an optical "afterglow" (van Paradijs et al. 1997) and a radio glow (Frail et al. 1997) were detected (for more on radio glows see Weiler et al. 2002). Beppo-SAX also found the *GRB afterglow* in X-rays. Numerous such γ-ray bursts were followed up. For an example of three of them see Fig. 18.5, where for two of these also spectral energy distributions are given. For a review of afterglows see van Paradijs et al. (2000). The bursts got named after their "birthday-burstday", like the first one GRB 970228. A later bright one, GRB 970508, allowed to take spectra which resembled supernova spectra. GRB 970508, having a redshift of $z = 0.8$,

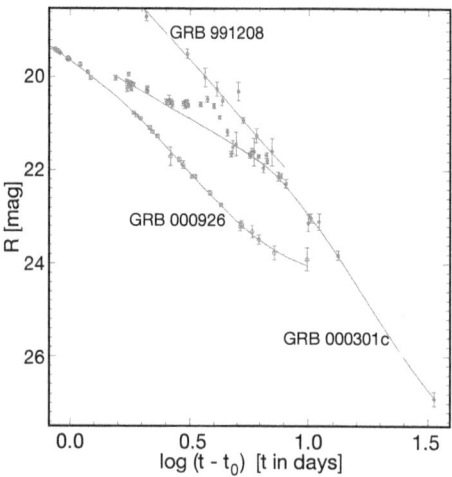 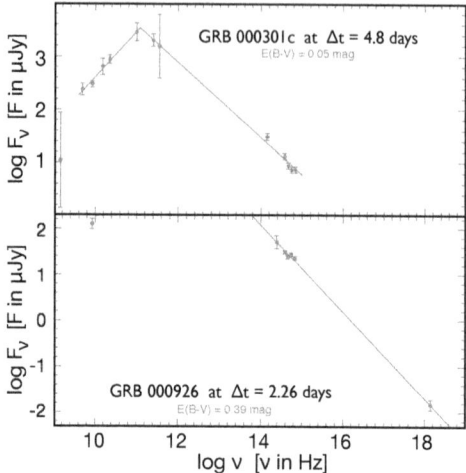

Figure 18.5: Observational data for gamma ray bursts (GRBs) are shown.
Left: Optical light curves of the "afterglow" of three well observed GRBs. The apparent R magnitude is plotted versus $\log(t - t_o)$. Note the difference in the exponential decay slopes as well as the "knee" in the light curve of GRB 000301c, signifying an effect of the medium near the GRB.
Right: Spectral energy distribution of the afterglows of GRB 000301c and GRB 000926 in $\log F_\nu$. Figures adapted from Sagar (2002).

supported the extragalactic nature. Given the brightness, both in γ-rays and in the visual, the event was dubbed "hypernova", because of the large energy involved. A later one, GRB 971224, was found to have even $z = 3.4$. The total energy, based on the observed levelness of f_ν over a large range in frequencies and the z-based distance, indeed would have to be gigantic, of the order of 10^{53} to 10^{54} erg. Such "explosions" thus would momentarily "outshine the big bang".

Expansion velocities of the material expelled seemed to be "super-luminal" (larger than c) and so alternative explanations came up. Today one sees the bright flashes as due to material expelled into a cone, leading also to the apparent superluminal velocities. This concept was supported by the sudden change in the brightness decline when the flash must have left the cone-like structure (becoming more diffuse, dispersing). It implied that the total energy involved was overestimated. The current best value is an GRB explosive energy of $\simeq 10^{51}$ to 10^{52} erg. The spectra resemble those of SN type Ib/Ic, so GRBs are nearly like supernovae of normal massive stars. A review on the theory of GRBs is available from Mészáros (2002) and on the SN nature from Woosley & Bloom (2006). A sketch of the possible schematics is given in Fig. 18.6.

The GRB supernovae at high redshift ($z > 6$ has been found!) show a low metal content in their spectra. These may be from Pop. III stars.

Supermassive stars (100 - 200 $M_\odot$; see Ch. 15) evolve much like a 60 $M_\odot$ star except that all fusion proceeds faster and is spatially (inside the star) and in time more blurred. The core region is quickly enriched with heavy elements and will, at the critical moment, come to Fe desintegration and electron capture, initiating the same chain of events as in hydrodynamic supernovae (as discussed in Sect. 18.3.1). However, given the much larger mass of the star, the expulsion of the shell is more difficult because of its larger mass. Although the thin outer layers will be much more accelerated (as with a whip where the speed is larger toward the thin end of its tip), the inner layers of the shell will eventually fall back on the already established neutron core. It is thus conceivable that, depending on how much mass is ejected and/or falling back, hypernovae may directly lead to the formation of a black hole (see Figs. 18.7 and 18.9).

Finally, GRBs of very short duration (< 1 s) have been found which also give off the highest energy photons ("hard" spectra). It is speculated that these come from two merging neutron stars. The neutron stars coalesce forming a black hole with the remaining material in an accretion disk.

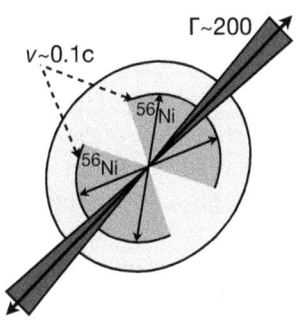

Figure 18.6: Possible spatial structure of the interior of a collapsing star becoming a supernova with a gamma ray burst (GRB). When the core of a massive star collapses (electron capture) material falls to the centre. With conservation of angular momentum the collapse is less pronounced in the equatorial plane, leading to an accretion disk. Along the polar axis a relativistic jet ($\Gamma \simeq 200$) develops from the core with an opening angle of less than a radian. Its radiation comes out in a flash, subsequently having an afterglow. GRBs apparently are so bright (at these very short wavelengths) because of the explosive energy ($\simeq 10^{51}$ erg) some is focussed into this cone-like structure. In a wider cone the actual explosion takes place ($v \simeq 0.1\ c$), making radioactive isotopes like ^{56}Ni which power the light curve ($\simeq 10^{52}$ erg). Explosive motions in the direction of the equatorial disk are blocked by the accretion disk. Figure after the model by Woosley & Heger (2006).

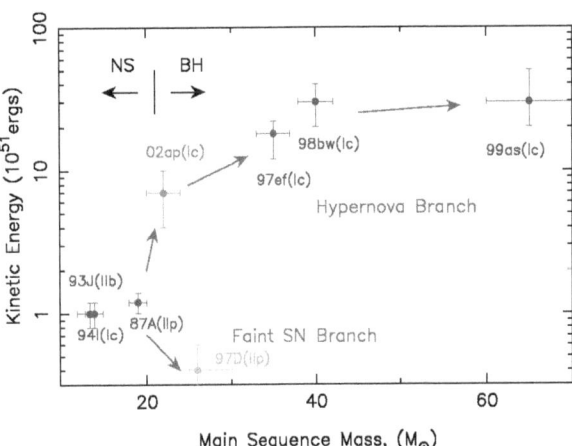

Figure 18.7: For several supernovae and hypernovae the type, the explosive energy and the likely M_{init} are shown. The limit marking the boundary between the nature of the remnant objects, a NS or a BH, is indicated. Figure from Nomoto et al. (2002).

18.6 Initial mass of stars becoming super- or hypernova

Calculations about the evolution of single massive stars have led to good insight in the possibility of stars to become supernova. Fig. 18.4 summarizes this knowledge for solar metallicity stars showing M_{init}, M_{final} and $M_{ejected}$ for the mass range of 3 to 100 $M_\odot$. Note the range of M_{init} which will lead to SNe Type II, and the one which (after heavy mass loss) becomes SN Type Ib/c. For metal poor stars similar results are available from Heger et al. (2002) as given in Fig. 18.9.

A correlation of observations of supernovae and hypernovae with their suspected M_{init} is shown in Fig. 18.7. Here the SNe are entered with their names and type on a M_{init} vs. explosive energy scale ($\frac{1}{2} M_{ejected} \times v_{ejecta}^2$).

Note that special kinds of SNe can emerge in binary systems (see, e.g., Fig. 19.16).

18.7 SN Type Ia and cosmology

The transition of a WD into a SN (see Sect. 18.3.2 above) is a transition of a star thought to be of almost exactly 1.4 $M_\odot$, so all such events should be physically the same.

The light curves of these supernovae have a characteristic temporal behaviour. This kind of SN was (long before their nature was understood) given the name SN Type Ia. These SNe are of great value for cosmology, since they allow to find the distance to the host galaxy from B_{max} of the light curve of the SN and the intrinsic Type Ia maximum M_B (see review by Branch 1998). This then leads to points in the very distant past on the curve of Hubble expansion. However, there is always doubt about the true nature of these SNe. Are they really so well defined (see Ch. 19.4.5)?

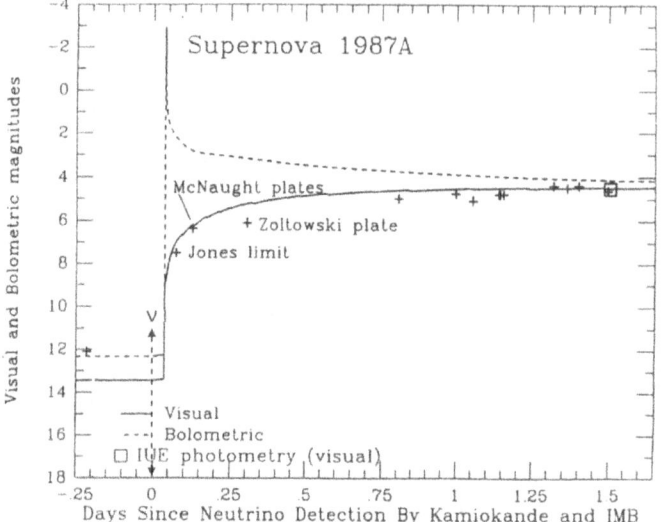

Figure 18.8: The first two days of SN 1987A. Shown are the curves of the visual brightness and of the bolometric magnitude. Brightness determinations in the visual are marked by +. Note the steep peak in $M_{\rm bol}$ at explosion.

Time $t = 0$ is choosen as the moment of the detection of neutrinos (ν) by Kamiokande and IMB. The Mont Blanc detector is said to have found neutrinos 5 hours earlier (+). Figure from Arnett et al. (1989).

18.8 SN 1987A in the LMC

18.8.1 SN 1987A itself

The 1987 SN in the LMC was a boon for astrophysics. It was nearby, satellites were available for observations in many wavelength ranges (including spectra), and (after a search of archives) the progenitor star was known: the blue supergiant Sk-69 202. It had type B3 with $V = 12.3$ mag, $T_{\rm eff} = 16000$ K, $L \simeq 10^5$ L$_\odot$ and $M \simeq 25$ M$_\odot$. Such a star would have had $M_{\rm init} \sim 50$ M$_\odot$. The low dispersion spectrum of Sk-69 202 available showed rather strong lines of nitrogen, indicating that perhaps a portion of the stellar outer layers had disappeared in the red giant wind and that convection had carried some of the products from fusion in the core to the exposed surface (see the [not quite applicable] Fig. 13.11).

SN 1987A was discovered on 24 Feb. 1987 on a photographic plate taken at 6 a.m. by Ian Shelton at the Las Campanas Observatory (Chile).

Very important was the detection of neutrinos (for neutrino experiments see Ch. 5.3.3), once people looked back into the measurements of the automatic neutrino detectors Kamiokande (Japan), IMB near Cleveland (Ohio), and Baksan (USSR). They helped pin down the moment of implosion to be 23 Feb. 1987, 7h 35m 35s (see Fig. 18.8).

The photosphere, defined as the layer with $\tau \simeq 1$, cooled from a temperature of $T \simeq 14000$ K at 1 day after outburst to $T \sim 6000$ K at day 10. It stabilized there at the location of the hydrogen recombination front (see McCray 1993) which moves inward into the gas, gas which itself is still moving away from the explosion centre.

The plateau in the lightcurve of SN 1987A was shown to really be due to β-decay of ^{56}Ni (half life 8.8 days) into ^{56}Co and on into ^{56}Fe releasing some 10^3 MeV per atom. The Solar Maximum Mission detected γ line radiation of ^{56}Co at 847 and 1238 keV in Aug.-Oct. 1987, with the Kuiper Airborne Observatory infrared emission lines of Ni II and Co II were detected.

The light curve of SN 1987A will slowly decay, being powered (as mentioned in Sect. 18.3.1) initially by radioactive decay of ^{56}Co but later by decay of other created nuclei such as ^{57}Co, ^{44}Ti, and ^{22}Na with much longer half life (see e.g. Arnett et al. 1989). After such a SN event a neutron star remains. The pulsar of SN 1987A has not been found (yet).

18.8.2 Effects of SN 1987A on its environment

Around SN 1987A *light echos* were discovered. The "flash" of the explosion (flash: a SN is considered to be bright at most 100 days; see Fig. 18.1) illuminates reflecting material (dusty

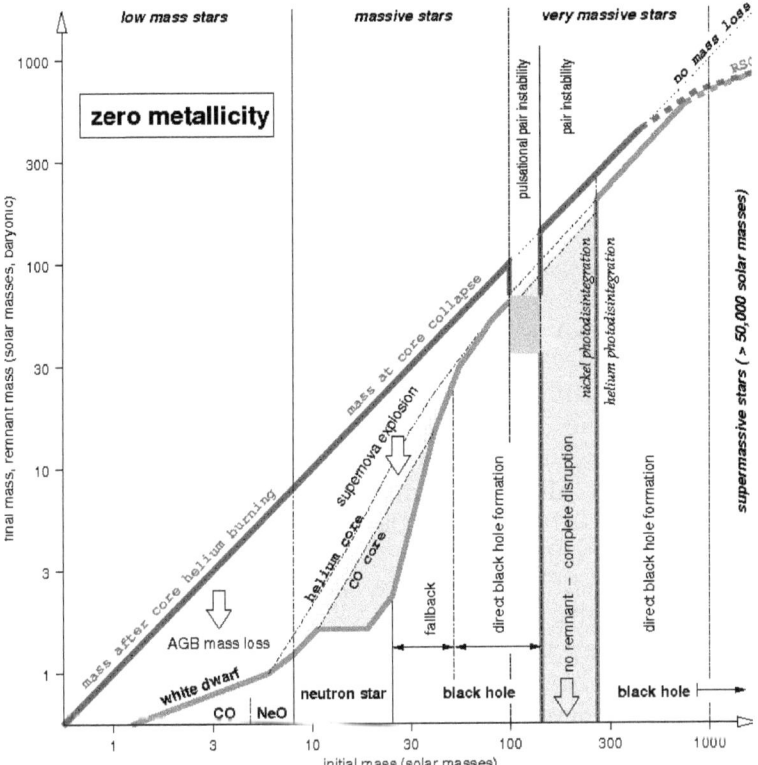

Figure 18.9: The fate of first stars for the full mass range of 1 - 1000 M$_\odot$.
Three lines run from the lower left to the upper right. The 45° line is the line of no mass loss. For zero-metallicity stars no or very little mass loss is expected. The heavy line marks the mass of the collapsed remnant. The nature of this remnant is given at the bottom in the diagram; if it is a WD, the nature of the core is given at the bottom. The dot-dashed line gives the amount of He produced.
Figure from Heger & Woosley (2002).
Compare with Fig. 18.4 for solar metallicity stars.

clouds) in the vicinity. The total travel time of light reflected depends, of course, on the location of the cloud (see e.g., Spyromilio et al. 1995).

After about a year the first signs of *dust* having formed in the SN 1987A ejected material became evident. Total extinction increased to $A_V \simeq 1$ mag after 2 years.

After three years a *luminous ring* appeared around SN 1987A (see Fig. 14.7). This ring (with diameter of $\simeq 0.3$ pc) is believed to be circumstellar gas having been ejected during the red giant wind phase of the progenitor. The two other rings are perhaps related with the overall rotation related mass-loss profile of the progenotor star.

The explosion led to material rushing out with $v_{\rm exp} \simeq 2 \cdot 10^4$ km s^{-1}. This material was predicted to hit the circumstellar ring in roughly 17 years. Indeed, the ring brightened considerably in 2003 due to the collisional interaction of the ejecta wind the wind blown ring (see, e.g., Park et al. 2005).

For a review of the numerous aspects of SN 1987A see Arnett et al. (1989) and McCray (1993).

18.9 Endproduct of first stars: $M_{\rm init}$ to $M_{\rm final}$

First stars (Ch. 15) have no metals and thus start out with limited fusion processes. In the course of evolution, He and C are formed which get distributed through the star because of the strong convection and the merian circulation (see Ch. 14).

The endproducts of the evolution of first stars range from WDs through NSs to BHs. Which star becomes what is shown in Fig. 18.9.

References

Arnett W.D., Bahcall J.N., Kirshner R.P., & Woosley S.E., 1989, ARAA 27, 629; *Supernova 1987A*
Branch, D. 1998, ARAA 36, 17; *Type Ia Supernovae and the Hubble Constant*
Clark, D.H., & Stephenson, F.R. 1977, *The Historical Supernovae*, Pergamon Press
de Boer, K.S., & Richtler, T. 1987, Sterne und Weltraum 26, 388

Filippenko, A.V. 1991, *The Progenitors and Explosion Mechanisms of Supernovae* in "Supernovae and stellar evolution", A. Ray & T. Velusamy (eds.); World Scientific, p. 58;

Filippenko, A.V. 1997, ARAA 35, 309; *Optical Spectra of Supernovae*

Frail, D.A., Kulkarni, S.R., Costa, E., Frontera, F., Heise, J., Feroci, M., Piro, L., dal Fiume, D., Nicastro, L., Palazzi, E., & Jager, R. 1997, ApJ 483, L91

Harkness, R.P., & Wheeler, J.C. 1990, *Classification of Supernovae*, in "Supernovae", A.G. Petschek (ed.); Springer, Heidelberg; p. 1

Heger, A., & Woosley, S.E. 2002, ApJ 567, 532

Heger, A., Woosley, S.E., Fryer, C.L., & Langer, N. 2003, in "From Twilight to Highlight: the Physics of Supernovae", W. Hillebrandt & B. Leibundgut (eds.); Springer, Heidelberg; p. 3

Hillebrandt, W. & Niemeyer, J.C. 2000, ARAA 38, 191; *Type Ia Supernova Explosion Models*

Kitaur, F.S., Janka, H.-T., & Hillebrandt, W. 2006, A&A 450, 345

McCray, R. 1993, ARAA 31, 175; *Supernova 1987A Revisited*

Mészáros, P. 2002, ARAA 40, 137; *Theories of Gamma-Ray Bursts*

Nomoto, K., Maeda, K., Umeda, H., Ohkubo, T., Deng, J., & Mazzali, P. 2002, IAU Symp. 212, "A massive star Odyssee", K.A. van der Hucht, A. Herrero & C. Esteban (eds.); p. 395

Ruiz-Lapuente, P., Canal, R., & Isertn, J. (eds.). 1995, "Thermonuclear Supernovae", Kluwer, Dordrecht

Park, S., Zhekov, S.A., Burrows, D.N., & McCray, R. 2005, ApJ 634, L73

Sagar, R. 2002, Bull. Astr. Soc. India, 29, 215

Spyromilio, J., Malin, D.F., Allen, D.A., Steer, C.J., & Couch, W.J. 1995, MNRAS 274, 256

Travaglio, C., Hillebrandt, W., Reinecke, M., & Thielemann, F.-K. 2004, A&A 425, 1029

van Paradijs, J., Groot, P.J., Galama, T., Kouveliotou, C., & 27 al. 1997, Nature 386, 686

van Paradijs, J., Kouveliotou, C., & Wijers, R.A.M.J. 2000, ARAA 38, 379; *Gamma-Ray Burst Afterglows*

Weiler, K., Panagia, N., Montes, M.J., & Sramek, R.A. 2002, ARAA 40, 387; *Radio Emission from Supernovae and Gamma-Ray Bursts*

Woosley, S.E. 1990, in "Supernovae", A.G. Petschek (ed.); Springer, Heidelberg; p. 182; *Type I supernovae: Carbon Deflagration and Detonation*

Woosley, S.E., & Bloom, J.S. 2006, ARAA 44, 507; *The Supernova-Gamma-Ray Burst Connection*

Woosley, S.E., & Heger, A. 2006, in "The Supernova Gamma-Ray Burst Connection", AIP Conf. Ser. 836, 398; Holt, S.-S., Gehrels, N., & Nousek, J.A. (eds.), p. 398; astro-ph/0604131

Woosley, S.E., & Janka, H.-T. 2005, Nature Physics 1, 147

Chapter 19

Evolution of binary stars

19.1 Introduction

Phenomena of a rather large collection of objects, such as novae, certain kinds of supernovae, aspects of Wolf-Rayet stars, pulsars with stellar companions and extremely X-ray bright (and sometimes variable) point sources did not simply fit in the schedules of normal stellar evolution. With time it became clear that evolution in a binary system could explain many of these phenomena. Since the majority of stars exists in binary systems and since stars in binary systems influence each other, the study of binary stars has developed to a very important part of stellar astrophysics. This is in particular true because of the large range of mass combinations possible for such systems.

Apart from the "strange objects" and the "unexplained phenomena" mentioned above, there are several classic reasons why investigating binarity is important:

- More than half of all stars seen in the sky are members of binary or multiple stellar systems.

- Binary stars are present among all types of non-variable and variable stars.

- Observations of binaries and their evaluation is indispensable for the determination of stellar masses. In addition, one can derive stellar radii, mean densities, luminosity ratios, centre-to-limb variations, atmospheric parameters, stellar rotation.

Of concern are thus the physical systems (nowadays called *binary stars*, or *binaries*, for short), which are gravitationally bound, and not the optical systems which are the product of accidental close projection onto the sky (called *double stars*).

Research in binary star evolution developed in the 1960s. First calculations of binary star evolution were made by Bogdan Paczynski (1966) in Warsaw and by Rudolf Kippenhahn and collaborators in Göttingen (1967 and later). The Wolf-Rayet star scenario was explored notably in Warsaw (Paczynski 1967) and in Amsterdam (1976) by Ed van den Heuvel.

In this chapter aspects of *evolution of stars in binary systems* (the differences with respect to single star evolution), and of the evolution of the system as binary are presented. Several reviews on (parts of) the topic are available, e.g., by Paczynski (1971b), Pringle & Wade (1985), De Loore & Doom (1992), Iben (1991), van den Heuvel (1994) and Eggleton (2006). On massive stars and on interacting and common envelope objects there are reviews by Podsiadlowski et al. (1992), Iben & Livio (1993), Livio (1994), Vanbeveren et al. (1998), and Taam & Sandquist (2000); on degenerate stars in binaries by Canal et al. (1990), Cowley (1992), Verbunt (1993), Verbunt & van den Heuvel (1995), Kahabka & van den Heuvel (1997) and Tauris & van den Heuvel (2006). Edited volumes are those by van Lewin et al. (1995) and Lewin & van der Klis (2006).

Because of the large variety of combinations possible for binary stars (mass, mass ratios, seperation, orbit ellipticity) there is no simple way to describe the phenomena of binary stars. In Sects. 19.2 and 19.3 general principles will be presented, using two examples from early binary research. In the sections thereafter a few examples are discussed which further illustrate the mechanisms and provide explanations for many previously unexplainable stellar states.

19.2 Equipotential surfaces

19.2.1 Mathematical formulation

The total potential of a binary system is the sum of the gravitational and the rotational potential.

A binary system has two components (called *primary component*, or *primary*, and *secondary component*, or *secondary*) with masses M_1 and M_2. The definition includes that initially $M_1 > M_2$. (Sometimes the components are just called after their mass, M_1 and M_2.)

The co-rotating cartesian coordinate system XYZ has its origin in the centre of gravity S. The binary (the stars are assumed to be point masses) is rotating around the axis through the centre of gravity (see Fig.19.1a). Let a be the distance between the components, then the position of the center of gravity is defined by

$$a_1 M_1 = a_2 M_2 \quad \text{or} \quad \frac{a_1}{a_2} = \frac{M_2}{M_1} \quad \text{with} \quad a = a_1 + a_2 \quad . \tag{19.1}$$

Some point $Q(x, y, z)$ is at distance r_1 and r_2 from M_1 and M_2, respectively, and at distance s from the rotation axis. Note that

$$r_1^2 = x^2 + y^2 + z^2, \quad r_2^2 = (a-x)^2 + y^2 + z^2, \quad \text{and thus} \quad s^2 = (a_1 - x)^2 + y^2 \quad . \tag{19.2}$$

Then the effective potential in point Q is

$$\Omega(x, y, z) = -G \frac{M_1}{r_1} - G \frac{M_2}{r_2} - \Omega_s \tag{19.3}$$

with Ω_s the potential of the centrifugal acceleration, $\Omega_s = \frac{1}{2}\omega^2 s^2$, in which the orbital frequency of the system is from Kepler's 3rd law, $\omega^2 = (2\pi/P)^2 = G(M_1 + M_2)/a^3$.

Combining these relations leads to an expression for the gravitational potential in $Q(x, y, z)$

$$\frac{\Omega(Q)}{-G} = \left(\frac{M_1}{r_1} + \frac{M_2}{r_2}\right) + \frac{1}{2} \frac{M_1 + M_2}{a^3} s^2 \quad . \tag{19.4}$$

An **equipotential surface** is a surface with the condition $\Omega = $ constant.

19.2.2 Graphical representation of equipotential surfaces

The graphical representation of a binary system can be as a

→ **2-dimensional cut** (contourlines of Ω) in the **meridional plane** (the X,Y-plane, see Fig. 19.1a) and in the **equatorial plane** (the X, Z-plane, see Fig. 19.1b), or as

→ **3-dimensional structure** (the X, Y-plane with Ω in perspective, see Fig. 19.3).

Of special interest are:

→ the **Lagrangian points** L_1 to L_3 where all forces cancel out, i.e. $d\Omega/dr = 0$, and

→ the **critical lobe** or **Roche lobe**, the surface through the *inner* Lagrangian point L_1.

The Roche lobe just includes the binary components whose individual lobes are in contact in L_1 (see Fig. 19.1), or, 3-dimensionally, the saddle of the double potential well (see Fig. 19.3).

The **volume** of the Roche lobe has been evaluated by different authors. A simple but sufficient approximation was given by Kopal (1959). The mean radius R_{L1} of the sphere approximating the Roche volume around M_1 (similar around M_2) is

$$R_{L1}/a = 0.38 + 0.2 \, \log(M_1/M_2) \quad . \tag{19.5}$$

valid in the range $0.3 < \frac{M_1}{M_2} < 20$. This equation will be used in Sect. 19.3.1.

An equation valid for all values of $M_1/M_2 = q$ to within a few percent (Eggleton 1983) is

$$\frac{R_{L1}}{a} = \frac{0.49 \, q^{2/3}}{0.6 \, q^{2/3} + \ln(1 + q^{1/3})} \quad . \tag{19.6}$$

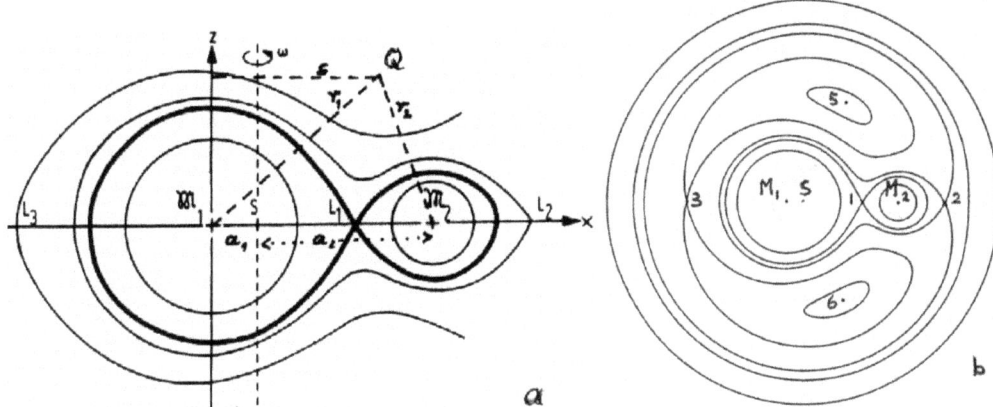

Figure 19.1: Equipotential surfaces: (a) meridional cross section (x, z plane), (b) equatorial cross section (x, y plane). The stars have mass M_1 and M_2, the centre of gravity is S. The Lagrangian points are indicated as L_1 - L_3 in (a), by their number 1,2,3, 5,6 in (b). L_1 is the point relevant for mass transfer. The geometry of point Q is indicated to help in the derivation of Eq. 19.3.

19.3 Mass exchange

19.3.1 General case

Binaries can be divided into close and wide systems. Individual stellar components in **wide binaries** evolve as if they were single stars (as in Chs. 10 and 13). But, what is a wide binary in comparison to a close binary?

Stellar systems are called **close binaries** when *one of the components will fill its individual Roche lobe in the course of stellar evolution*. Close binaries are supposed to be separated (*"detached"*) systems when they start their evolution on the main sequence. But, as demonstrated in the chapters on evolution of single stars, stellar radii may increase quite a lot during subsequent phases of evolution. As a consequence, the more massive component (which has the more rapid evolution) may start to fill its Roche lobe and then mass must (and will) flow to the less massive component via the inner Lagrangian point L_1, over the saddle of Fig. 19.3.

The surface brightness of a star is, of course, given by $F = \sigma T_{\text{eff}}^4$ (Stefan-Boltzmann law) and its total luminosity $L = 4\pi R^2 F = 4\pi R^2 \cdot \sigma T_{\text{eff}}^4$. Thus

$$R^2 = \frac{L}{4\pi \sigma T_{\text{eff}}^4}\ .$$

Using this equation one can easily calculate the change of stellar radii during evolution in relation with changes in L and T_{eff}. The size of the radius can be read from the theoretical HRD of Fig. 1.3.

Kippenhahn & Weigert (1967a) distinguish three binary evolution states which may lead to filling of the limited volume.

- Case A: expansion of the component during hydrogen burning close to the main sequence. The total expansion during this phase is, however, relatively small, but in a very close system the evolving star may reach the Roche lobe.

- Case B: expansion after the onset of hydrogen shell-burning and leading to crossing the Hertzsprung-Russell diagram on the way to the red-giant branch. Compared to Case A, this case is a dramatic increase of the stellar radius and with a high expansion rate.

- Case C: all subsequent phases of increase of the stellar radius. Here, the most important phase is the expansion at the onset of helium shell burning.

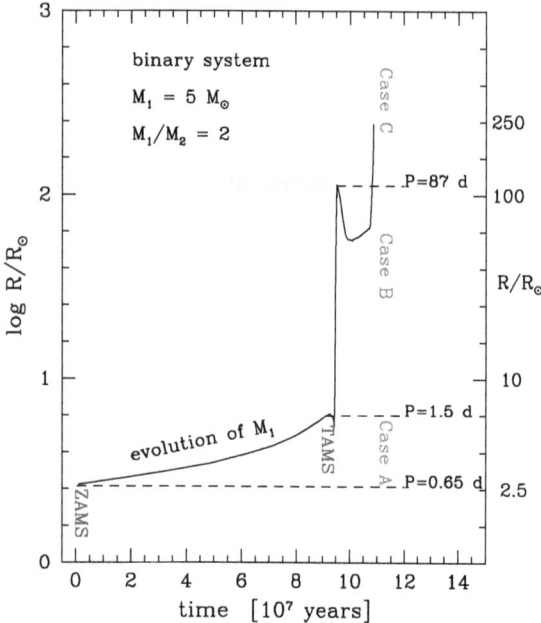

Figure 19.2: **Binary Example 1.** In a close binary system mass may be transferred because the primary star will, in the course of its evolution, increase in radius so that material may move to the secondary star through the Langrangian point L_1. The graph shows the evolution of the radius of a 5 $M_\odot$ star. H-ignition (ZAMS), transition to H-shell burning (TAMS) and He-ignition are labelled.
The figure indicates $M_1/M_2 = 2$, but the behaviour shown is only for a single 5 $M_\odot$ star whithout companion. The figure is similar to that in the early treatise on double star evolution by Paczynski (1971a), now using new stellar evolution data. Indicated are the three possible binary evolution states (mass transfer) Case A, B, and C.

These cases are illustrated in Fig. 19.2 for the radius evolution of a 5 $M_\odot$ star. In later evolutionary calculations some authors distinguished between *early* und *late* Cases B and Case C, resp., i.e.
– early Case B/Case C, if the outermost stellar layers are still radiative,
– late Case B/Case C, if the outer layers are convective.

As binary Example 1, consider the evolution of a close binary with a primary of $M_1 = 5$ $M_\odot$ and the mass ratio $M_2/M_1 = 0.5$. This case was discussed by Paczynski (1971a) the pioneer in the field. From Eq. 19.5 one derives that the primary fills its Roche lobe as soon as

$$R_{L1} = 0.44\, a \quad . \tag{19.7}$$

Kepler's 3$^{\rm rd}$ law, normally written as

$$\frac{a^3}{P^2} = M_1 + M_2 \quad (a \text{ in AU}, P \text{ in y and } M \text{ in } M_\odot),$$

can be expressed in logarithmic form as

$$\log a = \frac{2}{3} \log P + \frac{1}{3} \log(M_1 + M_2) + 0.624$$

with a in $R_\odot$, P in days, and M in $M_\odot$. Using this equation one thus gets the following condition for Roche lobe filling of this special binary system

$$\log P = 1.5 \log R_{L1} - 0.84 \quad (\text{for } M_1 = 5\, M_\odot \text{ and } M_2/M_1 = 0.5) \quad . \tag{19.8}$$

From the evolution of the radius of the 5 $M_\odot$ primary (Fig. 19.2) one can calculate the orbital period P (and the separation a) for which at any phase the primary fills its Roche lobe. On the main sequence the star will soon reach $R_1 = 2.7\, R_\odot$. Therefore, the initial condition for Roche lobe filling on the main sequence must be an orbital period of $P = 0.65$ d (and a separation of $a = 6.18\, R_\odot$). In fact, Case A is realized up to an initial period of $P = 1.5$ d. Initial periods between 1.5 d and 87 d (larger separation a) lead not to Case A but to Case B. Up to the large period of 4300 d (or nearly 12 yr and separation exceeding 2000 $R_\odot$) Case C will occur. We conclude that even binary systems with large dimensions must be called "close binaries", i.e., they are systems in which ultimately mass exchange (mass transfer) takes place via the inner Lagrangian point.

19.3. MASS EXCHANGE

Table 19.1: Rates $\dot{M}$ for mass transfer in the Kelvin-Helmholtz time scale (using Eq. 19.10)

Sp.type	M [$M_\odot$]	L [$L_\odot$]	R [$R_\odot$]	$\dot{M}$ [$M_\odot$ yr^{-1}]
B0 V	18	$5 \cdot 10^4$	7.5	$6.7 \cdot 10^{-4}$
A0 V	3	50	2.5	$1.3 \cdot 10^{-6}$
G2 V	1	1	1	$3.2 \cdot 10^{-8}$

In this example, the **time scale** of mass exchange apparently depends on the structure of the outer stellar layers.

– For stars with deep outer convective zones (like in later evolutionary states of stars with $M_{\text{init}} > 1.4$ $M_\odot$ which may become binaries of Cases B/C, or stars of the lower main sequence if Case A occurs), the mass loss may be approximated by

$$\dot{M} = f \frac{M}{P} \left(\frac{\Delta R}{R} \right)^3 \tag{19.9}$$

where the parameters refer to the primary. ΔR is the radius increase of the primary, and the factor f (in the order of 10) depends mainly on the density profile in the stellar interior.

– If the envelope of the star is radiative the mass transfer occurs in the thermal or Kelvin-Helmholtz time scale (see Ch. 4.2.2)

$$t_{\text{KH}} = \frac{G M^2}{R L} = 3.1 \cdot 10^7 \frac{M^2}{R L} \quad \text{yr}$$

with all variables in solar units. It follows for the mass loss rate

$$\dot{M} = \frac{M}{t_{\text{KH}}} = 3.2 \cdot 10^{-8} \frac{R L}{M} \quad M_\odot \text{ yr}^{-1} \quad . \tag{19.10}$$

Tab. 19.1 lists some mass exchange rates calculated from this equation.

– Finally, one assumes that, after equality of the stellar masses, mass transfer continues on the nuclear time scale, i.e., as long as the "primary" continues to evolve (expand) as it is doing for its evolutionary phase.

19.3.2 Conservative mass exchange

Mass exchange is called *conservative* if **no mass and no angular momentum is lost by the binary system**. Conservation of the orbital angular momentum J means

$$J = a_1 M_1 V_1 + a_2 M_2 V_2 = a \frac{M_1 M_2}{M_1 + M_2} V = \text{const} \;,$$

where for the second equation one uses: $a = a_1 + a_2$, the separation of the components, $a_1/a_2 = M_2/M_1$, and $V = V_1 + V_2$, their orbital velocities.

Under the assumption of circular orbits, the orbital acceleration is

$$\frac{V^2}{a} = G \frac{M_1 + M_2}{a^2} \;,$$

and therefore

$$J = \sqrt{\frac{G a}{M_1 + M_2}} M_1 M_2 \quad . \tag{19.11}$$

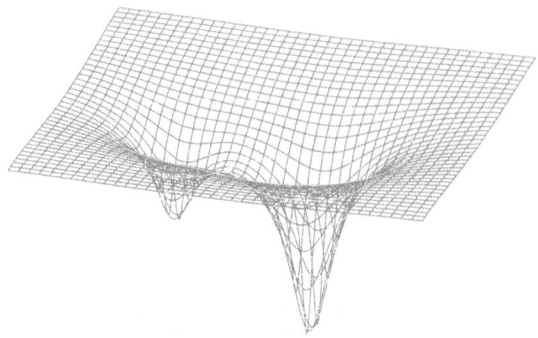

Figure 19.3: A double star system has a double gravitational potential well. The star at right has more mass than the star at left. Inside the potential well contours of equal gravitational force are indicated, shown in projection below the potential wells, too (compare with Fig. 19.1). When the more massive star evolves into a RG, it will fill its well and material will flow over the "saddle" into the well of the less massive star (as in Fig. 19.4), initiating the mass transfer. Then the deep well gets less deep, the shallower well deeper. With conservation of angular momentum the separation of the stars becomes smaller making the saddle deeper thus enhancing the mass transfer. In double stars the **mass transfer is driven by the evolution of the stars**, the *consequences for the system are determined by the shape of the potential well*. (Figure courtesy Manuel Metz.)

Note, that this formula can also be derived from $J = M \omega a^2$, where $\omega = 2\pi/P$. Introducing the mass ratio $\mu = M_2/M_1$ (!), the previous equation can be transformed into

$$a \sim \frac{1}{(M_1 M_2)^2} \sim \frac{(1+\mu)^4}{\mu^2} \quad . \tag{19.12}$$

Eq. 19.12 shows that the **separation runs into a minimum for** $\mu = 1$, i.e., as soon as the system reaches equal mass of the components; the value of a is larger for both $\mu > 1$ and $\mu < 1$. Accordingly, the dimension of the Roche lobe changes following Eq. 19.5 and the orbital period increases following Kepler's 3$^{\rm rd}$ law, or, after small transformations according to

$$\frac{P}{P_{\rm init}} = \left(\frac{M_{1,\rm init} \, M_{2,\rm init}}{M_1 \, M_2} \right)^3 , \tag{19.13}$$

where the subscript "init" denotes the initial case, i.e., the beginning of the mass-tranfer phase.

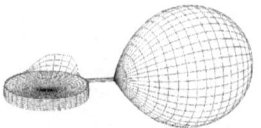 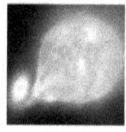

Figure 19.4: When the more massive star in a close double system evolves into a red giant, it will at some point in time fill its potential well and material will flow over to the less massive star, it fills and overflows the "Roche lobe" (the critical contour visible in Fig. 19.3 and Fig. 19.1). The material flows toward the less massive star and is subject to the Coriolis force, thus leading to an accretion disk (Sect. 19.3.4.2). The spot on which fresh material falls will react by violent motion and likely produces X-rays. Figure at left from Lederle & Kimeswenger (2003), the figure at right is an artist impression of such a transfer (from ESO PR 31-07).

19.3. MASS EXCHANGE

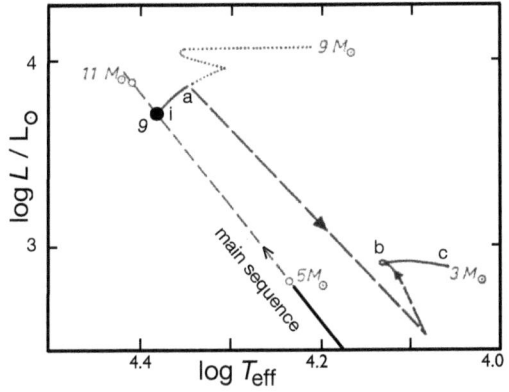

Phase	Component 1	Component 2	M_1	M_2	a	R_1	P [d]	t [My]
i			9	5	13.2	3.8	1.5	0
a			9	5	13.2	5.7	1.5	10
b			3.7	10.3	20	5.3	3.7	10.1
c			3	11	20	6.7	3.8	20

Figure 19.5: **Binary Example 2**. Simplified evolution of a close (9 $M_\odot$ + 5 $M_\odot$) binary system with Case A mass exchange according to Kippenhahn & Weigert (1967a). Near the end of the MS evolution the Roche lobe in the close binary system is filled.
Left: Evolutionary tracks showing how the evolution is changed due to the mass transfer (heavy dashed line for the primary 9 $M_\odot$ star with dotted extension for the evolution of a single 9 $M_\odot$ star). The secondary in this binary system evolves up the MS. **Right:** Actual shape of the binary system with tabulated values with M in $M_\odot$, a (the semi major axis of the system) and R in $R_\odot$.
· Phase i: The system starts with both stars on the main sequence with masses 9 $M_\odot$ and 5 $M_\odot$ and a radius of the primary of 3.8 $R_\odot$. The orbital period and major axis match (see tabular data). Both components are well *detached* within their individual Roche lobes.
· Phase a: After 10^7 yr the primary fills its Roche lobe of 5.7 $R_\odot$. The binary is called a *semi-detached* system. A rapid mass transfer starts via the inner Lagrangian point with the mass transfer (mass loss) rate of $\dot{M} = 8.8 \cdot 10^{-5} M_\odot \, \mathrm{yr}^{-1}$. The phase lasts only about $6 \cdot 10^4$ yr.
· Phase b: At the end of the rapid mass transfer only 3.7 $M_\odot$ are left for the primary. The star continues with a slow mass transfer of only $\dot{M} = 4.4 \cdot 10^{-8} M_\odot \, \mathrm{yr}^{-1}$. In the mean time the now more massive secondary star fills its Roche lobe.
· Phase c: When the slow mass transfer stops the primary has no more than 3 $M_\odot$ and the secondary has increased to 11 $M_\odot$. Both stars are filling their limited volumes, and the binary is called a *contact* system. The period has increased to $P = 3.8$ d, the separation to $a = 20$ $R_\odot$.

The **example of a 9 $M_\odot$ + 5$M_\odot$ star** as investigated by Kippenhahn & Weigert (1967a) was a revelation for the scientific community. For the first time it could be understood why the more evolved member of a binary system had less mass than the less evolved component. The fate of a 9 $M_\odot$ + 5 $M_\odot$ system is sketched in Fig. 19.5 ("Binary Example 2").

19.3.3 Classification scheme for close binary systems

These early calculations led to a new classification of binary systems. The configuration of the stars in the Roche lobe is divided into
 – D, detached systems,
 – SD, semi-detached systems, and
 – C, contact systems.
Specific stellar examples are mentioned with Fig. 19.6.

A natural continuation of this sequence leads to the over-contact systems which develop a common envelope filling a higher-order equipotential surface:
 – CE, common envelope systems (Sect. 19.3.4.3).

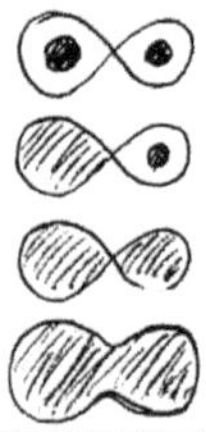

Class	Configuration	Type of eclipsing binary
D	detached	Algol
SD	semi detached	Algol, β Lyrae
C	contact	W UMa
	over-contact, common envelope	W UMa

Figure 19.6: Definition of configurations of close binary systems. The star types mentioned are rather of the lower mass range. Algol types of eclipsing binaries, e.g., β Persei, belong to classes D and SD. β Lyrae types are semi-detached (see Fig. 19.7, "Binary Example 3"). W UMa types belong to class C or are in over-contact.

19.3.4 Complications

19.3.4.1 Non-conservative mass exchange

Mass exchange is called *non-conservative* if mass and angular momentum is lost by the binary system. This can be described by two free parameters:
- β, the fraction of the mass loss from star 1 captured by star 2,
- α, the loss of specific angular momentum (i.e. angular momentum per unit of mass).

Then, the change of specific angular momentum J of the primary can be written as

$$\frac{dJ}{dM_1} = \alpha(1-\beta)\frac{2\pi a^2}{P} \quad . \tag{19.14}$$

Using Kepler's 3rd law in the form

$$\frac{2\pi\, a^2}{P} = \sqrt{G(M_1+M_2)a}$$

Eq. 19.14 gets as final form

$$\frac{dJ}{dM_1} = \alpha(1-\beta)\sqrt{G(M_1+M_2)a} \quad . \tag{19.15}$$

This equation together with Eq. 19.11 is the differential equation for J or a as function of M_1. (But note that the angular momentum of the axial rotation of the components and the angular momentum of the transferred mass has been neglected.)

Solutions of Eq. 19.15 have been given e.g. by Podsiadlowski et al. (1992) of the form

$$\frac{a}{a_{\text{init}}} = \begin{cases} \frac{M_1+M_2}{M_{1,\text{init}}+M_{2,\text{init}}}\left(\frac{M_1}{M_{1,\text{init}}}\right)^{2(\alpha-1)}\exp\left[\frac{2\alpha(M_1-M_{1,\text{init}})}{M_{2,\text{init}}}\right] & (\beta=0) \\ \frac{M_1+M_2}{M_{1,\text{init}}+M_{2,\text{init}}}\left(\frac{M_1}{M_{1,\text{init}}}\right)^{C_1}\left(\frac{M_2}{M_{2,\text{init}}}\right)^{C_2} & (\beta>0) \end{cases} \tag{19.16}$$

where the superscript 'init' refers again to the initial conditions. The secondary's mass M_2 is, of course,

$$M_2 = \beta(M_{1,\text{init}} - M_1) + M_{2,\text{init}}$$

and the constants C_1 and C_2 are

$$C_1 = 2\alpha(1-\beta) - 2 \quad \text{and} \quad C_2 = -\frac{2\alpha}{\beta}(1-\beta) - 2 \quad .$$

Generally, one supposes that
- conservative mass exchange occurs in Case A and early Case B/Case C, i.e. when energy transport in the outer layers of the mass-losing star is radiative,
- non-conservative mass exchange occurs in late Case B/Case C, when convection dominates.

Solutions for the orbital evolution of binares for a large variety of types of angular momentum loss are given by Soberman et al. (1997).

19.3. MASS EXCHANGE

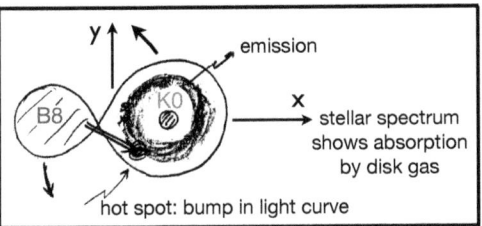

Figure 19.7: **Binary Example 3**: Model of a β Lyrae type star with Roche lobe overflow, accretion disk and hot spot. The typical spectral type of a β Lyrae type eclipsing binary is B8 Ve + K0 IV. The hot early-type primary is the exposed core of the once more massive star. It fills its Roche lobe and transfers mass via the inner Lagrangian point to the cool subgiant secondary component. The matter is stored in the accretion disk rotating around the secondary. The spectrum contains the absorption lines of the stars plus emission lines from the accretion disk (and possibly of the hot star wind). Moreover, the disk superimposes absorption lines in the spectrum when it is (from the observers point of view) projected onto the B8 type star close to the primary eclipse. The interaction of the mass stream with the disk causes a hot spot which is seen as additional light (a "bump") in the light curve. Note the direction to the observer as well as the direction of orbital motion.

19.3.4.2 Accretion disks

Conservation of (orbital) angular momentum of the transferred mass particles together with the Coriolis force leads to the formation of an *accretion disk* (see Fig. 19.4, and Fig. 19.5 "Binary Example 3") around the secondary component. Detailed calculations show that the disk forms if the radius R_2 of the secondary is smaller than about $0.05\,a$ (as before, a is the semi-major axis of the orbit which is also the separation or distance of the components).

If the radius of the secondary is large the gas stream will directly fall upon the stellar surface.

The disk consists of circular rings because such shapes contain the smallest amount of energy. Equilibrium between centrifugal acceleration V^2/r and gravitational acceleration $(GM_2)/r^2$ (with r the radius in a ring and V the orbital velocity) leads to "Keplerian disks" with

$$V \sim \frac{1}{\sqrt{r}} \quad \text{or} \quad \omega = \frac{V}{r} \sim \frac{1}{\sqrt{r^3}}$$

(with ω the angular velocity). Thus within the disk differential rotation prevails with shear of the velocity field. This leads to friction within the disk which is enhanced by the radial component of the particles' thermal motion and of the turbulent motion of blobs of matter. As a consequence there is efficient heating with subsequent radiation and thus radiative cooling. The loss of energy, finally, induces the matter to settle in the disk at smaller and smaller radii, and eventually to fall onto the secondary. Since this is based on cooling, the process runs on a thermal time scale.

In a stable, *steady-state* disk we will have conservation of energy which requires that the energy dE radiated by a ring (of radius a and thickness da) in a time interval t is equal to the difference in energy that passes through the ring's inner and outer boundaries. Thus

$$dE = \frac{dE}{da}\,da = \frac{d}{da}\left(-G\frac{M_2 m}{2a}\right)dr = G\frac{M_2 \dot{M} t}{2a^2}\,da \quad,$$

where $m = \dot{M}t$ is the orbiting mass entering and leaving the ring. Let dL_{ring} be the luminosity of the ring. Then the energy radiated by the ring in time t is related to this luminosity by

$$dL_{\text{ring}}\,t = dE = G\frac{M_2 \dot{M} t}{2a^2}\,da \quad \text{or} \quad dL_{\text{ring}} = G\frac{M_2 \dot{M}}{2a^2}\,da \quad . \tag{19.17}$$

On the other hand, the Stefan-Boltzmann law gives for that luminosity

$$dL_{\text{ring}} = 4\pi a \sigma T^4 da \quad, \tag{19.18}$$

where $4\pi a\,da$ is the surface area of the ring and T is the disk temperature at radius a. Integrating Eq. 19.17 over the disk from $a = R$ to $a = \infty$ gives the total luminosity of the disk

$$L_{\text{disk}} = G\frac{M_2 \dot{M}_1}{R_2} \quad . \tag{19.19}$$

Table 19.2: Parameters of accretion disks around typical white dwarfs and neutron stars in *semi-detached binary systems*

Parameter	white dwarf accretion disk	neutron star accretion disk
mass M	0.85 $M_\odot$	1.4 $M_\odot$
radius R	0.0095 $R_\odot$	10 km
mass exchange rate $\dot{M}$	$1.6 \cdot 10^{-10}$ $M_\odot$ yr^{-1}	$1.6 \cdot 10^{-9}$ $M_\odot$ yr^{-1}
max. temperature $T_{\max}$	$2.6 \cdot 10^4$ K	$6.9 \cdot 10^6$ K
wavelength of $I_{\max}$, $\lambda(I_{\max})$	1110 Å	4.23 Å
disk luminosity L_{disk}	$8.6 \cdot 10^{32}$ erg s^{-1} = 0.22 $L_\odot$	$9.3 \cdot 10^{36}$ erg s^{-1} = 2400 $L_\odot$

Combining Eqs. 19.17 and 19.18 one gets for the temperature at radius r

$$T = \left(\frac{GM_2\dot{M}_1}{8\pi\sigma R_2^3}\right)^{1/4} \left(\frac{R_2}{r}\right)^{3/4}. \qquad (19.20)$$

A more detailed analysis leads to the maximum temperature of the disk

$$T_{\max} = 0.488 \left(\frac{3GM_2\dot{M}_1}{8\pi\sigma R_2^3}\right)^{1/4}. \qquad (19.21)$$

Eqs. 19.19 and 19.21 have been used to calculate temperature and luminosity of disks around typical white dwarfs und neutron stars, respectively (see Table 19.2). The mass-exchange rates are chosen to be very small. Also, the mass of the WD and NS must be known (for dynamically derived masses see Fig. 17.3). In addition to temperature and luminosity the wavelength of maximum intensity disk radiation can be estimated assuming black-body radiation.

For more on accretion disks, see e.g., Papaloizou & Lin (1995) and Lin & Papaloizou (1996).

Conclusion: The data demonstrate that the accretion disks around white dwarfs have low luminosities with peak radiation in the ultraviolet, while those around neutron stars are strong X-ray sources.

Note that if the material in the disk loses angular momentum only moderately, the material falling onto the star will accelerate the star's rotation. Some material may be expelled in polar jets (Sect. 19.4.3). For effects in the very important case of WDs (novae, SN Ia) see Sect. 19.4.5.

19.3.4.3 Common envelopes; merging stars

In some cases the time scale of the transfer of mass from primary to secondary may be considerably shorter than the time scale of accretion, which (as discussed above) is the thermal scale. This may happen in systems with small separation a. In that case, more mass enters the accretion disk than is removed and the accreted layer will heat up, expand and fill the Roche lobe of the accreting star. The mass loss of the primary star will thereafter flow into what becomes a *common envelope* (CE) which encompasses both stars within a higher-order equipotential surface.

The frictional interaction between the secondary and the CE produces drag forces and removes angular momentum from the orbital motion. This has two consequences:

(1) The orbits of the stars, or rather the orbits of the stellar cores moving in the CE, shrink and the cores spiral-in toward one another. They get closer and closer and ultimately merge to form a single core with the CE now as envelope, in short, a star.

(2) The friction releases energy which heats the envelope. This process may continue until enough energy is added to the envelope to expel it. No single star will form.

The change of the binary's dimension can be calculated in the following simple way (see, e.g., Verbunt 1993). The primary star with mass M_1 is composed of a core with mass M_{core} and an

19.4. EVOLUTION OF BINARY STARS

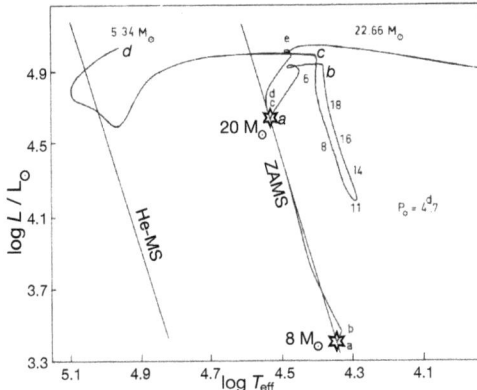

Figure 19.8: Evolutionary tracks of stars of a close binary system (20 M$_\odot$ + 8 M$_\odot$, further discussed in Sect. 19.4.1 and Fig. 19.9, "Binary Example 4") according to van den Heuvel (1976).
The normal evolution of each star is changed due to mass transfer. The various stages are labelled with letters: for the 20 M$_\odot$ star with a, b, c, d, for the 8 M$_\odot$ star with a, b, .., e. Numbers along the tracks indicate the mass of the star at that location. The primary star is on the last portion of the evolutionary track a WR star (compare with the location of WR stars in Fig. 13.9). Note the He MS, as in Fig. 6.8.

envelope of mass M_env. We may compare the binding energy of the envelope with the difference in total energy of the binary before and after the spiral-in:

$$\frac{GM_1 M_\text{env}}{\gamma R_1} = \alpha \left(\frac{GM_\text{core} M_2}{2a_\text{final}} - \frac{GM_1 M_2}{2a_\text{init}} \right) \quad , \tag{19.22}$$

where α is the loss of specific angular momentum (as used in Eq. 19.14), γ is a weighting factor for the gravitational binding of the envelope to the core, R_1 is the radius of the Roche lobe, and a_init and a_final are the distances between the binary stars before and after the spiral-in, respectively. The equation can be rewritten as

$$\frac{a_\text{final}}{a_\text{init}} = \frac{M_\text{core}}{M_1} \left(1 + \frac{2a_\text{init}}{\alpha\gamma R_1} \frac{M_\text{env}}{M_2} \right)^{-1} \quad . \tag{19.23}$$

The ratio R_1/a_init is given by Eq. 19.5. If all the energy released by the shrinking of the binary is efficient in unbinding the envelope, then $\alpha = 1$, and we can calculate the final distance a_final from Eq. 19.23. If, on the other hand, 90% (say) of the energy released is radiated away without contributing to the expansion of the envelope, then $\alpha = 0.1$, and a_final will accordingly be smaller. If a_final is smaller than the sum of the radii of the two stars, $R_\text{core} + R_2$, after spiral-in, then the companion of course merges with the core of the mass donor.

The spiralling in time may be long (years) or very short (like within 10 to 100 days). That time is set by the initial separation as well as by the mass of the two stars forming a common envelope. For an example see, e.g., Terman et al. (1994).

The concept of common envelope evolution is capable to explain the origin of a wide variety of close binary stars, e.g., cataclysmic variables (containing a white dwarf), or low-mass X-ray binaries (containing a neutron star). This will be worked out in more detail in Sect. 19.4. An observed case is given in "Binary Example 5" (Fig. 19.10). Reviews on common envelopes have been given by Iben & Livio (1993) and Taam & Sandquist (2000).

19.4 Evolution of binary stars

A wealth of studies on binary star evolution has been published since the early explorations. In this section three stellar pairs will be evaluated (a high-mass binary, a high-mass - low-mass binary, a low-mass binary), which explain many of the fascinating phenomena among binary stars. Note that in most case studies mass exchange is traeted as conservatory.

19.4.1 Towards massive X-ray binaries and beyond

Consider a massive binary system of 20 M$_\odot$ + 8 M$_\odot$ stars. Such stars will evolve quickly and have the potential that large amounts of mass are being transferred. Also, the large mass indicates these stars may end up as SN. The evolutionary tracks for this binary are shown in Fig. 19.8 and the shape and behaviour in the different phases are illustrated in "**Binary Example 4**" (Fig. 19.9).

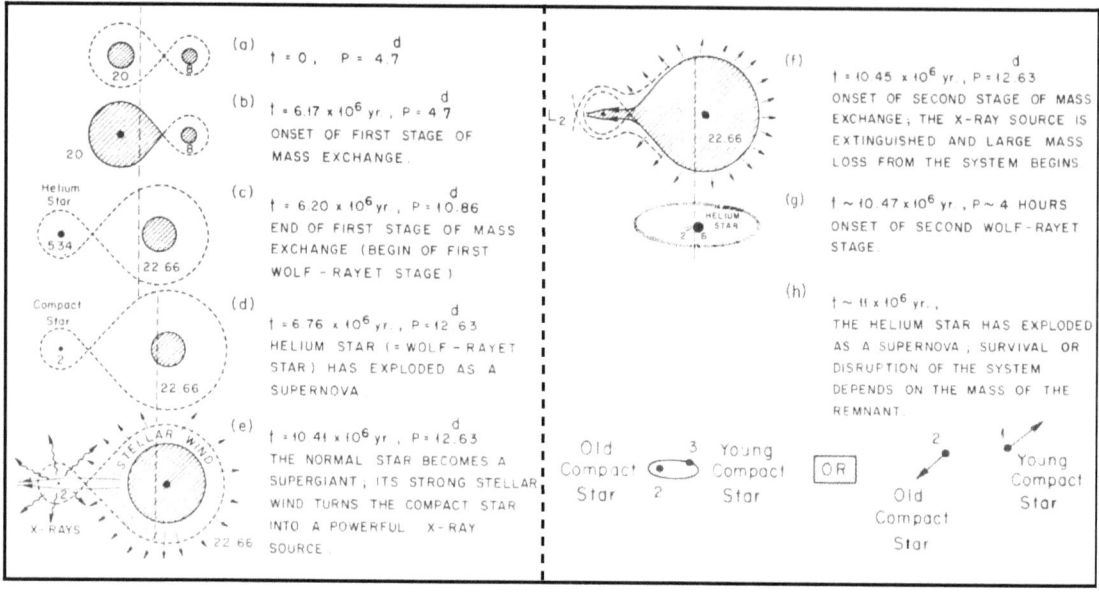

Figure 19.9: **Binary Example 4**. Phases of the evolution of a binary system of 20 $M_\odot$ + 8 $M_\odot$. The stages a - e are labelled along the evolutionary tracks of Fig. 19.8. Figure adapted from van den Heuvel (1976).

· Phase (a): The binary is assumed to start with the orbital period of $P = 4.7$ days. The spectral type is about O9 V + B3 V.

· Phase (b): After only 6 million years a rapid mass exchange starts which lasts only 30 000 years. The primary transfers 14.66 $M_\odot$ to its companion. The mass loss ends when core helium burning is ignited (and the star is no longer a red giant).

· Phase (c): As a result, the former less-massive secondary is now an O8-type main-sequence star of 22.66 $M_\odot$. The primary, now having only 5.34 $M_\odot$, has lost its hydrogen-rich layers and burns helium on the helium main-sequence (compare with Fig. 6.8). This helium star is a typical Wolf-Rayet (WR) star (in the first Wolf-Rayet stage).

· Phase (d): Only half a million years later the WR star explodes as a **Type Ib supernova** (no hydrogen lines!) and a collapsed compact object such as a **neutron star** (see Ch. 17.2) is left as pulsar, perhaps even a **black hole** (see Ch. 17.4). Depending on the width of such systems they may even be dynamically disrupted, thereby ejecting the O8 type star, which would become one of the galactic **high-speed OB stars** (Blaauw 1961).

· Phase (e): After a quiet phase of 4 million years in a stable system, the massive secondary evolves to supergiant. Its strong stellar wind is accreted by the compact star and turns it into a powerful X-ray source. This stage is known as **HMXB** = *high mass X-ray binary* star.

· Phase (f): After a life-time of 10^7 yr the secondary fills its Roche lobe and a second stage of mass transfer begins. The X-ray source is extinguished because the opacity is too high, and the X-rays will be degraded to photons of lower energy which finally escape as optical or ultraviolet radiation. A common envelope develops which might be expelled in all directions, or is partly lost via the Lagrangian point L_2. In any case, a large amount of mass and angular momentum is lost by the system. The stars spiral-in and the orbital period of the system shrinks to $P \sim 4$ hours only.

· Phase (g): The secondary loses its hydrogen rich layers and a (helium) WR star remains.

· Phase (h): The WR star finally explodes as a **supernova Type Ib** (the second) and two compact objects are left. Apparently, mass ejection from supernovae is asymmetric and so it depends on the direction of the kick of this asymmetric supernova explosion whether the system survives (maybe as a double pulsar) or will be disrupted.

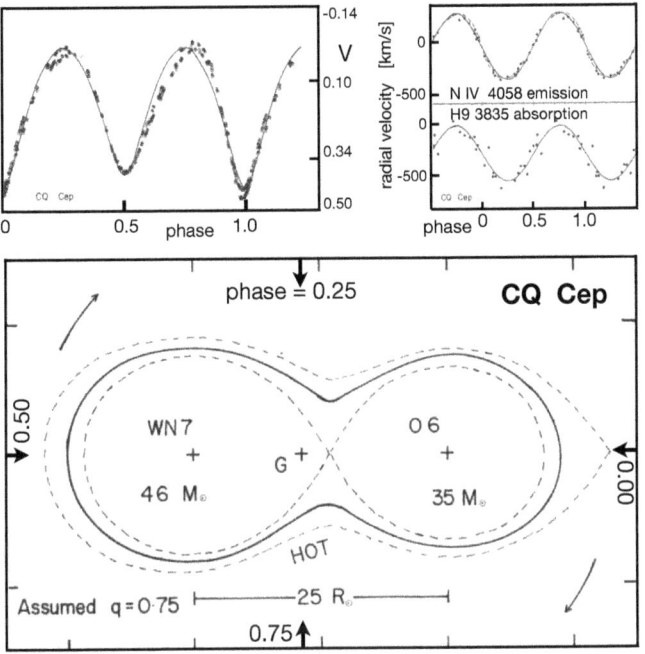

Figure 19.10: **Binary Example 5**.
The Wolf-Rayet (WN7 + O6) over-contact eclipsing binary CQ Cep.
The *upper left* diagram shows the light curve of the system in Johnson's V band. The line represents the fit to the best model for the system.
The *upper right* panel gives the radial velocity curves of the N IV λ 4058 emission and of the Balmer H9 λ 3835 absorption together with model fits.
The *lower* panel displays the configuration of the system in the orbital plane. A mass ratio of $q = 0.75$ (O-type star of now 35 $M_\odot$ and Wolf-Rayet star of now 47 $M_\odot$) has been assumed. The full line circumscribes the (contact) binary, dashed lines are two equipotential surfaces, G is the centre of gravity, the phases (aspect directions, arrows) are indicated at the edge of the box.

CQ Cephei has a period of 1.641 days and is a SB1 spectroscopic binary. The typical WR emission lines and the Balmer absorption lines are in phase (see upper right diagram) which means that also the hydrogen lines arise in the WR atmosphere. Thus the hydrogen-rich layers have not yet been tranferred completely to the companion, as is usual for late-type WR stars of the nitrogen sequence (WN7 to WN9). Analyses of the light and radial velocity curves lead to the interpretation as a binary system in over-contact (Fig. 19.8). CQ Cep is in its evolution between phases (b) and (c) of Fig. 19.9, but has developed an additional common envelope which might be expelled later on. Figure adapted from Leung et al. (1983).

Comments on different types of binary stars apparent in the evolutionary process of Example 4.

– The behaviour of this binary system is, of course, determined by the initial masses of the stars and by their separation. Depending on a, mass transfer is modest or heavy.

– **Wolf-Rayet stars, first stage**, phase (c). About half of all known WR stars are in orbit with a massive OB-type secondary. Prime types are the brightest WR stars γ Velorum and θ Muscae in the southern sky. But see also **Binary Example 5** (Fig. 19.10). Note that there is another mechanism for the origin of WR stars (see Ch. 13.4.1.3): mass loss by strong stellar winds which blow the outer hydrogen-rich stellar layers into the circumstellar space and expose the nuclear processed helium core as WR star.

– **Supernovae of Type Ib**, phases (d) and (h). WR stars end as hydrodynamical supernovae. WR's are helium stars, therefore they should generate Type I supernovae (no hydrogen but helium emission), probably of subtype Ib (additional oxygen emission). This is supported by the fact that SNe Ib have always been found in late-type galaxies (generally Sc), and usually near H II regions. This suggests that SNe Ib have massive progenitors, perhaps even more massive than SNe II (see Filippenko 1991 and Ch. 18.2).

– **HMXBs**, phase (e). These are systems in which the secondary usually is an OB star ($M_2 \geq 10\,M_\odot$). The luminosity is dominated by the secondary. The ratio between X-ray emission and emission in the optical region is $L_X/L_{opt} \simeq 10^{-3}$ to 10. Orbital periods range from about 2 to 600 d. The strong magnetic field of the compact object (10^8 - 10^9 Tesla) may play an important role: it guides the matter flow to the pole of the star. The temperature of the hot spot at the pole reaches values of $T \simeq 10^8$ K. Therefore, the emitted hard X-rays may be pulsed by the rotational period of the compact star. Observed values range from 0.7 s to several 100 s. About half of the

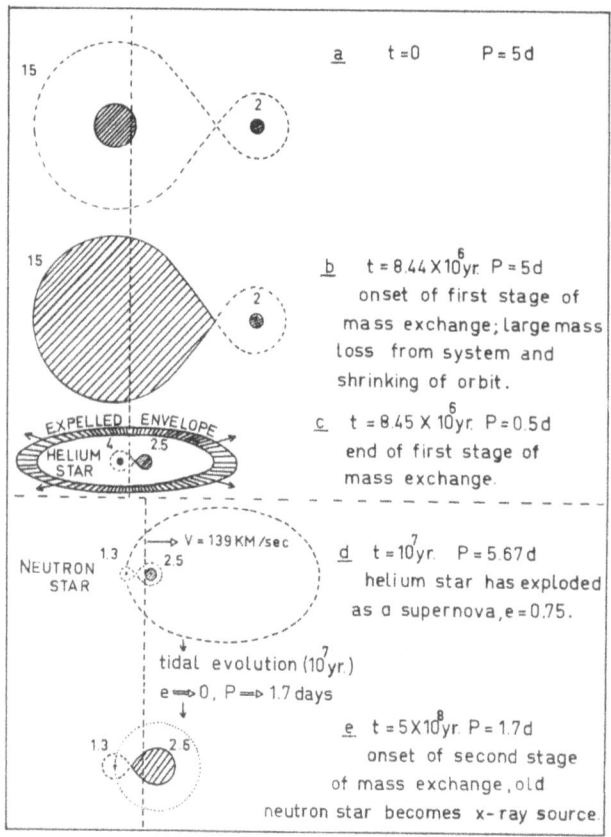

Figure 19.11: **Binary Example 6.** Evolution of a binary system with mass 15 M$_\odot$ + 2 M$_\odot$ (van den Heuvel 1976).

· Phase (a): The system starts with $P \simeq 5$ days corresponding to $a = 30\,\mathrm{R}_\odot$. The stars are well separated within their individual Roche lobes.

· Phase (b): After 8.4 million years the primary fills its Roche lobe and starts the first stage of mass exchange. The system experiences large loss of mass and angular momentum within only 10^4 years and the orbit shrinks dramatically to $a \sim 5\,\mathrm{R}_\odot$ with $P \sim 0.5$ days.

· Phase (c): The helium-rich core of the primary survives as a typical low-mass Wolf-Rayet star. The secondary has gained only about 0.5 M$_\odot$. In view of the high luminosity of the WR star and relatively low luminosity of its companion the system will appear as a single WR star surrounded by the expelled hydrogen-rich envelope.

· Phase (d): The WR star explodes as **SN Type Ib**. There is a small chance that the system survives (as in phase (h) of Fig. 19.9), but then with a large orbital eccentricity (perhaps $e = 0.75$) and a period of $P = 5.67$ days. In a long-lasting phase the orbit will be circularized by tidal effects, reducing the period to $P = 1.7$ days. If case of the SN Ib the system does not survive and the result will be a high speed neutron star and a high speed low mass star.

· Phase (e): After an even longer inactive stage the secondary starts modest mass transfer which switches the compact primary into an X-ray source. The system is observed as a typical *low-mass X-ray binary* (**LMXB**).

· Phase (f), not shown. The secondary ends as a white dwarf. This might be a **helium white dwarf** (perhaps with a hydrogen layer) or a carbon-oxygen one, depending on the secondary's mass at phase (e) and the total amount of mass exchanged.

HMXBs are **X-ray pulsars**. The most interesting HMXB is Cyg X-1 (or HDE 226868, O9.7 I). Reasonable assumptions for the binary system lead to the conclusion that the mass of the compact object is between 11 M$_\odot$ and 21 M$_\odot$ which is strong evidence for a black hole (see Tab. 17.1). For more on X-ray properties of BH binaries see Remillard & McClintock (2006).

– **Wolf-Rayet stars, second stage**, phase (g). This second WR stage has been detected and discussed in a series of papers by Moffat & Seggewiss (1979-1982). It was surprising that no harder X-rays have been observed coming from accretion of the WR star wind onto the compact star. But it has been shown (Moffat & Seggewiss 1979) that in most cases the WR wind is still sufficiently optical thick near the compact object that the X-rays will be degraded.

– **Double neutron stars**, phase (h). In 1975 the first double neutron star PSR 1913+16 was detected by Hulse & Taylor (1975). One of the stars is a pulsar with the pulse period of $P_{\mathrm{pulse}} = 59$ millisec which is modulated by the orbital period of $P_{\mathrm{orb}} = 0.32$ days. The analysis of the radial velocity curve suggests that the unseen component is a compact object, too. Several double neutron stars are known, 8 in the galactic disk and one in the globular cluster M 15 (van den Heuvel 2007).

19.4.2 Towards low-mass X-ray binaries

A second binary pair with interesting evolution is that of a massive system but with a large mass ratio $M_1/M_2 = 15/2$ ("**Binary Example 6**", Fig. 19.11). Since the more massive star evolves much quicker than the low mass star while it has the capability (M_{init}) to become a supernova of type II, the mass loss leads to stripping of the hydrogen layers and the star becomes SN Type Ib.

Once the secondary becomes RG, it will shed mass back on the post SN Ib neutron star. This leads to an energetic accretion disk and thus to X-ray emission.

Comments on different types of binary stars apparent in the evolutionary process of Example 6.

– **Wolf-Rayet stars with expelled envelopes**, phase (c). Indeed, a number of WR stars have been observed which are surrounded by a massive expanding shell; see e.g. the nebula NGC 6888 around the WR star HD 192168 in the Cyg OB1 association.

– **Binary pulsars**, phases (d) to (f). A small fraction of pulsars are members of binary systems. A most fascinating system is the pulsar PSR 1620-26. It lies near the core of the globular cluster M 4. In fact, the object is a triple system. The inner binary contains the $\simeq 1.4\,M_\odot$ neutron star (a pulsar with a pulse period of $P_p = 11$ millisec) and the $\simeq 0.3\,M_\odot$ white dwarf in an orbit with the period of $P_1 = 191$ days. Pulsar timing data (Ford et al. 2000) allow to determine the mass and the orbital parameters of the third component of the system. The best fit parameters correspond to a component of planetary (!) mass, $M_3 \sin i_2 \simeq 7 \cdot 10^{-3}\,M_\odot$ (i_2 is the unknown inclination of this orbit), in an orbit with semi-major axis $a_2 \simeq 60$ AU and period $P_2 \simeq 300$ yr.

– **LMXBs**, phase (e). LMBXs belong to the old bulge and globular cluster population of our Galaxy. They have luminosity ratios of $L_X/L_{\text{opt}} \simeq 10...10^4$. The periods are in the range from 40 minutes to about 10 days. Normally, LMXBs do not display X-ray pulses like the HMXB because the magnetic fields have considerably weakened in these old objects. Instead, they show irregular X-ray bursts. Those could be the consequence of explosive thermonuclear reactions on the surface of the neutron stars. About one quarter of the LMBXs are found in globular clusters. Here, another mechanism for the origin of close binary systems might operate, namely the gravitational capture of a neutron star by a normal cluster member.

19.4.3 Microquasars

When compact objects (black hole, neutron star) acquire mass from a companion star, the material assembles in an accretion disk (Sect. 19.3.4.2). When matter spirals toward the compact object its orbital speed increases (conservation of angular momentum). However, magneto-rotational instabilities propell part of the disk material into collimated polar jets (Fig. 19.12), taking angular momentum away from the disk. This allows most disk matter to fall onto the compact object.

When the receiving star establishes such relativistic jets (in particular seen in high-resolution radio interferometry observations) or exhibits a flat radiospectrum (being indirect evidence for an expanding continuous jet) such objects are called "microquasar". They are, sort-of, miniature versions of quasars, the nuclei of active galaxies. For more see, e.g., Mirabel & Rodríguez (1999).

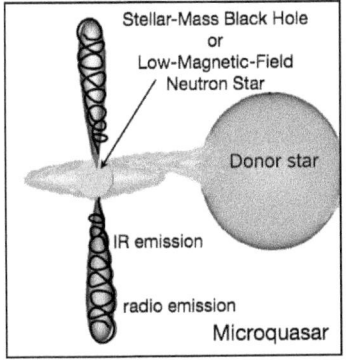

Figure 19.12: Sketch of the accreting / ejecting system called microquasar. The accretion disk emits thermal radiation in X-rays (this gives the name to the class "X-ray binaries"). The polar jets, typical of microquasars, are orthogonal to the disk and highly collimated. They have a twisted magnetic field and emit strong non-thermal radiation, mainly observed at radio wavelengths. Depending on the viewing angle, one of the jets may point toward the observer which then may cause the sometimes observed "superluminal" effect. Some microquasars have been detected at very high energy (i.e. γ-rays). Figure adapted from Massi & Kaufman Bernadó (2008).

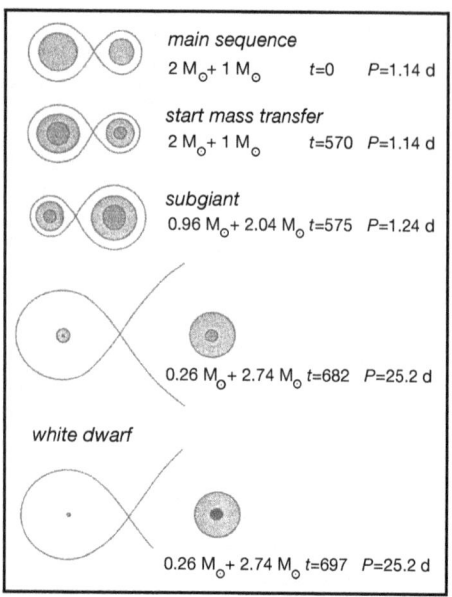

Figure 19.13: **Binary Example 7**.
Evolution of a binary system of 2 $M_\odot$ + 1 $M_\odot$ stars (Kippenhahn & Weigert 1967b; De Loore & Doom 1992). Time t is in units of 10^6 yr.

· Phase (a): The system starts on the main sequence with a period of $P = 1.14$ days.

· Phase (b): After $5.7 \cdot 10^8$ yr the primary star ends core hydrogen burning (the helium core then has 0.23 $M_\odot$) and begins its transition to the RG phase. This evolution soon slows due to Case B mass transfer as result of the increase of the stellar radius.

· Phase (c): The primary does become a red giant with a hydrogen burning shell and slows down the mass loss more and more.

· Phase (d): Mass transfer ends. The primary now has a mass of only $M_1 = 0.264$ $M_\odot$. 96% of the total mass is in the helium core which has a degenerate centre. 4% of the stellar mass is still hydrogen-rich and constitutes the huge outer layer of the star.

· Phase (e): The hydrogen layers of the primary shrink and collapse onto the surface of the helium core reducing the size of the object by a factor of about 130 and a **helium white dwarf** develops (the tiny point at the left) surrounded by a thin hydrogen-rich layer (0.9% of the total mass only).

· Phase (f), not shown. The now more massive secondary (2.74 $M_\odot$) sits inside a huge Roche lobe which is, according to Eq. 19.5, the consequence of the high mass ratio $M_2/M_1 = 10.4$ (note that now interchanged indices have been used in Eq. 19.5). The star will ignite helium in the core and fill this critical surface about $1.3 \cdot 10^8$ yr after phase (e). The mass transfer that follows forms a luminous ultraviolet-bright accretion disk around the white dwarf (see Tab. 19.2). The disk's matter falls inward and is collected at the surface of the white dwarf until nuclear hydrogen ignites explosively. This leads to a typical **nova outburst** of radiation (see also Ch. 17.1.5) and to the ejection of a shell of matter. The collection of hydrogen-rich matter on the white dwarf and the subsequent nuclear ignition may be a repeatable process. Such stars are observed as **recurrent novae**. If it comes to a steady flow and to steady burning on the surface, then the object manifests itself as a **luminous supersoft X-ray source** (see Kahabka & van den Heuvel 1997).

· Phase (g), not shown. The final outcome of this evolutionary scenario is **two white dwarfs** in a binary orbit. It depends on the initial mass which of the white dwarf types (from hydrogen over helium to the CO class, see Ch. 17.1.2), is be left. If the remaining pair is sufficiently narrow, the stars may merge and perhaps become SN (see Ch. 18.3.3).

19.4.4 Low mass binary systems: towards cataclysmic binaries, SN Ia

Low mass binaries evolve like low mass stars. When the more massive one becomes red giant, it transfers mass to the secondary. The primary may thus lose its envelope and turn into a WD.

Later, the now more massive secondary will lose mass back. The mass transfered may assemble in an accretion disk and fall at (ir)regular intervals onto the WD. Such material may come to explosive burning at the surface of the WD.

This kind of binary evolution was explored in detail by Kippenhahn et al. (1967b). Their example is of a 2 $M_\odot$ + 1 $M_\odot$ system sized to undergo Case B mass exchange. It is described in **"Binary Example 7"** (Fig. 19.13).

Comments on different types of binary stars apparent in the evolutionary process of Example 7.

– **Cataclysmic variables and Novae**, phase (f). Novae belong to a group of eruptive variables which are called cataclysmic variables. Common to all stars is the same mechanism of origin: Roche lobe overflow from a low-mass star onto a white dwarf as demonstrated in the evolutionary scheme

19.4. EVOLUTION OF BINARY STARS

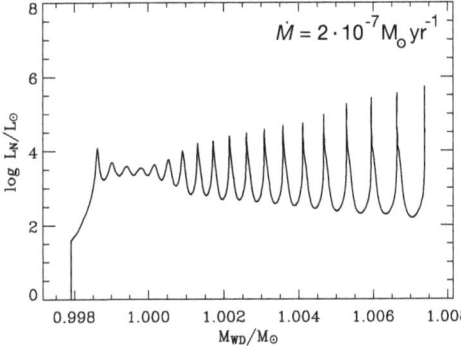
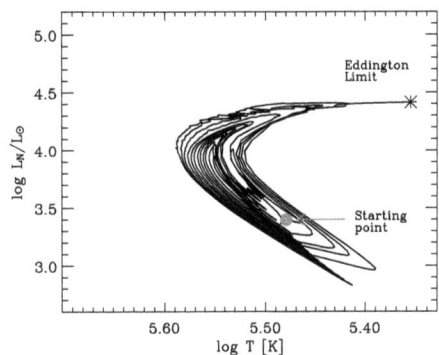

Figure 19.14: Helium shell burning in a rotating-accreting WD. Figures from Yoon et al. (2004). Left: Evolution of L_{He} in the shell of a rotating WD. The x-axis gives the actual mass (accretion, see $\dot{M}$), so being linearly representative of time. The He shell has variable brightness much like the thermal pulsing of AGB stars (Ch. 10.4.3). The pulses become stronger the more matter has been accreted. Modelling ends when the surface luminosity reaches the Eddington limit. With stronger chemical mixing the WD can continue accreting matter and more than 100 pulses may occur.
Right: The evolution shown in the HRD. The loops are due to the He shell pulses. Because of the opacity of the envelope gas the L_{He} comes out filtered and averaged out.

of Fig. 19.13. The subtypes of this class of variables differ in amplitude of the outbursts, their frequency (which may be periodic or non-periodic), strength of possible magnitic fields and many other respects. Subtypes are, e.g., novae (see also Ch. 17.1.5), dwarf novae, U Gem and Z Cam stars, or the magnetic AM Her and DQ Her systems. The latter four groups are named after their prototype stars. For more see Bode & Evans et al. (2008). For effects of angular momentum transfer see Sect. 19.4.5.

– **Supernovae of Type Ia**, phases (f) and (g). Several mechanisms have been proposed for stars to reach the state of becoming SNe Type Ia (for the explosion mechanism see Ch. 18.3.2).
○ A CO white dwarf in phase (f) might collapse and undergo explosive carbon ignition if rapid mass exchange lifts the white dwarf's mass over the Chandrasekhar limit of $M \simeq 1.4 \, M_\odot$.
○ On the other hand (phase (g)), two CO white dwarfs may spiral-in either in a common envelope or by gravitational wave energy loss. They then will come to carbon detonation.
○ When a He WD accumulates mass, the non-degenerate He shell may come to explosive nuclear burning triggering explosive core CO burning and deflagration of the WD.

However, rotation and angular momentum play an important role (see Sect. 19.4.5).

19.4.5 WDs and rotation: Nova and SN Ia phenomena

WDs have long been regarded as the objects resulting in novae and SNe of Type Ia. Various aspects from observations have led to reconsider the evolution of WDs. E.g., cataclysmic variables showed to have high rotational velocities of $50 < v \sin i < 1200$ km s^{-1} (Sion 1999). Also, novae with mass accretion should soon explode, while they do not seem to do that. The light curves of Type Ia SNe show some diversity, casting doubt on their unique origin. Moreover, the use of SN Ia as cosmological probes led to problems for the canonical models of the universe (requiring accelerated Hubble expansion). Further research in the binary WD scenario made clear that rotation of WDs is an important aspect of their evolution.

For SN Ia three scenarios exist of which two deal with close to centre C ignition. In one, two WDs merge due to angular momentum loss by gravitational wave radiation. When the meger object surpasses the Chandrasekhar limit the C ignites and the star deflagrates. In the other, a CO WD accretes H or He (Ch. 18.3.2) at a high rate leading with steady shell burning and ultimately (when enough mass has been accreted) a SN Ia follows. The third scenario has a moderate accretion rate

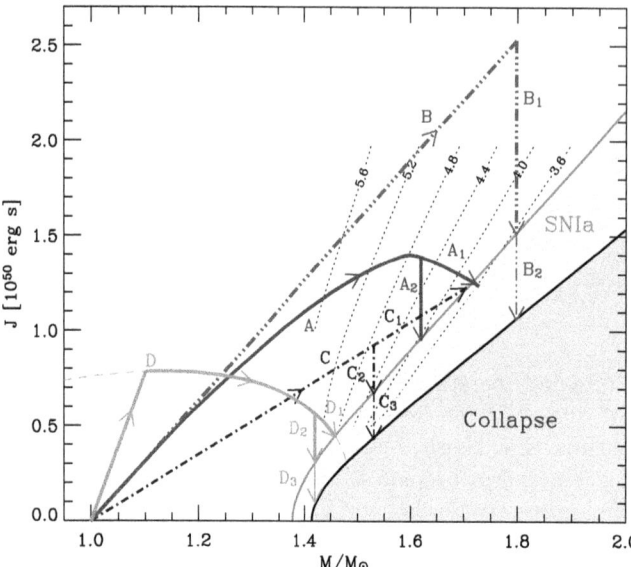

Figure 19.15: WDs in binary systems gain (with accretion) angular momentum. The figure (from Yoon & Langer 2005) shows which mass a WD of $M_{WD} = 1\,M_\odot$ may reach due to accretion of mass and angular momentum J and evolution of J before the SN instability limit is reached. The mass can perhaps reach as high as 2 $M_\odot$. If an accreting CO WD has a J smaller than the line bounding the hatched region, a SN Ia will occur. The modelling shows that SNe type Ia need not come from WDs with $M \equiv 1.4\,M_\odot$! If an accreting ONeMg WD (see Fig. 10.24) has a J smaller than the upper envelope of the coloured region it will come to an electron capture SN. The labels on the curves signify: A= accretion and loss of J; B= accretion of J but no loss of J; C= inefficient accretion of J; D= maximum internal angular momentum transport efficiency. In each case sub-possibilities are labeled with subscripts. The thin dashed line (merging with curve D) gives the M, J relation of WDs rigidly rotating at $v_{\rm crit}$. Dashed lines mark constant growth with times [log years] for the r-mode instability, which will lead to energy loss by gravitational radiation.

(Ch. 18.3.2) followed by ignition of He in the shell et the edge of the CO core. This then sends a shock wave through the star so that C ignites, deflagrating the star. All these scenarios were considered for more or less symmetric WDs and all scenarios have problems like rate of occurrence, too much mass loss, or instabilities of the modelled stars.

When a WD accretes matter in a binary system through a Keplerian disk (Sect. 19.3.4.2), it will also increase its angular momentum (see, e.g., Langer et al. 2000). The rate of increase depends on the size of the system. The separation of the stars, the mass of the donor and the rate of mass transfer are of relevance.

When a star rotates, its structure will change. Some negative $\log g$ has to be added to the structural description (see Fig. 14.1) so the internal gas pressure in the star is reduced. Rotation will also induce internal circulation (see Ch. 14). The models of Figs. 19.14 and 19.15 include the effects of the various rotation-induced instabilities leading to shellular rotation.

One consequence is, that rotating-accreting WDs will have He shell burning in a pulsating manner (Fig. 19.14), becoming brighter the more mass has been accreted. They will have a prolonged nova phenomenon phase (Yoon et al. 2004).

A further consequence is that mass accumulating WDs may become heavier than the Chandrasekhar limit (Ch. 17.1.4) and still do not explode. Thus such WDs may become SN Ia with M_{WD} of up to 2 $M_\odot$ (see Fig. 19.15). SN Ia are probably not equal enough to be useable for the derivation of reliable cosmological distances.

19.5 Variety of binary evolution; special objects explained

19.5.1 Multiple branching in binary evolution

A multitude of binary mass combinations is possible in which each individual star has an evolution foremost depending on its initial mass. Yet in closer binaries everything may become different. The outcome of binary evolution is also rather diverse because the factor of relevance for what really will happen is the initial separation of the stars of the system.

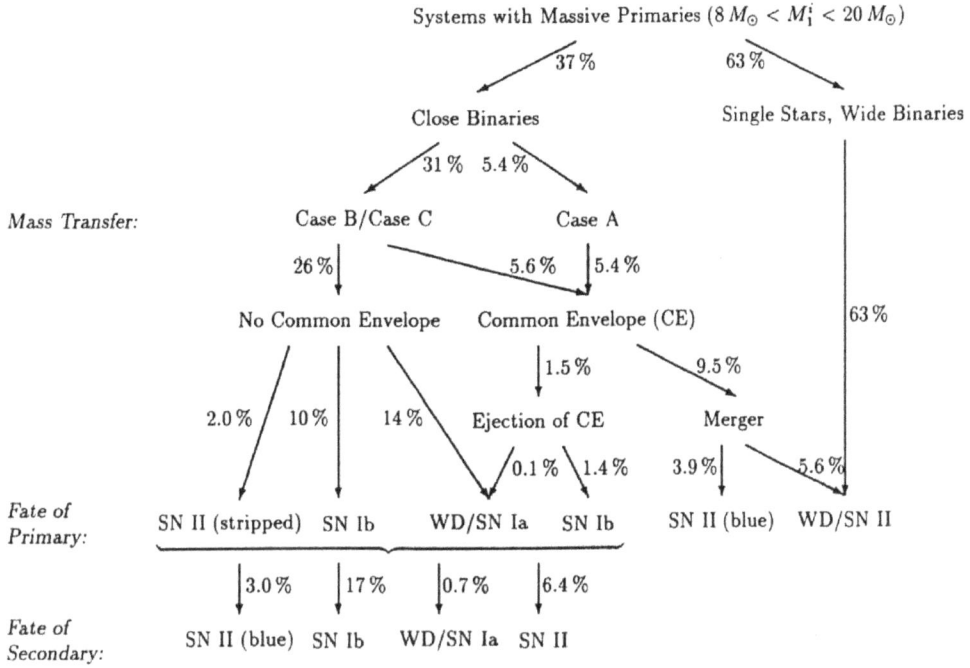

Figure 19.16: A statistical estimate is given of the final products of binary evolution starting with high mass primaries (8 M$_\odot$ < M$_{\mathrm{prim}}$ < 20 M$_\odot$). Figure from Podsiadlowski et al. (1992).

Several schemes based on statistical approaches are available in the literature. Fig. 19.16 gives an example, showing that the nature of the final products is rather diverse indeed. If the input would have been different, the outcome would have been different, too.

19.5.2 Special objects now explained by binary evolution

The investigations of binary stars and the transfer of mass between them has led to paths of stellar evolution which can explain several special kinds of stars and phenomena. Exemplary binary combinations were discussed in the previous sections.

19.5.2.1 Cataclysmic variables; Novae; Supersoft X-ray sources

In a pair of stars of the lower mass range, the primary will evolve first and ultimately become a WD. The secondary will in the course of its evolution transfer mass which collects in an accretion disk around the WD. This material then falls onto the WD. If it falls intermittently in small amounts ($10^{-10} < \dot{M} < 10^{-8}$ M$_\odot$ yr^{-1}) it leads to explosive nuclear fusion on the WD's surface, brightening the star considerably to be a Nova or a recurrent Nova (Sect. 19.4.4); only supernovae are brighter.

If $\dot{M} > 10^{-7}$ M$_\odot$ yr^{-1} the material accumulates at such a rate that continuous burning near the surface is possible, the system becomes a supersoft X-ray source (Fig. 19.11).

At intermediate accretion rates it produces the features of cataclysmic variables or recurrent novae (Sect. 19.4.4).

19.5.2.2 Type Ia supernovae

There are two possible explanations for the SN Type Ia phenomenon.

In a pair of stars of the lower mass range, the primary will evolve first and ultimately become a WD. The secondary will in the course of its evolution transfer mass. Ultimately, the WD may have accumulated so much mass that it surpasses the Chandrasekhar limit (see Ch. 17.1.4) leading to core collapse and a supernova explosion (SN Type Ia) leaving *no* stellar remnant.

The other scenario is that two CO white dwarfs may spiral-in and coalesce to form one 'star'. They then will come to carbon detonation being a SN Type Ia.

However, the fact that two scenarios are proposed as explanation for the SNe Type Ia as well as that sped-up rotation may lead to an explosion at $1.4 < M_{WD} < 2\,M_\odot$ (Sect. 19.4.5), indicates that the use as 'standard candle' is perhaps not well founded.

19.5.2.3 Type Ib and Ic supernovae

Supernovae of Type Ib are characterised by having helium but *no* hydrogen absorption in their spectrum. This means that the progenitor had no H left in its atmosphere. There are two explanations possible for that.

One is, that a single massive star had such a strong stellar wind in the luminous phases that all the shell material has been carried away (see Ch. 13.4.1). The second possibility is the case of a binary star, in which the massive star loses its outer layers by Roche overflow, thus ridding the star of its hydrogen (see Sect. 19.4.1).

Supernovae Type Ic have neither H nor He in their spectra. The progenitor is thus a WR star of type C (a WC star). A subgroup, the SNe Type Ic peculiar, give the long duration γ-ray bursts.

19.5.2.4 X-ray binaries (HMXB, LMXB)

There are two kinds of X-ray binaries, depending on their mass.

In a massive double system, the primary evolves and explodes as SN Type Ib (Sect. 19.5.2.3). When the massive secondary becomes a supergiant its strong stellar wind is accreted by the compact star which turns it into a powerful X-ray source, and the system is called a high mass X-ray binary (HMXB, Sect. 19.4.1).

In a system with a massive and a lower mass star, the massive becomes SN Ib first. Much later, the low mass secondary evolves and starts mass transfer onto the 'primary' neutron star. The accretion is erratic and leads to the low-mass X-ray binary (LMXB) phenomenon (Sect. 19.4.2).

19.5.2.5 Binary pulsars

If in a double system, the game of evolution and mass exchange results ultimately in two supernovae, the system ends as a double pulsar (Sect. 19.4.2). The mass transferred by the star becoming SN last, will in that time have led to a short X-ray binary phase in which the first born neutron star is brought to very fast spin due to accretion through a disk. One observes that in the 8 known double neutron stars in the galactic disk one of the pulsars spins very fast.

19.5.2.6 High speed OB stars

When in massive binary systems the first object turns supernova, the binary system may be dynamically disrupted. The remaining star, which has become massive and blue, is ejected and so turns into a high-speed OB star (Ch. 19.4.1).

"Hyper velocity stars" are those OB stars, that are faster than allowed from the binary star scenario. Several have been discovered (see e.g. Edelmann et al. 2005). These stars likely gained velocity by zipping by the central BH of the Milky Way (or another galaxy).

19.5.2.7 Merged stars

The nature of merged stars will depend on the initial conditions of the binary system, such as M_1 and M_2 and the initial separation a_{init}. Merged stars likely have a chemical composition and structure different from what a single star would obtain in the course of normal evolution.

Depending on initial conditions, the merging itself could lead to a supernova explosion when the core of the merged star would surpass the Chandrasekhar limit.

Stars known to be mergers are the blue-straggler stars (Ch. 10.8.4). One mechanism is that of normal low-mass binary stars evolving with mass transfer and spiral-in. The other is that stellar gravitational interactions in a globular cluster may bring stars into a collision course leading to a merger. The optimum mechanism may depend on environment conditions (Moretti et al. 2008).

19.6 Summary

After the presentation of dynamical effects in binary systems the effects of evolution of the stars (notably radius change) have been presented. Evolution leads in the not wide binary systems to mass transfer and this to substantial deviations in the evolutionary path of the individual stars compared to those of single stars.

By far not all phenomena have been presented in exhaustive detail. The most notable aspects of the evolution of binary systems are:
- The increase in stellar radius in the course of stellar evolution may lead to filling of the *Roche lobe* and thus to mass transfer.
- Since the stars in a binary system are normally different, the more massive one grows first, the secondary gains mass but in turn may transfer mass back.
- The mass being transferred may assemble in *accretion disks* which often are dusty and can obscure the collecting star.
- Mass collected will plunge onto the star often leading to eruptive phenomena, sometimes to nuclear fusion in the stellar envelope: the *Nova* phenomenon.
- Mass exchange may be conservative, in which the receiving star gains angular momentum. If mass excahnge is not conservative, angular momentum is lost from the system.

Several examples of binaries were elaborated:
- A massive pair may lead to *high-mass X-ray binaries*, to Type Ib supernovae (no H absorption lines) and thus to binary pulsars, and possibly to *run-away OB stars*.
- A combination of a massive and a low mass star may lead to *low-mass X-ray binaries*, to WR-stars exploding as SN Type Ib, and thus to pulsar-WD combinations.
- A combination of two lower mass stars leads to *cataclismic variables*, to *Novae*, to supernovae Type Ia, or to just double WDs.
- Binary systems may, provided there is enough mass to go around, also lead to a *Black Hole*.
- A WD accreting matter accretes also angular momentum. Its rotation will speed up, reducing the internal pressure, which allows the accretion of mass to $M_{WD} > 1.4$ $M_\odot$ before it leads to a SN Ia explosion, thus a SN Ia of a star heavier than the classical Chandrasekhar limit.
- In tight binary systems stars may merge after a common envelope phase. In globular star clusters the blue stragglers are proof of the existence of merged stars.

References

Blaauw, A. 1961, Bull. Astron. Inst. Neth. 15, 265
Bode, M.F., & Evans, A. (eds.) 2008, "Classical Novae", 2nd edition. Cambridge Univ. Press
Canal, R, Isern, J., & Labay, J. 1990, ARAA 28, 183; *The Origin of Neutron Stars in Binary Systems*
Cowley, A.P. 1992, ARAA 30, 287; *Evidence for Black Holes in Stellar Binary Systems*
De Loore, C.W.H., & Doom, C. 1992, "Structure and Evolution of Single and Binary Stars", Kluwer, Dordrecht
Edelmann, H., Napiwotzki, R., Heber, U., Christlieb, N., & Reimers, D. 2005, ApJ 634, L181
Eggleton, P.P. 1983, ApJ 268, 368
Eggleton, P.P. 2006, "Evolutionary Processes in Binary and Multiple Stars", Cambridge Univ. Press
Filippenko, A.V. 1991, in "Supernovae and stellar evolution", A. Ray & T. Velusamy (eds.); World Scientific, p. 58

Ford, E.B., Joski, K.J., Rasio, F.A., & Zbarsky, B. 2000, ApJ 528, 336
Hulse, R.A., & Taylor, J.H. 1975, ApJ 195, L51
Iben, I., Jr. 1991, ApJS 76, 55, *Single and Binary Star Evolution*
Iben, I., Jr. Livio M., 1993, PASP 105, 1373, *Common Envelopes in Binary Star Evolution*
Kahabka, P., & van den Heuvel, E.P.J. 1997, ARAA 35, 69; *Luminous Supersoft X-Ray Sources*
Kippenhahn, R., & Weigert, A. 1967a, ZfA 65, 251
Kippenhahn, R., & Weigert, A. 1967b, ZfA 66, 58
Kopal, Z. 1959, "Close binary systems", The Int. Astrophys. Series; Chapman & Hall, London
Langer, N., Deutschmann, A., Wellstein, S., & Höflich, P. 2000, A&A 362, 1046
Lederle, C., & Kimeswenger, S. 2003, A&A 397, 951
Leung, K.C., Seggewiss, W., & Moffat, A.F.J. 1983, ApJ 265, 961
Lewin, W.H.G., & van der Klis, M. (eds.). 2006, "Compact Stellar X-Ray Sources", Cambridge Univ. Press
Lewin, W.H.G., van Paradijs, J.A., & van den Heuvel, E.P.J. (eds.). 1995, "X-Ray Binaries", Cambridge Univ. Press
Lin, D.N.C., & Papaloizou, J.C.B. 1996, ARAA 34, 703; *Theory of Accretion Disks II: Application to Observed Systems*
Livio, M. 1994, in "Interacting Binaries", H. Nussbaumer & A. Orr (eds.); Saas-Fee Adv. Course 22, Springer, Heidelberg; p. 135
Massi, M., & Kaufman Bernadó, M. 2008, A&A 477, 1
Mirabel, I.F., & Rodríguez, L.F. 1999, ARAA 37, 409; *Sources of Relativistic Jets in the Galaxy*
Moffat, A.F.J., & Seggewiss, W. 1979, A&A 77, 128
Moretti, A., De Angeli, F., & Piotto, G. 2008, A&A 483, 183
Paczynski, B. 1966, Acta Astron. 16, 231
Paczynski, B. 1971a, Acta Astron. 21, 417
Paczynski, B. 1971b, ARAA 9, 183; *Evolutionary Processes in Close Binary Systems*
Papaloizou, J.C.B., & Lin, D.N.C. 1995, ARAA 33, 505; *Theory of Accretion Disks I: Angular Momentum Transport Processes*
Podsiadlowski, P., Joss, P.C. & Hsu, J.J.L. 1992, ApJ 391, 246
Pringle, J.E., & Wade, R.A. (eds.). 1985, "Interacting Binary Stars", Cambridge Univ. Press
Remillard, R.A., & McClintock, J.E. 2006, ARAA 44, 49; *X-Ray Properties of Black-Hole Binaries*
Sion, E.M. 1999, PASP 111, 532
Soberman, G.E., Phinney, E.S., & van den Heuvel, E.P.J. 1997, A&A 327, 620
Taam, R.E., & Sandquist, E.L. 2000, ARAA 38, 113; *Common Envelope Evolution of Massive Binary Stars*
Tauris, T.M., & van den Heuvel, E.P.J. 2006, in "Compact Stellar X-Ray Sources", W.H.G. Lewin, & M. van der Klis (eds.), Cambridge Univ. Press, p. 623
Terman, J.L., Taam, R.E., & Hernquist, L. 1994, ApJ 422, 729
Vanbeveren, D., De Loore, C., & Van Rensbergen, W. 1998, A&A Rev. 9, 63; *Massive Stars*
van den Heuvel, E.P.J. 1976, in IAU Symp. 73, "Structure and Evolution of Close Binary Systems", P. Eggleton, S. Mitton, J. Whelan (eds.), Reidel, Dordrecht; p. 35
van den Heuvel, E.P.J. 1994, "Interacting Binaries", H. Nussbaumer & A. Orr (eds.); Saas-Fee Adv. Course 22, Springer, Heidelberg; p. 263
van den Heuvel, E.P.J. 2007, "Double Neutron Stars: Evidence For Two Different Neutron-Star Formation Mechanisms", in "The Multicolored Landscape of Compact Objects and their Explosive Origins", AIP Conf. 924, p. 598; astro-ph/0704.1215
Verbunt, F. 1993, ARAA 31, 93; *Origin and Evolution of X-ray Binaries and Binary Radio Pulsars*
Verbunt, F., & van den Heuvel, E.P.J. 1995, in "X-Ray Binaries", W.H.G. Lewin, J.A. van Paradijs, & E.P.J. van den Heuvel (eds.); Cambridge Univ. Press, p. 457
Yoon, S.-C., & Langer, N. 2005, A&A 435, 967
Yoon, S.-C., Langer, N., Scheithauer, S. 2004, A&A 425, 217

Chapter 20

Luminosity and mass function

An important aspect to understand star formation and stellar populations is the comparison of observations with models. For that one needs to know what kind of stars are formed (see Ch. 7) and how they evolve (see Chs. 10 and 13), in short, how they and their star systems look like after some evolutionary time.

On the observational side one has CMDs (and perhaps HRDs) which one can relate with theoretical ones (see Synthetic CMDs, Ch. 21.4). On the theoretical side one has stellar models and models for stellar evolution. The models may be used to arrive at a time- and origin-dependent distribution of stars over brightness, in other words, to a *luminosity function*, LF.

One also needs to know the statistics of the kinds of stars (mass!) formed when stars form. This latter information has to be arrived at from observations of star groups in relation to the expected behaviour with time. The statistics of the stars against mass is given by a *mass function*, MF.

20.1 The luminosity function

The LF describes the number of stars per unit of luminosity. A luminosity function can be derived from observations by just counting the stars per luminosity interval. Since luminosities require the determination of the distance as well as information about all other wavelengths, one mostly uses the statistics of brightness in the V-band. It has become common astronomical slang to use in this case the word *"luminosity"* even when only M_V is meant.

The luminosity of stars changes with evolution. Thus the LF of a star group is not something eternal, but it is changing with time. Since red giant stars represent only a relatively short phase of stellar evolution, and because their luminosity is (at most) poorly related to the luminosity of the progenitor main-sequence stars, one prefers to work with the LF of just the main-sequence stars in a population.

The expected LF can be modelled using some mass function for the stars in the population and the stellar evolution information (e.g., a library of isochrones; see Ch. 21). Such theoretical luminosity functions then can be compared with observed luminosity functions to retrieve information about the nature and history of the stellar group under consideration.

When just one group of stars formed at time t_0, which stayed together and to which no stars have been added (from elsewhere or due to additional star formation at a later time) the LF should be smooth. If one recognizes in the LF a jump in the number statistics then this points at additional star formation at a later moment in time than t_0. (Note that if all stars of an old population are included the HB may cause a legitimate jump, which is easily recognizable by its M_V and which can even be used for distance determination.)

20.2 The stellar initial mass function

The most relevant stellar mass function is the *initial mass function* (IMF). It describes how many stars were formed per mass unit in the range available for star formation. One important question is, whether the IMF is the same for all star forming regions in all places, for all kinds of gas (composition), and at all times (galactic evolution).

Knowledge of the IMF is important for the modelling of present day star groups, for the modelling of the field population (e.g the Solar neighbourhood), and clearly for the modelling of entire galaxies, all in terms of spatial distribution, but also to model the present and the past emission characteristics (spectral synthesis, spectral flux, ionizing power, etc.) of star groups. Furthermore, the ratio of high to low mass stars in any stellar ensemble plays a role in the evolution of that system. How does it develop dynamically, how fast does the production of heavy elements proceed, what is the impact of the system on its surrounding matter? An estimate of the number of low mass stars could give indications for the number and total mass of brown dwarfs, perhaps leading to the solution of the problem of dark matter.

After normalisation, the stellar mass function can be interpreted as the probability density function describing the probability for a star of a given mass m (more precisely: a mass interval dm) to be created within a molecular cloud.

In practice one has to consider for an arbitrary star group two kinds of mass function: the present day mass function (PDMF), $N_P(m)$, as observed in present day systems, and the IMF, $N(m, Z, T, ...)$, as valid at the time of the formation of the stellar system. These two functions are related through the effects of additional star formation taking place over time (the star formation rate $\Psi(t)$) and the effects of stellar evolution ($E(t, m)$ giving the time over which a star of mass m remains in the main-sequence phase) as

$$N_P(m) = \int_0^{t_0} N(m, Z, t, \ldots) \, E(t, m) \, \Psi(t) dt \qquad (20.1)$$

with t_0 the age of the oldest stars in the system. Implicitly, $N_P(m)$ is the PDMF of only main-sequence stars. The IMF may depend, as indicated, on other parameters than just the mass (m), like the abundance of the heavy elements in the gas, Z, the point in time (t) of formation (evolution of the IMF with the evolution of the universe), or on further parameters. Assuming that the IMF is 'universal', i.e. independent from all parameters but the mass itself, we can conclude that

$$N_P(m) = N(m) \int_0^{t_0} E(t, m) \, \Psi(t) dt \qquad . \qquad (20.2)$$

The star formation rate $\Psi(t)$ can be a very complicated function. In the case of *star clusters*, we can assume that all stars formed at more or less the same time, i.e., in general the duration of star formation is much smaller than the age of the object. One then can show that Eq. 20.2 is reduced to

$$N_P(m) = N(m) \cdot E(t, m) \qquad . \qquad (20.3)$$

This means that the PDMF and the IMF of star clusters only differ by the effects of stellar evolution, $E(t, m)$. Thus, star clusters (and possibly associations) are ideal objects for studies of mass functions since the cluster IMF can (after correction for stars having already evolved away from the main sequence) be derived from its PDMF. In addition, members of a cluster have – besides the same age – the same distance, reddening, and chemical composition, so that adjustments of the individual measurements are not required.

So far, we have not discussed the shape of the IMF, which was not necessary for the above general consideration.

The birthplace of IMF studies – although the term 'initial mass function' itself is not mentioned in the paper – was the publication of Salpeter (1955) who studied the luminosity function including effects of stellar evolution of the stars in the Solar neighbourhood. Two major reviews published in more recent years were written by Scalo (1986, 1998).

20.2. THE STELLAR INITIAL MASS FUNCTION

20.2.1 Power law mass functions; equivalences

In most studies it was found that the IMF has – at least in the investigated mass intervals – the shape of a power law:
$$N(m) \propto m^\alpha \qquad (20.4)$$
or, more precisely,
$$dN(m) \propto m^\alpha \cdot dm \qquad (20.5)$$
with a negative exponent α. Qualitatively this means: few high mass stars, lots of low mass stars. With this formalism the IMF – understood as a probability density function – is entirely described by giving the value of the exponent α. However, there are various different ways to describe the IMF which lead to different characteristic values with which the IMF is being described, for example:
$$d\log N = \Gamma \cdot d\log m \qquad (20.6)$$
$$dN \propto m^{-x} \cdot d\log m \qquad (20.7)$$
where Eq. 20.6 is used by Salpeter (1955). *The three notations (20.5), (20.6), (20.7) are equivalent.*

One way to deal with this lack of a 'standard' is to also give the value found by Salpeter (1955) for the Solar neighbourhood in the notation used. For the three descriptions this value would be $\alpha = -2.35$, $\Gamma = -1.35$ and $x = +2.35$, based on the equivalences $\Gamma - 1 = \alpha$ and $\alpha = -x$.

The equivalence mentioned above can be demonstrated as follows. The transition (20.5) $\iff$ (20.7) is evident with
$$d\log m = \frac{1}{m} \cdot dm \quad . \qquad (20.8)$$
The transition (20.5) $\implies$ (20.6) follows from
$$N = \int dN \stackrel{(20.5)}{=} \rho \cdot \int m^\alpha \cdot dm = \frac{\rho}{\alpha+1} \cdot m^{\alpha+1} \qquad (20.9)$$
with a constant of proportionality ρ and with
$$\begin{aligned}
d\log N &\stackrel{(20.8)}{=} \frac{1}{N} \cdot dN \\
&\stackrel{(20.9)}{=} \frac{1}{\frac{\rho}{\alpha+1} \cdot m^{\alpha+1}} \cdot dN \\
&\stackrel{(20.5)}{=} \frac{1}{\frac{\rho}{\alpha+1} \cdot m^{\alpha+1}} \cdot \rho \cdot m^\alpha \cdot dm \\
&= (\alpha+1) \cdot \frac{1}{m} \cdot dm \\
&\stackrel{(20.8)}{=} (\alpha+1) \cdot d\log m \\
&= \Gamma \cdot d\log m \quad .
\end{aligned}$$
Similarly, one can prove the transition (20.6) $\implies$ (20.5) with
$$\log N = \int d\log N \stackrel{(20.6)}{=} \Gamma \cdot \int d\log m = \Gamma \cdot \log m \quad . \qquad (20.10)$$
It then follows that
$$N = m^\Gamma \qquad (20.11)$$
and thus
$$\begin{aligned}
dN &\stackrel{(20.8)}{=} N \cdot d\log N \\
&\stackrel{(20.6)}{\propto} N \cdot d\log m
\end{aligned}$$

$$\stackrel{(20.8)}{=} N \cdot \frac{1}{m} \cdot \mathrm{d}m \quad (20.8)$$
$$\stackrel{(20.11)}{=} m^{\Gamma-1} \cdot \mathrm{d}m \quad (20.11)$$
$$= m^{\alpha} \cdot \mathrm{d}m \quad,$$

completing the proof of equivalence of the various representations of the mass function.

20.2.2 Salpeter mass function

We will use the nomenclature of Eq. 20.6 which is equivalent to the Salpeter one. More precisely, we define:

> The initial mass function is the differential probability distribution of stellar mass m per unit *logarithmic* mass interval, denoted $N = N(\log m)$.

$$\Gamma := \frac{\mathrm{d}\,\log N(\log(m))}{\mathrm{d}\,\log m} \qquad (20.12)$$

is called the index of the IMF.

With this definition, Γ can be determined for any differentiable representative of the IMF, i.e., the definition is independent from the shape of the IMF. In general, Γ is a function of the mass. Only in the case the IMF is (at least within certain mass intervals) represented by a power law, Γ seems to be constant within these intervals and stands for the exponent of the IMF. In the case of a power law IMF, this nomenclature has the advantage that in a log–log plot the IMF results in a line with a slope of Γ. In addition, no further constant of proportionality (like ρ in the above proof) is required. As stated before, in this notation, the Salpeter value would be $\Gamma = -1.35$.

20.3 Relation between the luminosity and mass functions

Since the stellar mass is a parameter which cannot be directly measured, stellar models are required to transform a measured stellar magnitude in a given filter into a stellar mass. A function which performs this task is called a mass-luminosity relation (**MLR**).

The distribution of stars over mass is thought to be smooth and is represented by the mass function. The reverse need not be true: a smooth mass function may, also on the MS, imply a luminosity function with structure.

A good example of LF structure is based in the transition region in the MS from the lower mass range with only central PP-chain burning to the slightly more massive CNO-cycle hydrogen burners. Stars below this mass boundary have energy production which depends on central temperature like $\epsilon \sim T_c^{4.5}$ (Ch. 5.1.4) while stars above the mass limit have $\epsilon \sim T_c^{19.9}$ (Ch. 5.1.5). The consequence is that in/near the transition region a star more massive by small Δm may be more luminous by a large ΔL. The statistical distribution in L of stars will have there a "discontinuity", or will not be a power law. In other parts of the MS there is most likely a monotonous relation of Δm and ΔL.

As a consequence of this, difficulties occur when stars of the stargroup under investigation have evolved away from the main sequence: in that case the MLR is no longer a function, i.e. there is no unique relation between luminosity and mass. Therefore IMF studies have to be restricted to parts of the CMD where the MLR still is definite, which is true for the main-sequence region.

20.4 Determinations of the mass function

To determine the mass function from photometric data one proceeds as follows:

1) compare the data with isochrones and select the best;

2) decide for each data point which isochrone (Ch. 21) applies (due to photometric errors, etc., many data points lie *not* exactly on an isochrone);

3) identify the mass associated with the individual data points;

4) build the statistics of the stellar masses so found in mass intervals (according to the favoured mathematical representation of the IMF);

5) determine the slope of the statistics to find the exponent of the assumed exponential mass function, perhaps using multiple fitting functions;

6) possibly, before step 5, make a correction for the incompleteness (see Sect. 20.4.3, below) in the faintest bins.

Note that a transformation of observed brightness (M_V) into luminosity L is rather inaccurate because isochrones in the MS part of a CMD form a rather tight set of curves. Phrased differently, evolution in the main-sequence phase leads to brightening of the stars, which for massive stars does not (in the visual; see Ch. 1.4.2) go along with a change in colour.

Star groups of different nature require in part different treatments. Investigating different groups leads to results for different mass ranges.

20.4.1 Star clusters

At first glance, star clusters are ideal objects for IMF determinations. With all members of a cluster formed at (more or less) the same time in the same cloud in the same small volume, studies of clusters might reveal variations of the IMF with parameters like time, location in the Galaxy, or chemical composition. A field star IMF would have to be *assumed* to be independent from these parameters, since the 'birthplace' and the circumstances of star formation in general are unknown for individual stars and the field consists of objects of different ages.

20.4.1.1 Open clusters

Open star clusters are ensembles of usually several hundred to a few thousand member stars, which are located within a volume of roughly 5 to 20 pc across. Their ages range from some 10^6 yr up to 10^{10} yr. Open clusters are mainly located within or near the Galactic plane, mostly hidden behind interstellar dust. Over 1000 open clusters are known in the visible Milky Way, half of which have been observed at least in one photometric system.

The small amount of stars in open clusters leads to small number statistics, especially if one considers the heavy contamination of the observational data with field stars due to the location of most clusters in or close to the disk of the Galaxy.

20.4.1.2 Globular clusters

Galactic globular clusters are generally old, with ages of 10 to $15 \cdot 10^9$ yr. The metal content Z in globular clusters ranges in [M/H] from -2.3 to -0.5. Each contains (even now) more stars than an open cluster. Given their age and present mass ($3 \cdot 10^4$ to 10^6 M$_\odot$) they must have formed in really huge (low metallicity) gas clouds.

About 150 globular clusters are known in the galaxy forming a roughly spherical halo. Half of the globular clusters are well studied.

The high age of globular clusters limits IMF studies to masses below 1 M$_\odot$.

20.4.1.3 Mass segregation

All clusters may show mass segregation, i.e., spatial distributions of the stars differing for different masses. One usually finds that high mass stars are concentrated towards the centre of the cluster, whereas the distribution of the lower mass stars has a higher density in the outer parts. As a consequence, it has to be taken into account that the apparent size of in particular an open star cluster might be underestimated when based on images with a too bright limiting magnitude, since the low mass stars might not be detected and thus low-mass stars may be underrepresented.

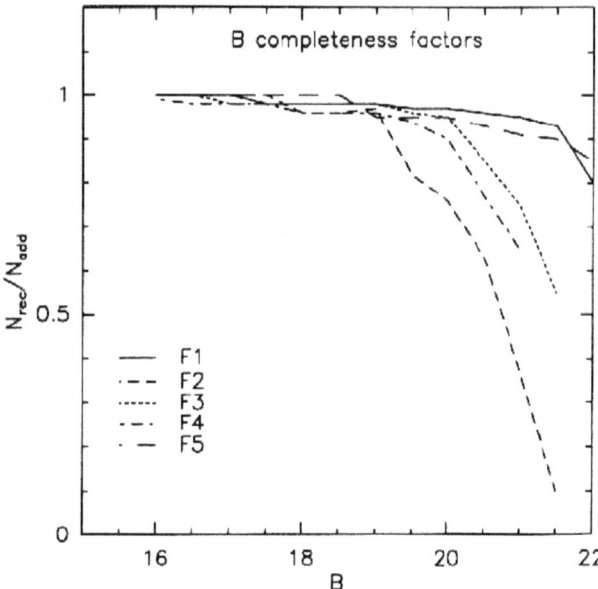

Figure 20.1: Factors of *completeness* $N_{\text{recovered}}/N_{\text{added}}$ (see Sect. 20.4.3) for 5 fields (F1 to F5) in an OB association in the LMC, plotted against the B magnitude of each star. Toward the fainter end more and more stars stayed apparently undected, but by amounts which differ between the fields (likely due to differing numbers of stars per unit surface area). Figure from Vallenari et al. (1993).

Mass segregation was found even for very young clusters (Pandey et al. 1990, Raboud & Mermilliod 1998a, 1998b). This suggests it already takes place at the time of cluster formation. Mass segregation is a process which possibly consists of three steps. The early type stars already form mainly in the central region of the emerging cluster ('initial mass segregation'), or, even vice versa, massive stars mainly form together while lower mass stars surround them. Due to the evolution of the most massive stars in the early stages of a cluster lifetime, the entire cluster may be mixed in such a way that a phase may follow in which all stars are distributed homogeneously. After that the dynamics of the cluster leads to another concentration of the massive stars around the gravitational centre of the object (core collapse), whereas the lower mass stars form a corona around this region, which may extend as far out as the tidal radius of the cluster.

These points have to be carefully taken into consideration. However, since one can handle them, star clusters still are a good source for IMF studies.

20.4.2 Field stars

Investigation of the field stars may lead to a meaningless result, since one deals with a mix of stars of different ages and thus histories, or from (so to say) different IMFs. However, the stars in the disk of the Milky Way form apparently a mix from very similar original conditions (except time) so that Salpeter (1955) was able to derive an IMF for the field stars close to the 'real IMF' as found studying various star clusters.

20.4.3 Completeness of the photometry

One important aspect for the determination of the luminosity or mass functions is the quality of the input data. Photometry reaches to a certain faint limit in brightness and stars fainter cannot be recognized in the photometry. Luminosity or mass functions derived do not reach beyond this limit. But some stars are seemingly brighter due to noise (see Fig 21.6).

Photometric values for stars are derived from a CCD frame by using a software package which finds 'stars' and determines their brightness (in instrumental units). The software looks for pixels with values elevated over the background level. It then determines the full signal of the object, either by summing up pixel values in a well defined area around the peak or by fitting a point-spread-function to such a peak. Well known photometric packages are, for each method, SEXTRACTOR or DAOPHOT, respectively. The software produces lists of objects with coordinates

20.4. DETERMINATIONS OF THE MASS FUNCTION

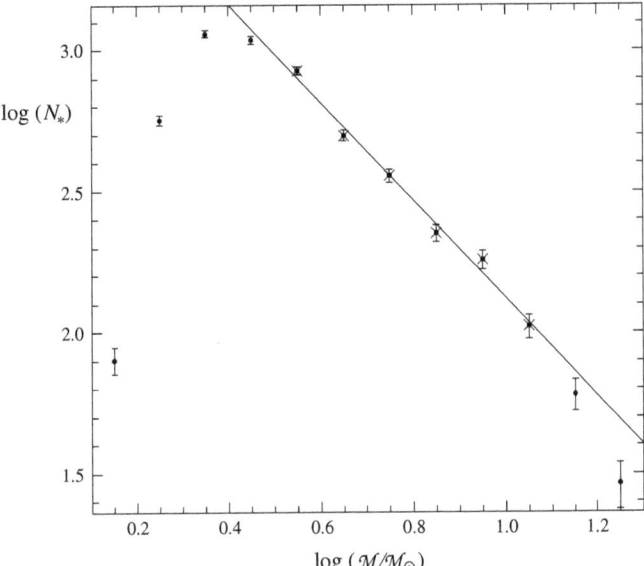

Figure 20.2: The mass function for the stars inside the supergiant shell No. 4 in the LMC has a slope $\Gamma = -1.73$ in the mass range of $2.5 < m < 16$ $M_\odot$. Note the *incompleteness* toward the smaller masses. At the high mass end possibly effects of small number statistics play a role or the IMF for those stars has a different slope. Figure from Braun et al. (1997).

X, Y and instrumental intensity J. Using exposures of calibration stars one now can transform the J-values to magnitudes. Using such lists from, e.g., a B and a V exposure, one can crosscorrelate the lists and identify the stars of one frame with that on the other, in order to arrive at values for $B - V$ for each star.

When approaching the fainter limit of the photometry, stars will be missing randomly, but progressively so toward the fainter bins. In order to determine where the *incompleteness* of the photometry sets in, experiments are performed with the data.

With a random number generator positions in the CCD frame are selected at which *artificial stars* (digital images with a noisy intensity distribution over pixels like that of the real stars) are added to the digital values of the CCD frame. The artificial stars may be chosen to lie in well defined brightness ranges. The software to reduce CCD data to photometric values for the stars is used again but now on the frame containing also the artificial stars. The goal is to explore how many of the inserted stars are recovered. If one recovers, e.g., 84% of the inserted stars, one can assume that the original photometry also 'saw' 84% of the stars, thus was 84% complete. (Note that if one adds too many stars, the original statistics is disturbed.) Such tests are run several times (with each time new artificial stars in the original frame) to eliminate statistical errors. An example of the results of such an investigation is given in Fig. 20.1.

With such a result one may 'correct' a luminosity function for the stars missing at the faint end, the so-called *incompleteness correction*. (Note that often the faulty phrase 'completeness correction' is used.) This allows the determination of the luminosity and mass functions toward fainter and less massive stars, respectively.

The incompleteness test is very important in the case of dense star fields. There, the photometry will be seriously incomplete (also for the brighter stars!) due to the larger probability of star images overlapping mutually. But especially the fainter stars are affected which will be drowned in the signal of the brighter ones.

An example of a mass function derived from photometry inside supergiant shell LMC 4 is shown in Fig. 20.2. Effects of incompleteness can be recognized there, too.

An example of the very low end of the MS, in fact showing the cut-off toward the lower mass limit for stars, is shown in Fig. 20.3.

20.4.4 Results for mass functions

The results of recent IMF studies is (in most cases) well represented by power laws within different mass intervals (Scalo 1998). This clearly means that **the IMF is not** a power law, only that

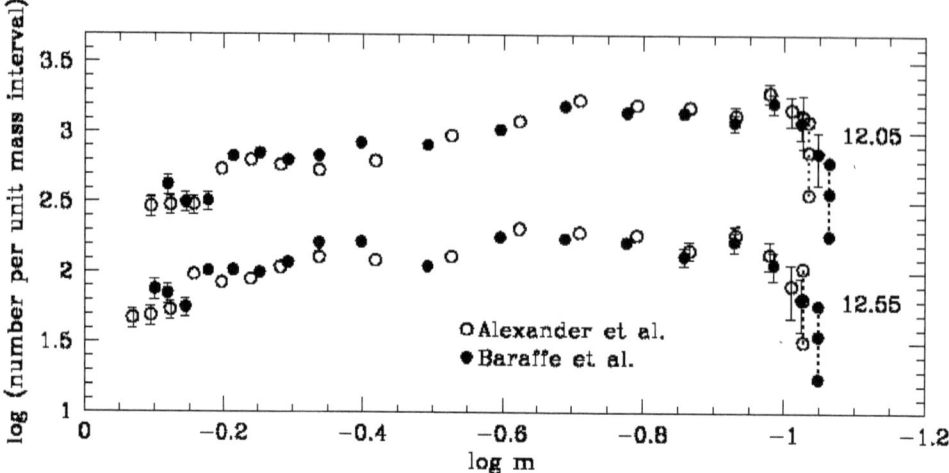

Figure 20.3: The mass function for the stars of the low end of the MS, as derived from HST V, $V-I$ data for the globular cluster NGC 6397 (see Fig. 6.3) is shown. Note the shallow slope (cf. Sect. 20.4.4). Since for this cluster the distance is not exactly known the mass function is given for the distance moduli 12.05 and 12.55. The transformation of I-bins to mass-bins leads for the same magnitude interval to different statistics. Also different input stellar models (Alexander et al., Baraffe et al.) have been used. Figure from King et al. (1998).

it locally can be approximated by such a function. From studies of different mass ranges typical exponents are found:

$$\begin{aligned} \Gamma &= -1.3 \pm 0.5 & \text{for} \quad & m > 10 \text{ M}_\odot, \\ \Gamma &= -1.7 \pm 0.5 & \text{for} \quad & 1 \text{ M}_\odot < m < 10 \text{ M}_\odot, \\ \Gamma &= -0.2 \pm 0.3 & \text{for} \quad & m < 1 \text{ M}_\odot, \end{aligned} \qquad (20.13)$$

where the $\pm$ values refer to a rough range of Γ values derived for the indicated mass intervals, caused by empirical uncertainties or probable real IMF variations. Some authors distinguish only between masses above and below $\simeq 1 M_\odot$, whereas others divide the sub-Solar mass interval into two parts with, e.g., $0.3 > \Gamma > -0.85$ for $0.08 \text{ M}_\odot \leq m \leq 0.5 \text{ M}_\odot$ and $\Gamma = -1.2$ for $0.5 \text{ M}_\odot \leq m \leq 1 \text{ M}_\odot$. Clearly, an IMF as power law does not have one unique slope.

As noted before, star clusters do change due to dynamical interactions. E.g., lower mass stars, which normally are on wider orbits in a cluster, may be more easily stripped off by tidal forces. This probably plays a more important role the older the studied object is. This has to be kept in mind when using the term 'initial' mass function for objects of an age of, say, 100 Myr.

20.4.5 The high-mass end of the IMF

How far does the IMF extend to the high mass end? Is there a limit or does the IMF continue in case of star formation in gigantic clouds? Since the high mass end will have fewer and fewer stars, it may well be that the chances to form very-high mass stars are so slim that there is an upper limit. Moreover, once stars are forming they take up mass no longer available for further star formation.

Modelling all these aspects, Weidner & Kroupa (2006) show that in practice the *maximum mass of stars* is $M_{\text{max}} \simeq 150 \text{ M}_\odot$ (in gas of roughly solar metallicity). They also show that the mass function of clusters with a modest total mass (say $10^2 \text{ M}_\odot$) cuts off at lower mass than the mass function of a single cluster of $10^6 \text{ M}_\odot$. The reason is that the small number statistics in small mass clusters prevents the formation of high-mass stars there, while the high-mass cluster does have that possibility. This result also means that the "integrated galactic IMF" (the integral of all the cluster IMFs of a galaxy) will for $M > 1\,M_\odot$ be always steeper than an individual cluster IMF. And thus that modelling of, e.g., the UV output of a galaxy to estimate the number of young stars or the recent star formation rate has to be done with an IMF *steeper* than the one given in Eq. 20.13.

20.5 The IMF and its universality

With the help of a – more or less – 'universal' mass function, it would be sufficient to know about a limited (e.g. due to the range of data available) mass range of a certain object to extrapolate over the entire mass spectrum, so that the total mass of a stellar system and its dynamics could be easily estimated. The data available so far show that approximately 20% of the total mass of a stellar system is bound in the stars with more than $10\,M_\odot$, 75% in objects with masses between 0.1 and $10\,M_\odot$, and only 5% in stars less massive than $0.1\,M_\odot$ (Larson 1999). If this distribution could be generally confirmed, this would mean that brown dwarfs and low mass stars would not contribute much to the total mass of a stellar system, a point which would be essential for the discussion of the nature of dark matter (Chabrier & Méra 1997, Kroupa 1998).

One defines three types of universality:

- **Strong universality.** IMF studies always lead to (within the errors) exactly the same shape described by the same parameter(s).

- **Intermediate universality.** The shape of the IMF is basically the same, but its parameter(s) vary. However, these variations can be explained, e.g., by certain circumstances of star formation or other cluster parameters.

- **Weak universality.** IMFs of different objects show more or less the same shape and features, but significant variations occur randomly.

Many IMF studies indeed lead to similar results, most studies are consistent with a 'universal' IMF at least in the weak sense (Kennicutt 1998). The most obvious parameter which varies with time and could lead to IMF variations is the metallicity and thus the cooling rate of a molecular cloud. But there may also be an influence of the stars formed early, in particular if they are massive, by way of the flux of Lyman-continuum photons which will tend to ionize the birth cloud from whithin. In general, the initial mass function may be uniform in spite of the fact that various star systems show otherwise considerable variations (Kroupa 2002).

Cluster formation is only poorly understood in detail. Some authors speculate that two different, physically unrelated mechanisms produce stars of different masses which contribute to the total IMF. One leads to the power law shaped IMF of stars with masses above roughly $1\,M_\odot$, the other to the more or less flat(ter) part below $1\,M_\odot$. One attempt to explain the high mass end of the IMF is the theory of a *universal fractal structure* of star forming clouds, which is caused by turbulence (Larson 1992, 1995). The fractal dimension of the cloud, then, would be equal to $-\Gamma$. Fractal structure would indeed lead to a power law for the distribution over mass of stars formed.

20.6 The mass function for the first stars

In the big bang model, the universe expanded from a very hot phase into a cool phase, in which mass assembled gravitationally to form galaxies. In those galaxies the first stars (Ch. 15) must have formed.

The material after the big bang contained only H, D, He and some Li. There was no dust in the ISM and therefore H_2, essential for the quick formation of stars, was absent. It is very likely that the mass function of the first stars being formed at that time is, in the absence of H_2, different from the 'normal' IMF. It has been speculated, that the mass function of the first generation of

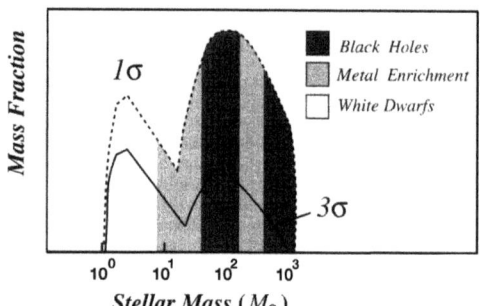

Figure 20.4: Possible mass function for Pop. III stars (the first star generation) differs from the normal IMF because the conditions for star formation were completely different then. In particular an 'excess' of mass rich stars was likely. The $1\,\sigma$ and $3\,\sigma$ curves represent the model margin of the gas density in the early galaxy. The shadings indicate mass ranges together with the nature of the stellar remnant for that mass range.
Figure after Nakamura & Umemura (2001).

stars (frequently called the Population III stars) had an excess of mass rich stars. Fig. 20.4 shows the possible shape of that function.

Acknowledgement. A large portion of the text of this chapter was adapted from the introductory chapter of the doctoral thesis of Sanner (2001).

References

Braun, J.M., Bomans, D.J., Will, J.-M., & de Boer, K.S. 1997, A&A 328, 167
Chabrier, G., & Méra, D. 1997, A&A 328, 83
Kennicutt, R.C. 1998, ASP Conf. Ser. 142, G. Gilmore & D. Howell (eds.); p. 1
King, I.R., Anderson, J., Cool, A.M., & Piotto, G. 1998, ApJ 492, L 37
Kroupa, P. 1998, ASP Conf. Ser. 134, R Rebolo et al. (eds.); p. 1
Kroupa, P. 2002, Science 295, 82, *The Initial Mass Function of Stars; Evidence for Uniformity in Variable Systems*
Larson, R.B. 1992, MNRAS 256, 641
Larson, R.B. 1995, MNRAS 272, 213
Larson, R.B. 1999, in "Star Formation 1999", T. Nakamoto (ed.); p. 336
Nakamura, F., & Umemura, M. 2001, ApJ 548, 19
Pandey, A.K., Paliwal, D.C., & Mahra, H.S. 1990, ApJ 362, 165
Raboud, D., & Mermilliod, J.-C. 1998a, A&A 329, 101
Raboud, D., & Mermilliod, J.-C. 1998b, A&A 333, 897
Salpeter, E.E. 1955, ApJ 121, 161
Sanner, J. 2001, "Photometric and Kinematic Studies of Open Clusters", PhD Thesis, Sternwarte Univ. Bonn
Scalo, J.M. 1986, Fundamentals of Cosmic Physics, 11, 1
Scalo, J.M. 1998, ASP Conf. Ser. 142, G. Gilmore & D. Howell (eds.); p. 201
Vallenari, A., Bomans, D.J., & de Boer, K.S. 1993, A&A 268, 137
Weidner, C., & Kroupa, P. 2006, MNRAS 365, 1333

Chapter 21

Isochrones

21.1 Definition

A *colour magnitude diagram (CMD)* gives the colours and magnitudes of the members of a group of stars. If the group is a star cluster, of which one may assume that all stars are of the same age (and at the same distance), it identifies a *population of stars with a certain age*. Models for the evolution of stars of different masses make it possible to calculate for a given group of coeval stars the distribution of their data points (colour and magnitude). *The line connecting the datapoints of equal time elapsed in the evolution of the full range of stellar masses is called* **isochrone**.

The age of a star cluster can then be determined by comparing its CMD with different isochrones. When making such a comparison several factors are of importance.

- The quality of the stellar evolution and structure models with respect to:
  - the chemical composition (X, Y, Z) and its effects (mainly Z; in certain cases the abundance of the different elements must be taken into account individually),
  - the completeness of the functions of opacity,
  - the completeness of the possible reactions for nuclear fusion,
  - the completeness of other important internal processes such as convection,
  - the completeness of atmosphere physics (chemical composition, stratification, convection, magnetic fields),
  - the completeness of overall stellar structure aspects (e.g., effects of stellar rotation on colour).

  Most of these aspects have been discussed in other Chapters.

- The quality of the observational data with respect to:
  - the accuracy of the photometry and its completeness,
  - the presence in CMD of fore- & background stars or background galaxies ($V > 22$ mag),
  - the assumptions for $[M/H]$, distance and reddening.

  These aspects will be addressed below.

21.2 Examples

Isochrones come in different shapes related with the parameters chosen for the representation.

In Fig. 21.1 isochrones are plotted as a function of $\log L$ and $\log T_{\text{eff}}$. But since isochrones are meant to be used in relation with observational data, other parametrizations are suitable.

Fig. 21.2 shows isochrones as a function of $\log g$ and $\log T_{\text{eff}}$. These parameters can be determined from spectroscopic and photometric observations which makes them independent of the availability of the distance.

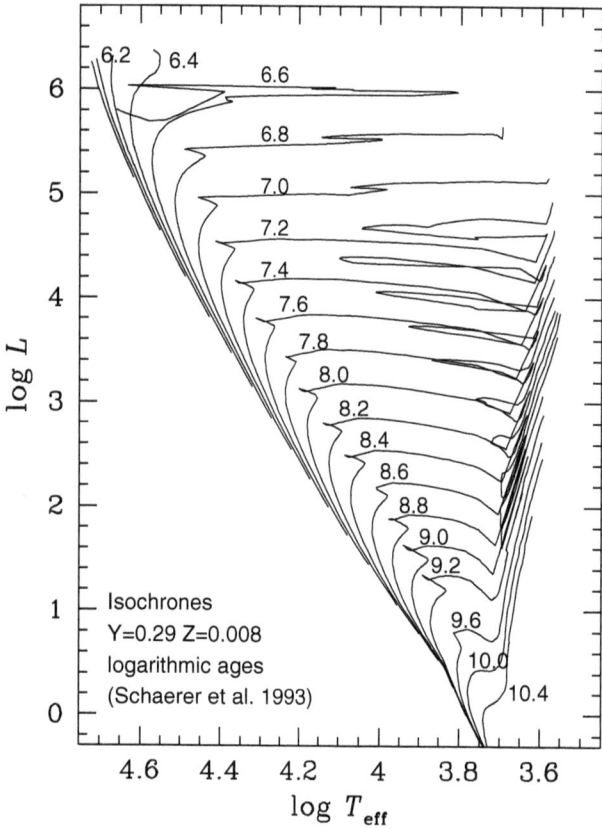

Figure 21.1: Isochrones as a function of $\log L$ and $\log T_{\text{eff}}$ (with the logarithm of the age in years).
Note that this diagram is only relevant for observations if the distance of the stars is known.
The isochrones for log (age) = 9.4, 9.8 and 10.2 are not plotted. For stars in the lower mass range the post-He-flash evolution is not included. Thus the horizontal branch phase and other late stages are *not* represented!
Models from Schaerer et al. (1993) for $Z = 0.008$ ($\simeq 1/3$ solar metal content).

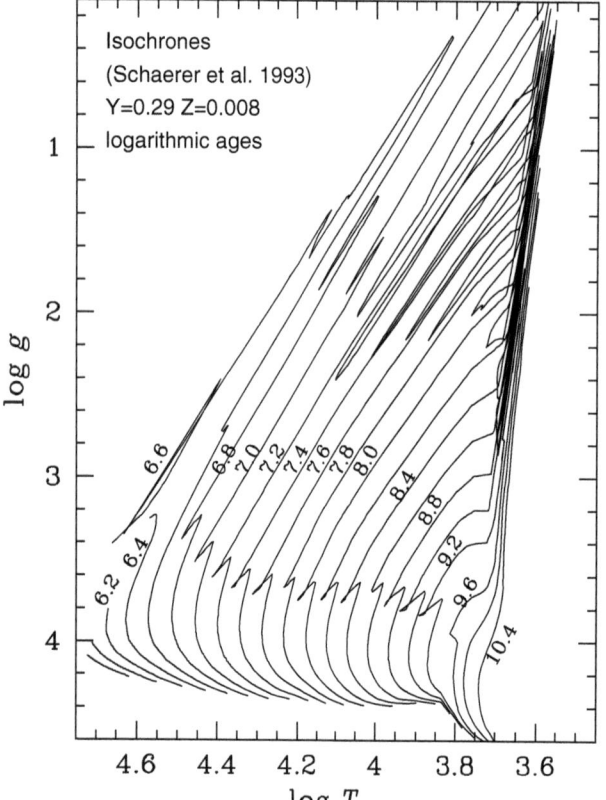

Figure 21.2: Isochrones as a function of $\log g$ and $\log T_{\text{eff}}$ (with the logarithm of the age in years).
This diagram uses only parameters which can be derived from photometry and spectroscopy without the need to know the distance of the stars.
The isochrones for log (age) = 9.4, 9.8 and 10.2 are not plotted and a few isochrones are not labelled (but compare with Fig. 21.1). For stars in the lower mass range the post-He-flash evolution is not included. Thus the horizontal branch phase and other late stages are *not* represented!
Models from Schaerer et al. (1993) for $Z = 0.008$ ($\simeq 1/3$ solar metal content).

21.2. EXAMPLES

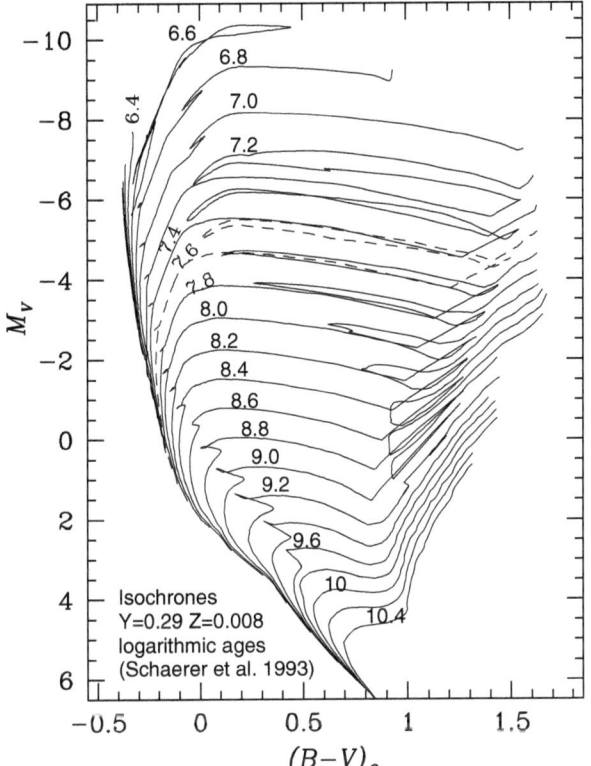

Figure 21.3: Isochrones as a function of M_V and $(B-V)_0$ (with the logarithm of the age in years).

Note that, on the blue side, due the limiting value of $(B-V)_0$, a small change in $(B-V)_0$ gives a large change in age. The use of the M_V versus $B-V$ ioschrones can result in large age uncertainties for star groups younger than $10^{8.5}$ years. And the use of M_V implies that the distance must be known.

The isochrone for log (age) = 7.6 is plotted as a dashed line for better recognition in the part of the diagram where the isochrones have considerable overlap. For stars in the lower mass range the post-He-flash evolution is not included. Thus the horizontal branch phase and other late stages are *not* represented! For isochrones with HB see the lower panels of Fig. 21.4. Models from Schaerer et al. (1993) for $Z = 0.008$ ($\simeq 1/3$ solar metal content).

In Fig. 21.3 the same isochrones are plotted as a function of M_V and $(B-V)_0$. Photometry only provides a $(V, B-V)$-diagram so that here knowledge of distance as well as colour excess is required for the transformation from V and $B-V$ to M_V and $(B-V)_0$.

The isochrones can be used to *determine the age of the stars* of the particular star group. Once the best fitting isochrone is found one also can translate the stellar data into, e.g., the mass of the stars. These aspects are discussed in Ch. 20.

The *age of a field star* can only be determined if its observational parameters allow accurate placement on an isochrone. This would require a very accurate determination of the luminosity (distance + spectral energy distribution) and the temperature. Moreover, many isochrones overlap so that often no unique age can be found. For double stars this is, of course, easier since there are two stars of the same age. In general, ages of field stars cannot be determined reliably.

21.2.1 Effects of metal content of stars

Fig. 21.4 illustrates how isochrones differ if the initial metallicity of the stars is different. At lower Z the opacity of stellar gas is lower so that mass-rich stars evolve with blue loops (see Chs. 10.4.4 & 13.4.1.2) up to higher T_{eff} than in the case of metal-rich stars (see also Fig. 16.3). Note that X, Y, Z are values by weight: $Z = 0.020$ is about solar metallicity. With HB stars the blueness in the metal-poor case is caused by a thinner envelope than with less metal-poor HB stars (see Ch. 10.4.2.2).

21.2.2 Transforming (L,T)-isochrones to $(M_V, B-V)$-isochrones

The comparison with observational data requires a transformation of theory to observable parameters. The transformation is based on data of some well known stars, for which distances, radii, and metalicities in the atmospheres are known and fluxes are measured accurately in many spectral ranges. However, transformations of (L,T) to $(M_V, B-V)$ are often not perfect, as stellar atmospheres are not always taken into account in detail or are incomplete (e.g., molecular opacity).

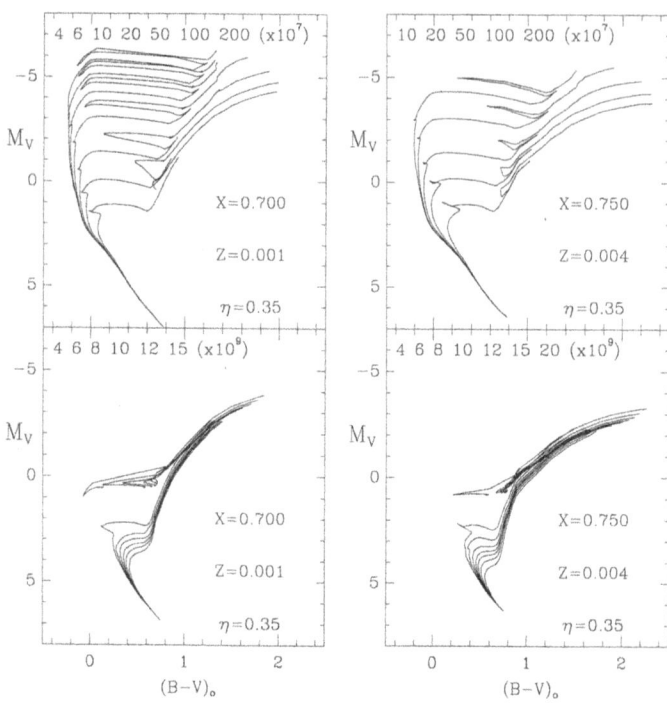

Figure 21.4: Isochrones for metal poor star clusters.
Left: very metal-poor population.
Right: less metal-poor population.
Upper row: for small ages.
Lower row: for larger ages.
The range in ages (years) is given in the top of each box.
The blue loops in the evolution of high-mass stars are for metal-poor populations longer than with less metal-poor stars, and thus also in isochrones (see also Fig. 21.3).
The HB stretches further blueward for metal-poorer populations (metal-poor stars have thinner shells). Solar-metallicity stars become RHB stars (Table 10.1) with thick shells. Compare with the globular cluster CMDs of Figs. 10.7 and 10.8. Figures from Bertelli et al. (1990).

The imperfectness of the transformation almost always leads to slight mismatches between observed CMDs and the isochrones used for the fit.

21.2.3 Difference between isochrones and evolutionary tracks

It is important to dwell on the difference between an evolutionary track and an isochrone. For high-mass stars the evolutionary track clearly starts at the given mass on the ZAMS, and leads away directly from it toward the TAMS. An isochrone of moderate age has, away from the ZAMS, almost the same shape as an evolutionary track. At lower L it asymptotically approaches the ZAMS. For stars with low masses the isochrone is the curve along the lower main sequence and the giant branch, not differing a lot from the evolutionary track of a single star, but again, the isochrone diverges from the lower main sequence asymptotically.

21.3 Using isochrones in CMDs

Isochrones give the expected locations for the datapoints of stars with the same age in the (L,T or $M_V, B-V$) plane depending on the time the stars have evolved.

If only a few stars are present, a very accurate photometry and a good determination of both the colour excess $E(B-V)$ and the distance (or the distance modulus $m-M$) is required in order to obtain reasonable results.

If many members of a star cluster are measured (especially if also red giants are present), a fit of the isochrones can yield $(m-M)$, $E(B-V)$ and age, as well as $[M/H]$ because the location of the RGB depends on metallicity.

Fig. 21.5 shows the CMDs for the open star clusters Pleiades and M 67 with isochrones fitted. The diagram of M 67 obviously also contains field stars. CMDs for globular clusters are given in Figs. 10.7 and 10.8, which may be compared with Fig. 21.3. For further information concerning parameters derived from the CMDs see Ch. 20.

In spite of the well defined location of an isochrone, the isochrones do **not** show **how many stars** can be found **where** along the isochrone (but see below, Sect. 21.4).

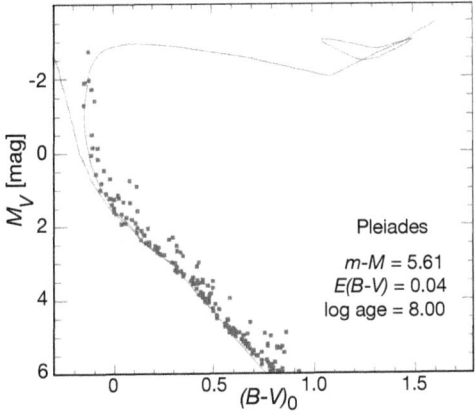 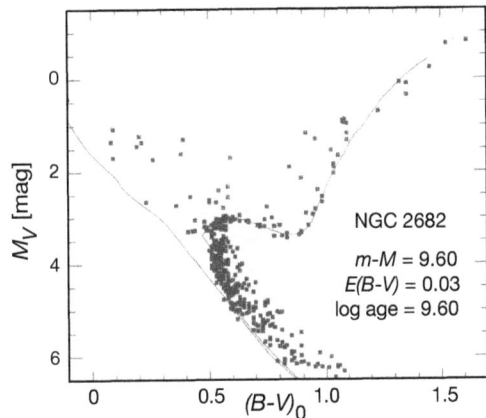

Figure 21.5: Comparison of star cluster main sequences with isochrones (models without overshooting). *Left*: the Pleiades. There are no RGs in the data. *Right*: M 67 (= NGC 2682). The fit is made to the TO; the model does not represent the RG well. From Meynet et al. (1993).

21.4 Synthetic CMDs

To overcome the problem of not knowing how many stars lie where on an isochrone, synthetic CMDs have been constructed. This requires models for stellar evolution and models for stellar atmospheres (the material to create an isochrone), and some knowledge about the stellar-mass statistics (how many stars exist at what mass). For the latter one assumes the so called *mass function* (see Ch. 20).

The "initial mass function", mostly of the form $dN = M^\alpha dM$, is used to generate a synthetic stellar population. For all these stars the evolution is followed. This basically is a job using tables, since evolutionary tracks can be stored as a library. At the desired time (of evolution) the surface parameters of the stars of the synthetic population are taken from the library and plotted in an HRD or CMD. Since the starting point was the chosen mass function, the data points are spread through the HRD or CMD according to their true likelihood.

Starting with known parameters of the IMF for a newborn stargroup with given chemical composition, and using stellar evolution models and models for stellar atmospheres, one can follow the change in time of the values of colour, magnitude, surface metallicity, gravity and luminosity for each star in the modelled group. The CMD thus constructed is unique for that star group at the state of evolution choosen.

In order to make *a realistic comparison with observed CMDs* further modifications are required. These are given below and deal with (1) photometric errors, (2) binary stars, (3) the effect of both errors nad binaries, (4) optical confusion and completeness and (5) mixing of objects at different distances. Finally, (6) stellar rotation has effects on colours.

1) observation errors:
To simulate observation errors the colour and brightness of the stars arrived at in the modelling are changed by adding randomly brightness dependent errors of a size as expected in real observations. The effect is visible in Fig. 21.6, Panel **c**.

2) presence of binary stars:
In the field of the Milky Way most stars form binary systems, several form multiple systems. It is reasonable to assume that also in larger groupings (up to open clusters) many stars are binary objects. (In globular clusters the frequency of dynamical interactions is large so that many pairs will be disrupted with time.) In binary systems, the mass of the two stars is mostly similar.

Observations containing a binary will normally not allow to resolve the binary spatially. In photometry, the brightness and colour found for the sum of the the two stars can only be treated as if it was one. The effect of binaries on a CMD is shown in Fig. 21.6, Panel **a** and Panel **b**. Note

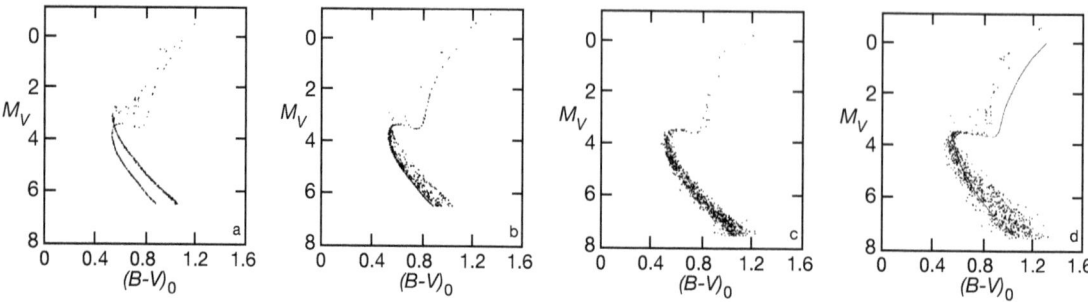

Figure 21.6: Synthetic CMDs for a young star group, showing separately and combined the effects of binary stars and of photometric errors.
Panel a: Synthetic CMD with 50% of the stars in binaries, with mass ratios between 0.9 and 1.0. Since in this case the MS stars are almost identical, their M_V values are almost the same so that the MS binary is $\simeq 0.75$ mag brighter than a single star. A MS + RG combination essentially has the M_V of the RG, in the case of two nearly equal RGs the M_V of the binary is again 0.75 mag brighter than of the single RG.
Panel b: Synthetic CMD with 50% of the stars as binaries, with mass ratios between 0.3 and 1.0. Note that now the sequence of binaries has much more scatter than in Panel a because of the more different M_V of primary and secondary due to their more different mass.
Panel c: Synthetic CMD for single stars, but with observing errors added.
Panel d: Synthetic CMD with 50% of the stars in binaries with mass ratio between 0.3 and 1.0 as in Panel **b** and with photometric errors as in Panel **c**. Note that Panel d looks like a "normal" CMD (as observed). The fiducial line represents the observations of the star cluster Kron 2. Figures adapted from Apparicio et al. (1990).

that one can recognize in Fig. 21.5 (Pleiades) physical binaries above the true main sequence.

3) binaries plus errors:
The effect of both binaries and errors is shown in Fig. 21.6, Panel **d**.

4) optical confusion and completeness:
When the spatial resolution of the observational data is limited (either due to seeing or because of the instrumentation) stars with very small angular separation are seen by the photometry as one star. Since the superposition is random in terms of stellar types it can only be modelled in a statistical sense. It causes noise similar to that of Fig. 21.6b but likely with even more scatter.

At the faint limit of observation, noise prevents seeing all stars. Some faint ones above the detection threshold disappear due to noise, while other fainter ones may reach above the detection threshold because of noise.

One accounts for both problems through the determination of the *completeness* of the photometry. This aspect is discussed in Ch. 20.4.3.

5) intermixing distance:
Different distances of the stars observed in a field clearly cause confusion for the interpretation. When one observes a star cluster, the foreground and background field stars can be easily recognized since they do not lie near the 'well established sequences' in the CMD. However, MS stars cover the full range in colour and these may turn up anywhere in the diagram, sometimes masquerading as evolved cluster stars.

In particular crowded regions of the CMD, like the red clump (Ch. 10.10, Fig. 10.23), spread recognizably due to distance effects. The distance spread of the red clump is sometimes called "the red plume". Distance effects can simply be modelled but that is in practice seldom a feature of synthetic CMDs.

6) stellar rotation:
A final issue to be considered in relation with colour is the effect of stellar rotation on colour (apart from effects on evolution, see Chs. 14 and 16). Rotation deforms stars and the equatorial zone has

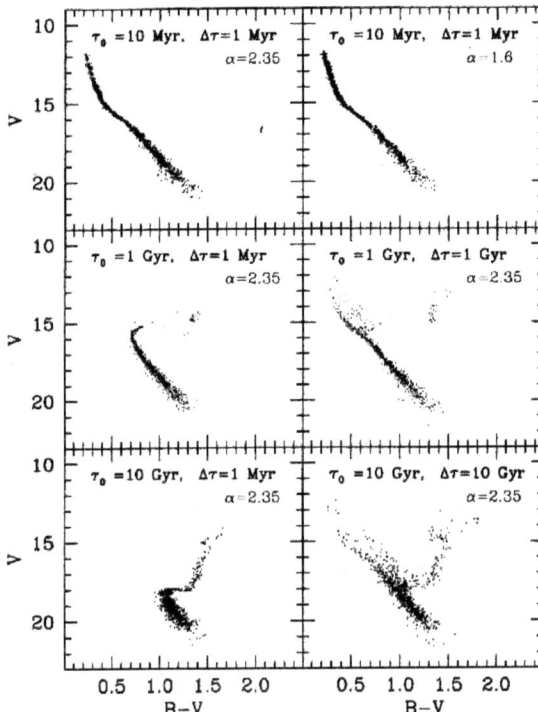

Figure 21.7: Synthetic CMDs are shown for particular star fields, containing stars of different ages (a field with a complicated star formation history). The time of onset of star formation τ_0, the duration of the star formation period $\Delta\tau$, and the slope α of the mass function used (in the notation $\alpha = -1 - \Gamma$; see Ch. 20.2) are given in the various panels. Figure from Tosi (2001).

a temperature different from that of the poles (see Ch. 3.7). Depending on the viewing angle, the colours of rotating stars are different from the reference values, leading to a shift in and spread of, e.g., the main-sequence location (see Fig. 3.14).

21.5 Special CMD-regions to find the age of star groups

Certain special CMD regions are useful to make a good first guess at the age of a star group.

TO. If stars in a group form at the same time, the stars of the upper MS will with time evolve away from the MS ($\tau_{\rm MS} \simeq M^{-2}$). After a while, the MS reaches less high in the CMD and the isochrone (Fig. 21.3) bends away from the ZAMS. This bending away location is called the "tunrn-off" (TO).

For young star groups the ages found from CMDs rely heavily on the "location of the TO". This boils down to just finding the brightest stars still being on the MS, but the nature of the IMF (see Ch. 20) leads often to the problem of small number statistics. Also the red supergiant stars present can be used to find the applicable isochrone.

Blue Loop. The core-He-burning stars represent a special group for age determination. Fig. 9.6 shows the duration of time spent on the MS and as core-He burner (in the blue loop). The core-He-burning phase has a duration of about 10% (with only small variation) of the core-H-burning phase. In particular the sequence of blue loop stars forms, from bright to faint, a time/age sequence (inspect Fig. 21.3).

Δ(HB-TO). For very old groups, like globular clusters, the TO is important. Here a very useful (*distance independent*) indicator is the vertical separation of the HB and the TO. For examples of CMDs showing that separation clearly, see Figs. 10.7 and 10.8. The HB in a LF can also be used for distance determination (see Ch. 20.1).

Red Clump. The red clump (see Ch. 10.10), although almost always a mix of stars from various evolutionary stages, can be very useful to sort the various populations present.

21.6 Star formation history (SFH)

Often observations contain data of stars belonging to a range of ages. This is the case, e.g., for galaxies. It is also the case for the field of the Large Magellanic Cloud (LMC), a galaxy seen almost face on, with all stars at the same distance (no distance-brightness confusion), but with a mix of ages from very old (from the beginnings of the LMC) up to very young (recent star formation). Reaching a model for the star formation history (SFH) with a good fit to the observations is in such cases much more difficult.

21.6.1 Photometric SFH

The star formation episodes have to be identified (in time and duration). This can be done either by just iterating the model until the synthetic CMD matches the observations, but better by using other information, too. Among that extra information can be
1) Inspection of the luminosity function of the main sequence. When looking along the main sequence, a jump in the observed luminosity function may signify the presence of an additional population of MS stars of just that (turn-off) age.
2) Inspection of the red clump domain. The red clump region contains red HB stars, the red giant clump stars, and possibly He-core burning stars of moderate mass. Each of these kinds of stars indicates the presence of stars of a roughly defined age range.

Fig. 21.7 shows examples of synthetic CMDs for stellar populations with special age structure. It is clear that there are large numbers of free parameters (time, frequency and intensity of bursts of star formation or continuous star formation) so that a unique answer is not possible in automated analyses (but see the model of Fig. 10.23). Additional information is almost always required.

21.6.2 SFH and synthetic spectral energy distributions

In the case of a system which cannot be (spatially) resolved by photometry (normally then being at a large distance) one may attempt to reconstruct the star formation history using observed integrated spectra. The goal is to match a model with the observed spectral energy distribution.

Consider the star groups of the lower panels in Fig. 21.7 for which different star formation histories were assumed (the left panel has a burst of $\Delta\tau = 1$ Myr starting τ_0 years ago, the right panel a continuous formation, with $\Delta\tau = \tau_0$). In each case one can synthesise the integrated spectrum by taking for each star in the modelled CMD the appropriate spectral energy distribution (like the ones of Figs. 1.4, 2.6, 3.10, 10.21, 13.1). Summing such spectra up produces the synthetic spectral energy distribution which will have the particular age and SFH relevant characteristics. E.g., the spectral energy distribution may just be bright in the red (due to a large number of RG stars and few MS stars), or it may have a low level but rather flat energy distribution in the UV (due to the presence of blue HB stars), or could be bright in both the UV and the red (due to young MS stars and old globular cluster populations).

References

Apparicio, A., Bertelli, G., Chisoi, C., & García-Pelayo, J.M. 1990, A&A 240, 262
Bertelli, G., Betto, R., Bressan, A., Chiosi, C., Nasi, E., & Vallenari, A. 1990, A&AS 85, 845
Meynet, G., Mermilliod, J.-C., & Maeder, A. 1993, A&AS 98, 477
Schaerer, D., Meynet, G., Maeder, A., & Schaller, G. 1993, A&AS 98, 523
Tosi, M. 2001, in "Dwarf galaxies and their environment", K.S. de Boer, R.-J. Dettmar, U. Klein (eds.); Shaker Verlag, p. 67

Chapter 22

Stars influence their environment

Stars and their very diverse ways of evolution have effects on their immediate environment but far reaching cosequences are present, too. The effects come from the return to interstellar space of the heavy elements produced in stellar interiors, because heavy elements influence the way new stellar generations are born. Thus stars influence the way the universe evolves.

22.1 Star formation and IS cloud metal content

For interstellar clouds to form stars, the clouds must become dense and cool, eventually surpassing some critical limit.

The most important *cooling mechanism* is based on the cooling capacity of interstellar atoms and ions (see Ch. 7 with 7.2.1 and 7.3.2). Cooling proceeds mostly through collisional excitation of metals into "forbidden" levels, which then deexcite and emit a photon. The photon thus has an energy (wavelength) for which normally absorption is not possible. Energy is lost from the cloud. Each temperature phase has its most efficient "cooling lines", e.g., lines from O^{2+} in H II regions, from C^+ and Fe^+ in neutral gas, and by various molecules in molecular gas.

Molecule formation will *make gas denser* because for every multi-atom molecule formed, the number of particles is reduced. New molecules form in the gas phase, but only if molecules of some sort are present already. The one molecule really needed for gas chemistry is H_2, and it forms essentially only with the help of dust grains. Stars produce, in the course of their evolution, heavy elements which are returned to the ISM through winds and explosions. The seeds of interstellar dust form in particular in metal rich stellar winds (like that of an AGB star).

The metal richer the gas, the better it can cool. And more (star-borne) dust makes molecule formation easier. Thus star formation became easier in the course of the evolution of the universe.

22.2 Effects of first stars

Without metals in the gas stars must have formed differently. One speculation is, that only gigantic stars formed because star formation is more difficult in metal free gas (no cooling). Thus in this case, the full Jeans criterium (Ch. 7.3.1) must have been operating in relatively warm gas. These first stars then contain no metals and thus have initially no CNO-cycle burning (there is no C in the universe yet). Once the He formed in the core becomes so hot that C is produced, this C is, through convection, spread up into the H-burning zones, allowing then for CNO-burning there. For more see Ch. 15.

One important effect first stars have is that they produce photons. Photons in the 11-13 eV range will readily dissociate the H_2. And photons with energy > 13.6 eV will ionize the surroundings of the star creating the first H II regions. This marks the "epoch of reionization" of the at that moment neutral gas of the early universe. They provide the first photons for the radiative brilliance of the galaxies to come.

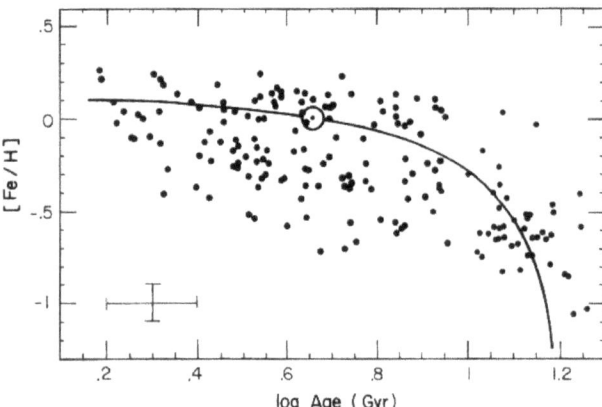

Figure 22.1: The chemical evolution of the Milky Way (and thus of the universe, for that matter) can be seen in the plot of the abundance of Fe against the age of the star. Note that the age is plotted in a *logarithmic* scale and that the present is at the *left* side. Younger generations of stars are formed in gas enriched in metals by previous generations. Figure from Pagel (1997).

Then, through supernova explosions, the first stars will bring metals created in their interior into the ISM. The ISM will therefore now have the possibility to form dust as well as to form H_2 much more easily. And the enriched material will allow easier formation of stars. First stars mark the onset of a transformation of the extreme mass function of the early universe into the mass function as observed currently in the Solar vicinity.

Finally, the black holes formed after the evolution of these very massive stars are massive, they are in the category of "intermediate mass black holes". These IMBHs may become the central objects of larger amounts of stars, perhaps catching some of the stars (Baumgardt et al. 2006a), throwing some stars out as hyper-velocity stars (Baumgardt et al. 2006b; see also Ch. 19.5.2.6), or coalescing other IMHBs and so becoming the central objects of larger entities like dwarf galaxies.

22.3 Chemical evolution

22.3.1 Consumption of primordial D, Li, He

In the first stars, primordial species like D and Li will also be used up in the nuclear fusion (see Chs. 5 and 8). Since these first stars likely are huge and have deep convection zones, D and Li will be destroyed. However, in later generations of stars, in particular in stars without (or with at most shallow) outer convection zones, which are the stars with $M_{\rm init} > 1.2$ $M_\odot$, D and Li will not be burned up und thus may be returned to interstellar space during the late (super-)giant stellar-wind phases. This means that the D and Li content of the universe will decrease, but only very slowly.

Once stars come to conditions in which He comes to fusion, it will fuse foremost the He from the hydrogen fusion, but of course primordial He will be consumed as well. On the other hand, stars in the lower mass range will make He without coming to He fusion. These stars will return He to the ISM through the red-giant wind. Will the He abundance in the universe change?

22.3.2 Metal production and yield

Due to the production of heavy elements by stars and their liberation the chemical composition of the interstellar medium changes with time. This represents the *chemical evolution* of the ISM, in fact of the entire universe.

In modelling this enrichment process, one has to study the yield, Y, of nucleosynthesis in stars (see also Ch. 5.2.5). Various stellar types ($M_{\rm init}$) contribute in different ways. In general, the high mass stars contribute in particular α elements (Ch. 5.2.3), the lower mass stars contribute predominantly the elements formed through the s-process, by neutron capture (Ch. 5.2.4). Overviews can by found in, e.g., Trimble (1991), Arnett (1995) and Pagel (1997). A general summary of which elements come predominantly from what kind of star is given in Table 22.1.

Values for yields vary largely between different studies, since the yields depend not only on the

Table 22.1: Stellar mass range and fusion products returned to the ISM

stellar mass range	process	elements to ISM
high mass	WR-star wind	He, CNO, Mg
(see Ch. 13)	SN type II	CNO, Fe, >Fe
	SN type Ib	Mg, Fe, Ni, Cr, Co, >Fe
middle mass	RG-star winds	He
(see Ch. 10)	AGB-star winds	He, CNO, Mg, Al, Si, etc.
	SN type Ia	Mg, >Mg
lower mass	RG-star winds	He, CNO?,
(see Ch. 10)	AGB-star winds	He, CNO, s-process elements (Eu, Ba, etc.)
	SN type Ia	He, CNO, Fe, Ni

knowledge about the possible fusion processes and their rates, but also on how much of the fusion products is expelled in winds and explosions.

The abundance of the elements in stars is in the literature normally not related to H, since a comparison with spectral lines of similar strength (see Ch. 3) is more reliable. Moreover, in cooler stars there are no H absorption lines visible. The element of reference is **Fe**, because it is abundant and because it can be seen in almost all stars in one or more of the various ionization stages present related with the particular T_{eff}. In the Sun the Fe abundance is $\log N(\text{Fe})/N(\text{H}) \simeq -4.5$.

The metals produced by stars and returned to the ISM cause enrichment of the ISM. New generations of stars are formed in this enriched material and thus contain more heavy elments than very old stars. For an example see the relation of [Fe/H] with stellar age (Fig. 22.1).[1] Note that $[\text{Fe/H}] \equiv \log(N(\text{Fe})/N(\text{H}))_* - \log(N(\text{Fe})/N(\text{H}))_\odot$, as defined in Ch. 1.4.3.

In metal poor stars, the abundance of elements in relation to Fe is nearly constant, irrespective of the abundance of Fe. That is another reason to take the abundance of Fe to be representative for all metals. For examples of very metal-poor stars studied see Sect. 15.1.

The yields of stars depend also on stellar rotation (Ch. 14.7). In particular in low metallicity stars (low Z) rotation has an effect on the fusion processes taking place as well as on the mass loss. For a study of that kind, arriving at overall yields for low Z stars, see e.g., Meynet et al. (2006).

In the early phases of the evolution of galaxies, when [Fe/H] is small, the enrichment of the ISM with heavy elements was erratic and spatially irregular. Also, the relative abundance of the elements shed by stars depends on stellar type so that the enrichment of the ISM was variable for each element. This explains the large scatter of the [X/Fe] abundance values[2] in Fig. 22.3. As of $[\text{Fe/H}] \simeq -1$ the scatter disappears and enrichment runs parallel for all elements. At that state of (time in) the evolution, SNe of Type Ia (which can occur once stars in low mass binaries have evolved to WDs) add in particular Fe, thus reducing [X/Fe].

To model the chemical evolution of a galaxy one has to combine the applicable IMF with the appropriate yield function. The result of such a model for the Milky Way is shown by the solid line in Fig. 22.3 matching the observational data.

22.3.3 Radioactive decay and nucleochronometry

Elements heavier than Fe are radio-active and they deacy with time. Each isotope has a specific half-life. This decay can be used to derive the age of stars using the current ratio of the abundance of such elements.

To use that method one needs to know what the abundance (ratio) of these elements was in the gas of the birth cloud. In particular the stars formed early in the universe are of interest. The

---

[1] Use of a logarithmic scale is justified because age determinations have an accuracy which is roughly similar for *logarithmic* ages. Graphs with log age on an axis are somewhat misleading in the appreciation of time elapsed.

[2] Note that "abundance" is a logarithmic value, so that already for that reason scatter decreases toward larger [Fe/H]. The effect of adding 1 (linear) unit of Fe to 2 available is very different from that of adding 1 to 20 of Fe.

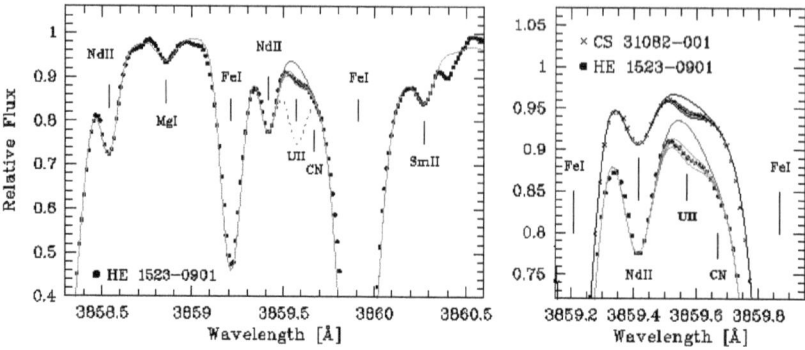

Figure 22.2: The spectrum of the metal-poor star HE 1523-0901 (and CS 31082-001) reveals absorption by U II. Because the star is metal poor (actually, [Fe/H] = −2.95), the opacity in the normal spectral lines is low such that U II can be seen. The thin lines indicate a set of spectral model fits. Using abundance ratios of radio-active elements (like U, Th, Ir, Eu) with different half-lives, the actual age of the star (rather the "age" of the chemical composition of the birth cloud) can be derived. The method is called stellar nucleochronometry. Figure from Frebel et al. (2007).

first supernovae in the universe were those of the first massive stars (see "first" stars; Ch. 15), and the amount of radionucleides produced can be modelled. Those heavy elements are added to the at that time almost pristine interstellar gas. Thus one can estimate the composition of the birth cloud, i.e., the ratio of the abundance of heavy radionucleides.

In Fig. 22.2 the spectrum around a line of U II is shown. Using the column density ratio of elements with different half-lives, such as U/Th or U/Ir one can derive how long ago the birth-cloud matter of the star got its composition. For the star of Fig. 22.2 it turned out to be $\simeq 13.2$ Gyr, pretty close to the age cosmologists derive for the universe.

Such ages are "absolute" and can be derived for individual stars. The accuracy of these ages depends (apart from the usual aspects of the observations like spectral noise, the required T_{eff} and $\log g$) foremost on the accuracy of the modelled chemical composition of the brith-cloud gas. The other way to derive stellar ages is through fitting an isochrone to a CMD of a star group (Ch. 21).

22.4 What comes of all evolution?

22.4.1 Stars and their light

Stars produce photons which, in principle, can travel infinite distances. Nuclear fusion is, thus, the transformation of nuclear mass into photons.

After the big bang, the universe was (and is) filled with a radiation field being the same (but relativistically diluted) as the Planck-function intensity distribution immediately after the first recombination in the universe (at $T \simeq 3000$ K). This radiation field now has $T_{\text{eff}} \simeq 2.79$ K.

The stellar radiation adds to the radiation density in the universe. It has a spectral distribution roughly equal to that of an A to F star, so of $T_{\text{eff}} \simeq 7000$ K. However, near young stars in dusty environments a considerable amount of radiation is absorbed by the dust, and reemitted. Such regions show an excess emission in the near to far IR. Thus entire galaxies, in particular the dusty ones (so not elliptical galaxies), have a radiation spectrum with an IR peak somewhere in the 20 to 100 μm range.

22.4.2 Stellar remnants

The ultimate fate of stars is to end as white dwarfs, neutron stars, or black holes (see Ch. 10, Ch. 13, Ch. 17, Ch. 19). Cooling is all these blobs of degenerate matter can do.

22.4. WHAT COMES OF ALL EVOLUTION?

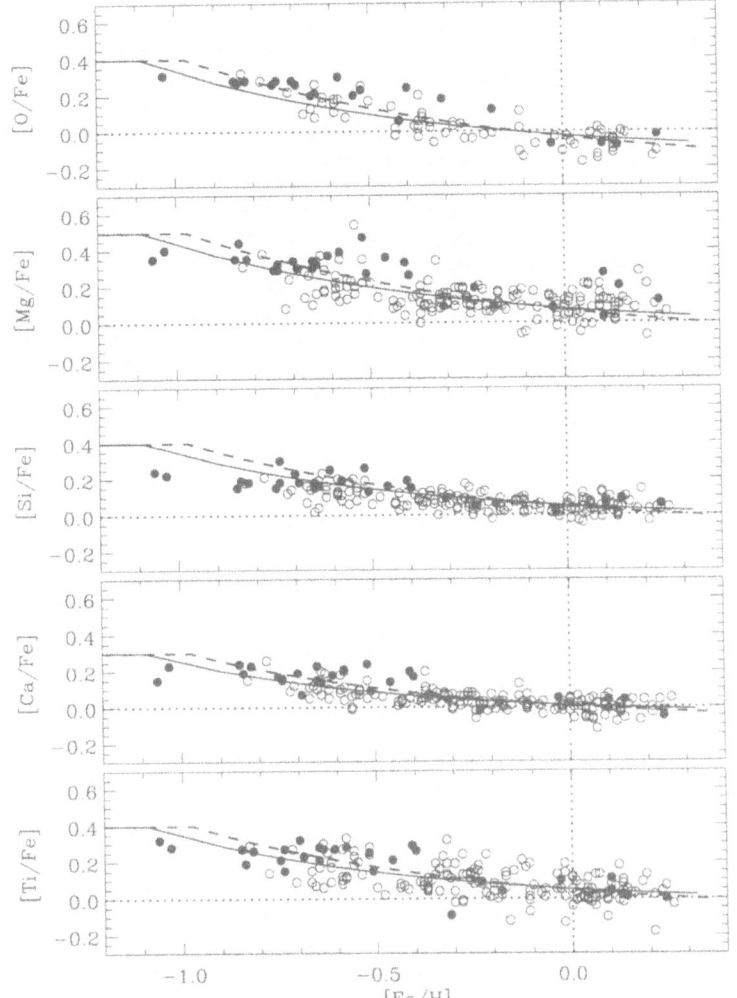

Figure 22.3: The chemical evolution of the Milky Way visible in a plot of the abundance of various elements against Fe. Note that [Fe/H] itself changes (evolves) with time (Fig. 22.1). The horizontal axis can therefore be treated as a 'transformed' time axis, with the *present at right*. Data were taken from various sources (the various symbols) and pertain to disk stars as well as halo and metal-weak thick disk stars. Chemical enrichment was 'inhomogeneous' at first, later all abundances concurr (for [Fe/H] > −1). The line represents a model of chemical evolution. Figure from Pagel & Tautvaisiené (1995).

22.4.3 Gas returned to IS space

During stellar evolution, lots of gas is returned to interstellar space, which then can proceed to again form molecular clouds, and new stars. Since also lots of H is returned from stars to the ISM, this process can go on for many Hubble times. More and more gas will in this way end up in stellar remnants, until the gaseous content of galaxies is so low, that new clouds surpassing the Jeans criterium have ceased to exist.

Looking at distant galaxies one also looks back in time. At such distances stars cannot any more be recognized but emission or absorption lines due to gas can. The measurements of, e.g., quasar absorption lines allows to probe the chemical composition of their IS gas at the particular redshift. It has become clear that out onto $z \simeq 2$ (equivalent to a look-back time of $\simeq 7$ Gyr) those objects appear to be chemically similar to stars of that age in the Milky Way.

References

Arnett, D. 1995, ARAA 33, 115; *Explosive Nucleosynthesis Revisted: Yields*
Baumgardt, H., Hopman, C., Portegies Zwart, S.F., & Makino, J. 2006a, MNRAS 372, 467
Baumgardt, H., Portegies Zwart, S.F., & Gaulandris, A. 2006b, MNRAS 372, 174
Frebel, A., Christlieb, N., Norris, J.E., Thom, C., Beers, T.C., & Rhee, J. 2007, ApJ 660, L117
Meynet, G., Ekström, S., & Maeder, A. 2006, A&A 447, 623
Pagel, B.E.J. 1997, "Nucleosynthesis and Chemical Evolution of Galaxies", Cambridge Univ. Press

Pagel, B.E.J., & Tautvaisiené, G. 1995, MNRAS 276, 505
Trimble, V. 1991, A&A Rev 3, 1; *The Origin and Abundances of the Chemical Elements Revisited*

Chapter 23

Summary; Questions, Constants, Acronyms, Lists

An attempt is made to recapitulate the material by focussing on the most important more general aspects of the theories about stellar structure and stellar evolution. Four summarizing diagrams are presented. Statements are global. And of course, when looking at details, reality is always more complicated.

Reference is made to chapters with most of the particular relevant information.

23.1 Stars and their structure

- Stars form in interstellar gas clouds consisting of $\simeq$75% hydrogen, $\simeq$25% helium (by mass), and traces of heavier elements. Cooling, compaction (through molecule formation) and gravity dominate star formation.
  (Ch. 7, Ch. 20, Ch. 22)

- The actual nature of any star is foremost determined by its initial mass, M_{init}, and its age; elements heavier than He play rather a secondary role (in the main-sequence phase).
  (Ch. 4, Ch. 6)

- The initial mass and the age determine what the luminosity, L, is at any time in the evolution of a star. Imprecise input physics leads (naturally) to variations in model results.
  (Ch. 4, Ch. 6, Ch. 16)

- Stellar evolution proceeds in an orderly manner; lighter elements are transformed into heavier ones in the core thereby releasing energy leading to the luminosity of the star.
  (Ch. 4, Ch. 5)

- The surface parameters of a star (T_{eff} and $\log g$) establish themselves based on the luminosity and on the capacity of the layers without fusion (the envelope) to accommodate L; here the chemical composition does play a role through opacity.
  (Ch. 2, Ch. 3, Ch. 9)

- The radius of a star follows from the total mass (gravitationally) and the luminosity together with the actual condition of the gases (equation of state) in the stellar interior as well as in the envelope.
  (Ch. 4, Ch. 6)

- Stars formed are distributed over mass intervals according to a mass function. This function may be "universal" and is close to a power law.
  (Ch. 20)

- Stars may vibrate according their structural properties. Pulsation shows itself is a pronounced manner in surface brightness variations. Stars may have hot outer zones, coronae.
  (Ch. 11; Ch. 10, Ch. 13, Ch. 12)

- Stars may have convetive zones. They augment energy transport and mixing and thus influence evolution.
  (Ch. 4, Ch. 6, Ch. 9, Ch. 14)

- Stars rotate and stars have magnetic fields. Rotation influences internal structure and thus evolution. Internal meridional circulation enhances mixing. Magnetic fields influence internal structure as well as surface structure.
  (Ch. 14; Ch. 12)

- Depending on their luminosity and their extent (radius) stars will have little or much mass loss. Mass loss is driven by radiation and opacity; rotation helps, too.
  (Ch. 4; Ch. 13; Ch. 14)

23.2 Stars and their evolution

- Evolution is simple or complex depending on M_{init} (and binarity):
  (Ch. 5, Ch. 6, Ch. 7, Ch. 8, Ch. 9, Ch. 10, Ch. 13, Ch. 15, Ch. 17, Ch. 18, [Ch. 19])

  – Stars having nuclear fusion in the core are rather compact, like main-sequence stars.
  – Stars having fusion in a shell are, in general, extended (giants or super-giants).
  – Evolution on the MS and in further stages is driven mainly by the fusion produced change in mean molecular weight in the interior.
  – The luminosity of a star is foremost set by the fusion reactions possible; the temperature dependence is steep.
  – Objects starting with $0.012 < M_{\text{init}} < 0.08$ $M_\odot$ will become BDs (Ch. 8).
  – Stars starting with $M_{\text{init}} < 0.5$ $M_\odot$ evolve (after times comparable to or longer than the age of the universe) from MS stars into WDs (Ch. 10).
  – Stars starting with $0.5 < M_{\text{init}} < 8$ $M_\odot$ evolve into RG stars, then proceed to become core-He burners as HB stars. Stars with $M_{\text{HB}} < 0.6$ $M_\odot$ turn into WDs, the more massive ones evolve into AGB and pAGB stars, all finally become WDs (Ch. 10).
  – Stars with $8 < M_{\text{init}} < 15$ $M_\odot$ evolve from H-burning MS stars into core-He-burning MS stars (blue loop stars) and ultimately become mostly supernovae of Type II (Ch. 13).
  – Stars with $15 < M_{\text{init}} < 25$ $M_\odot$ evolve from H-burning MS stars into stars going through various fusion processes, and ultimately become supernovae of Type II (Ch. 13).
  – Stars with $M_{\text{init}} > 25$ $M_\odot$ evolve from H-burning MS stars into stars with many fusion stages, lots of mass loss, turn into WR stars, and ultimately become mostly supernovae of Type Ib (Ch. 13).
  – Stars end their life as WDs, through supernovae as NSs, in binaries possibly as BHs (Ch. 17).
  – Very massive (and first) stars may become "hypernova" with a γ-ray burst.

- Nuclear burning of the primary elements takes place in stars in certain parts of the HRD:
  (Ch. 6, Ch. 5, Ch. 8, Ch. 10, Ch. 13 [Ch. 22])

  – Burning of D without further fusion stages takes place in BDs.
  – H-burning stars are found in the HRD on as well as anywhere above and to the right of the MS.
  – He-burning stars are found in the HRD on the HB as well as on and to the upper right of the blue-loop and perhaps in WR stars.
  – C-burning stars are found only in the upper part of the HRD.

- Gravity holds a star together, fusion leads to higher interior matter density. The central regions of stars become with time progressively denser and hotter (perhaps degenerate), the outer layers become more diffuse and are lost with time. The higher the initial mass, the higher the central density reached.
  (Ch. 6, Ch. 17, Ch. 10, Ch. 13)

- Most stars are part of binary systems; this may influence the evolution of stars, as summarized above for single ones, considerably:
  (Ch. 19, Ch. 17)

  - Close binaries have various forms of mass exchange which leads to many stellar types not explained by simple stellar evolution (Novae, SN Type Ia, Black Holes).
  - A star starting with $M_{init} > 8$ $M_\odot$ will rather end as a SN Type Ib and may ultimately become a Black Hole.
  - Stars starting with less mass may end up as SN Type Ia.

23.3 Stellar evolution in comparison

As a summary of all of stellar evolution, evolutionary tracks are shown in Figs. 23.1 & 23.2 for stars with 0.15, 1, 10, and 100 $M_\odot$.

The figures need extensive explanations which are given here for all of them, instead of in a figure caption.

The diagrams show "3D"-views of the evolution of T_{eff} and $\log L$ with age. The difference with most of the diagrams presented previously is, that age is in Figs. 23.1 & 23.2 more recognizably visible, plotted along the base-plane of the figures. It shows clearly that evoolution *after* the MS-phase is only a small part of the life of a star.

Of each track, projections are shown onto the planes of age-L, age-T_{eff}, and on the familiar L-T_{eff} plane. Note that, in contrast to what is shown in most theoretical HRDs, T_{eff} is here *linear* (not logarithmic). Between the plots the scales in T_{eff} and age differ while the logarithmic luminosity axis is only shifted in level. For the evolution up to the end of the RG, vertical projection lines onto the age-T_{eff} plane are added. The evolutionary tracks for the 0.15, 1, and 10 $M_\odot$ are from Salasnich et al. (2000), the one for 100 $M_\odot$ is from Fagotto et al. (1994).

Spectral energy distributions are shown for the points in evolution labelled A,B,C with each 3D evolutionary track. These spectral energy distributions are taken from the Basel stellar library (BaSeL2.2; see Lejeune et al. 1997). The spectral models choosen are for [M/H] = −0.4 dex (since at other metallicities the database is less complete). This small underabundance hardly affects the overal evolution compared to evolution at solar abundance.

Notes on the individual models:

0.15 $M_\odot$: The star evolves very slowly (model from Salasnich et al. 2000) and stays in the main-sequence phase (inside the MS band) in the 15 Gyr shown, this being the (currently believed) time since the beginning of the Universe.

1 $M_\odot$: The evolution is the familiar one, except for the fact that [M/H] = −0.4 dex and thus that T_{eff} for the model differs from the Solar value. Beyond the onset of He-core burning there is a gap in the track. The track continues starting at the HB for an HB star of 0.8 $M_\odot$, then evolves up to the first significant thermal pulse near the tip of the AGB. The track is from models by Salasnich et al. (2000). The discontinuities in the evolution shown are due to the difficulties in modelling the fast phases of evolution, such as the end of RGB and transition into an HB star, as well as the end of the AGB phase. Beyond the AGB tip a simple pAGB cooling track was added, which extends beyond the figure limits to $\simeq 10^5$ K and then descends into the WD cooling track (not shown to not overextend the T_{eff} axis).

298 CHAPTER 23. SUMMARY; QUESTIONS, CONSTANTS, ACRONYMS, LISTS

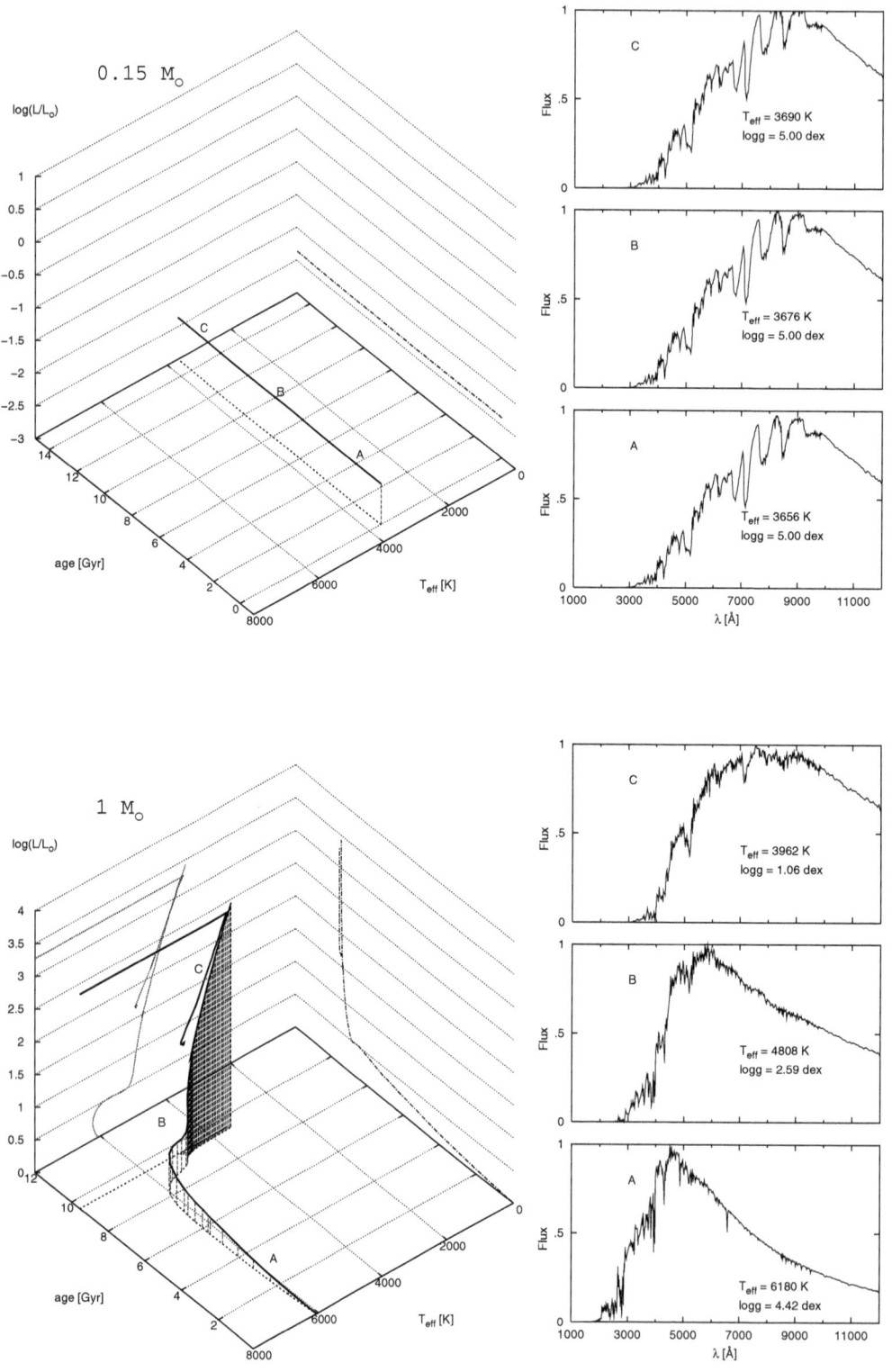

Figure 23.1: Overview of evolutionary tracks for stars of 0.15 and 1.0 $M_\odot$; explanation in Sect. 23.3.

23.3. STELLAR EVOLUTION IN COMPARISON

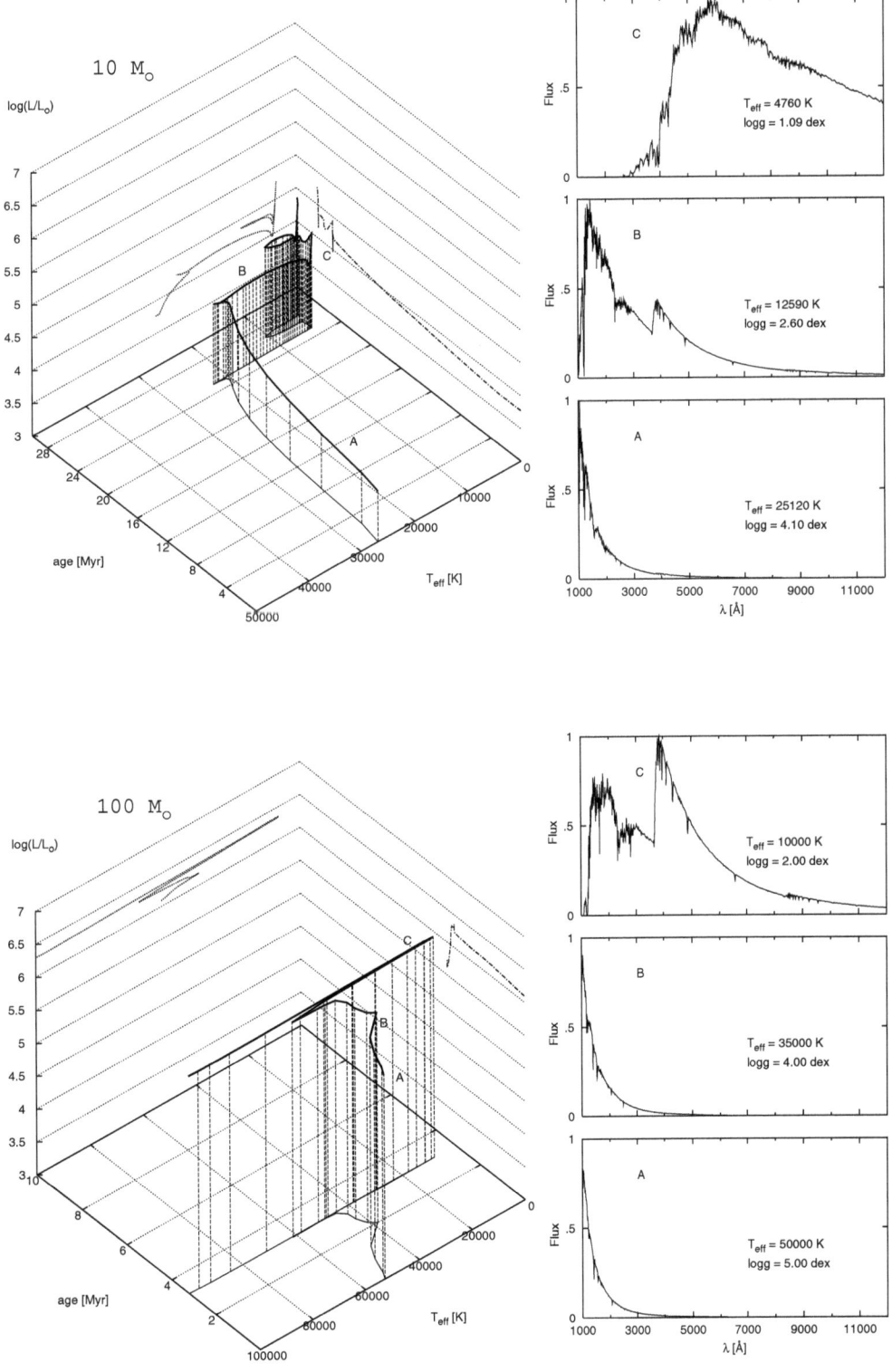

Figure 23.2: Overview of evolutionary tracks for stars of 10 and 100 $M_\odot$; explanation in Sect. 23.3.

10 $M_\odot$: The track (model from Salasnich et al. 2000) shows the evolution only up to the onset of C-burning. Note that carbon ignition in stars of this mass is very difficult to model because of the closeness to the 8 $M_\odot$ limit (the risk for such stars to come to explosive burning and disruption). For further evolution see Ch. 13.

100 $M_\odot$: The track (model from Fagotto et al. 1994) shows the full evolution including the heavy mass loss (WR-stage) leading to blueing and a reduction in luminosity. For further evolution see Ch. 13.

23.4 Stars and effects for their environment

- The first stars form in material without heavy elements. These stars are "supermassive" and end their short life as very high to middle energy supernovae (BHs, GRBs?).
  (Ch. 15, Ch. 18)

- Stars more massive than 8 $M_\odot$ have high enough T_{eff} that they can create regions of ionized gas around them. Supernovae induce a sudden burst of ionization of the circumstellar gas, stars with mass of $2.5 < M < 8$ $M_\odot$ may create planetary nebulae.
  (Ch. 13, Ch. 18, Ch. 10)

- Fusion produces elements heavier than H, which in later stages of stellar evolution are, through stellar winds and supernova explosions, in part returned to interstellar space. These elements from nucleosynthesis become part of secondary generations of stars.
  (Ch. 5, Ch. 22)

- Most stars return most of their mass to the ISM.
  (Ch. 10, Ch. 13)

- Stars contribute to the metal enrichment of the ISM depending on M_{init}.
  (Ch. 5, Ch. 10, Ch. 13, Ch. 18)

    – The elements C, N, and O, as well as heavier s-process elements (n-capture), come (mostly) from stars with $M_{\text{init}} < 8$ $M_\odot$, being returned to the ISM by stellar winds, after processed material has been dredged-up in deep convection zones
    – Elements around the Fe peak come (mostly) from more massive stars through SN explosions.

- Investigating stellar populations requires the use of synthetic CMDs and leads to the derivation of the initial mass function. Knowledge of the IMF is required for studies of the structure of star systems (from clusters up to entire galaxies).
  (Ch. 21, Ch. 20)

- Star formation will proceed as long as sufficient gas is available to assemble into neutral interstellar clouds.
  (Ch. 22)

- A rather uniform universe containing essentially only hydrogen and helium is, through the formation and evolution of stars, slowly enriched in heavy elements. It is ultimately transformed into a universe containing innumerable small spheres of degenerate heavy matter which are embedded in a sea of photons.
  (Ch. 22)

References

Fagotto, F., Bressan, A., Bertelli, G., & Chiosi, C. 1994, A&AS 105, 29
Lejeune, Th., Cuisinier, F., & Buser, R. 1997, A&AS 125, 229
Salasnich, B., Girardi, L., Weiss, A., & Chiosi, C. 2000, A&A 361, 1023

23.5 List of questions

What is the formula collecting all stellar surface parameters?

Sketch the derivation of the simplest form of the equation of radiative transfer.

Why does a hot gaseous nebula exhibit emission lines but a normal star absorption lines?

Explain opacity. Give the various sources/processes of opacity considering different temperature regimes.

Explain reasons for the simplifications applied to the full radiative transfer equation leading to the description of limb darkening, the concept of grey atmosphere, the Rosseland mean, the Eddington approximation.

Explain the Boltzmann statistics and the Saha equation.

What is a curve of growth? Explain how one can use this concept to derive from relevant line observations temperature, ionization structure, electron pressure and element abundances in a stellar atmosphere.

Explain how the size of the Balmer jump and the strength of the Balmer lines change with temperature (from 6000 K to 30 000 K).

Describe why one can derive the surface $\log g$ and surface T_{eff} from the profile of a spectral line.

What is the effect of stellar rotation on the shape of a spectral line? And on the colour of (and the derived astrophysical parameters of) a star?

Give the basic formulae needed to describe the structure of a star.

Give the essentials for the functions kappa (κ), epsilon (ϵ), rho (ρ), and of the concepts of degeneration and polytrope.

Why is opacity in general larger for temperatures $10^7 < T < 10^5$ K and again small for $T < 10^4$ K? Is there a frequency dependence?

Why is it so difficult to build a neutrino telescope?

Explain the so-called solar neutrino problem. What is the solution?

What makes an object a star? Why is it stable? Why can one see a star as a "leaky ball" of gas?

Explain how one arrives at first estimates for the maximum and the minimum value for the mass of an object called star.

How does one make stellar models? What is homology? Why have main-sequence stars $L \simeq M^3$?

Why does the He-MS lie in a different location in the HRD than the H-MS? And the MS of metal poor stars?

Why do low-mass stars have surface convection and high-mass stars a convective core? How come convection reaches so deep in the lowest mass MS stars?

What is the Jeans criterion? What is the effect of a thermal instability?

Which processes help star formation, which aspects hamper it?

Describe the internal structure of an object in the process of becoming a MS-star.

Explain the line spectrum and the spectral energy distribution of the T Tauri stars.

What is a brown dwarf? What is its source of energy?

Why does a star become brighter in the MS phase?

Why and how does a star become red giant? What is "stellar thermal equilibrium"?

What happens to a star in the RG-phase? How is a stellar wind driven?

Give criteria for convection. What is convective overshoot?

What is the lifespan of MS stars? Why is it shorter for more massive stars?

What is a blue loop? What do stars do at the left end of the blue loop?

What is the Hertzsprung gap? What is the RR Lyrae gap?

How is it that a pulsating star stops pulsating (think of Polaris)?

Why does an RR Lyrae star become "redder" in Strömgren c_1 when its atmosphere gets hotter?

What happens at the end of the RG phase?

Describe the structure of HB stars. Which fusion processes are active? What mass do they have? Why does the HB have the shape it has (in CMD and in HRD)?

Why is a red HB star red and a blue one blue?

Which fusion processes take place in AGB stars? What is their structure? What is "dredge-up"?

What are OH/IR stars? And planetary nebulae?

What is asteroseismology? Which pulsation modes are known?

What has one learned from stellar vibrations? And from helioseismology?

What are stellar coronae?

What are the essential aspects of white dwarfs? Which stars evolve into WDs?

Give the mass ranges of the stars which become WDs and give for each range the approximate time it takes to become a WD. How long is a star WD?

Give the criteria for the limits between WDs, NSs and BHs.

Describe in which way a WD can become NS and a NS can become a BH.

Which phase transitions are known for matter in stars?

How massive are the stars of the upper mass range? Why is their lower limit at SpT B2?

Explain the shape of the P Cygni profiles.

Give processes of nuclear fusion. Over what time span do they take place?

What is a Wolf-Rayet star?

Which effects does rotation have on stars? What is the importance of magnetic fields in relation with rotation?

Why are "first stars" different? What is the endproduct of the evolution of first stars?

Describe the processes in massive stars leading to a supernova explosion.

Explain the difference between thermonuclear supernovae and hydrodynamical supernovae.

What is the difference between SN Ia and SN Ib? Which evolution leads to each of these kinds?

What is a hypernova?

Sketch into your own HRD or CMD the evolutionary tracks for stars of 0.1, 1.0, 1.4, 2.5, 5, 10, 20, 50, 80 solar masses. Mark the essential phase transitions.

Where in the HRD does one find core-H burning stars? And generally H burning stars?

Where in the HRD does one find core-He burning stars? And generally He burning stars?

Where in the HRD are pulsating stars found?

Describe the physics leading to a Roche Lobe.

What happens to close binary stars?

Why does matter flowing to a companion assemble in an accretion disk?

Which process makes the material fall toward the star? And what may happen then? Accretion and angular momentum? What is a merged star?

What is a Nova, a recurrent Nova, a cataclysmic variable? What is the mass of an exploding WD?

Describe how a double pulsar may come about.

How does one detect a black hole?

What is a luminosity function? And a mass function? How are they derived? What is the relevance of these functions?

List the various problems in the transformation from CMD to IMF.

What is an isochrone? Is it different from an evolutionary track? Do the shapes of isochrone and evolutionary track differ?

What is the yield of stars? Which elements are produced through what kind of fusion in which kinds of star?

How come galaxies evolve chemically?

23.6 Acronyms, Constants, Abbreviations

Table 23.1: List of acronyms and location where first mentioned

acronym		defined in Chapter
AGB	asymptotic giant branch	10
BD	brown dwarf	8
BH	black hole	17
BHB	blue horizontal branch	10
CMD	colour-magnitude diagram	1
CNO	carbon nitrogen oxygen	5
GB	giant branch	10
GC	globular cluster	10
GRB	gamma ray burst	18
HB	horizontal branch	1
HH	Herbig-Haro (object)	7
HRD	Hertzsprung-Russell diagram	1
IMF	initial mass function	20
LF	luminosity function	20
LTE	local thermodynamic equilibrium	2
MF	mass function	20
MKK	Morgan Keenan Kellman spectral classification	3
MS	main sequence	1
NS	neutron star	17
OH/IR	OH / infrared (star)	10
pAGB	post asymptotic giant branch	10
PMS	pre main sequence	7
PN	planetary nebula	10
Pop.	Population (I, II, III)	7
PP	proton-proton	5
RG	red giant	10
RGB	red giant branch	1
RHB	red horizontal branch	10
SN	supernova	18
TAMS	terminal-age main sequence	9
TE	thermodynamic equilibrium	2
WD	white dwarf	17
WR	Wolf-Rayet (star)	13
X	hydrogen fraction by mass	1
Y	helium fraction by mass	1
Z	heavy element fraction (non X,Y) by mass	1
ZAMS	zero-age main sequence	9

Table 23.2: **Constants of physics and astronomy**

Parameter		mks	cgs
Physics			
speed of light	c	$2.998 \cdot 10^8$ m s^{-1}	$2.998 \cdot 10^{10}$ cm s^{-1}
constant of gravity	G	$6.673 \cdot 10^{-11}$ m^3 kg^{-1} s^{-2}	$6.673 \cdot 10^{-8}$ cm^3 g^{-1} s^{-2}
mass of hydrogen atom	m_H	$1.674 \cdot 10^{-27}$ kg	$1.674 \cdot 10^{-24}$ g
mass of electron	m_e	$9.109 \cdot 10^{-31}$ kg	$9.109 \cdot 10^{-28}$ g
Planck constant	h	$6.626 \cdot 10^{-34}$ J s	$6.626 \cdot 10^{-27}$ erg s
Boltzmann constant	k	$1.381 \cdot 10^{-23}$ J K^{-1}	$1.381 \cdot 10^{-16}$ erg K^{-1}
radiation constant	a	$7.566 \cdot 10^{-16}$ J m^{-3} K^{-4}	$7.566 \cdot 10^{-15}$ erg cm^{-3} K^{-4}
Stefan-Boltzmann constant	σ	$5.67 \cdot 10^{-8}$ W m^{-2} K^{-4}	$5.67 \cdot 10^{-5}$ erg cm^{-2} s^{-1} K^{-4}
gas constant	$\Re$	8.314 J mol^{-1} K^{-1}	$8.314 \cdot 10^7$ erg mol^{-1} K^{-1}
Astronomy			
Sun			
solar luminosity	$L_\odot$	$3.845 \cdot 10^{26}$ W	$3.845 \cdot 10^{33}$ erg s^{-1}
solar mass	$M_\odot$	$1.989 \cdot 10^{30}$ kg	$1.989 \cdot 10^{33}$ g
solar radius	$R_\odot$	$6.960 \cdot 10^8$ m	$6.960 \cdot 10^{10}$ cm
solar surface gravity	$g_\odot$	$2.736 \cdot 10^2$ m s^{-2}	$2.736 \cdot 10^4$ cm s^{-2}
solar effective temeperature	T_eff	5780 K	
conversions			
electronvolt	1 eV	$1.602 \cdot 10^{-19}$ J	$1.602 \cdot 10^{-12}$ erg
seconds in 1 year	1 yr	$3 \cdot 10^7$ s	$3 \cdot 10^7$ s
Gyr, Myr		10^9 yr, 10^6 yr	
distance	1 pc	$3 \cdot 10^{16}$ m	$3 \cdot 10^{18}$ cm
visual brightness [mag]	$V=0$	$10^{-10.42}$ W m^{-2} nm^{-1}	$10^{-8.42}$ erg cm^{-2} s^{-1} Å^{-1}

Table 23.3: **Abbreviations of the names of common journals**

Abbreviation	Name
A&A	Astronomy & Astrophysics
A&A Rev	Astronomy & Astrophysics Review
A&AS	Astronomy & Astrophysics Supplements
AJ	Astronomical Journal
ApJ	Astrophysical Journal
ApJS	Astrophysical Journal Supplements
ARAA	Annual Reviews of Astronomy & Astrophysics
MNRAS	Monthly Notices of the Royal Astronomical Society
PASP	Publications of the Astronomical Society of the Pacific

23.7 List of Figures

1.1	Planck function, double logarithmic	8
1.2	Three types of diagram: HRD, CMD, HRD	9
1.3	HRD with relations of L, T_{eff}, and R	10
1.4	Spectra of stars (visual) covering spectral types O through M	11
1.5	Relationship of m_V, M_V, M_{bol}, L	12
2.1	Geometry of definition of intensity	16
2.2	Geometry to define absorption coefficient	18
2.3	Geometry for definition of radiation transport equation	20
2.4	Geometry of opacity in atmosphere	22
2.5	Absorption coefficient of hydrogen; Balmer, Paschen edge, etc.	26
2.6	Spectral energy distribution of metal poor and metal rich star	27
2.7	Abundance of molecules in cool atmospheres in relation with temperature	28
2.8	Total absorption coefficient κ; hot and cool	29
2.9	Comparing giant and main-sequence star: gas density and Balmer jump	30
2.10	Two colour diagram and Planck colour	31
3.1	Spectral line shape and pressure (gravity)	34
3.2	Voigt function for spectral lines	35
3.3	Curve of growth for absorption lines; dependence on α	37
3.4	Hydrogen: levels of excitation and ionization	41
3.5	Strömgren $b-y$ versus c_1 diagram	42
3.6	Metallicity from m_1 vs. $b-y$	43
3.7	Curve of growth of the Solar atmosphere	44
3.8	The G-band in spectra	45
3.9	Quasi molecular H_2 absorption	46
3.10	Molecular absorption in RG atmosphere	47
3.11	Shifting of Balmer lines due to stellar magnetic field	47
3.12	Stellar magnetic field splits Na D lines	48
3.13	Change of line shape due to stellar rotation	49
3.14	Rotational effects on stellar colours	50
4.1	Convective cell and Schwarzschild's criterium	59
4.2	Contour lines of adiabatic gradients in atmosphere gas	61
4.3	Sketch of geometry leading to convective overshoot	63
4.4	Rosseland mean opacity (OPAL); $\overline{\kappa}$ vs. T, ρ	64
4.5	Phase diagram for stellar gas: ρ vs. T	67
5.1	Binding energy per nucleon	71
5.2	p-p interaction: Coulomb energy barrier	72
5.3	Graphic representation of fusion probability; Gamow peak	73
5.4	The three branches of the PP chain	74
5.5	CNO cycle reaction network with two extra branches	75
5.6	Total energy production rate ϵ_{H}, ϵ_{He}	76
5.7	Proton capture reaction network of Ne, Na, Mg, Al to Si	78
5.8	Neutron capture through s- and r-process for high mass elements	79
5.9	The predicted spectrum of neutrinos from the Sun	81
5.10	Atmospheric electron neutrinos and muon neutrinos with Kamiokande	84
6.1	Balance between gas and radiation pressure	90
6.2	Central temperature and radius for $n=3/2$ polytropes; hydrogen ignition	92
6.3	Low end of main sequence in NGC 6397	92
6.4	MS stars: run of T, L, ρ, κ, ϵ, r vs. m/M	95

6.5	MS stars: run of T_c and ρ_c	96
6.6	MS stars: internal structure; plot against $\log M_{\rm MS}$	97
6.7	MS stars: location of MS for metallicities other than Solar	99
6.8	Location of the MS of H-stars, He-stars, C-stars: HRD and T_c, ρ_c	100
7.1	Discovery of H_2: UV spectrum of the O7-type star	103
7.2	Star forming regions: The Orion GMC; Phenomena in Taurus	105
7.3	Infrared sources in NGC 2024	106
7.4	Bipolar molecular outflow in IRAS 04166+2706 and in L 1448	107
7.5	Sketch of thermal instability	108
7.6	Effects of magnetic field in interstellar plasma; winding	111
7.7	Structure of a protostar (schematic)	112
7.8	Mid- to far-IR spectrum of protostar NGC 1333-IRAS 4	112
7.9	HRD: evolution tracks of accreting PMS stars	113
7.10	HRD: evolution of a PMS star at constant mass; birth-line	114
7.11	The outflow activity around L1551–IRS 5; CO map	116
7.12	Disk of YSO Orion 114 and jet of HH 111 in Orion B	117
7.13	Images of the HL Tau - HH 30 region; CO and S II	118
7.14	Spectra of the classical T Tau star DR Tau	119
7.15	Spectra of four late-K early-M T Tau stars	120
7.16	HRD of pre-MS stars of the Taurus-Auriga T association	121
7.17	CMDs to find Herbig Be and Ae stars	122
7.18	Schematic view of a bipolar nebula	123
8.1	Evolution of T_c and fusion in BDs and low mass MS stars	126
8.2	Brown dwarf evolution of deuterium	127
8.3	Evolution tracks of brown dwarfs with isochrones	127
8.4	Brown dwarf near-IR spectrum	128
8.5	CMD (near IR) with MS stars and BDs	129
8.6	Mass and radius of planets, brown dwarfs, low-mass stars	130
9.1	Location of ZAMS and TAMS; convection / no-convection	133
9.2	Gedankenexperiment: MS to RG; radius reacts on luminosity	134
9.3	Gedankenexperiment: MS to RG; temperature reacts on luminosity	135
9.4	Gedankenexperiment: MS to RG; comparison with reality	136
9.5	Gedankenexperiment: deviations from stellar thermal equilibrium	138
9.6	Life time in years for stellar H- and He-burning phases	140
10.1	HRD: evolutionary tracks (L and $T_{\rm eff}$) of lower mass range stars	142
10.2	Phase diagram: evolution of T_c and ρ_c	142
10.3	HRD: evolution of stars with sub-solar mass	143
10.4	Evolution of the internal structure of a 1 $M_\odot$ star	144
10.5	HRD: evolution from RG phase to ZAHB; He flashes	146
10.6	Evolution of RG star due to He flash	146
10.7	CMD of the stars in M 3: all low-mass evolutionary phases	150
10.8	CMDs of two globular clusters showing different morphology (47 Tuc & M 15)	150
10.9	HRD: evolution of metal-poor HB stars, ZAHB, TAHB, AGB, pAGB	151
10.10	AGB phase: the compact He and H shell burning zones	152
10.11	Evolution of the internal structure of a 7 $M_\odot$ star	153
10.12	Evolution of the internal structure of a 3 $M_\odot$ star	154
10.13	End of AGB phase: He shell flashes, pulses	155
10.14	Spectral energy distribution of an OH/IR star	156
10.15	Planetary nebula NGC 6751 in Aquila	157
10.16	CMD of stars in the solar vicinity showing WDs	157
10.17	Cooling of WDs: H_2 in the atmosphere and spectral change	158
10.18	HRD: born-again stars	159

10.19	Relation between initial mass, M_{init}, and final mass as WD, M_{final}	160
10.20	HRD: PG 1159 star pulsation on the WD cooling sequence	161
10.21	Spectra of cool subdwarf stars	162
10.22	Gaps in the main sequence of the Pleiades	163
10.23	CMD: synthetic diagram with red clump stars	164
10.24	Summary of evolution of stars in the low mass range	165
11.1	Radial pulsation (schematic)	168
11.2	Seismic nodes in a sphere (schematic)	168
11.3	Possible g-mode and p-mode oscillations; Lamb and Brunt-Väisälä	169
11.4	HRD with locations of pulsation	170
11.5	The strip with "Cepheids" as presented by Shapley (1927)	171
11.6	Temporal behaviour of amplitude of Polaris	173
11.7	Three RR Lyrae types: amplitude and period	173
11.8	RR Lyrae stars: light and colour curves; T_{eff} $\log g$, R/R_0 curves of SS Leo	174
11.9	Hysteresis in RR Lyr variable: colour and T_{eff}, $\log g$, R, v_{rad} loop of RR Gem	174
11.10	RR Lyra: observed curve of atmospheric radial velocity	175
11.11	Acoustic ray paths of two solar p modes	176
11.12	Power spectrum of solar brightness variation	177
11.13	3-D image of Solar 5 min p_{15} oscillation mode	177
11.14	Depth of the solar convection zone from helioseismology	178
11.15	Model showing how starspots cause features in spectra	179
11.16	Surface structure of a spotted star reconstructed from Doppler imaging	179
11.17	Asteroseismology: model of spectral line changes due to stellar vibration	180
12.1	The corona of the Sun as seen with SoHO	183
12.2	The Sun at 195 Å as seen with SoHO	184
12.3	Model of actual magnetic field lines of the Sun	184
12.4	Collection of sunspots as seen with the DOT	185
12.5	Sunspot of the Sun with surge as seen with the DOT	185
12.6	A prominence and a flare of the Sun as seen with SoHO	186
13.1	Spectra of WR and Of stars	188
13.2	Image of η Carinae	191
13.3	HRD showing location of LBVs	191
13.4	Spectral lines of red supergiants with signs of stellar wind	192
13.5	Model structure of a red supergiant atmosphere	193
13.6	HRD: stars with mass loss rates	194
13.7	Expanding shells and the P-Cyg profile	195
13.8	Velocity-radius relation for WR star HD 151932	196
13.9	HRD: evolutionary tracks for stars with $12 < M_{\text{init}} < 120$ $M_\odot$	198
13.10	Evolution of T_c and ρ_c for stars with $1 < M_{\text{init}} < 120$ $M_\odot$	198
13.11	Evolution of a 15 $M_\odot$ star	199
13.12	Evolution and atmospheric chemical composition of a 60 $M_\odot$ star	201
13.13	Internal structure of a $M_{\text{init}} = 60$ $M_\odot$ star at end of C burning	202
13.14	Brightness evolution of P Cyg 1710 - 1990	203
14.1	Vector addition of gravity components of a rotating star	206
14.2	Meridional circulation in rotating stars	207
14.3	Solar rotation of envelope as derived from SoHO data	208
14.4	Evolution of rotation frequency $\Omega/2\pi$ inside a 1 $M_\odot$ star	209
14.5	Rotating star and formation of mass loss disk	210
14.6	Rotation causes latitude dependent mass loss	211
14.7	Two nebulae with shapes based on mass-loss profile	211
14.8	Abundance ratios Na/Fe and O/Fe in globular cluster red giants	212
14.9	Abundance variations of Na/Fe and O/Fe due to rotation	212

14.10 Evolution of stellar rotation for 12 to 60 $M_\odot$ stars 213
14.11 HRD: evolutionary tracks for rotating high mass stars 214

15.1 Cooling behaviour of gas in relation with metal content 218
15.2 HRD: evolution of stars of 120 to 1000 $M_\odot$ 219
15.3 Evolution of T_c and ρ_c for stars of 300, 500 and 1000 $M_\odot$ 220
15.4 Internal composition of a 500 $M_\odot$ star before exploding 220
15.5 Internal temperature versus radius in a 25, 500, and 1000 $M_\odot$ star 220

16.1 Evolutionary tracks (7 $M_\odot$) with effects of convective overshoot 224
16.2 Location of ZAMS and TAMS in various model grids 224
16.3 Comparison of evolutionary tracks for different mass-loss rates and metallicities . 225
16.4 Evolution: effects of metallicity and rotation 226

17.1 Spectral energy distribution of a hot WD 228
17.2 Behaviour of dense and degenerate gas: crystallization 229
17.3 Mass derived for several neutron stars in binary systems 231
17.4 Speculative internal structure of a neutron star 232
17.5 Phase diagram, mass-radius relation: planets, WDs, NSs 233
17.6 Mass derived for several stellar black holes in binary systems 235

18.1 Examples of supernova light curves 238
18.2 Visible evolution of supernova 1987A, day 3 to 60 241
18.3 Thermonuclear supernova: model of deflagration of SN Ia 242
18.4 Fate of solar metallicity stars of $M_{\text{init}} = 3$ to 100 $M_\odot$ 243
18.5 Lightcurves and spectral energy distributions of Gamma-Ray Bursts 244
18.6 Gamma Ray Bursts due to cone-focussed radiation 245
18.7 Energy and M_{init} of hypernovae; NS or BH 245
18.8 First two days of SN 1987A 246
18.9 Fate of first stars of $M_{\text{init}} = 1$ to 1000 $M_\odot$ 247

19.1 Equipotential surfaces for binary objects 251
19.2 Binary Example 1: evolution of the radius of a 5 $M_\odot$ star in a binary system . . . 252
19.3 Potential well of double star system: 3-dimensional 254
19.4 Roche overflow in a double star system: 3-dimensional 254
19.5 Binary Example 2: evolution of a close 9 $M_\odot$ + 5 $M_\odot$ binary 255
19.6 Definition of configurations of close binary systems 256
19.7 Binary Example 3: model of a β Lyrae type binary, Roche lobe & accretion disk . 257
19.8 Evolutionary tracks of stars in a massive 20 $M_\odot$ + 8 $M_\odot$ close binary 259
19.9 Binary Example 4: phases of the evolution of a binary system of 20 $M_\odot$ + 8 $M_\odot$. 260
19.10 Binary Example 5: the over-contact eclipsing WR+O binary system CQ Cephei . 261
19.11 Binary Example 6: evolution of a binary system of 15 $M_\odot$ + 2 $M_\odot$ 262
19.12 Sketch of structure of an accreting compact object, a microquasar 263
19.13 Binary Example 7: evolution of a binary system of 2 $M_\odot$ + 1 $M_\odot$ 264
19.14 Helium shell burning in a rotating-accreting WD 265
19.15 SN Ia progenitors - rotating-accreting WDs with $M > 1.4 M_\odot$ 266
19.16 Estimate of the final products of binary evolution 267

20.1 Example of factors of photometric completeness 276
20.2 Mass function for the stars inside the supergiant shell LMC 4 277
20.3 Mass function for the low end of the MS in NGC 6397 278
20.4 Possible mass function for Population III stars 280

21.1 Isochrones as a function of $\log L$ and $\log T_{\text{eff}}$ 282
21.2 Isochrones as a function of $\log g$ and $\log T_{\text{eff}}$ 282
21.3 Isochrones as a function of M_V and $(B - V)_0$ 283
21.4 Isochrones for populations of two low levels of metal content 284

23.7. LIST OF FIGURES

21.5	Comparison of the MS of the Pleiades and of M 67 with isochrones	285
21.6	Synthetic CMDs for young star group: effects of binaries, photometric errors . . .	286
21.7	Synthetic CMDs for star fields: effects of age and SF period	287
22.1	Chemical evolution; iron abundance against age of the star	290
22.2	Nucleochronometry of metal-poor stars: example spectrum	292
22.3	Chemical evolution of the Milky Way; various elements vs. iron	293
23.1	Overview evolutionary tracks 0.15 and 1.0 $M_\odot$ (L, T_{eff}, time)	298
23.2	Overview evolutionary tracks 10 and 100 $M_\odot$ (L, T_{eff}, time)	299

23.8 List of Tables

1.1	Photometric filter bands (Johnson, Strömgren)	6
2.1	Limb darkening values of the Sun	23
2.2	Parameters for outer solar atmosphere; depth	25
3.1	Nomenclature hydrogen lines	41
4.1	Opacity coefficient for conduction	65
7.1	Known interstellar molecules	103
7.2	Parameters of molecular clouds	104
10.1	Parameters and (spectral) type of HB core He burning stars	148
10.2	Stars in lower mass range: timescale of evolution and mass lost	155
10.3	Summary of mass limits relevant for lower mass star evolution	166
13.1	Definition of " f " in O star spectral classification	189
13.2	Properties of main-sequence OB-type stars	189
13.3	Definition of WR star types	190
13.4	Lifetimes of H-, He-, and C-burning in massive stars	204
17.1	Observational evidence for stellar black holes	234
18.1	Historical visible supernovae	238
18.2	Supernova classification scheme	239
19.1	Binary star mass exchange rates	253
19.2	Temperature, maximum wavelength, and luminosity of accretion disks	258
22.1	Stellar mass range and fusion products returned	291
23.1	Summary of acronyms	303
23.2	List with constants of physics and astronomy	304
23.3	Abbreviations of common astronomy journal names	304

Index

absorption coefficient .. 18, 23ff, 33, 63, 95, 172
accretion disk,   see disk
ambipolar diffusion 109ff
asteroseismology 178
asymptotic giant branch (AGB) stars .. 46, 70,
         ...... 77, 78, 79, 147, 151ff, 192, 289, 291
AGB-manqué star 149, 151
atmosphere 15ff, 33ff
—, definition of 94
—, pressure structure of Solar 24
α-capture   see nuclear fusion
balance gravity-pressure 3, 90, 227
Balmer jump 31, 41
baroclinic instability 207
binary stars 230, 231, 249ff, 285
—, examples . 252, 255, 257, 260, 261, 262, 264
bipolar outflow 115, 123, 263
birth line 96
black hole 219, 233ff, 243, 245, 247, 290
blanketing 31
blue loop 138, 142ff, 153, 173, 197, 200,
         .... 202, 213, 283, 284
born-again star 159
brown dwarfs 80, 92, 125ff, 143
Brunt-Väisälä frequency 169, 207
buoyancy 59
buoyancy frequency 61
C burning,   see nuclear fusion
C main sequence 99, 198
C star 99, 200
cataclysmic variables 230, 264, 267
Chandrasekhar limit 89, 230, 265
chemical composition
—, effects of 98, 108, 139, 145, 149, 162,
         .... 200, 202, 217, 221, 224, 283, 290
—, and evolution 48, 217ff, 290
chromosphere 183, 192
colour index 6
common envelope 258
conduction 56, 64, 228
contraction,   see gravity
convection 59, 132
—, opacity driven 62, 97
—, radiation driven 62, 97
—, due to flash 145, 151, 152
—, in core 97, 132, 146, 199
—, in envelope 97, 144, 146, 153, 178, 199
—, in Sun 178
—, and Ledoux criterion 61, 223
—, and modelling uncertainty 223
—, and overshoot 62, 200

—, and Schwarzschild criterion 59, 223
—, and turbulence 208
convective stars 89
cooling line 229
core, definition of 94
—, contraction 202
—, driving evolution 140
—, He burning 147ff
—, isothermal 136, 139
corona 68, 183
cosmology 5, 128, 245, 266, 290
critical rotation velocity 211, 266
crystallization 229
curve of growth 36, 43
deflagration 230, 241
degeneracy parameter 66, 96, 100, 142, 198
degenerate gas 66, 90, 91, 96, 145, 151, 228
degenerate stars 227ff, 292
deuterium 113, 125, 128, 290
—, buring of 113, 125
disk
—, accretion 110, 230, 257ff
—, circumstellar 116
—, Keplerian 257, 266
—, mass loss 210
dissociation 27, 28, 289
doppler imaging 179
dredge-up 147
—, first 148
—, second 148
—, third 148, 152
dust 70, 144, 156, 192, 217
dynamical time scale 58, 109
dynamo 209
Eddington 3, 22, 89, 98, 194, 265
emission coefficient 17
energy transport
—, convective 56, 97
—, radiative 55, 97
—, by neutrinos   see neutrinos
energy production 67, 71ff, 201, 202
—, from fusion,   see nuclear fusion
—, from gravity,   see gravity
envelope, definition of 95
—, expanding 192ff
equation of state 65ff, 87ff, 167
equipotential surfaces 250, 254
escape velocity 68, 193
Fermi pressure,   see degenerate gas
first stars 75, 217, 247, 279, 289
flash ignition,   see H flash, He flash

flash supernova 246
fractal structure 104
fusion,   see nuclear fusion
G-band 45
g-mode oscillation 168ff
γ ray burst 243
Gamow peak 72
Gauss function 35, 94
gaps along branches in CMD/HRD 163
gas pressure 90
globular clusters 148ff, 275, 284
gravity
—, energy from 68, 137, 143, 153ff, 201ff
—, importance of 3
—, instability (Jeans) 106
—, surface 13
—, waves, g-mode oscillations 168ff
—, and balance with pressure 3, 90
—, and broadening of lines 34, 228
—, and contraction 3, 134, 202
—, and levitation, settling 48
—, and mass loss 68
gravothermal hysteresis 134, 200
gray atmosphere 21, 23
grid points 94
H burning 92, 113, 126, 131
    see also nuclear fusion
H excitation 38, 41
H flash 151, 159
H ionization 25, 40, 289
H shell burning 136, 143, 151, 153, 176
H$^-$ ion 26, 29, 31
H II region 101, 141, 187, 289
Hayashi 114
He burning 75, 147
    see also nuclear fusion
He flash 145, 151
He ignition 136, 142, 145
He main sequence 99, 146, 198
He shell pulses 265
He star 99, 200, 260, 262
helioseismology 177
Herbig-Haro objects 115, 119
Hertzsprung-Russell Diagram (definition) 9
homology 12, 96
horizontal branch 14, 99, 142, 148ff
hot bottom,   see nuclear fusion
Hoyle 3, 76
hydrodynamic supernova,   see supernova
hydrostatic equilibrium 15, 53
hypernovae 243
hysteresis 134ff, 174, 200
initial mass function (IMF),   see mass function

initial to final mass 160, 165, 202, 243, 247,
    .... 266, 292
ionization, photon absorption 25
isochrone 274, 281ff
Jeans instability/criterion 106, 289
jets 263
    see also ambipolar diffusion
Jupiter mass 125
Kelvin-Helmholtz time scale 57, 113, 253
Keplerian disk 257, 266
κ-mechanism 160, 169, 172
Lamb frequency 169
leaky star; leakyness 4, 96ff, 197, 200, 202, 218
Ledoux criterion,   see convection
levitation 48
lifetimes of phases .140, 155, 165, 189, 203, 204
limb darkening 21
line profile 33ff
lithium 73, 79, 120, 128, 155, 221, 290
—, burning of 80, 126
local thermodynamic equilibrium 19
Lorentz function 33
luminosity
—, and mass 98, 140
—, and metallicity 98, 99
—, and molecular weight 98, 201
luminosity function 271ff
magnetic field 47, 109ff, 183, 209, 232
main sequence
—, definition 2, 95, 132
—, duration of 139, 203
mass (of a star)
—, maximum 90
—, minimum 91, 126
mass exchange 253ff
mass function 102, 143, 271ff
mass limits 141, 166, 187, 218, 230
—, of convection 97, 132
mass loss 68, 144, 152, 154, 194, 195, 200,
    .... 201, 210, 219, 225
—, disks 210
mass transfer 230, 251ff
merged stars 258, 268
meridional circulation 206
metallicity 8, 43
    see also chemical composition
metal poor stars 79, 98, 102, 139, 149, 217, 290
microquasar 263
mixing 63, 211, 213
—, length 63, 223
molecular weight, mean 56, 61, 65, 98, 155
—, and luminosity 98, 201

molecules 47, 103, 112, 128, 143, 156, 158,
..... 217, 228
molecular clouds 102ff
molecular outflows 105
neutrino 80ff, 131
—, mean free path 80
—, oscillations 83
—, production in supernovae 81, 239, 246
—, stellar core cooling 145, 153, 201, 230
neutron star 66, 230ff, 243, 247, 260, 262
novae 230, 264ff, 267
—, recurrent 264ff
nuclear fusion 71ff
—, energy production 67, 72ff, 95, 137
—, hot-bottom burning 147, 152, 155, 165
—, ideas about 2ff
—, minimum stellar mass for 91
—, C burning 76, 152, 199, 200
—, D burning 113, 125, 290
—, H burning 92, 197
—, He burning 75, 92, 147ff, 197, 200, 201
—, Li burning 80, 126, 290
—, N burning 77, 213
—, O burning 77, 152
—, CNO cycle 74, 152, 200, 274, 289
—, pp chain 73, 132, 274
—, p-process 79
—, r-process 77, 79, 240
—, s-process 77, 78, 147, 152, 290
—, triple α process 75
—, α-capture 76, 78, 213
nuclear statistical equilibrium 78, 219
nuclear time scale 58
nucleochronometry 291
nucleosynthesis 79, 290
OH/IR stars 155ff, 192
opacity 25, 64, 87
—, definition of 18
—, processes of 25
—, and convection 62
—, and gray atmosphere 23
optical depth 18, 36, 148, 184
oscillations 168ff
outflows 115ff
P-Cyg profile 119, 188, 195
—, inverse 119
p-mode oscillation 168ff
pair production 219
period-luminosity relation 171
Planck function 7, 19, 21, 30
planetary nebula 157, 159
planets 125, 129, 177
polytropes 87ff, 232

post AGB star 156ff
Population I, II, III 102, 221, 244, 284
pp chain,   see nuclear fusion
pre main sequence 113
pressure scale height 25, 61, 178, 207, 226
protostars 112
pulsar 232, 262
—, binary pulsar 263, 268
—, X-ray pulsar 262
pulsation 167ff
—, and atmosphere velocities 175
—, and κ-mechanism 172
pulsational variables 160
pulses (thermal) 147, 145, 155, 265
quark star,   see strange star
quasi molecules 46
radiation
—, pressure 17, 65, 90
—, transport 16ff, 55
—, and particle acceleration 193
recurrent novae,   see novae
red giant 14, 134, 143, 200
—, branch 14, 150, 153, 163
—, from MS to 134
—, H shell burning 143ff
—, luminosity evolution 138, 144
—, pulsators 172, 176
—, second stage,   see AGB star
—, spectral energy 14, 30, 43
red supergiant 192, 197, 199, 200
Roche lobe 250ff, 257ff, 260ff
Rosseland mean opacity, definition of 23
rotation
—, braking of 209
—, effects of 49, 202, 203, 205ff
—, white dwarf, increase of 265
—, and colour index 49
—, and deformation 49, 205
—, and magnetic field 209
—, and mass loss 210
—, and spectral lines 49
—, and star formation (clouds) 109
—, and stellar wind 70
Saha equation 39
Salpeter function, value 102, 273, 274
scale height,   see pressure
scattering 27, 64, 194
Schönberg-Chandrasekhar limit 138, 145
Schwarzschild criterion,   see convection
Schwarzschild radius 234
settling 48, 228
shear instability 208
shell, definition of 95

Sirius A B 227
SN 1987A 246
solar neutrinos 81ff
Solberg-Høiland instability 207
sound speed 169, 171, 178
spectral classification 1, 9, 50, 128
spots (star-) 179, 185
s-process,   see nuclear fusion
star
—, as leaky box 4, 96ff
—, as thermostat 4, 96ff
star (types)
—, Ae 121, 215
—, AGB,   see asymptotic giant branch
—, Algol 256
—, Ap 48
—, B 101, 178ff
—, Be 121, 190, 215
—, blue straggler 150, 162
—, Cepheids 161, 173, 197
—, DA variables 175
—, HBA 148
—, HBB 148
—, LBV 176, 187, 191
—, Mira 2, 176
—, O 101, 178ff
—, Oe 190
—, P Cygni 187, 191, 195, 203
—, PG 1159 .... 161, 181
—, RHB 148
—, RR Lyrae 148, 160, 172ff
—, subdwarf 14, 148, 157, 162
—, T Tau 105, 118
—, WC, WN, WO,   see Wolf-Rayet
—, Wolf-Rayet 187, 190, 198, 261, 262, 263
—, W UMa 255
—, W Vir 172
—, ZZ Ceti 161, 175
—, β Cep 176
—, β Lyrae 256ff
—, δ Sct 172, 175
—, λ Boo 161
star clusters 275, 285
star formation 3, 101ff, 217
—, history (SFH) 288
stellar thermal equilibrium 4, 95, 137ff
strange star 233
subdwarf stars 14, 148, 157, 162
Sun 1, 7, 13, 23, 25, 44, 63, 64, 67, 81,
.... 144, 177, 183, 208
supernova 2, 219, 237ff
—, Type Ia 230, 237ff, 265ff, 291
—, Type Ib 237ff, 260ff, 268, 291

—, Type Ic 239, 268
—, Type II 154, 237, 291
—, hydrodynamic 239
—, pair instability 219
—, thermonuclear 241
—, and neutrinos 239
—, and r-process fusion 78, 79, 240
supersoft X-ray source 264, 267
temperature
—, central 95, 96, 142, 198
—, effective (definition) 10
thermal equilibrium:
    see stellar thermal equilibrium
    see thermodynamic equilibrium
thermal instability (clouds) 108
terminal age horizontal branch 151
terminal age main sequence 132, 198, 252
thermodynamic equilibrium 19
thermonuclear supernova,   see supernova
thermostat 4, 96ff
time scale
—, dynamical 58
—, free fall 57, 113
—, nuclear 58
—, C burning 203
—, H burning 140, 203
—, He burning 140, 203
variable stars 107ff
vibration 167, 176ff
Virial theorem 57
Voigt function 35
white dwarf .. 66, 77, 92, 143, 157ff, 180, 227ff,
.... 233, 264
—, cooling 143, 151, 158, 161, 229
—, fast rotating 265
—, He 158, 165, 262, 264
—, and accretion of matter 230, 241, 265
—, and nuclear fusion 77, 230, 241
—, and supernova,   see supernova Type Ia
wind (stellar),   see also mass loss 68
—, continuum driven 69
—, dust driven 70, 156
—, line driven 69
—, line profile 195
—, and pulsation 70, 155
—, and stellar rotation 70, 210
—, and velocity profile 68, 196
X-ray binaries 259, 268
yield 80, 290
Zeeman 42, 158
   see also magnetic field
zero age horizontal branch 151
zero age main sequence 95, 115, 132, 198

www.ingramcontent.com/pod-product-compliance
Ingram Content Group UK Ltd.
Pitfield, Milton Keynes, MK11 3LW, UK
UKHW051259180426
11947UKWH00020B/1807